Beginning Algebra, Form A

Other Books in the Streeter Series

Also Available from McGraw-Hill

Schaum's Outline Series in Mathematics & Statistics

Most outlines include basic theory, definitions, and hundreds of solved problems and supplementary problems with answers.

Titles on the Current List Include:

Advanced Calculus
Advanced Mathematics
Analytic Geometry
Beginning Calculus
Boolean Algebra
Calculus, 3d edition
Calculus for Business, Economics, & the Social and Life Sciences
Calculus of Finite Differences & Difference Equations
College Algebra
College Mathematics, 2d edition
Complex Variables
Descriptive Geometry
Differential Equations
Differential Geometry
Discrete Math
Elementary Algebra
Essential Computer Math
Finite Mathematics
Fourier Analysis
General Topology
Geometry, 2d edition

Group Theory
Laplace Transforms
Linear Algebra, 2d edition
Mathematical Handbook of Formulas & Tables
Matrix Operations
Modern Abstract Algebra
Modern Elementary Algebra
Modern Introductory Differential Equations
Numerical Analysis, 2d edition
Partial Differential Equations
Probability
Probability & Statistics
Projective Geometry
Real Variables
Review of Elementary Mathematics
Set Theory & Related Topics
Statistics, 2d edition
Technical Mathematics
Tensor Calculus
Trigonometry, 2d edition
Vector Analysis

Schaum's Solved Problems Books

Each title in this series is a complete and expert source of solved problems containing thousands of problems with worked out solutions.

Titles on the Current List Include:

3000 Solved Problems in Calculus
2500 Solved Problems in College Algebra and Trigonometry
2500 Solved Problems in Differential Equations
2000 Solved Problems in Discrete Mathematics

3000 Solved Problems in Linear Algebra
2000 Solved Problems in Numerical Analysis
3000 Solved Problems in Precalculus

Available at your College Bookstore. A complete list of Schaum titles may be obtained by writing to:

Schaum Division
McGraw-Hill, Inc.
Princeton Road S-1
Hightstown, NJ 08520

Beginning Algebra, Form A

Third Edition

James Streeter
Late Professor of Mathematics
Clackamas Community College

Donald Hutchison
Clackamas Community College

Louis Hoelzle
Bucks County Community College

McGraw-Hill, Inc.
New York   St. Louis   San Francisco   Auckland   Bogotá   Caracas
Lisbon   London   Madrid   Mexico   Milan   Montreal   New Delhi   Paris
San Juan   Singapore   Sydney   Tokyo   Toronto

Beginning Algebra, Form A

 This book is printed on recycled, acid-free paper containing a minimum of 50% recycled de-inked fiber.

4 5 6 7 8 9 0 DOW DOW 9 0 9 8 7 6 5

ISBN 0-07-031718-6

This book was set in Times Roman by York Graphic Services, Inc.
The editors were Michael Johnson, Karen Minette, and Jack Maisel;
the design was done by Caliber/Phoenix Color Corp.;
the production supervisor was Al Rihner.
R. R. Donnelley & Sons Company was printer and binder.

Library of Congress Cataloging-in-Publication Data

Streeter, James (James A.)
    Beginning algebra. Form A/James Streeter, Donald Hutchison,
Louis Hoelzle. —3rd ed.
       p.   cm.
    Includes index.
    ISBN 0-07-031718-6 —ISBN 0-07-063020-8 (teacher's ed.)
    1. Algebra.   I. Hutchison, Donald (date).   II. Hoelzle, Louis
F.   III. Title.
QA152.2.S779   1993
512′.9—dc20   92-11715

While a graduate student at the University of Washington, James Streeter paid for his education as a math tutor. It was here that he began to formulate the ideas that would eventually become this package. Upon graduation he taught for two years at Centralia Community College. In 1968 he moved on to Clackamas Community College to become their first mathematics chair.

At the community college, Jim recognized that he faced a very different population than the one for whom he had tutored at UW. Jim was convinced that, in order to reach the maximum number of these students, he would have to utilize every medium available to him. Jim opened a math lab that included CAI, original slides and tapes (which were eventually published by Harper and Row) and original worksheets and text materials. With the assistance of the people at McGraw-Hill, that package has been refined to include media and supplements that did not even exist when this project began.

Donald Hutchison spent his first 10 years of teaching working with disadvantaged students. He taught in an intercity elementary school and an intercity high school. He also worked for two years at Wassaic State School in New York and two years at the Portland Habilitation Center. He worked with both physically and mentally disadvantaged students in these two settings.

In 1982 he was hired by Jim Streeter to teach at Clackamas Community College. It was here that he discovered the two things that, along with his family, serve as a focus for his life. Jim introduced him to the joy of writing (with the first edition of *Beginning Algebra*) and Jack Scrivener converted him to a born-again environmentalist.

Don is also active in several professional organizations. He is a member of the ACM committee that has undertaken the writing of computer curriculum for the two-year college. Since 1990 he has chaired the Technology in Mathematics Education committee for AMATYC.

In 1989 Don became Chair of the Mathematics department at Clackamas Community College.

Louis Hoelzle has been teaching at Bucks County Community College for 23 years. He has taught the entire range of courses from Arithmetic to Calculus. This gives him the perspective of the current and future needs of developmental students.

Over the past 30 years Lou has also taught Physics courses at four year colleges. This gives him the perspective of the practical applications of mathematics.

In addition, Lou has done extensive reviewing of manuscripts and writing of several solutions manuals for major texts. In these he has focused on writing for the student.

Lou is also active in professional organizations and has served on the Placement and Assessment Committee for AMATYC since 1989.

In 1989, Lou became Chair of the Mathematics Department at Bucks County Community College.

A teacher affects eternity; he can never tell where his influence stops.

—Henry Brooks Adams 1838–1918

This series is dedicated to the memory of James Arthur Streeter, an artisan with words, a genius with numbers, and a virtuoso with pictures from 1940 until 1989.

There is a German proverb that has been translated, ''More is to be got from one teacher than from two books.'' This reflects the essence of what I have learned about writing from Jim Streeter. Jim was always a teacher who wrote, and never a writer who taught. Jim was a teacher in the classroom, in his office, during meetings, and behind his word processor. From Jim's perspective, it seemed that life's single challenge was to find the components that, when properly assembled, would reveal the concept that he wanted to explain. One night Jim and I went out to celebrate the completion of a manuscript. I taught him a dart game called ''301.'' He was fascinated. I doubt that he ever played again, but he used the game in the classroom and in his writing. He found it an entertaining medium for teaching subtraction, multiplication, and critical thinking skills (e.g., ''If you need 15 points, and you must finish with a double score, how many different ways can you win with two darts?'')

Jim did not believe in platitudes, but he did convince me that ''writing makes one a better teacher and teaching makes one a better writer.'' Because I knew Jim, I have become a better writer, a better teacher, and a better person.

Donald Hutchison

CONTENTS

PREFACE

When we study mathematics, we study questions at three levels:

What is mathematics?

What can we do with mathematics?

How do we do it?

The first question is reserved for higher-level mathematics and philosophy courses (where one can pursue the debate about whether mathematics is a product of human endeavor or rather the discovery of a natural system).

The remaining two questions are at the core of every lecture in every mathematics class we teach. The purpose of this textbook is to help the student discover answers to the last two questions.

The focus in teaching this course is to make our students better problem-solvers. Our emphasis as writers is to communicate this to our audience. The language must be at the appropriate level. The topics must be broad enough to include all of the concepts that the student will need both in future mathematics classes and future employment. And the problems must relate enough to student experience to motivate the topic.

One of our goals has been to create a set of applications that would be more relevant to the student. A greater part of our goal was reached by incorporating a theme.

The *environmental theme* is an important new element of this edition of *Beginning Algebra.* The reason for this focus is twofold: our own, and the mathematics community's concern for the state of the earth and our realization that most students share this concern. As a result, students become more interested in learning how to use mathematics when they are interested in what kinds of problems can be solved.

Each chapter of this text opens with an *environmental essay.* These essays were written by Jack Scrivener, who teaches both beginning algebra and environmental science at Clackamas Community College in Oregon City, Oregon. The essays have been written specifically for these students at this level. Professor Scrivener's ten years of teaching these two courses makes him uniquely qualified to reach this audience.

Within each chapter are exercises that specifically relate to the environmental essay for that chapter. These were also written by Professor Scrivener. They are easily identified by the tree logo that appears over the exercise number. Our goal was to produce a set of noncontrived problems that students could solve with the skills they have just learned. We are very pleased with the results. Occasionally one of the problems will be a stretch for all of the students, but it's a reachable stretch. It is also a stretch that motivated students are more likely to attempt.

Several other features new to the third edition of this text are designed to better teach students how to do the mathematics. Among those features, we have

A *worktext format* printed on *recycled paper* that incorporates *four colors* to guide students through critical steps and increase clarity in graphs.

Also new to this edition are the *expression screens*. In teaching this course (and intermediate algebra as well) we have discovered that many students are unable to identify whether they are working on an expression or an equation. When this is the case, they do not know what form the answer should take. We have tried to address this problem in two ways. First, we have expanded the discussion about the difference, and we have included several exercises in which the student is to do no more than identify whether each item is an expression or an equation. Then, to avoid confusion, we have *screened the expression* in any example in which the student is asked to simplify. By screening the expression, it is clear to the student just what it is they are simplifying.

Think about These are new exercises that require the student to extend or generalize from the skills just learned. These exercises promote and utilize critical thinking skills.

This new edition also includes the *Retain Your Skills* exercise sets. These twenty-question review quizzes appear at the end of chapters 5, 7, and 9. They remind the student of skills learned and encourage regular cumulative review of the material.

Other features retained from the second edition include

Check Yourself   These exercises accompany the text examples and are designed to involve the student actively in the learning process. Answers are provided at the end of the section for immediate feedback.

Build Your Skills   Exercises that allow the student to practice and master the skills of the section.

Skillscan   Draws problems from previous sections of the text. Designed to aid the student in the process of reviewing concepts that will be applied to the following section.

Chapter Summaries and Summary Exercises   These sections give the student an opportunity to practice and review at the end of each chapter. Answers are provided with section references to aid in the process of review.

Self-Tests   Each chapter concludes with a self-test to give students guidance in preparing for in-class tests. Answers are provided at the back of the book.

Cumulative Test   These are designed to give the student further opportunity for building skills which should be especially useful for students preparing for mid-term and final exams.

We continue to encourage the participation of students in the learning process by directing our attention to readability and student involvement. Each topic is developed in a straightforward fashion with numerous examples to clarify the subject being developed. All important definitions, rules, and algorithms are enhanced with color for easy identification and reference.

Pattern recognition and problem-solving strategies are two themes retained in this edition. One way in which we encourage pattern recognition is in the matching exercises incorporated into various exercise sets. For instance, in Chapter 10 (''Quadratic Equations'') we include linear graphs in the exercise set to encourage more discriminant thinking on the part of the students.

All the features are designed to encourage, facilitate, and motivate problem solving among the students. This is not just the nature of this text, it is the primary basis for the inclusion of mathematics in virtually every curriculum.

Supplements

There are a number of supplements available for both the student and instructor. The *Teacher's Edition* includes answers to all exercises and tests. These answers are printed in a second color for easier use by the instructor. The *Instructor's Solutions Manual* contains worked-out solutions and answers to all of the exercises in the text. The *Student's Solutions Manual* contains worked-out solutions to the odd-numbered exercises in the text. *The Instructor's Resource Manual* contains multiple-choice placement tests for three levels, a diagnostic pretest for each chapter, three forms of multiple-choice and open-ended chapter tests, two forms of multiple-choice and open-ended cumulative tests, two forms of multiple-choice and open-ended final tests, an answer section, and appendixes. The *video series* provides the student with additional instructional and visual support of the lessons. *The Professor's Assistant* is a computerized test generator that allows the instructor to create tests using algorithmically generated test questions and those from a standard testbank. This testing system enables the instructor to choose questions either manually or randomly by section, question type, difficulty level, and other criteria. This system is available for IBM, IBM compatible, and Macintosh computers. The *Print Test Bank* is a printed and bound copy of the questions found in the standard testbank. The *Interactive Tutorial Software* is a self-paced interactive tutorial specifically linked to the text. This tutorial reinforces selected topics and provides unlimited opportunities to review concepts and to practice problem solving. It requires virtually *no* computer training on the part of the student and is available for IBM, IBM compatible, and Macintosh computers. The *Calculator Enhancement Manual* presents an integrated approach that utilizes calculator-based graphing to enhance understanding and development. It includes calculator exercises and examples as well as appendixes on how to use the most popular calculators. For further information about these supplements, please contact your local college division sales representative.

Acknowledgments

The people at McGraw-Hill have made this a better and more exciting project. Special thanks go to each of the following: Michael Johnson, our editor, constantly asked good questions, looked for ways to improve the text, and found excellent reviewers and problem checkers; Denise Schanck challenged us to effectively use an incredible array of resources; Jack Maisel provided communications between the marvelous graphic design people at York Graphic Services, Inc. and the authors; Karen Minette was the one person with more answers than questions; and Marni Centor provided a much needed contact between authors and programmers.

We would also like to thank the students and faculty at Bucks County Community College and Clackamas Community College. They have helped us in the development of the material by asking the right questions. Special thanks to Barry Bergman, Susan Hopkirk, Betsy Farber, and Maryann Klicka.

The copy-editors and problem-checkers have done a remarkable job of helping us work towards the elusive goal of an error-free text. Particular thanks go to Patti Scott and Gail Santner.

As mentioned, we had many contributions to the development of this edition and the previous edition. Thanks to

Kathleen J. Bavelas, Manchester Community College

Robert J. Blain, Salt Lake Community College

Ruth A. Koelle, Roger Williams College

Susann Mathews, The Ohio State University

John Pazdar, Greater Hartford Community College

Kathryn Pletsch, Antelope Valley Community College

Debbie Singleton, Lexington Community College

Barbara Jane Sparks, Camden Community College

Ann Bartholomay, Southwest Virginia Community College

Henri Feiner, West Los Angeles College

Susan L. Friedman, Bernard M. Baruch College

Maria Kelly, Fresno City College

Jack W. Kotman, Lansing Community College

Virginia Lee, Brookdale Community College

Shirley Markus, University of Louisville

Linda Murphy, Northern Essex College

Nancy K. Nickerson, Northern Essex College

Carol O'Loughlin, Northern Essex College

Elise Price, Tarrant County Junior College

Peggy Rejto, Normandale Community College

Sylvester Roebuck, City College of Chicago, Olive-Hardy College

Jack Rotman, Lansing Community College

Alexa Stiegemeier, Elgin Community College

Eleanor Strauss, Community College of Philadelphia

Tommy Thompson, Brookhaven College

The ancillary material has been provided in an unusually timely and efficient manner. Along with Karen Minette, the following people deserve recognition:

Richard & Beth Morel, Strong House, Inc.

John R. Martin, Tarrant County Junior College—Northeast

Carolyn Meitler, Concordia University

John Garlow, Tarrant County Junior College—Northwest

Linda Dodge, Greenfield Community College

Finally, none of this would happen without the support of our families. We especially want to thank our wives, Claudia Hutchison and Rose Hoelzle. Their lessons in patience by example were among the most important that we have learned.

Donald Hutchison

Louis Hoelzle

TO THE STUDENT

You are about to begin a course in algebra. We have made every attempt to provide a text that will help you understand what algebra is about and how to effectively use it. We have made no assumptions about your previous experience with algebra. Your rate of progress through the course will depend both upon the amount of time and effort that you give to the course and to your previous background in mathematics. There are some specific features in this textbook that will aid you in your studies. Here are some suggestions about how to use those features.

Keep in mind that a review of *all* of the chapter material will further enhance your ability to grasp later topics and to move more effectively through the following chapters.

1. If you are in a lecture class, make sure that you take the time to read the appropriate text section *before* your instructor's lecture on the subject. Then take careful notes on the examples that your instructor presents during class.

2. After class, work through similar examples in the text, making sure that you understand each of the steps shown. Examples are followed by *Check Yourself* exercises. Algebra is best learned by being involved in the process and that is the purpose of these exercises. Always have a pencil and paper at hand and work out the problems that are presented and check your results immediately. If you have difficulty, go back and carefully review the previous exercises. Make sure that you understand what you are doing and why. The best test of whether you do understand a concept lies in your ability to explain that concept to one of your fellow students. Try working together.

3. At the end of each chapter section you will find a set of exercises. Work these carefully in order to check your progress on the section you have just finished. You will find the solutions for the odd-numbered exercises following the problem set. If you have had difficulties with any of the exercises, review the appropriate parts of the chapter section. If your questions are not completely cleared up, by all means do not become discouraged. Ask your instructor or an available tutor for further assistance. A word of caution: Work the exercises on a regular (preferably daily) basis. Again, learning algebra requires becoming involved. As is the case with learning any skill, the main ingredient is practice.

4. When you have completed a chapter, review by using the *Chapter Summary.* You will find all the important terms and definitions in this section, along with examples illustrating all the techniques that have been developed in the chapter. Following the summary are *Summary Exercises* for further practice. The exercises are keyed to chapter sections, so you will know where to turn if you are still having problems.

5. When done with the *Summary Exercises,* try the *Self-Test* that appears at the end of each chapter. This will give you an actual practice test to work as you review for in-class testing. Again, answers with section references are provided.

6. Finally, an important element of success in studying algebra is the process of regular review. We have provided a series of *Cumulative Tests* throughout the textbook (they are located after Chapters 3, 6, 8, and 10) and *Retain Your Skills* exercises (Chapters 5, 7, and 9). These will help you review not only the concepts of the chapter that you have just completed, but those of previous chapters. Use these tests in preparation for any midterm or final examinations. If it appears that you have forgotten some concepts that are being tested, don't worry. Go back and review the sections where the idea was initially explained, or the appropriate chapter summary. That is the purpose of these cumulative tests.

We hope that you will find our suggestions helpful as you work through this material, and we wish you the best of luck in the course.

Donald Hutchison
Louis Hoelzle

CHAPTER 1

The Language of Algebra

The ENVIRONMENT

Introduction to Environmental Problems

The earth's population has more than doubled in the last 40 years. It is highly likely that it will double again in the next 40 years. Almost 61,000 square kilometers of desert are formed each year. We are destroying 202,000 square kilometers of tropical forest every year. One-third of the world's cropland is losing soil faster than it is being formed. One-half of the world's wetlands have been lost to development or pollution. Thousands of lakes in the northern hemisphere contain no fish because of their high acid content. Tons of waste wash up onto the world's beaches daily.

The environment is no longer the concern of only a few scientists and environmental activists. Environmental pollution and disruption has become front-page news for the citizens of the planet. Understanding how we are affected by these problems is important to our individual lives. Knowing how individual people cause environmental problems helps us make decisions about our personal actions and lifestyles.

Each chapter of this text will begin with an environmental essay. These essays will show how people are causing rapid changes on this planet. They will also discuss how these changes are affecting the people of the planet. Many people are not aware that their behavior can cause problems for other living things. These essays will show a connection between the actions of individuals and the problems that affect all of us.

Exercises within the problem sets will relate to the theme of that chapter. These exercises will provide practice with the mathematics necessary to grasp the importance of the ideas discussed in each essay. Although the algebra has occasionally been simplified, it is never contrived. These are real problems facing all of us.

Environmental problems are not confined to one part of the globe or to one aspect of the environment. The air, water, land, plants, animals, and humans have all been affected by these problems. Problems such as acid rain, ozone depletion, and overpopulation make the news regularly. We also hear about world hunger, extinction of plants and animals, and global warming. These are not local problems. They affect the entire planet.

When we first look at the wide range and size of these problems, we can become overwhelmed. These essays will show that individual action can both avoid creating future problems and help cure some of those that already exist. The exercises will show that basic algebra can help us analyze and solve these problems.

From Arithmetic to Algebra

OBJECTIVES

1. To represent the operations of addition, subtraction, multiplication, and division by using the notation of algebra
2. To identify algebraic expressions

In arithmetic you learned how to do calculations with numbers by using the basic operations of addition, subtraction, multiplication, and division.

In algebra you will still be using numbers and the same four operations. However, you will also be using letters to represent numbers. Letters such as *x*, *y*, *L*, or *W* are called *variables* when they represent numerical values.

Here we see three rectangles whose lengths and widths are labeled with numbers.

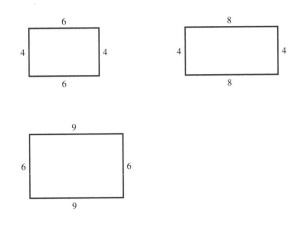

If we need to represent the length and width of *any* rectangle, we can use the variables *L* and *W*.

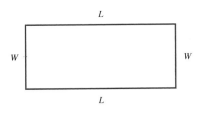

In arithmetic: + denotes addition, − denotes subtraction, × denotes multiplication, ÷ denotes division.

You are familiar with the symbols used to indicate the four fundamental operations of arithmetic.

Let's look at how these operations are indicated in algebra.

ADDITION

$x + y$ means the *sum* of *x* and *y* or *x plus y*.

Example 1

Some other words that tell you to add are "more than" and "increased by."

(a) The *sum* of a and 3 is written as $a + 3$.
(b) L *plus* W is written as $L + W$.
(c) 5 *more than* m is written as $m + 5$.
(d) x *increased by* 7 is written as $x + 7$.

CHECK YOURSELF 1*

Write, using symbols.

1. The sum of y and 4   **2.** a plus b
3. 3 more than x **4.** n increased by 6

Let's look at how subtraction is indicated in algebra.

SUBTRACTION

$x - y$ means the *difference* of x and y or x *minus* y.

Example 2

Some other words that mean to subtract are "decreased by" and "less than."

(a) r *minus* s is written as $r - s$.
(b) The *difference* of m and 5 is written as $m - 5$.
(c) x *decreased by* 8 is written as $x - 8$.
(d) 4 *less than* a is written as $a - 4$.

CHECK YOURSELF 2

Write, using symbols.

1. w minus z
2. The difference of a and 7
3. y decreased by 3
4. 5 less than b

You have seen that the operations of addition and subtraction are written exactly the same way in algebra as in arithmetic. Multiplication presents a slight difficulty because the sign $\times$ looks like the letter x. So in algebra we use other symbols to show multiplication to avoid any confusion. Here are some ways to write multiplication.

*Check Yourself Answers appear at the end of each section throughout the book.

Note: x and y are called the *factors* of the product xy.

MULTIPLICATION

A raised dot $x \cdot y$

Parentheses $(x)(y)$ These all indicate the *product* of x and y.

Writing the letters next to each other xy

Example 3

(a) The product of 5 and a is written as $5 \cdot a$, $(5)(a)$, or $5a$. The last expression, $5a$, is the shortest and the most common way of writing the product.

Note: You can place two letters next to each other or a number and a letter next to each other to show multiplication. But you *cannot* place two numbers side by side to show multiplication: 37 means the number "thirty-seven," not 3 times 7.

(b) 3 times 7 can be written as $3 \cdot 7$ or $(3)(7)$.
(c) Twice z is written as $2z$.
(d) The product of 2, s, and t is written as $2st$.
(e) 4 more than the product of 6 and x is written as $6x + 4$.

CHECK YOURSELF 3

Write, using symbols.

1. m times n
2. The product of h and b
3. The product of 8 and 9
4. The product of 5, w, and y
5. 3 more than the product of 8 and a

Before we move on to division, let's look at how we can combine the symbols we have learned so far.

An *expression* is a meaningful collection of numbers, variables, and signs of operation.

Not every collection of symbols is an expression.

Example 4

(a) $2m + 3$ is an expression. It means that we multiply 2 and m, then add 3.

(b) $x + \cdot + 3$ is not an expression. The three operations in a row have no meaning.

(c) $y = 2x - 1$ is not an expression. The equals sign is not an operation sign.

(d) $3a + 5b - 4c$ is an expression. Its meaning is clear.

CHECK YOURSELF 4

Identify which are expressions and which are not.

1. $7 - \cdot x$ **2.** $6 + y = 9$
3. $a + b - c$ **4.** $3x - 5yz$

To write more complicated products in algebra, we need some "punctuation marks." Parentheses ( ) mean that an expression is to be thought of as a single quantity. Brackets [ ] and braces { } are used in exactly the same way as parentheses in algebra. Look at the following example showing the use of these signs of grouping.

Example 5

(a) 3 times the sum of a and b is written as

This can be read as "3 times the quantity a plus b."

$$3(a + b)$$

The sum of a and b is a single quantity, so it is enclosed in parentheses.

(b) The sum of 3 times a and b is written as

No parentheses are needed here since the 3 multiplies *only* the a.

$$3a + b$$

(c) 2 times the difference of m and n is written as

$$2(m - n)$$

(d) The product of s plus t and s minus t is written as

$$(s + t)(s - t)$$

(e) The product of b and 3 less than b is written as

$$b(b - 3)$$

CHECK YOURSELF 5

Write, using symbols.

1. Twice the sum of p and q
2. The sum of twice p and q
3. The product of a and the quantity $b - c$
4. The product of x plus 2 and x minus 2
5. The product of x and 4 more than x

Now let's look at the operation of division. Think of the variety of ways you showed division in arithmetic. Besides the division sign $\div$, you used the long division symbol $\overline{)}$ and the fraction notation. For example, to indicate the quotient when 9 is divided by 3, you could write

$$9 \div 3 \qquad \text{or} \qquad 3\overline{)9} \qquad \text{or} \qquad \frac{9}{3}$$

In algebra the fraction form is usually used.

DIVISION

$\dfrac{x}{y}$ means x *divided by* y or the *quotient* of x and y.

Example 6

Write, using symbols.

(a) m divided by 3 is written as $\dfrac{m}{3}$.

(b) The quotient of a plus b divided by 5 is written as

$$\frac{a + b}{5}$$

(c) The sum p plus q divided by the difference p minus q is written as

$$\frac{p + q}{p - q}$$

CHECK YOURSELF 6

Write, using symbols.

1. r divided by s
2. The quotient when x minus y is divided by 7
3. The difference a minus 2 divided by the sum a plus 2

CHECK YOURSELF ANSWERS

1. (1) $y + 4$; (2) $a + b$; (3) $x + 3$; (4) $n + 6$.
2. (1) $w - z$; (2) $a - 7$; (3) $y - 3$; (4) $b - 5$.
3. (1) mn; (2) hb; (3) $8 \cdot 9$ or $(8)(9)$; (4) $5wy$; (5) $8a + 3$.
4. (1) Not an expression; (2) not an expression; (3) an expression; (4) an expression.
5. (1) $2(p + q)$; (2) $2p + q$; (3) $a(b - c)$; (4) $(x + 2)(x - 2)$; (5) $x(x + 4)$.
6. (1) $\dfrac{r}{s}$; (2) $\dfrac{x - y}{7}$; (3) $\dfrac{a - 2}{a + 2}$.

Name _____

Date _____

Build Your Skills

Write each of the following phrases, using symbols.

1. The sum of *c* and *d*

2. *a* plus 7

3. *w* plus *z*

4. The sum of *m* and *n*

5. *x* increased by 2

6. 3 more than *b*

7. 10 more than *y*

8. *m* increased by 4

9. *a* minus *b*

10. 5 less than *s*

11. *b* decreased by 7

12. *r* minus 3

13. 6 less than *r*

14. *x* decreased by 3

15. *w* times *z*

16. The product of 3 and *c*

17. The product of 5 and *t*

18. 8 times *a*

19. The product of 8, *m*, and *n*

20. The product of 7, *r*, and *s*

21. The product of 3 and the quantity *p* plus *q*

22. The product of 5 and the sum of *a* and *b*

23. Twice the sum of *x* and *y*

24. 3 times the sum of *m* and *n*

25. The sum of twice *x* and *y*

26. The sum of 3 times *m* and *n*

ANSWERS

1. _____
2. _____
3. _____
4. _____
5. _____
6. _____
7. _____
8. _____
9. _____
10. _____
11. _____
12. _____
13. _____
14. _____
15. _____
16. _____
17. _____
18. _____
19. _____
20. _____
21. _____
22. _____
23. _____
24. _____
25. _____
26. _____

27. _____

28. _____

29. _____

30. _____

31. _____

32. _____

33. _____

34. _____

35. _____

36. _____

37. _____

38. _____

39. _____

40. _____

41. _____

42. _____

43. _____

44. _____

45. _____

46. _____

47. _____

48. _____

49. _____

27. Twice the difference of x and y

28. 3 times the difference of c and d

29. The quantity a plus b times the quantity a minus b

30. The product of x plus y and x minus y

31. The product of m and 3 less than m

32. The product of a and 7 more than a

33. x divided by 5

34. The quotient when b is divided by 8

35. The quotient of a plus b, divided by 7

36. The difference x minus y, divided by 9

37. The difference of p and q, divided by 4

38. The sum of a and 5, divided by 9

39. The sum of a and 3, divided by the difference of a and 3

40. The difference of m and n, divided by the sum of m and n

Write each of the following phrases, using symbols. Use the variable x to represent the number in each case.

41. 5 more than a number

42. A number increased by 8

43. 7 less than a number

44. A number decreased by 10

45. 9 times a number

46. Twice a number

47. 6 more than 3 times a number

48. 5 times a number decreased by 10

49. Twice the sum of a number and 5

50. 3 times the difference of a number and 4

51. The product of 2 more than a number and 2 less than that same number

52. The product of 5 less than a number and 5 more than that same number

53. The quotient of a number and 7

54. A number divided by 3

55. The sum of a number and 5, divided by 8

56. The quotient when 7 less than a number is divided by 3

57. 6 more than a number divided by 6 less than that same number

58. The quotient when 3 less than a number is divided by 3 more than that same number

Identify which are expressions and which are not.

59. $2(x + 5)$

60. $4 + (x - 3)$

61. $4 + \div m$

62. $6 + a = 7$

63. $2b = 6$

64. $x(y + 3)$

65. $2a + 5b$

66. $4x + \cdot 7$

50. _____

51. _____

52. _____

53. _____

54. _____

55. _____

56. _____

57. _____

58. _____

59. _____

60. _____

61. _____

62. _____

63. _____

64. _____

65. _____

66. _____

67. _____

68. _____

69. _____

a. _____

b. _____

c. _____

d. _____

e. _____

f. _____

Write an algebraic expression for each of the following.

67. The earth's population has doubled in the last 40 years. If we let x represent the earth's population 40 years ago, what is the population today?

68. It is estimated that the earth is losing 4000 species of plants and animals every year. If S represents the number of species living last year, how many species are on earth this year?

69. One-third of the world's cropland is losing its topsoil. Let L represent the world's cropland in square kilometers. How many square kilometers of cropland are losing topsoil?

Skillscan (Appendix 1)

Perform each of the indicated operations.

a. $3 + 3 + 3$

b. $3 \cdot 3 \cdot 3$

c. $5 + 5 + 5 + 5$

d. $5 \cdot 5 \cdot 5 \cdot 5$

e. $4 + 4 + 4 + 4 + 4$

f. $4 \cdot 4 \cdot 4 \cdot 4 \cdot 4$

ANSWERS

We provide the answers for the odd-numbered exercises at the end of each exercise set.

1. $c + d$ **3.** $w + z$ **5.** $x + 2$ **7.** $y + 10$ **9.** $a - b$ **11.** $b - 7$ **13.** $r - 6$
15. wz **17.** $5t$ **19.** $8mn$ **21.** $3(p + q)$ **23.** $2(x + y)$ **25.** $2x + y$
27. $2(x - y)$ **29.** $(a + b)(a - b)$ **31.** $m(m - 3)$ **33.** $\dfrac{x}{5}$ **35.** $\dfrac{a + b}{7}$ **37.** $\dfrac{p - q}{4}$
39. $\dfrac{a + 3}{a - 3}$ **41.** $x + 5$ **43.** $x - 7$ **45.** $9x$ **47.** $3x + 6$ **49.** $2(x + 5)$
51. $(x + 2)(x - 2)$ **53.** $\dfrac{x}{7}$ **55.** $\dfrac{x + 5}{8}$ **57.** $\dfrac{x + 6}{x - 6}$ **59.** Expression   **61.** Not an
expression   **63.** Not an expression   **65.** Expression   **67.** $2x$ **69.** $\dfrac{1}{3}L$ or $\dfrac{L}{3}$ **a.** 9
b. 27   **c.** 20   **d.** 625   **e.** 20   **f.** 1024

1.2

Exponents and the Order of Operations

OBJECTIVES

1. To write a product of factors in exponential form
2. To determine the order in which operations should be done

In Section 1.1, we showed how symbols are used to denote the four basic operations. Often in mathematics we define other symbols that will allow us to write expressions in a more compact or "shorthand" form. This is an idea that you have encountered before. For example, given an expression with repeated addition, such as

$$5 + 5 + 5$$

how could this be rewritten? You might respond by noting that the expression

$$3 \cdot 5$$

has the same meaning. You have learned that multiplication is shorthand for repeated addition.

In algebra we frequently have a *factor* that is repeated in an expression several times. For instance, we might have

$$5 \cdot 5 \cdot 5$$

To abbreviate this product, we write

$$5 \cdot 5 \cdot 5 = 5^3$$

This is called *exponential notation* or *exponential form*. The exponent or power, here 3, indicates the number of times that the factor or base, here 5, appears in a product.

$$5 \cdot 5 \cdot 5 = 5^3 \quad \nearrow \text{Exponent or power}$$
$$\searrow \text{Base}$$

Be careful: 5^3 is *not* the same as $5 \cdot 3$. Notice that $5^3 = 5 \cdot 5 \cdot 5 = 125$ and $5 \cdot 3 = 15$.

Example 1

(*a*) Write $3 \cdot 3 \cdot 3 \cdot 3$, using exponential form. The number 3 appears 4 times in the product, so

$$3 \cdot 3 \cdot 3 \cdot 3 = 3^4 \quad \nwarrow \text{Four factors of 3}$$

This is read "3 to the fourth power."

Margin notes:

$5 + 5 + 5 = 15$
and
$3 \cdot 5 = 15$

A factor is a number or a variable that is being multiplied by another number or variable.

Since an exponent represents repeated multiplication, 5^3 is an expression.

⟨CAUTION⟩

13

(*b*) Write $x \cdot x \cdot x$, using exponential form. The same idea works for letters or variables, so since x appears 3 times in the product

$$x \cdot x \cdot x = x^3$$

This is read "*x* to the third power" or "*x* cubed."

CHECK YOURSELF 1

Write in exponential form.

1. $4 \cdot 4 \cdot 4 \cdot 4 \cdot 4 \cdot 4$ **2.** $y \cdot y \cdot y \cdot y$

If a product involves a combination of numbers and letters or different letters, the exponential form can also be used.

Example 2

Write each of the following, using exponents.

(*a*) $5 \cdot m \cdot m$

Note that the exponent applies *only* to *m*, *not* to 5.

$$5 \cdot m \cdot m = 5m^2$$
Two factors
of *m*

This is read "5 times *m* to the second power" or "5 times *m* squared."

(*b*) $a \cdot a \cdot b \cdot b \cdot b \cdot b$

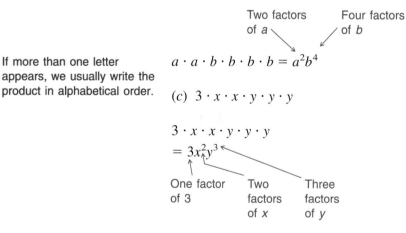

Two factors   Four factors
of *a*   of *b*

If more than one letter appears, we usually write the product in alphabetical order.

$$a \cdot a \cdot b \cdot b \cdot b \cdot b = a^2 b^4$$

(*c*) $3 \cdot x \cdot x \cdot y \cdot y \cdot y$

$$3 \cdot x \cdot x \cdot y \cdot y \cdot y$$
$$= 3x^2 y^3$$

One factor   Two   Three
of 3   factors   factors
of *x*   of *y*

CHECK YOURSELF 2

Write in exponential form.

1. $6 \cdot b \cdot b \cdot b$ **2.** $p \cdot p \cdot p \cdot q \cdot q \cdot q \cdot q \cdot q$ **3.** $5 \cdot m \cdot m \cdot m \cdot n \cdot n$

There may also be situations in which it will be useful to change an expression from exponential form to an expanded form (written as a product of factors).

Example 3

Write each expression in expanded form.

(*a*) $4x^3$

The exponent 3 applies *only* to *x*.

$$4x^3 = 4 \cdot x \cdot x \cdot x$$

(*b*) $(4x)^3$

Now the exponent applies to 4*x* because of the parentheses. So 4*x* is repeated as a factor 3 times.

$$(4x)^3 = (4x) \cdot (4x) \cdot (4x)$$

(*c*) $5x^2y^3$

$$5x^2y^3 = 5 \cdot x \cdot x \cdot y \cdot y \cdot y$$

CHECK YOURSELF 3

Write each expression in expanded form.

1. $5x^4$ **2.** $(5x)^4$ **3.** $3a^3b^3$

An expression will have different values depending on the number assigned to the variables in the expression. Finding the value of an expression is called *evaluating an expression*. This involves replacing each variable with a given value, as illustrated in our next example.

Example 4

Evaluate each expression.

(*a*) $5x^2$ when $x = 2$

We replace *x* with the value 2.

$$5x^2 = 5(2)^2 = 5 \cdot 2 \cdot 2 \qquad \text{2 appears as a factor 2 times}$$
$$= 20$$

(*b*) $2xy^3$ when $x = 2$, $y = 3$

Now let *x* be 2 and *y* be 3. Then

$$2xy^3 = 2(2)(3)^3$$
$$= 2 \cdot 2 \cdot 3 \cdot 3 \cdot 3$$
$$= 108$$

CHECK YOURSELF 4

Evaluate each expression.

1. $3a^3$ when $a = 2$
2. $5m^2n$ when $m = 3$ and $n = 4$

You have now seen all the parts that make up arithmetic expressions. In order to evaluate those expressions, you need to know the order in which the operations are done. To see why, simplify the expression $5 + 2 \cdot 3$.

Method 1	or	Method 2
$\underbrace{5 + 2} \cdot 3$		$5 + \underbrace{2 \cdot 3}$
Add first.		Multiply first.
$= 7 \cdot 3$		$= 5 + 6$
$= 21$		$= 11$

CAUTION

Only one of these results can be correct.

Since we get different answers depending on how we do the problem, the language of algebra would not be clear if there were no agreement on which method is correct. The following rules tell us the order in which operations should be done.

Parentheses and brackets are both grouping symbols. Later we will see that fraction bars and radicals are also grouping symbols.

THE ORDER OF OPERATIONS

STEP 1   Evaluate all expressions inside grouping symbols first.

STEP 2   Evaluate all expressions involving exponents.

STEP 3   Do any multiplication or division in order, working from left to right.

STEP 4   Do any addition or subtraction in order, working from left to right.

Example 5

Evaluate $5 + 2 \cdot 3$.

There are no parentheses or exponents, so start with step 3. You should multiply and then add.

$5 + 2 \cdot 3$
      ↖——— Multiply first.
$= 5 + 6$
    ↖——— Then add.
$= 11$

Note: Method 2 shown above is the correct one.

CHECK YOURSELF 5

Evaluate the following expressions.

1. $20 - 3 \cdot 4$ **2.** $9 + 6 \div 3$

When there are no parentheses, evaluate the exponents first.

Example 6

Evaluate $5 \cdot 3^2$.

$5 \cdot 3^2$ Evaluate the power first.

$= 5 \cdot 9$ Multiply.

$= 45$

CHECK YOURSELF 6

Evaluate $4 \cdot 2^4$.

If an expression contains no parentheses do multiplication and division in order from left to right, then add and subtract.

Example 7

Evaluate $7 \cdot 6 - 5 \cdot 4$.

$7 \cdot 6 - 5 \cdot 4$
$= 42 - 20$
$= 22$

CHECK YOURSELF 7

Evaluate $8 \div 4 + 3 \cdot 2$.

Operations inside grouping symbols are done first.

Example 8

Evaluate $(5 + 2) \cdot 3$.

Do the operation inside the parentheses as the first step.

$(5 + 2) \cdot 3$ Add.

$= 7 \cdot 3$
$= 21$

CHECK YOURSELF 8

Evaluate $4(9 - 3)$.

The principle is the same when more than two or more "levels" of operations are involved.

Example 9

(*a*) Evaluate $5 \cdot 2^4 + 3$.

— Evaluate the power.

$= 5 \cdot 16 + 3$

— Multiply.

$= 80 + 3$

— Add.

$= 83$

(*b*) Evaluate $4(2 + 3)^3$.

— Add inside the parentheses first.

$= 4(5)^3$

— Evaluate the power.

$= 4 \cdot 125$

— Multiply.

$= 500$

(*c*) Evaluate $5(7 - 3)^2 - 10$.

— Evaluate the expression inside the parentheses.

$= 5(4)^2 - 10$

— Evaluate the power.

$= 5 \cdot 16 - 10$

— Multiply.

$= 80 - 10$

— Subtract.

$= 70$

CHECK YOURSELF 9

Evaluate

1. $4 \cdot 3^3 - 8 \cdot 11$. **2.** $12 + 4(2 + 3)^2$.

CHECK YOURSELF ANSWERS

1. (1) 4^6; (2) y^4.
2. (1) $6b^3$; (2) $p^3 q^5$; (3) $5m^3 n^2$.
3. (1) $5 \cdot x \cdot x \cdot x \cdot x$; (2) $5x \cdot 5x \cdot 5x \cdot 5x$; (3) $3 \cdot a \cdot a \cdot a \cdot b \cdot b \cdot b$.
4. (1) 24; (2) 180.
5. (1) 8; (2) 11.
6. 64. **7.** 8. **8.** 24. **9.** (1) 20; (2) 112.

Name _____

Date _____

Build Your Skills

Write each expression, using exponential form.

1. $4 \cdot 4 \cdot 4$

2. $3 \cdot 3 \cdot 3 \cdot 3 \cdot 3$

3. $5 \cdot 5 \cdot 5 \cdot 5 \cdot 5 \cdot 5$

4. $a \cdot a \cdot a \cdot a$

5. $m \cdot m \cdot m \cdot m \cdot m$

6. $2 \cdot 2 \cdot 2 \cdot 2 \cdot 2 \cdot 2 \cdot 2$

7. $2 \cdot x \cdot x \cdot x$

8. $3 \cdot y \cdot y \cdot y \cdot y$

9. $a \cdot a \cdot a \cdot a \cdot b \cdot b$

10. $m \cdot m \cdot n \cdot n \cdot n$

11. $9 \cdot r \cdot r \cdot r \cdot s \cdot s \cdot s$

12. $6 \cdot a \cdot b \cdot b \cdot b \cdot b$

13. $5 \cdot w \cdot z \cdot z \cdot z \cdot z \cdot z$

14. $7 \cdot x \cdot x \cdot x \cdot x \cdot y \cdot y \cdot y$

Write each expression in expanded form.

15. $5x^3$

16. $2y^5$

17. $2x^2y$

18. $3ab^3$

19. $12a^3b^4$

20. $9x^5y^2$

21. $(2x)^3$

22. $(3y)^4$

23. $(5p)^5$

24. $(8s)^3$

Answer	
1.	_____
2.	_____
3.	_____
4.	_____
5.	_____
6.	_____
7.	_____
8.	_____
9.	_____
10.	_____
11.	_____
12.	_____
13.	_____
14.	_____
15.	_____
16.	_____
17.	_____
18.	_____
19.	_____
20.	_____
21.	_____
22.	_____
23.	_____
24.	_____

ANSWERS

25. _____

26. _____

27. _____

28. _____

29. _____

30. _____

31. _____

32. _____

33. _____

34. _____

35. _____

36. _____

37. _____

38. _____

39. _____

40. _____

41. _____

42. _____

43. _____

44. _____

45. _____

46. _____

47. _____

48. _____

49. _____

50. _____

51. _____

52. _____

Evaluate each expression if $x = 3$ and $y = 4$.

25. $3x^2$

26. $2y^2$

27. $(3x)^2$

28. $(2y)^2$

29. $3xy^2$

30. $2x^3y$

Evaluate each expression if $a = 2$ and $b = 5$.

31. a^2b^2

32. a^3b^3

33. $(ab)^2$

34. $(ab)^3$

Evaluate each of the following expressions.

35. $4 + 3 \cdot 5$

36. $8 - 2 \cdot 3$

37. $(4 + 3) \cdot 5$

38. $(8 - 2) \cdot 3$

39. $12 - 8 \div 4$

40. $10 + 20 \div 5$

41. $(12 - 8) \div 4$

42. $(10 + 20) \div 5$

43. $8 \cdot 7 + 2 \cdot 2$

44. $48 \div 8 - 4 \div 2$

45. $8 \cdot (7 + 2) \cdot 2$

46. $48 \div (8 - 4) \div 2$

47. $3 \cdot 5^2$

48. $5 \cdot 2^3$

49. $(3 \cdot 5)^2$

50. $(5 \cdot 2)^3$

51. $4 \cdot 3^2 - 2$

52. $3 \cdot 2^4 - 8$

53. $4(3^2 - 2)$

54. $3(2^4 - 8)$

55. $2 \cdot 4^2 - 8 \cdot 3$

56. $3 \cdot 2^3 - 7 \cdot 3$

57. $(2 \cdot 4)^2 - 8 \cdot 3$

58. $(3 \cdot 2)^3 - 7 \cdot 3$

59. $4(2 + 6)^2$

60. $3(8 - 4)^2$

61. $(4 \cdot 2 + 6)^2$

62. $(3 \cdot 8 - 4)^2$

63. $3(4 + 3)^2$

64. $5(4 - 2)^3$

65. $3 \cdot 4 + 3^2$

66. $5 \cdot 4 - 2^3$

67. $4(2 + 3)^2 - 25$

68. $8 + 2(3 + 3)^2$

69. $(4 \cdot 2 + 3)^2 - 25$

70. $8 + (2 \cdot 3 + 3)^2$

Evaluate each of the following expressions if $a = 2$, $b = 3$, $c = 5$, and $d = 6$.

71. $6bd$

72. $3ac$

73. $a + d$

74. $c - a$

75. $5a + 2b$

76. $3d - 7a$

53. _____

54. _____

55. _____

56. _____

57. _____

58. _____

59. _____

60. _____

61. _____

62. _____

63. _____

64. _____

65. _____

66. _____

67. _____

68. _____

69. _____

70. _____

71. _____

72. _____

73. _____

74. _____

75. _____

76. _____

A series of doublings may be written as $2 \times 2 \times 2 \times 2 \times 2 \ldots$ or 2^n, where n is the number of doublings.

77. The earth's population has doubled approximately 5 times since the birth of Christ. Write this factor in exponential form.

78. Sulfur dioxide (SO_2) is an important part of acid rain. Natural sources add about 10^8 metric tons (t) of sulfur dioxide to the atmosphere each year ($1\ t = 10^3$ kilograms). Express, in expanded form, the number of metric tons of SO_2 added to the atmosphere each year.

Skillscan

Evaluate each expression.

a. $6(2 + 3)$ **b.** $6 \cdot 2 + 6 \cdot 3$

c. $(4 + 5) + 7$ **d.** $4 + (5 + 7)$

e. $2 \cdot (3 \cdot 8)$ **f.** $(2 \cdot 3) \cdot 8$

ANSWERS

1. 4^3 **3.** 5^6 **5.** m^5 **7.** $2x^3$ **9.** a^4b^2 **11.** $9r^3s^3$ **13.** $5wz^5$ **15.** $5 \cdot x \cdot x \cdot x$
17. $2 \cdot x \cdot x \cdot y$ **19.** $12 \cdot a \cdot a \cdot a \cdot b \cdot b \cdot b \cdot b$ **21.** $2x \cdot 2x \cdot 2x$
23. $5p \cdot 5p \cdot 5p \cdot 5p \cdot 5p$ **25.** 27   **27.** 81   **29.** 144   **31.** 100   **33.** 100   **35.** 19
37. 35   **39.** 10   **41.** 1   **43.** 60   **45.** 144   **47.** 75   **49.** 225   **51.** 34
53. 28   **55.** 8   **57.** 40   **59.** 256   **61.** 196   **63.** 147   **65.** 21   **67.** 75
69. 96   **71.** 108   **73.** 8   **75.** 16   **77.** 2^5 **a.** 30   **b.** 30   **c.** 16   **d.** 16
e. 48   **f.** 48

The Properties of Addition and Multiplication

OBJECTIVE
To recognize applications of the commutative, associative, and distributive laws

All the integers, decimals, and fractions that we see in this course are real numbers.

Everything that we do in algebra will be based on certain rules for the operations introduced in Section 1.1. We call these rules *properties of the real numbers*. In this section we consider those properties that we will be using in the remainder of this chapter.

The first two properties tell us that we can add or multiply in any order.

THE COMMUTATIVE PROPERTIES

If a and b are any numbers,

1. $a + b = b + a$ Commutative property of addition
2. $a \cdot b = b \cdot a$ Commutative property of multiplication

---

These rules should come as no surprise. In arithmetic you naturally add or multiply in whichever order is most convenient.

Example 1

(*a*) $5 + 9 = 9 + 5$ and $x + 7 = 7 + x$

These are applications of the commutative property of addition.

(*b*) $5 \cdot 9 = 9 \cdot 5$

This is an application of the commutative property of multiplication.

CHECK YOURSELF 1

Identify the property being applied.

1. $7 + 3 = 3 + 7$
2. $7 \cdot 3 = 3 \cdot 7$
3. $a + 4 = 4 + a$
4. $x \cdot 2 = 2 \cdot x$

We will also want to be able to change the grouping in simplifying expressions. This is possible because of the associative properties. Numbers or variables can be grouped in any manner to find a sum or a product.

> **THE ASSOCIATIVE PROPERTIES**
>
> If a, b, and c are any numbers,
>
> **1.** $a + (b + c) = (a + b) + c$ Associative property of addition
> **2.** $a \cdot (b \cdot c) = (a \cdot b) \cdot c$ Associative property of multiplication

Example 2

(a) Show that $2 + (3 + 8) = (2 + 3) + 8$.

Remember, as we saw in Section 1.2, we always do the operation in the parentheses first.

$2 + \underbrace{(3 + 8)}$
Add first.
$= 2 + 11$
$= 13$

$\underbrace{(2 + 3)} + 8$
Add first.
$= 5 + 8$
$= 13$

So

$2 + (3 + 8) = (2 + 3) + 8$

(b) Show that $\dfrac{1}{3} \cdot (6 \cdot 5) = \left(\dfrac{1}{3} \cdot 6\right) \cdot 5$.

$\dfrac{1}{3} \cdot \underbrace{(6 \cdot 5)}$
Multiply first.
$= \dfrac{1}{3} \cdot (30)$
$= 10$

$\underbrace{\left(\dfrac{1}{3} \cdot 6\right)} \cdot 5$
Multiply first.
$= (2) \cdot 5$
$= 10$

So

$\left(\dfrac{1}{3} \cdot 6\right) \cdot 5 = \dfrac{1}{3}(6 \cdot 5)$

CHECK YOURSELF 2

Show that the following statements are true.

1. $3 + (4 + 7) = (3 + 4) + 7$
2. $3 \cdot (4 \cdot 7) = (3 \cdot 4) \cdot 7$
3. $\left(\dfrac{1}{5} \cdot 10\right) \cdot 4 = \dfrac{1}{5} \cdot (10 \cdot 4)$

Another important property involves addition and multiplication together. We can illustrate this property with an application. Suppose that we want to find the total of the two areas shown.

Remember: The area of a rectangle is the product of its length and width:

$A = L \cdot W$

We can find the total area by adding first to find the overall width. Then we multiply by the length.

[or]

We can find the total area as a sum of the two areas.

Length	Overall width		(Area 1) Length · Width		(Area 2) Length · Width
30	$\cdot (10 + 15)$		$30 \cdot 10$		$+ \; 30 \cdot 15$
$= 30$	$\cdot 25$		$= 300 + 450$		
$= 750$			$= 750$		

So

$$30 \cdot (10 + 15) = 30 \cdot 10 + 30 \cdot 15$$

This leads us to the following property.

Note the pattern.

$a(b + c) = a \cdot b + a \cdot c$

We "distributed" the multiplication "over" the addition.

THE DISTRIBUTIVE PROPERTY

If a, b, and c are any numbers,

$a(b + c) = a \cdot b + a \cdot c$

Example 3

Use the distributive property to simplify the following.

(*a*) $5(3 + 4)$

$5(3 + 4) = 5 \cdot 3 + 5 \cdot 4$

Note: $5(3 + 4) = 5 \cdot 7 = 35$

or

$5 \cdot 3 + 5 \cdot 4 = 15 + 20 = 35$

(b) $8(x + y)$

Because the variables are different, $8x + 8y$ cannot be simplified further.

$$8(x + y) = 8x + 8y$$

(c) $2(3x + 5)$

$$2(3x + 5) = 2 \cdot 3x + 2 \cdot 5$$
$$= 6x + 10$$

(d) $\dfrac{1}{3}(9 + 12) = \dfrac{1}{3} \cdot 9 + \dfrac{1}{3} \cdot 12$
$$= 3 + 4 = 7$$

Note: It is also true that

$$\frac{1}{3}(9 + 12) = \frac{1}{3}(21) = 7$$

CHECK YOURSELF 3

Use the distributive property to simplify (remove the parentheses).

1. $4(6 + 7)$ **2.** $9(m + n)$

3. $3(5a + 7)$ **4.** $\dfrac{1}{5}(10 + 15)$

We will use the distributive property in this fashion in Section 1.4.

Note: The distributive property can also be used in a "right-hand" manner:

$$(2 + 3)x = (2 + 3)x = 2x + 3x = 5x$$

CHECK YOURSELF ANSWERS

1. (1) Commutative property of addition; (2) commutative property of multiplication; (3) commutative property of addition; (4) commutative property of multiplication.

2. (1) $3 + (4 + 7) = 3 + 11 = 14$ (2) $3 \cdot (4 \cdot 7) = 3 \cdot 28 = 84$
      $(3 + 4) + 7 = 7 + 7 = 14.$ $(3 \cdot 4) \cdot 7 = 12 \cdot 7 = 84.$

     (3) $\left(\dfrac{1}{5} \cdot 10 \right) \cdot 4 = 2 \cdot 4 = 8$

        $\dfrac{1}{5} \cdot (10 \cdot 4) = \dfrac{1}{5} \cdot 40 = 8.$

3. (1) $4 \cdot 6 + 4 \cdot 7$, or 52; (2) $9m + 9n$; (3) $15a + 21$; (4) 5.

Name

Date

Build Your Skills

Identify the property that is illustrated by each of the following statements.

1. $5 + 9 = 9 + 5$

2. $6 + 3 = 3 + 6$

3. $2 \cdot (3 \cdot 5) = (2 \cdot 3) \cdot 5$

4. $3 \cdot (5 \cdot 6) = (3 \cdot 5) \cdot 6$

5. $10 \cdot 5 = 5 \cdot 10$

6. $8 \cdot 4 = 4 \cdot 8$

7. $8 + 12 = 12 + 8$

8. $6 + 2 = 2 + 6$

9. $(5 \cdot 7) \cdot 2 = 5 \cdot (7 \cdot 2)$

10. $(8 \cdot 9) \cdot 2 = 8 \cdot (9 \cdot 2)$

11. $9 \cdot 8 = 8 \cdot 9$

12. $6 \cdot 4 = 4 \cdot 6$

13. $2(3 + 5) = 2 \cdot 3 + 2 \cdot 5$

14. $5 \cdot (4 + 6) = 5 \cdot 4 + 5 \cdot 6$

15. $5 + (7 + 8) = (5 + 7) + 8$

16. $8 + (2 + 9) = (8 + 2) + 9$

17. $(10 + 5) + 9 = 10 + (5 + 9)$

18. $(5 + 5) + 3 = 5 + (5 + 3)$

19. $7 \cdot (3 + 8) = 7 \cdot 3 + 7 \cdot 8$

20. $5 \cdot (6 + 8) = 5 \cdot 6 + 5 \cdot 8$

Verify that each of the following statements is true by evaluating each side of the equation separately and comparing the results.

21. $7 \cdot (3 + 4) = 7 \cdot 3 + 7 \cdot 4$

22. $4 \cdot (5 + 1) = 4 \cdot 5 + 4 \cdot 1$

23. $2 + (9 + 8) = (2 + 9) + 8$

24. $6 + (15 + 3) = (6 + 15) + 3$

25. $5 \cdot (6 \cdot 3) = (5 \cdot 6) \cdot 3$

26. $2 \cdot (9 \cdot 10) = (2 \cdot 9) \cdot 10$

ANSWERS

1. _____
2. _____
3. _____
4. _____
5. _____
6. _____
7. _____
8. _____
9. _____
10. _____
11. _____
12. _____
13. _____
14. _____
15. _____
16. _____
17. _____
18. _____
19. _____
20. _____
21. _____
22. _____
23. _____
24. _____
25. _____
26. _____

27. $5 \cdot (2 + 8) = 5 \cdot 2 + 5 \cdot 8$

28. $3 \cdot (10 + 2) = 3 \cdot 10 + 3 \cdot 2$

29. $(3 + 12) + 8 = 3 + (12 + 8)$

30. $(8 + 12) + 7 = 8 + (12 + 7)$

31. $(4 \cdot 7) \cdot 2 = 4 \cdot (7 \cdot 2)$

32. $(6 \cdot 5) \cdot 3 = 6 \cdot (5 \cdot 3)$

33. $\frac{1}{2} \cdot (2 + 6) = \frac{1}{2} \cdot 2 + \frac{1}{2} \cdot 6$

34. $\frac{1}{3} \cdot (6 + 9) = \frac{1}{3} \cdot 6 + \frac{1}{3} \cdot 9$

35. $\left(\frac{2}{3} + \frac{1}{6}\right) + \frac{1}{3} = \frac{2}{3} + \left(\frac{1}{6} + \frac{1}{3}\right)$

36. $\frac{3}{4} + \left(\frac{5}{8} + \frac{1}{2}\right) = \left(\frac{3}{4} + \frac{5}{8}\right) + \frac{1}{2}$

37. $2.5 + (4.6 + 3.2) = (2.5 + 4.6) + 3.2$

38. $8.1 + (1.9 + 3.4) = (8.1 + 1.9) + 3.4$

39. $\frac{1}{2} \cdot (2 \cdot 8) = \left(\frac{1}{2} \cdot 2\right) \cdot 8$

40. $\frac{1}{5} \cdot (5 \cdot 3) = \left(\frac{1}{5} \cdot 5\right) \cdot 3$

41. $\left(\frac{3}{5} \cdot \frac{5}{6}\right) \cdot \frac{4}{3} = \frac{3}{5} \cdot \left(\frac{5}{6} \cdot \frac{4}{3}\right)$

42. $\frac{4}{7} \cdot \left(\frac{21}{16} \cdot \frac{8}{3}\right) = \left(\frac{4}{7} \cdot \frac{21}{16}\right) \cdot \frac{8}{3}$

43. $2.5 \cdot (4 \cdot 5) = (2.5 \cdot 4) \cdot 5$

44. $4.2 \cdot (5 \cdot 2) = (4.2 \cdot 5) \cdot 2$

Use the distributive property to remove the parentheses in each of the following expressions. Then simplify your result where possible.

45. $2(3 + 5)$

46. $5(4 + 6)$

47. $3(x + 5)$

48. $5(y + 8)$

49. $4(w + v)$

50. $7(c + d)$

51. $2(3x + 5)$

52. $3(7a + 4)$

53. $\frac{1}{3} \cdot (15 + 9)$

54. $\frac{1}{6} \cdot (36 + 24)$

Use the properties of addition and multiplication to complete each of the following statements.

55. $5 + 7 = \quad + 5$

56. $(5 + 3) + 4 = 5 + (\quad + 4)$

57. $(8)(3) = (3)(\quad)$

58. $8(3 + 4) = 8 \cdot 3 + \quad \cdot 4$

59. $7(2 + 5) = 7 \cdot \quad + 7 \cdot 5$

60. $4 \cdot (2 \cdot 4) = (\quad \cdot 2) \cdot 4$

Use the indicated property to write an expression that is equivalent to each of the following expressions.

61. $3 + 7$ (commutative property)

62. $2(3 + 4)$ (distributive property)

63. $5 \cdot (3 \cdot 2)$ (associative property)

64. $(3 + 5) + 2$ (associative property)

65. $2 \cdot 4 + 2 \cdot 5$ (distributive property)

66. $7 \cdot 9$ (commutative property)

Think About These

You will find that many of the exercise sets in this text have a set of problems labeled "Think About These." These are particularly challenging exercises which either introduce ideas that extend the material of the section or require you to generalize from what you have learned.

Evaluate each of the following pairs of expressions. Then answer the given question.

67. $8 - 5$ and $5 - 8$
Do you think subtraction is commutative?

68. $12 \div 3$ and $3 \div 12$
Do you think division is commutative?

ANSWERS

53. _____

54. _____

55. _____

56. _____

57. _____

58. _____

59. _____

60. _____

61. _____

62. _____

63. _____

64. _____

65. _____

66. _____

67. _____

68. _____

69. _____

70. _____

71. _____

72. _____

a. _____

b. _____

c. _____

d. _____

e. _____

f. _____

g. _____

h. _____

i. _____

69. $(12 - 8) - 4$ and $12 - (8 - 4)$
Do you think subtraction is associative?

70. $(48 \div 16) \div 4$ and $48 \div (16 \div 4)$
Do you think division is associative?

71. $3(6 - 2)$ and $3 \cdot 6 - 3 \cdot 2$
Do you think multiplication is distributive over subtraction?

72. $\dfrac{1}{2}(16 - 10)$ and $\dfrac{1}{2} \cdot 16 - \dfrac{1}{2} \cdot 10$
Do you think multiplication is distributive over subtraction?

Skillscan (Appendix 1)

a. $\dfrac{5}{2} + \dfrac{3}{2}$

b. $\dfrac{5}{6} + \dfrac{2}{3}$

c. $\dfrac{4}{3} + \dfrac{3}{5}$

d. $\dfrac{3}{8} + \dfrac{7}{12}$

e. $\left(\dfrac{5}{3}\right)\left(\dfrac{9}{2}\right)$

f. $\left(\dfrac{8}{5}\right)\left(\dfrac{5}{12}\right)$

g. $\left(\dfrac{12}{25}\right)\left(\dfrac{15}{20}\right)$

h. $\left(\dfrac{9}{16}\right)\left(\dfrac{24}{27}\right)$

i. $(3)\left(\dfrac{1}{3}\right)$

ANSWERS

1. Commutative property of addition   3. Associative property of multiplication
5. Commutative property of multiplication   7. Commutative property of addition
9. Associative property of multiplication   11. Commutative property of multiplication
13. Distributive property   15. Associative property of addition   17. Associative property of
addition   19. Distributive property   21. $49 = 49$ 23. $19 = 19$ 25. $90 = 90$

27. $50 = 50$ 29. $23 = 23$ 31. $56 = 56$ 33. $4 = 4$ 35. $\dfrac{7}{6} = \dfrac{7}{6}$ 37. $10.3 = 10.3$

39. $8 = 8$ 41. $\dfrac{2}{3} = \dfrac{2}{3}$ 43. $50 = 50$ 45. 16 47. $3x + 15$ 49. $4w + 4v$

51. $6x + 10$ 53. 8 55. 7 57. 8 59. 2 61. $7 + 3$ 63. $(5 \cdot 3) \cdot 2$

65. $2 \cdot (4 + 5)$ 67. No   69. No   71. Yes   a. 4 b. $\dfrac{3}{2}$ c. $\dfrac{29}{15}$ d. $\dfrac{23}{24}$

e. $\dfrac{15}{2}$ f. $\dfrac{2}{3}$ g. $\dfrac{9}{25}$ h. $\dfrac{1}{2}$ i. 1

Adding and Subtracting Algebraic Expressions

OBJECTIVES

1. To add algebraic expressions
2. To subtract algebraic expressions

To find the perimeter of (or the distance around) a rectangle, we add 2 times the length and 2 times the width. In the language of algebra, this can be written as

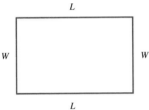

Perimeter $= 2L + 2W$

We call $2L + 2W$ an *algebraic expression,* or more simply an *expression.* Recall from Section 1.1 that an expression allows us to write a mathematical idea in symbols. It can be thought of as a meaningful collection of letters, numbers, and operation signs.

Some expressions are

(*a*) $5x^2$
(*b*) $3a + 2b$
(*c*) $4x^3 - 2y + 1$
(*d*) $3(x^2 + y^2)$

In these expressions, the addition and subtraction signs break the expressions into smaller parts called *terms.*

> A *term* is a number, or the product of a number and one or more variables, raised to a power.

In an expression, each sign ($+$ or $-$) is a part of the term that follows the sign.

Example 1

Note that each term "owns" the sign that precedes it.

(*a*) $5x^2$ has one term.
(*b*) $3a + 2b$ has two terms: $3a$ and $2b$.

   Term   Term

(*c*) $4x^3 - 2y + 1$ has three terms: $4x^3$, $-2y$, and 1.

   Term   Term   Term

CHECK YOURSELF 1

List the terms of each expression.

1. $2b^4$ **2.** $5m + 3n$ **3.** $2s^2 - 3t - 6$

Note that a term in an expression may have any number of factors. For instance, $5xy$ is a term. It has factors of 5, x, and y. The number factor of a term is called the *numerical coefficient*. So for the term $5xy$, the numerical coefficient is 5.

Example 2

(a) $4a$ has the numerical coefficient 4.
(b) $6a^3b^4c^2$ has the numerical coefficient 6.
(c) $-7m^2n^3$ has the numerical coefficient -7.
(d) Since $1 \cdot x = x$, the numerical coefficient of x is understood to be 1.

CHECK YOURSELF 2

Give the numerical coefficient for each of the following terms.

1. $8a^2b$ **2.** $-5m^3n^4$ **3.** y

If terms contain exactly the *same letters* (or variables) raised to the *same powers*, they are called *like terms*.

Example 3

(a) The following are like terms.

$6a$ and $7a$	Each pair of terms
$5b^2$ and b^2	has the same letters,
$10x^2y^3z$ and $-6x^2y^3z$	with each letter
$-3m^3$ and m^3	raised to the same power—the numerical coefficients can be any number.

(b) The following are *not* like terms.

Different letters

$6a$ and $7b$

Different exponents

$5b^2$ and b^3

Different exponents

$3x^2y$ and $4xy^2$

CHECK YOURSELF 3

Circle the like terms.

$$5a^2b \qquad ab^2 \qquad a^2b \qquad -3a^2 \qquad 4ab \qquad 3b^2 \qquad -7a^2b$$

Like terms of an expression can always be combined into a single term. Look at the following:

$$\underbrace{2x}_{x+x} + \underbrace{5x}_{x+x+x+x+x} = \underbrace{7x}_{x+x+x+x+x+x+x}$$

Rather than having to write out all those x's, try

Here we use the distributive property from Section 1.3.

$$2x + 5x = (2 + 5)x = 7x$$

In the same way,

$$9b + 6b = (9 + 6)b = 15b$$

You don't have to write all this out—just do it mentally!

and $10a - 4a = (10 - 4)a = 6a$

This leads us to the following rule.

COMBINING LIKE TERMS

To combine like terms, use the following steps.

STEP 1   Add or subtract the numerical coefficients.

STEP 2   Attach the common variables.

Example 4

Combine like terms.

(a) $8m + 5m = (8 + 5)m = 13m$

(b) $5pq^3 - 4pq^3 = 1pq^3 = pq^3$

(c) $7a^3b^2 - 7a^3b^2 = 0a^3b^2 = 0$

Remember, that when any factor is multiplied by 0, the product is 0.

CHECK YOURSELF 4

Combine like terms.

1. $6b + 8b$ **2.** $12x^2 - 3x^2$ **3.** $8xy^3 - 7xy^3$

4. $9a^2b^4 - 9a^2b^4$

Let's look at some expressions involving more than two terms. The idea is just the same.

The distributive property can be used over any number of terms.

Example 5

Combine like terms.

(a) $5ab - 2ab + 3ab$

$\qquad = (5 - 2 + 3)ab = 6ab$

Only like terms can be combined.

(b) $\overbrace{8x - 2x} + 5y$

$\qquad = 6x \qquad + 5y$

$\qquad\quad$ Like terms $\qquad$ Like terms

(c) $\overbrace{5m + 8n \qquad + 4m} - 3n$

$\qquad = (5m + 4m) + (8n - 3n)$

$\qquad = \quad 9m \qquad + \quad 5n$

Here we have used the associative and commutative properties.

With practice you won't be writing out these steps, but doing them mentally.

(d) $4x^2 + 2x - 3x^2 + x$

$\qquad = (4x^2 - 3x^2) + (2x + x)$

$\qquad = x^2 + 3x$

As these examples illustrate, combining like terms often means changing the grouping and the order in which the terms are written. Again all this is possible because of the properties of addition that we introduced in Section 1.3.

CHECK YOURSELF 5

Combine like terms.

1. $4m^2 - 3m^2 + 8m^2$ $\qquad\qquad$ **2.** $9ab + 3a - 5ab$

3. $4p + 7q + 5p - 3q$

CHECK YOURSELF ANSWERS

1. (1) $2b^4$; (2) $5m, 3n$; (3) $2s^2, -3t, -6$.

2. (1) 8; (2) -5; (3) 1.

3. The like terms are $5a^2b$, a^2b, and $-7a^2b$.

4. (1) $14b$; (2) $9x^2$; (3) xy^3; (4) 0.

5. (1) $9m^2$; (2) $4ab + 3a$; (3) $9p + 4q$.

Name _____

Date _____

ANSWERS

Build Your Skills

List the terms of the following expressions.

1. $5a + 2$

2. $7a - 4b$

3. $4x^3$

4. $3x^2$

5. $3x^2 + 3x - 7$

6. $2a^3 - a^2 + a$

Circle the like terms in the following groups of terms.

7. $5ab, 3b, 3a, 4ab$

8. $9m^2, 8mn, 5m^2, 7m$

9. $4xy^2, 2x^2y, 5x^2, -3x^2y, 5y, 6x^2y$

10. $8a^2b, 4a^2, 3ab^2, -5a^2b, 3ab, 5a^2b$

Combine the like terms.

11. $4m + 6m$

12. $5a^2 + 9a^2$

13. $5b^3 + 12b^3$

14. $8rs + 12rs$

15. $21xyz + 7xyz$

16. $4mn^2 + 15mn^2$

17. $9z^2 - 3z^2$

18. $7m - 6m$

19. $5a^3 - 5a^3$

20. $13xy - 9xy$

21. $19n^2 - 18n^2$

22. $7cd - 7cd$

23. $18p^2q - 3p^2q$

24. $15r^3s^2 - 6r^3s^2$

25. $7x^2 - 5x^2 + 4x^2$

26. $9uv + 5uv - 8uv$

1. _____

2. _____

3. _____

4. _____

5. _____

6. _____

7. _____

8. _____

9. _____

10. _____

11. _____

12. _____

13. _____

14. _____

15. _____

16. _____

17. _____

18. _____

19. _____

20. _____

21. _____

22. _____

23. _____

24. _____

25. _____

26. _____

27. _____

28. _____

29. _____

30. _____

31. _____

32. _____

33. _____

34. _____

35. _____

36. _____

37. _____

38. _____

39. _____

40. _____

41. _____

42. _____

43. _____

44. _____

45. _____

46. _____

27. $5a - 3a + 4b$

28. $7m^2 - 3m + 4m^2$

29. $5x + 3y - 2x - 2y$

30. $8a^2 + 9a + 5a^2 - 7a$

31. $4a + 7b + 3 - 2a + 3b - 2$

32. $5p^2 + 2p + 8 + 4p^2 + 5p - 6$

33. $\dfrac{2}{3}m + 3 + \dfrac{4}{3}m$

34. $\dfrac{1}{5}a - 2 + \dfrac{4}{5}a$

35. $\dfrac{13}{5}x + 2 - \dfrac{3}{5}x + 5$

36. $\dfrac{17}{12}y + 7 + \dfrac{7}{12}y - 3$

37. $2.3a + 7 + 4.7a + 3$

38. $5.8m + 4 - 2.8m + 11$

Perform the indicated operations.

39. The sum of $5a^4$ and $8a^4$ is

40. The sum of $9p^2$ and $12p^2$ is

41. Subtract $12a^3$ from $15a^3$.

42. Subtract $5m^3$ from $18m^3$.

43. Subtract $4x$ from the sum of $8x$ and $3x$.

44. Subtract $8ab$ from the sum of $7ab$ and $5ab$.

45. Subtract $3mn^2$ from the sum of $9mn^2$ and $5mn^2$.

46. Subtract $4x^2y$ from the sum of $6x^2y$ and $12x^2y$.

Think About These

Use the distributive property to remove the parentheses in each expression. Then simplify by combining like terms.

47. $2(3x + 2) + 4$

48. $3(4z + 5) - 9$

49. $5(6a - 2) + 12a$

50. $7(4w - 3) - 25w$

51. $4s + 2(s + 4) + 4$

52. $5p + 4(p + 3) - 8$

Evaluate each of the following expressions if $a = 2$, $b = 3$, and $c = 5$. Be sure to combine like terms where possible as the first step.

53. $5a^2 + 2a$

54. $7b^2 - 7b$

55. $6c^2 + 2c^2$

56. $6b^3 - 2b^3$

57. $5b + 3a - 2b$

58. $7c - 2b + 3c$

59. $5ac^2 - 2ac^2$

60. $5a^3b - 2a^3b$

Forest use in the United States can be categorized in three ways: wilderness, national parks, and national forests. Although the Pacific Northwest states of Oregon and Washington are known for their forests, much of this land is actually designated as wilderness or national park instead of as national forest. Oregon has 13 national forests, 1 national park, and 35 wilderness areas. Washington has 6 national forests, 3 national parks, and 25 wilderness areas.

61. Let F represent the area of forests, P represent the area of parks, and W represent the area of wilderness.
   (a) Write an expression representing the total area of the three types of forest categories in Oregon.

   (b) Write an expression representing the total area of three types of forest categories in Washington.

   (c) Find the sum of parts (a) and (b) to determine the total area of the three types of forest use in the Pacific northwest states of Oregon and Washington.

© 1993 McGraw-Hill, Inc.

47. _____

48. _____

49. _____

50. _____

51. _____

52. _____

53. _____

54. _____

55. _____

56. _____

57. _____

58. _____

59. _____

60. _____

61. _(a)_ _____

   _(b)_ _____

   _(c)_ _____

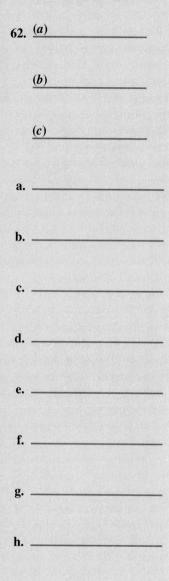

62. If C represents the average area cut on the national forests and R the average area of roads in the forests, then $C + R$ represents the total area of forest land that is cut or contains roads.

(a) How much of the Oregon's national forests has been cut or contains roads?

(b) How much of the Washington's national forests has been cut or contains roads?

(c) What is the total amount of national forest that has been cut or contains roads in Oregon and Washington?

Skillscan (Section 1.2)

Write each expression in expanded form.

a. 2^2 **b.** 2^3

c. 2^4 **d.** 2^5

e. $2^3 \cdot 2^2$ **f.** $2^2 \cdot 2^5$

g. $\dfrac{2^5}{2^2}$ **h.** $\dfrac{2^4}{2^3}$

ANSWERS

1. $5a, 2$ **3.** $4x^3$ **5.** $3x^2, 3x, -7$ **7.** $5ab, 4ab$ **9.** $2x^2y, -3x^2y, 6x^2y$ **11.** $10m$
13. $17b^3$ **15.** $28xyz$ **17.** $6z^2$ **19.** 0 **21.** n^2 **23.** $15p^2q$ **25.** $6x^2$ **27.** $2a + 4b$
29. $3x + y$ **31.** $2a + 10b + 1$ **33.** $2m + 3$ **35.** $2x + 7$ **37.** $7a + 10$ **39.** $13a^4$
41. $3a^3$ **43.** $7x$ **45.** $11mn^2$ **47.** $6x + 8$ **49.** $42a - 10$ **51.** $6s + 12$ **53.** 24
55. 200 **57.** 15 **59.** 150 **61.** $a.$ $13F + P + 35W$ $b.$ $6F + 3P + 25W$ $c.$ $19F + 4P + 60W$
a. $2 \cdot 2$ **b.** $2 \cdot 2 \cdot 2$ **c.** $2 \cdot 2 \cdot 2 \cdot 2$ **d.** $2 \cdot 2 \cdot 2 \cdot 2 \cdot 2$ **e.** $2 \cdot 2 \cdot 2 \cdot 2 \cdot 2$
f. $2 \cdot 2 \cdot 2 \cdot 2 \cdot 2 \cdot 2 \cdot 2$ **g.** $\dfrac{2 \cdot 2 \cdot 2 \cdot 2 \cdot 2}{2 \cdot 2}$ or $2 \cdot 2 \cdot 2$ **h.** $\dfrac{2 \cdot 2 \cdot 2 \cdot 2}{2 \cdot 2 \cdot 2}$ or 2

1.5 Multiplying and Dividing Algebraic Expressions

OBJECTIVES

1. To find the product of certain algebraic expressions
2. To find the quotient of certain algebraic expressions

In Section 1.2 we introduced the exponential notation. Remember that the exponent tells us how many times the base is to be used as a factor.

Exponent
↓
$$2^5 = 2 \cdot 2 \cdot 2 \cdot 2 \cdot 2 = 32$$
↑   ↖
Base   The fifth power of 2

The notation can also be used when you are working with letters or variables.

In general,

$$x^m = \underbrace{x \cdot x \cdot \cdots \cdot x}_{m \text{ factors}}$$

where m is a natural number. Natural numbers are the numbers we use for counting: 1, 2, 3, and so on.

$$x^4 = \underbrace{x \cdot x \cdot x \cdot x}_{4 \text{ factors}}$$

Now look at the product $x^2 \cdot x^3$.

$$x^2 \cdot x^3 = \underbrace{(x \cdot x)(x \cdot x \cdot x)}_{} = \underbrace{x \cdot x \cdot x \cdot x \cdot x}_{} = x^5$$

2 factors + 3 factors = 5 factors

So

Note that the exponent of x^5 is the *sum* of the exponents in x^2 and x^3.

$$x^2 \cdot x^3 = x^{2+3} = x^5$$

This leads us to the following property of exponents.

THE FIRST PROPERTY OF EXPONENTS

For any natural numbers m and n and any number x,

$$x^m \cdot x^n = x^{m+n}$$

In words, to multiply expressions with the same base, keep the base and add the exponents.

Example 1

If no exponent is written, it is understood to be 1. In symbols,

$$x = x^1$$

‹**CAUTION**›

(*a*) $a^5 \cdot a^7 = a^{5+7} = a^{12}$
(*b*) $x \cdot x^8 = x^1 \cdot x^8 = x^{1+8} = x^9$
(*c*) $3^2 \cdot 3^4 = 3^{2+4} = 3^6$

Be careful! The product is *not* 9^6. The base does not change.

39

(d) $y^2 \cdot y^3 \cdot y^5 = y^{2+3+5} = y^{10}$

You can extend the first property to find the product of any number of factors.

(e) $x^3 \cdot y^4$ *cannot* be simplified. The bases are not the same.

CHECK YOURSELF 1

Multiply.

1. $b^6 \cdot b^8$ **2.** $y^7 \cdot y$ **3.** $2^3 \cdot 2^4$ **4.** $a^2 \cdot a^4 \cdot a^3$

Suppose that numerical coefficients (other than 1) are involved in a product. For instance, look at the product

Note that although we have several factors, this is still a single *term*.

$2x^3 \cdot 3x^5$

To find the product, multiply the numbers and then use the first property of exponents to combine the variables.

Multiply.

$2x^3 \cdot 3x^5 = (2 \cdot 3)(x^3 \cdot x^5)$

Add the exponents.

$$= 6x^{3+5}$$
$$= 6x^8$$

You may have noticed that we have again changed the order and grouping. This method uses the commutative and associative properties of Section 1.3.

Example 2*

Multiply.

Again we have written out all the steps. You can do the multiplication mentally with practice.

(a) $5a^4 \cdot 7a^6 = (5 \cdot 7)(a^4 \cdot a^6) = 35a^{10}$
(b) $y^2 \cdot 3y^3 \cdot 6y^4 = (1 \cdot 3 \cdot 6)(y^2 \cdot y^3 \cdot y^4) = 18y^9$
(c) $2x^2y^3 \cdot 3x^5y^2 = (2 \cdot 3)(x^2 \cdot x^5)(y^3 \cdot y^2) = 6x^7y^5$

CHECK YOURSELF 2

Multiply.

1. $4x^3 \cdot 7x^5$ **2.** $3a^2 \cdot 2a^4 \cdot 2a^5$ **3.** $3m^2n^4 \cdot 5m^3n$

---

*When an example requires simplification of an expression, the expression will be screened. The simplification will then follow the equals sign.

What about dividing expressions when exponents are involved? For instance, what if we want to divide x^5 by x^2? We can use the following approach to division:

$$\frac{x^5}{x^2} = \frac{\overbrace{x \cdot x \cdot x \cdot x \cdot x}^{5 \text{ factors}}}{\underbrace{x \cdot x}_{2 \text{ factors}}} = \frac{x \cdot x \cdot x \cdot x \cdot x}{x \cdot x}$$

We can divide by 2 factors of x.

$$= \overbrace{x \cdot x \cdot x}^{3 \text{ factors}} = x^3$$

So

$$\frac{x^5}{x^2} = x^{5-2} = x^3$$

Note that the exponent of x^3 is the *difference* of the exponents in x^5 and x^2.

This leads us to a second property of exponents.

THE SECOND PROPERTY OF EXPONENTS

For any natural numbers m and n where m is greater than n and any number x not equal to zero,

$$\frac{x^m}{x^n} = x^{m-n}$$

In words, to divide expressions with the *same base*, keep the base and subtract the exponents.

Example 3

Divide the following.

(a) $\dfrac{y^7}{y^3} = y^{7-3} = y^4$

(b) $\dfrac{m^6}{m} = \dfrac{m^6}{m^1} = m^{6-1} = m^5$

(c) $\dfrac{a^3 b^5}{a^2 b^2} = a^{3-2} \cdot b^{5-2} = ab^3$ — Apply the second property to each variable separately.

CHECK YOURSELF 3

Divide.

1. $\dfrac{m^9}{m^6}$ 2. $\dfrac{a^8}{a}$ 3. $\dfrac{a^3 b^5}{a^2}$ 4. $\dfrac{r^5 s^6}{r^3 s^2}$

If numerical coefficients are involved, just divide the numbers and then use the second law of exponents to divide the variables. Look at the following example.

Example 4

Divide the following.

Subtract the exponents.

(a) $\dfrac{6x^5}{3x^2} = 2x^{5-2} = 2x^3$

6 divided by 3

20 divided by 5

(b) $\dfrac{20a^7b^5}{5a^3b^4} = 4a^{7-3} \cdot b^{5-4}$

Again apply the second property to each variable separately.

$= 4a^4b$

CHECK YOURSELF 4

Divide.

1. $\dfrac{4x^3}{2x}$ **2.** $\dfrac{20a^6}{5a^2}$ **3.** $\dfrac{24x^5y^3}{4x^2y^2}$

Be careful! Later in this text you will encounter expressions such as

$\dfrac{x^3 + 2}{x}$

Students are sometimes tempted to apply the second property of exponents (*incorrectly*) to the first *term* of the numerator and the denominator to write

$\dfrac{x^3 + 2}{x} = x^2 + 2$

Again this is *not correct*. The given expression cannot be further simplified with this approach.

CHECK YOURSELF ANSWERS

1. (1) b^{14}; (2) y^8; (3) 2^7; (4) a^9.
2. (1) $28x^8$; (2) $12a^{11}$; (3) $15m^5n^5$.
3. (1) m^3; (2) a^7; (3) ab^5; (4) r^2s^4.
4. (1) $2x^2$; (2) $4a^4$; (3) $6x^3y$.

Name _____

Date _____

Build Your Skills

Multiply.

1. $x^4 \cdot x^8$

2. $b \cdot b^5$

3. $5^3 \cdot 5^3$

4. $y^7 \cdot y^3$

5. $a^9 \cdot a$

6. $3^4 \cdot 3^5$

7. $z^{10} \cdot z^3$

8. $x^7 \cdot x$

9. $p^5 \cdot p^7$

10. $s^6 \cdot s^9$

11. $x^3y \cdot x^2y^4$

12. $m^2n^3 \cdot mn^4$

13. $w^5 \cdot w^2 \cdot w$

14. $x^5 \cdot x^4 \cdot x^6$

15. $m^3 \cdot m^2 \cdot m^4$

16. $r^3 \cdot r \cdot r^5$

17. $a^3b \cdot a^2b^2 \cdot ab^3$

18. $w^2z^3 \cdot wz \cdot w^3z^4$

19. $p^2q \cdot p^3q^5 \cdot pq^4$

20. $c^3d \cdot c^4d^2 \cdot cd^5$

21. $3a^5 \cdot 2a^4$

22. $5s^7 \cdot s^3$

23. $x^4 \cdot 3x^3$

24. $4m^5 \cdot 3m^6$

25. $5m^3n^2 \cdot 4mn^3$

26. $7x^2y^5 \cdot 6xy^4$

27. $6x^3y \cdot 9xy^5$

28. $5a^3b \cdot 10ab^4$

ANSWERS

1. _____
2. _____
3. _____
4. _____
5. _____
6. _____
7. _____
8. _____
9. _____
10. _____
11. _____
12. _____
13. _____
14. _____
15. _____
16. _____
17. _____
18. _____
19. _____
20. _____
21. _____
22. _____
23. _____
24. _____
25. _____
26. _____
27. _____
28. _____

29. $2a^2 \cdot a^3 \cdot 3a^7$

30. $4x^5 \cdot 2x^3 \cdot 3x^2$

31. $3c^2d \cdot 4cd^3 \cdot 2c^5d$

32. $5p^2q \cdot p^3q^2 \cdot 3pq^3$

33. $5m^2 \cdot m^3 \cdot 2m \cdot 3m^4$

34. $3a^3 \cdot 2a \cdot a^4 \cdot 2a^5$

35. $2r^3s \cdot rs^2 \cdot 3r^2s \cdot 5rs$

36. $6a^2b \cdot ab \cdot 3ab^3 \cdot 2a^2b$

Divide.

37. $\dfrac{a^8}{a^5}$

38. $\dfrac{m^7}{m}$

39. $\dfrac{y^7}{y}$

40. $\dfrac{b^8}{b^3}$

41. $\dfrac{p^{15}}{p^{10}}$

42. $\dfrac{s^{18}}{s^{12}}$

43. $\dfrac{x^5y^3}{x^2y^2}$

44. $\dfrac{s^5t^4}{s^3t^2}$

45. $\dfrac{6m^3}{3m}$

46. $\dfrac{8x^5}{4x}$

47. $\dfrac{12a^5}{3a^2}$

48. $\dfrac{20x^7}{4x^6}$

49. $\dfrac{18m^5n}{9m^3}$

50. $\dfrac{50a^4b^3}{10b^2}$

51. $\dfrac{28w^3z^5}{7wz}$

52. $\dfrac{48p^6q^7}{8p^4q}$

53. $\dfrac{18x^3y^4z^5}{9xy^2z^2}$

54. $\dfrac{25a^5b^4c^3}{5a^4bc^2}$

Simplify each of the following expressions where possible.

55. $2a^3b \cdot 3a^2b$

56. $2xy^3 \cdot 3xy^2$

57. $2a^3b + 3a^2b$

58. $2xy^3 + 3xy^2$

59. $2x^2y^3 \cdot 3x^2y^3$

60. $5a^3b^2 \cdot 10a^3b^2$

61. $2x^2y^3 + 3x^2y^3$

62. $5a^3b^2 + 10a^3b^2$

63. $\dfrac{8a^2b \cdot 6a^2b}{2ab}$

64. $\dfrac{6x^2y^3 \cdot 9x^2y^3}{3x^2y^2}$

65. $\dfrac{8a^2b + 6a^2b}{2ab}$

66. $\dfrac{6x^2y^3 + 9x^2y^3}{3x^2y^2}$

67. When one is using exponential notation, how many kilograms (kg) of SO_2 are added to the atmosphere each year (yr) from natural sources? (See question 78 in Section 1.2.)

68. In 1987, nearly 1.4 billion kilograms of air pollutants were produced by approximately 14,000 of the nation's top industrial polluters. What was the average amount of air pollution generated by each of these polluters?

$$\dfrac{1.4 \times 10^9}{1.4 \times 10^4}$$

51. _____

52. _____

53. _____

54. _____

55. _____

56. _____

57. _____

58. _____

59. _____

60. _____

61. _____

62. _____

63. _____

64. _____

65. _____

66. _____

67. _____

68. _____

ANSWERS

a. _____

b. _____

c. _____

d. _____

e. _____

f. _____

g. _____

h. _____

Skillscan (Section 1.2)

Evaluate each of the following expressions for the indicated value of the variable.

a. a^2 $(a = 4)$ **b.** x^3 $(x = 4)$

c. $2w^3$ $(w = 3)$ **d.** $3z^3$ $(z = 5)$

e. $5p^2$ $(p = 2)$ **f.** $4s^3$ $(s = 2)$

g. $(5p)^2$ $(p = 2)$ **h.** $(4s)^3$ $(s = 2)$

ANSWERS

1. x^{12} **3.** 5^6 **5.** a^{10} **7.** z^{13} **9.** p^{12} **11.** x^5y^5 **13.** w^8 **15.** m^9 **17.** a^6b^6
19. p^6q^{10} **21.** $6a^9$ **23.** $3x^7$ **25.** $20m^4n^5$ **27.** $54x^4y^6$ **29.** $6a^{12}$ **31.** $24c^8d^5$
33. $30m^{10}$ **35.** $30r^7s^5$ **37.** a^3 **39.** y^6 **41.** p^5 **43.** x^3y **45.** $2m^2$ **47.** $4a^3$
49. $2m^2n$ **51.** $4w^2z^4$ **53.** $2x^2y^2z^3$ **55.** $6a^5b^2$ **57.** Cannot simplify   **59.** $6x^4y^6$
61. $5x^2y^3$ **63.** $24a^3b$ **65.** $7a$ **67.** $10^8 \cdot 10^3 = 10^{11}$ kg/yr   **a.** 16   **b.** 64   **c.** 54
d. 375   **e.** 20   **f.** 32   **g.** 100   **h.** 512

1.6 Evaluating Algebraic Expressions

OBJECTIVE
To evaluate algebraic expressions

In applying algebra to problem solving, you will often want to find the value of an algebraic expression when you know certain values for the letters (or variables) in the expression. As we pointed out earlier, finding the value of an expression is called *evaluating the expression* and uses the following steps.

TO EVALUATE AN ALGEBRAIC EXPRESSION

STEP 1 Replace each variable by the given number value.

STEP 2 Do the indicated arithmetic, following the order of the operations.

Example 1

Suppose that $a = 5$ and $b = 7$.

(*a*) To evaluate $a + b$, we replace a with 5 and b with 7.

$$a + b = 5 + 7 = 12$$

(*b*) To evaluate $3ab$, we again replace a with 5 and b with 7.

$$3ab = 3 \cdot 5 \cdot 7 = 105$$

CHECK YOURSELF 1

If $x = 6$ and $y = 7$, evaluate.

1. $y - x$ **2.** $5xy$

We are now ready to evaluate algebraic expressions that require following the rules for the order of operations.

Example 2

Evaluate the following expressions if $a = 2$, $b = 3$, $c = 4$, and $d = 5$.

(*a*) $5a + 7b = 5 \cdot 2 + 7 \cdot 3$ Multiply first
$$= 10 + 21 = 31 \quad \text{Then add}$$

(*b*) $3c^2 = 3 \cdot 4^2$ Evaluate the power.
$$= 3 \cdot 16 = 48 \quad \text{Then multiply.}$$

◁ CAUTION ▷
This is different from
$(3c)^2 = (3 \cdot 4)^2$
$= 12^2 = 144$

47

(c) $7(c + d) = 7(4 + 5)$ Add inside the parentheses.

$\qquad\qquad = 7 \cdot 9 = 63$

(d) $5a^4 - 2d^2 = 5 \cdot 2^4 - 2 \cdot 5^2$ Evaluate the powers.

$\qquad\qquad\quad = 5 \cdot 16 - 2 \cdot 25$ Multiply.

$\qquad\qquad\quad = 80 - 50 = 30$ Subtract.

CHECK YOURSELF 2

If $x = 3$, $y = 2$, $z = 4$, and $w = 5$, evaluate the following expressions.

1. $4x^2 + 2$ **2.** $5(z + w)$ **3.** $7(z^2 - y^2)$

To evaluate algebraic expressions when a fraction bar is used, do the following: Start by doing all the work in the numerator, then do the work in the denominator. Divide the numerator by the denominator as the last step.

Example 3

If $p = 2$, $q = 3$, and $r = 4$, evaluate:

(a) $\dfrac{8p}{r}$

Replace p with 2 and r with 4.

As we mentioned in Section 1.2, the fraction bar is a grouping symbol, like parentheses. Work first in the numerator and then in the denominator.

$\dfrac{8p}{r} = \dfrac{8 \cdot 2}{4} = \dfrac{16}{4} = 4$ Divide as the last step.

(b) $\dfrac{7q + r}{p + q} = \dfrac{7 \cdot 3 + 4}{2 + 3}$ Now evaluate the top and bottom separately.

$\qquad\qquad = \dfrac{21 + 4}{2 + 3} = \dfrac{25}{5} = 5$

CHECK YOURSELF 3

Evaluate the following if $c = 5$, $d = 8$, and $e = 3$.

1. $\dfrac{6c}{e}$ **2.** $\dfrac{4d + e}{c}$ **3.** $\dfrac{10d - e}{d + e}$

CHECK YOURSELF ANSWERS

1. (1) 1; (2) 210.
2. (1) 38; (2) 45; (3) 84.
3. (1) 10; (2) 7; (3) 7.

1.6 Exercises

Build Your Skills

Evaluate each of the following expressions of $a = 2$, $b = 3$, $c = 5$, and $d = 6$.

1. $2c^3$

2. $3d^2$

3. $2b^2 + 3c^2$

4. $d^3 - 3b^2$

5. $3(a + c)$

6. $5(d - a)$

7. $3a + c$

8. $5d - a$

9. $6(a + 2d)$

10. $3(3b + c)$

11. $6a + 2d$

12. $3(3b) + c$

13. $5(4d - 3c)$

14. $7(2a + 3b)$

15. $c(4a + 3d)$

16. $d(4c - 3b)$

17. $2a^2 + b^2$

18. $5a^2 + c^2$

19. $2(a^2 + b^2)$

20. $5(a^2 + c^2)$

ANSWERS

1. _____
2. _____
3. _____
4. _____
5. _____
6. _____
7. _____
8. _____
9. _____
10. _____
11. _____
12. _____
13. _____
14. _____
15. _____
16. _____
17. _____
18. _____
19. _____
20. _____

21. _____

22. _____

23. _____

24. _____

25. _____

26. _____

27. _____

28. _____

29. _____

30. _____

31. _____

32. _____

33. _____

34. _____

35. _____

36. _____

50

21. $(2a^2 + b)^2$

22. $(5a^2 + c)^2$

23. $\dfrac{10a}{c}$

24. $\dfrac{9d}{b}$

25. $\dfrac{5cd}{b}$

26. $\dfrac{8ab}{d}$

27. $\dfrac{7b - d}{c}$

28. $\dfrac{3c + d}{b}$

29. $\dfrac{a + 2d}{2c - b}$

30. $\dfrac{2b + d}{c - b}$

31. $\dfrac{3a^3 - 2d}{c - a}$

32. $\dfrac{2b^2 + 3c}{2b + c}$

Think About These

Insert parentheses where necessary so that each of the following statements will be true.

33. $2 \cdot 3 + 4 = 14$

34. $4 \cdot 5 - 4 = 16$

35. $5 \cdot 4 - 2 = 18$

36. $4 \cdot 7 - 3 = 16$

37. $5 \cdot 6 - 2 \cdot 3 = 24$

38. $4 \cdot 5 + 2 \cdot 3 = 84$

39. $3 \cdot 6 + 2 \cdot 2 = 48$

40. $5 \cdot 3 - 2 \cdot 7 = 1$

41. $2 + 3^2 = 25$

42. $5 - 2^2 = 9$

43. $3 + 2 \cdot 4^2 = 80$

44. $5 + 3 \cdot 2^3 = 64$

Solve each of the following problems.

45. The perimeter of a rectangle is given by the formula $P = 2L + 2W$. Find the perimeter of a rectangle with $L = 8$ centimeters (cm) and $W = 5$ cm.

46. The area of a triangle is given by the formula $A = \dfrac{1}{2} \cdot b \cdot h$. Find the area in square feet (ft^2) of a triangle with $b = 4$ feet (ft) and $h = 6$ ft.

47. The volume of a square prism is given by $V = LW^2$. Find the volume in cubic centimeters (cm^3) of a prism with $L = 5$ cm and $W = 3$ cm.

48. The area of a trapezoid is given by $A = \dfrac{1}{2}h(B + b)$. Find the area in square inches (in^2) of a trapezoid with $h = 6$ inches (in), $B = 8$ in, and $b = 5$ in.

ANSWERS

37. _____

38. _____

39. _____

40. _____

41. _____

42. _____

43. _____

44. _____

45. _____

46. _____

47. _____

48. _____

49. _____

50. _____

51. _____

49. The surface area of a square prism is given by $S = 2W^2 + 4LW$. Find the surface area in square inches (in^2) of a prism with $L = 8$ in and $W = 3$ in.

50. The kinetic energy (KE) of a moving body is given by

$$KE = \frac{wv^2}{2g}$$

Find the kinetic energy, in foot pounds (ft · lb) of a moving body where $w = 20$ pounds (lb), $v = 60$ feet per second (ft/s), and $g = 32$ feet per second squared (ft/s^2).

51. In Section 1.4, problem 62, assume the average area cut on Oregon forests is 15,000 hectares and the average roaded area on Oregon forests is 250 hectares. How many hectares of Oregon forests have been cut and contain roads?

ANSWERS

1. 250   **3.** 93   **5.** 21   **7.** 11   **9.** 84   **11.** 24   **13.** 45   **15.** 130   **17.** 17
19. 26   **21.** 121   **23.** 4   **25.** 50   **27.** 3   **29.** 2   **31.** 4   **33.** $2 \cdot (3 + 4) = 14$
35. True   **37.** True   **39.** $3 \cdot (6 + 2) \cdot 2 = 48$ **41.** $(2 + 3)^2 = 25$
43. $(3 + 2) \cdot 4^2 = 80$ **45.** 26 cm   **47.** 45 cm^3 **49.** 114 in^2 **51.** 198,250

Summary

From Arithmetic to Algebra [1.1]

The sum of x and 5 is $x + 5$.
7 more than a is $a + 7$.
b increased by 3 is $b + 3$.

The difference of x and 3 is $x - 3$.
5 less than p is $p - 5$.
a decreased by 4 is $a - 4$.

The product of m and n is mn.
The product of 2 and the sum of a and b is $2(a + b)$.

n divided by 5 is $\dfrac{n}{5}$.

The sum of a and b, divided by 3, is $\dfrac{a + b}{3}$.

Addition $x + y$ means the *sum* of x and y or x plus y. Some other words indicating addition are ''more than'' and ''increased by.''

Subtraction $x - y$ means the *difference* of x and y or x minus y. Some other words indicating subtraction are ''less than'' and ''decreased by.''

Multiplication $\left.\begin{array}{r} x \cdot y \\ (x)(y) \\ xy \end{array}\right\}$ These all mean the *product* of x and y or x *times* y.

Division $\dfrac{x}{y}$ means x *divided by* y or the *quotient* when x is divided by y.

Exponents and the Order of Operations [1.2]

The Notation

$$\overset{\text{Exponent}}{\underset{\text{Base} \qquad \text{4 factors}}{a^4 = \underbrace{a \cdot a \cdot a \cdot a}}}$$

$5^3 = 5 \cdot 5 \cdot 5$
$\quad = 125$
$a^2 b^3 = a \cdot a \cdot b \cdot b \cdot b$
$6m^2 = 6 \cdot m \cdot m$

The number or letter used as a factor, here a, is called the *base*. The *exponent*, which is written above and to the right of the base, tells us how many times the base is used as a factor.

The Order of Operations

1. Do any operations in grouping symbols first.
2. Evaluate all expressions involving exponents.
3. Do any multiplication or division in order, working from left to right.
4. Do any addition or subtraction in order, working from left to right.

Evaluate the power.
$5 + 3 \cdot 2^2$
Multiply.
$= 5 + 3 \cdot 4$ Add.
$= 5 + 12$
$= 17$

The Properties of Addition and Multiplication [1.3]

The Commutative Properties   If a and b are any numbers,

1. $a + b = b + a$
2. $a \cdot b = b \cdot a$

The Associative Properties   If a, b, and c are any numbers,

1. $a + (b + c) = (a + b) + c$
2. $a \cdot (b \cdot c) = (a \cdot b) \cdot c$

The Distributive Properties   If a, b, and c are any numbers,

$$a(b + c) = a \cdot b + a \cdot c$$

Adding and Subtracting Expressions [1.4]

Term   A number, or the product of a number and one or more variables, raised to a power.

4a^2 and 3a^2 are like terms.
5x^2y and 2xy^2 are not like terms.

Like Terms   Terms that contain exactly the same variables raised to the same powers.

Combining Like Terms

$5a + 3a = 8a$
$7xy - 3xy = 4xy$

1. Add or subtract the numerical coefficients.
2. Attach the common variables.

Multiplying and Dividing Expressions [1.5]

The First Property of Exponents

$x^3 \cdot x^5 = x^{3+5}$
$\quad\quad = x^8$

$$x^m \cdot x^n = x^{m+n}$$

$3a^2b^3 \cdot 4a^3b^4 = 12a^5b^7$

Multiplying Expressions   Multiply the numerical coefficients and use the first property of exponents to combine the variables.

The Second Property of Exponents

$\dfrac{m^8}{m^4} = m^{8-4}$
$\quad\quad = m^4$

$$\frac{x^m}{x^n} = x^{m-n}$$

where x is not 0 and m is greater than n.

$\dfrac{28x^5y^3}{7xy^2} = 4x^4y$

Dividing Expressions   Divide the numerical coefficients and use the second property of exponents to combine the variables.

Evaluating Algebraic Expressions [1.6]

If $a = 2$ and $b = 4$,
$\quad 5a^2 - 3b$
$= 5 \cdot 2^2 - 3 \cdot 4$
$= 5 \cdot 4 - 3 \cdot 4$
$= 20 - 12$
$= 8$

To Evaluate an Expression

1. Replace each variable by the given number value.
2. Do the indicated arithmetic operations, following the rules for the order of operations.

Summary Exercises Chapter 1

This exercise set is provided to give you practice with each of the objectives of the chapter. Each exercise is keyed to the appropriate chapter section. The answers are provided in the instructor's manual. Your instructor will give you guidelines on how to best use these exercises.

[1.1] Write, using symbols.

1. 5 more than y

2. c decreased by 10

3. The product of 8 and a

4. The quotient when y is divided by 3

5. 5 times the product of m and n

6. The product of a and 5 less than a

7. 3 more than the product of 17 and x

8. The quotient when a plus 2 is divided by a minus 2

[1.2] Write each of these in expanded form.

9. x^3

10. $y^3 \cdot y^2$

11. $2x^3y^4$

12. $(2x)^3y^4$

[1.2] Evaluate each of the following expressions.

13. $18 - 3 \cdot 5$

14. $(18 - 3) \cdot 5$

15. $5 \cdot 4^2$

16. $(5 \cdot 4)^2$

17. $5 \cdot 3^2 - 4$

18. $5(3^2 - 4)$

19. $5(4 - 2)^2$

20. $5 \cdot 4 - 2^2$

21. $(5 \cdot 4 - 2)^2$

22. $3(5 - 2)^2$

23. $3 \cdot 5 - 2^2$

24. $(3 \cdot 5 - 2)^2$

[1.3] Identify the property that is illustrated by each of the following statements.

25. $5 + (7 + 12) = (5 + 7) + 12$

26. $2(8 + 3) = 2 \cdot 8 + 2 \cdot 3$

27. $4 \cdot (5 \cdot 3) = (4 \cdot 5) \cdot 3$

28. $4 \cdot 7 = 7 \cdot 4$

[1.3] Verify that each of the following statements is true by evaluating each side of the equation separately and comparing the results.

29. $8(5 + 4) = 8 \cdot 5 + 8 \cdot 4$

30. $2(3 + 7) = 2 \cdot 3 + 2 \cdot 7$

31. $(7 + 9) + 4 = 7 + (9 + 4)$

32. $(2 + 3) + 6 = 2 + (3 + 6)$

33. $(8 \cdot 2) \cdot 5 = 8(2 \cdot 5)$

34. $(3 \cdot 7) \cdot 2 = 3 \cdot (7 \cdot 2)$

[1.3] Use the distributive law to remove parentheses.

35. $3(7 + 4)$

36. $4(2 + 6)$

37. $4(w + v)$

38. $6(x + y)$

39. $3(5a + 2)$

40. $2(4x^2 + 3x)$

Write, using exponents.

41. $5 \cdot 5 \cdot 5 \cdot 5$

42. $6 \cdot 6 \cdot 6 \cdot 6 \cdot 6$

43. $y \cdot y \cdot y \cdot y \cdot y$

44. $b \cdot b \cdot b \cdot b$

45. $8 \cdot a \cdot a \cdot a$

46. $c \cdot c \cdot d \cdot d \cdot d \cdot d$

[1.4] List the terms of the expressions.

47. $4a^3 - 3a^2$

48. $5x^2 - 7x + 3$

[1.4] Circle like terms.

49. $5m^2, -3m, -4m^2, 5m^3, m^2$

50. $4ab^2, 3b^2, -5a, ab^2, 7a^2, -3ab^2, 4a^2b$

[1.4] Combine like terms.

51. $5c + 7c$

52. $2x + 5x$

53. $4a - 2a$

54. $6c - 3c$

55. $9xy - 6xy$

56. $5ab^2 + 2ab^2$

57. $7a + 3b + 12a - 2b$

58. $6x - 2x + 5y - 3x$

59. $5x^3 + 17x^2 - 2x^3 - 8x^2$

60. $3a^3 + 5a^2 + 4a - 2a^3 - 3a^2 - a$

61. Subtract $4a^3$ from the sum of $2a^3$ and $12a^3$.

62. Subtract the sum of $3x^2$ and $5x^2$ from $15x^2$.

[1.4] Multiply.

63. $a^7 \cdot a^{10}$

64. $x^6 \cdot x^4$

65. $2m^2 \cdot 5m^7$

66. $6x^5 \cdot 3x^2$

67. $3y^2 \cdot y^3 \cdot 8y^4$

68. $2a^2 \cdot 3a \cdot 5a^3$

69. $3p^2q^3 \cdot 7pq^5$

70. $6a^2b^3 \cdot 5a^3b^2$

[1.5] Divide.

71. $\dfrac{x^{10}}{x^3}$ **72.** $\dfrac{a^5}{a^4}$ **73.** $\dfrac{x^2 \cdot x^3}{x^4}$

74. $\dfrac{m^2 \cdot m^3 \cdot m^4}{m^5}$ **75.** $\dfrac{18p^7}{9p^5}$ **76.** $\dfrac{24x^{17}}{8x^{13}}$

77. $\dfrac{30m^7n^5}{6m^2n^3}$ **78.** $\dfrac{108x^9y^4}{9xy^4}$ **79.** $\dfrac{48p^5q^3}{6p^3q}$

80. $\dfrac{52a^5b^3c^5}{13a^4c}$

[1.6] Evaluate each of the following expressions if $x = 2$, $y = 3$, $z = 4$, and $w = 5$.

81. $x + y$ **82.** $x + w + z$ **83.** $8xw$

84. $3xyz$ **85.** $2xw$ **86.** $5xyzw$

87. $4w - 3z$ **88.** $3z - 2x$ **89.** $5x - 2y$

90. $3y - 2z$ **91.** $3z^2 - 4x^3$ **92.** $5x^2 - 2y^2$

93. $5(2z - w)$ **94.** $4(6x + y)$ **95.** $2(x + y)^2$

96. $(2x + y)^2$ **97.** $\dfrac{5yz}{2w}$ **98.** $\dfrac{wx^3y}{10z}$

99. $\dfrac{2w + 3z}{3x + w}$ **100.** $\dfrac{5x + 2y + z}{wz}$

Name

Section

Date

The purpose of this self-test is to help you check your progress and to review for a chapter test in class. Allow yourself about an hour to take the test. When you are done, check your answers in the back of the book. If you missed any problems, be sure to go back and review the appropriate sections in the chapter and the exercises that are provided.

ANSWERS

1. _____

2. _____

3. _____

4. _____

5. _____

6. _____

7. _____

8. _____

9. _____

10. _____

11. _____

12. _____

13. _____

14. _____

Write, using symbols.

1. 5 less than a

2. The product of 6 and m

3. 4 times the sum of m and n

4. The quotient when the sum of a and b is divided by 3

Write, using exponents.

5. $4 \cdot 4 \cdot 4 \cdot 4$

6. $7 \cdot b \cdot b \cdot b$

Evaluate the following expressions.

7. $23 - 4 \cdot 5$

8. $4 \cdot 5^2 - 35$

9. $4(2 + 4)^2$

Identify the property that is illustrated by each of the following statements.

10. $6 \cdot 7 = 7 \cdot 6$

11. $2(6 + 7) = 2 \cdot 6 + 2 \cdot 7$

12. $4 + (3 + 7) = (4 + 3) + 7$

Use the distributive law to remove parentheses. Then simplify your result.

13. $3(5 + 2)$

14. $4(5x + 3)$

59

Combine like terms.

15. $8a + 7a$

16. $8x^2y - 5x^2y$

17. $10x + 8y + 9x - 3y$

18. Subtract $9a^2$ from the sum of $12a^2$ and $5a^2$.

Multiply.

19. $a^5 \cdot a^9$

20. $3x^2y^3 \cdot 5xy^4$

Divide.

21. $\dfrac{4x^5}{2x^2}$

22. $\dfrac{20a^3b^5}{5a^2b^2}$

Evaluate the following expressions if $a = 3$, $b = 4$, and $c = 5$.

23. $2a^3 - 3b^2$

24. $5(4a - 2c)$

25. $\dfrac{4a + 3c}{b + c}$

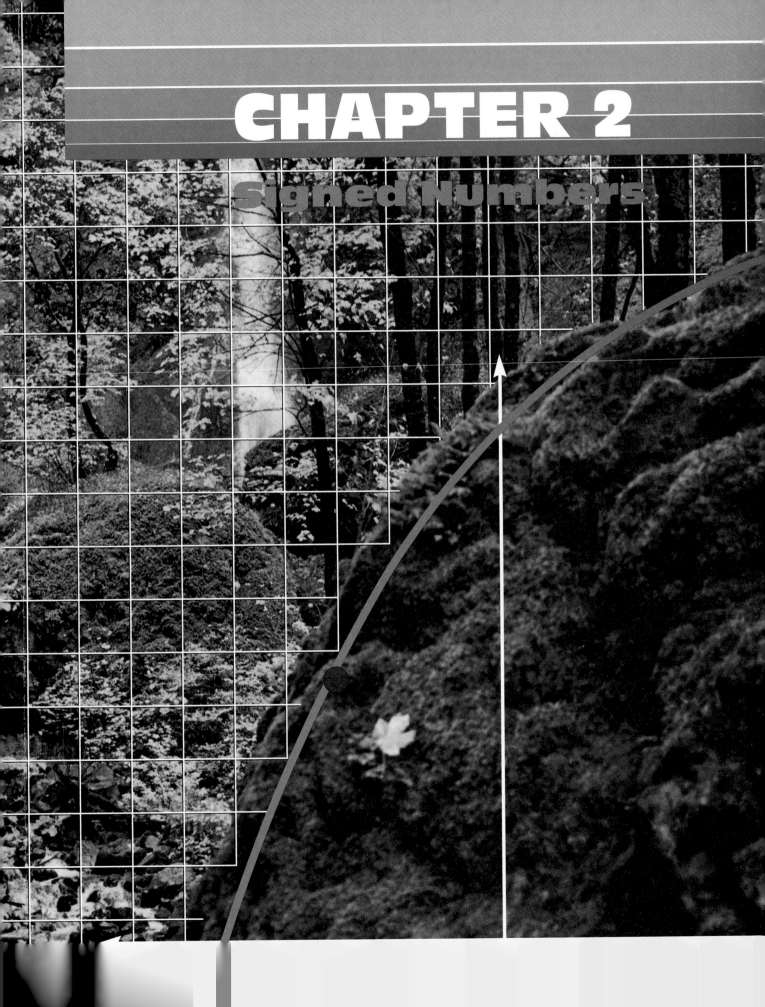

CHAPTER 2

Signed Numbers

The ENVIRONMENT

Soil Erosion

Although not commonly mentioned in the daily news, soil erosion is one of our major environmental problems. The food we eat needs healthy soil in which to grow. If we lose the soil, our ability to feed the world's growing population will be impaired. The health of most life on earth depends on the health of the soil.

Soil is classified as a renewable resource. It is continually being regenerated by natural processes. However, these processes are extremely slow. In most agricultural areas, it takes an average of 500 years to generate 2.5 centimeters (approximately 1 inch) of new soil. Most forest soils take 2 to 3 times this long to be renewed.

Worldwide we are losing more than 20 billion metric tons of topsoil every year (1 metric ton = 1000 kilograms). Most tropical and temperate agricultural soils are renewed at about 1 metric ton per hectare per year. The worldwide erosion rate for agricultural land is 18 to 100 times the natural rate of soil renewal, depending on the soil type, climate, wind, and other factors. Even though many agricultural areas have rich soils several centimeters deep, this rate of soil loss will deplete those soils rapidly.

According to the Food and Agricultural Organization (FAO) of the United Nations, this level of soil erosion may reduce agricultural production in Africa by 25 percent between 1975 and 2000. Studies of soil loss in the United States indicate that a loss of 2.5 centimeters of topsoil may reduce corn and wheat yields by 6 percent. Applying the U.S. figure to the world as a whole, we see that annual soil loss may cause an annual grain production loss of over 8 million metric tons.

The United States is often referred to as the world's breadbasket. The United States is blessed with some of the best agricultural soils on earth. These soils allow us to be the best food-producing nation in the world.

Soil surveys indicate that about one-third of the original topsoil on U.S. agricultural lands has eroded into rivers, lakes, and oceans. The average erosion rate on cultivated land in the United States is about 7 times the natural soil formation rate. However, this average erosion rate masks much higher rates of erosion in heavily farmed areas such as the midwest and California. Some agricultural lands in the midwest have lost half of their topsoil. The erosion rate in some regions of California is 80 times the rate of natural soil formation. Two-thirds of the eroded soils in the United States come from one-fourth of the cropland. This is some of the most productive cropland in the world.

High levels of erosion carry a high price tag. Plant nutrient losses from erosion of U.S. soils are worth $18 billion a year. Damages caused by silt, plant nutrients, and pesticides eroded into the nation's waters cost over $4 billion a year. Farmers are not the only people who need to be concerned with these costs. Anyone who uses food grown on eroding croplands is affected by these increased costs.

Several farming practices, both old and new, can be employed to slow the rate of soil erosion. For many years farmers have used contour farming and terracing on steeper slopes to slow runoff and erosion. In high-wind areas, shelterbelts or windbreaks can be planted to protect exposed soils. Although known for years, these practices are spreading as more and more people realize the need to protect the soil.

More modern techniques include land use zoning, strip and alley cropping, and no-till farming. Zoning is the designating of different areas for certain uses. Strip and alley cropping means planting a variety of crops in wide rows or alleys to protect the soils better. No-till farming is planting crops with little or no plowing of the soil. These practices all disturb the soil less than conventional farming, thus slowing the rate of erosion.

Signed Numbers—An Introduction

OBJECTIVES

1. To understand the meaning of a negative number
2. To recognize the set of integers
3. To represent integers on a number line
4. To evaluate a numerical expression involving the absolute value notation

The numbers you use to count things, 1, 2, 3, 4, 5, and so on, are called the *natural* (or *counting*) *numbers*. The *whole numbers* consist of the natural numbers and zero—0, 1, 2, 3, 4, 5, and so on. They can be represented on a number line like the one shown.

The number line continues forever in both directions.

The origin

When numbers are used to represent physical quantities (altitudes, temperatures, and amounts of money are examples), it may be necessary to distinguish between *positive* and *negative* quantities. It is convenient to represent these quantities with plus (+) or minus (−) signs. For instance,

The altitude of Mount Whitney is 14,495 feet (ft) *above* sea level (+14,495).

The altitude of Death Valley is 282 ft *below* sea level (−282).

The temperature in Chicago is 10° *below* zero (−10°).

The temperature in Miami is 60° *above* zero (+60°).

A *gain* of $100 (+100).

A *loss* of $100 (−100).

These numbers suggest the need to extend the whole numbers to include both positive numbers (like +100) and negative numbers (like −282).

To represent the negative numbers, we extend the number line to the *left* of the origin and name equally spaced points.

Numbers used to name points to the right of the origin, positive numbers, are written with a positive (+) sign or with no sign at all.

+6 and 9 are positive numbers

Numbers used to name points to the left of the origin are negative numbers. They are always written with a negative (−) sign.

−3 and −20 are negative numbers

Read "negative 3."

Positive and negative numbers considered together are *signed numbers.* Here is the number line extended to include negative numbers.

0 is not considered a signed number.

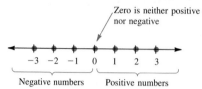

The numbers used to name the points shown on the number line above are called the *integers.* The integers consist of the natural numbers, their negatives, and the number 0. We can represent the set of integers by

The dots are called *ellipses* and indicate that the pattern continues.

$$\{ \ldots, -3, -2, -1, 0, 1, 2, 3, \ldots \}$$

Example 1

Represent the following integers on the number line shown.

$-3, -12, 8, 15, -7$

CHECK YOURSELF 1

Represent the following integers on a number line.

$-1, -9, 4, -11, 8, 20$

Integers are not the only kind of signed numbers. Decimals and fractions can also be thought of as signed numbers.

Example 2

(a) 145 is an integer.
(b) -28 is an integer.
(c) 0.35 is not an integer.
(d) $-\dfrac{2}{3}$ is not an integer.

Which of the following numbers are integers?

$$-23 \quad 1054 \quad -0.23 \quad 0 \quad -500 \quad -\frac{4}{5}$$

An important idea for our work in this chapter is the *absolute value* of a number. This represents the distance of the point named by the number from the origin on the number line.

The absolute value of 5 is 5. The absolute value of −5 is also 5.
     In symbols we write

The absolute value of a number does not *depend on whether the number is to the right or to the left of the origin, but on its* distance *from the origin.*

$$|5| = 5 \quad \text{and} \quad |-5| = 5$$

Read "the absolute value of 5."   Read "the absolute value of negative 5."

Example 3

(*a*) $|7| = 7$
(*b*) $|-7| = 7$
This is the negative, *or opposite, of the absolute value of negative 7.*
(*c*) $-|-7| = -7$

(*d*) $|-10| + |10| = 10 + 10 = 20$

Absolute value bars serve as another set of grouping symbols, so do the operation inside *first.*
(*e*) $|8 - 3| = |5| = 5$

Here, evaluate the absolute values, then subtract.
(*f*) $|8| - |3| = 8 - 3 = 5$

Evaluate.

1. $|8|$ **2.** $|-8|$ **3.** $-|-8|$
4. $|-9| + |4|$ **5.** $|9 - 4|$ **6.** $|9| - |4|$

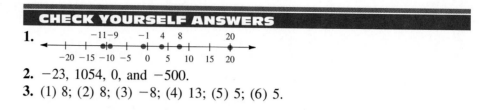

CHECK YOURSELF ANSWERS

2. -23, 1054, 0, and -500.

3. (1) 8; (2) 8; (3) -8; (4) 13; (5) 5; (6) 5.

Name _____

Date _____

Build Your Skills

Represent each quantity with a signed number.

1. An altitude of 300 ft above sea level

2. An altitude of 50 ft below sea level

3. A loss of $200

4. A profit of $200

5. A temperature of 30° above zero

6. A temperature of 5° below zero

Represent the integers on the number lines shown.

7. 5, −15, 18, −8, 3

$$\begin{array}{c} \hline \end{array}$$
-20 \quad -10 \quad 0 \quad 10 \quad 20

8. −18, 4, −5, 13, 9

$$\begin{array}{c} \hline \end{array}$$
-20 \quad -10 \quad 0 \quad 10 \quad 20

Which numbers in the following sets are integers?

9. $\left\{5, -\dfrac{2}{9}, 175, -234, -0.64\right\}$

10. $\left\{-45, 0.35, \dfrac{3}{5}, 700, -26\right\}$

Evaluate.

11. $|18|$

12. $|26|$

13. $|-10|$

14. $|-7|$

ANSWERS

1. _____

2. _____

3. _____

4. _____

5. _____

6. _____

7. _____

8. _____

9. _____

10. _____

11. _____

12. _____

13. _____

14. _____

15. _____

16. _____

17. _____

18. _____

19. _____

20. _____

21. _____

22. _____

23. _____

24. _____

25. _____

26. _____

27. _____

28. _____

29. _____

30. _____

31. _____

32. _____

33. _____

34. _____

35. _____

15. $-|2|$

16. $-|4|$

17. $-|-8|$

18. $-|-13|$

19. $|-2| + |3|$

20. $|4| + |-3|$

21. $|-9| + |9|$

22. $|11| + |-11|$

23. $|4| - |-4|$

24. $|5| - |-5|$

25. $|15| - |8|$

26. $|11| - |3|$

27. $|15 - 8|$

28. $|11 - 3|$

29. $|-9| + |2|$

30. $|-7| + |4|$

31. $|-8| - |-7|$

32. $|-9| - |-4|$

Think About These

Label each statement as true or false.

33. All whole numbers are integers.

34. All nonzero integers are signed numbers.

35. All integers are whole numbers.

36. All signed numbers are integers.

37. All negative integers are whole numbers.

38. Zero is neither positive nor negative.

Place absolute value bars in the proper location on the left side of the expression so that the statement is true.

39. $6 + (-2) = 4$

40. $8 + (-3) = 5$

41. $6 + (-2) = 8$

42. $8 + (-3) = 11$

For all exercises in Chapter 2 relating to soil erosion we will consider the formation of soil to represent positive quantities and the erosion of soil to represent negative quantities.

Example: The formation of 4 centimeters (cm) of new soil can be represented by the number 4. The erosion of 12 cm of soil can be represented by the number -12.

Represent each quantity with a signed number.

43. The erosion of 5 cm of topsoil from an Iowa corn field

44. The formation of 2.5 cm of new topsoil on the African savanna

ANSWERS

36. _____

37. _____

38. _____

39. _____

40. _____

41. _____

42. _____

43. _____

44. _____

a. _____

b. _____

c. _____

d. _____

e. _____

f. _____

g. _____

h. _____

i. _____

j. _____

Skillscan (Section 1.3)

Find each sum.

a. $3 + (8 + 9)$

b. $(6 + 12) + 3$

c. $(3 + 8) + 9$

d. $6 + (12 + 3)$

e. $\dfrac{2}{3} + \left(3 + \dfrac{1}{3}\right)$

f. $\left(\dfrac{3}{4} + 1\right) + \dfrac{5}{4}$

g. $\left(\dfrac{2}{3} + \dfrac{1}{3}\right) + 3$

h. $\left(\dfrac{3}{4} + \dfrac{5}{4}\right) + 1$

i. $\left(\dfrac{1}{2} + \dfrac{1}{3}\right) + \dfrac{1}{6}$

j. $\left(\dfrac{1}{4} + \dfrac{3}{8}\right) + \dfrac{1}{2}$

ANSWERS

We provide the answers for the odd-numbered exercises at the end of each exercise set. The answers for the even-numbered exercises are at the end of the book.

1. 300 or (+300)   **3.** −200   **5.** 30   **7.**   **9.** 5, 175, −234

11. 18   **13.** 10   **15.** −2   **17.** −8   **19.** 5   **21.** 18   **23.** 0   **25.** 7   **27.** 7
29. 11   **31.** 1   **33.** True   **35.** False   **37.** False   **39.** $|6 + (−2)| = 4$
41. $|6| + |−2| = 8$ **43.** −5   **a.** 20   **b.** 21   **c.** 20   **d.** 21   **e.** 4   **f.** 3
g. 4   **h.** 3   **i.** 1   **j.** $\dfrac{9}{8}$

2.2 — Adding Signed Numbers

OBJECTIVE
To find the sum of signed numbers

In the previous section, we introduced the idea of signed numbers. Now we will examine the four arithmetic operations (addition, subtraction, multiplication, and division) and see how those operations are performed when signed numbers are involved. We start by considering addition.

An application may help. As before, let's represent a gain of money as a positive number and a loss as a negative number.

If you gain $3 and then gain $4, the result is a gain of $7:

$$3 + 4 = 7$$

If you lose $3 and then lose $4, the result is a loss of $7:

$$-3 + (-4) = -7$$

If you gain $3 and then lose $4, the result is a loss of $1:

$$3 + (-4) = -1$$

If you lose $3 and then gain $4, the result is a gain of $1:

$$-3 + 4 = 1$$

The number line can be used to illustrate the addition of integers. Starting at the origin, we move to the *right* for positive numbers but to the *left* for negative numbers.

Example 1

Add $\dfrac{4}{3} + \dfrac{2}{3}$.

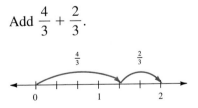

Start at the origin and move $\dfrac{4}{3}$ of a unit to the right. Then move $\dfrac{2}{3}$ more to the right to find the sum. So we have

$$\frac{4}{3} + \frac{2}{3} = \frac{6}{3} = 2$$

CHECK YOURSELF 1

Add.

1. $5 + 6$

2. $\dfrac{5}{4} + \dfrac{7}{4}$

The number line will also help you visualize the sum of two negative numbers. Remember, we move to the left for negative numbers.

Example 2

(*a*) Add $(-3) + (-4)$.

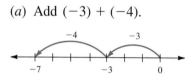

Start at the origin and move 3 units to the left. Then move 4 more units to the left to find the sum. From the graph we see that the sum is

$(-3) + (-4) = -7$

(*b*) Add $\left(-\dfrac{3}{2}\right) + \left(-\dfrac{1}{2}\right)$.

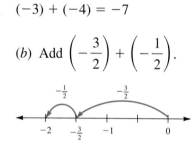

As before, we start at the origin. From that point move $\dfrac{3}{2}$ units left. Then move another $\dfrac{1}{2}$ unit left to find the sum. In this case

$$\left(-\dfrac{3}{2}\right) + \left(-\dfrac{1}{2}\right) = -2$$

CHECK YOURSELF 2

Add.

1. $(-4) + (-5)$

2. $(-3) + (-7)$

3. $(-5) + (-15)$

4. $\left(-\dfrac{5}{2}\right) + \left(-\dfrac{3}{2}\right)$

You have probably noticed some helpful patterns in the previous examples. These patterns will allow you to do the work mentally without having to use the number line. Look at the following rule.

This means that the sum of two positive numbers is positive and the sum of two negative numbers is negative.

> **ADDING SIGNED NUMBERS CASE 1: SAME SIGN**
>
> If two numbers have the same sign, add their absolute values. Give the sum the sign of the original numbers.

Example 3

(*a*) $(-8) + (-5) = -13$ Add the absolute values (8 + 5 = 13), and give the sum the sign (−) of the original numbers.

(*b*) $[(-3) + (-4)] + (-6)$ Add inside the brackets as your first step
$= (-7) + (-6)$
$= -13$

CHECK YOURSELF 3

Add mentally.

1. $7 + 9$
3. $(-5.8) + (-3.2)$

2. $(-7) + (-9)$
4. $[(-5) + (-2)] + (-3)$

Let's again use the number line to illustrate the addition of two numbers. This time the numbers will have *different* signs.

Example 4

(*a*) Add $3 + (-6)$.

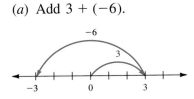

First move 3 units to the right of the origin. Then move 6 units to the left.

$3 + (-6) = -3$

(*b*) Add $-4 + 7$.

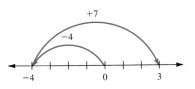

This time move 4 units to the left of the origin as the first step. Then move 7 units to the right.

$-4 + 7 = 3$

CHECK YOURSELF 4

Add.

1. $7 + (-5)$ **2.** $4 + (-8)$ **3.** $-4 + 9$ **4.** $-7 + 3$

You have no doubt noticed that in adding a positive number and a negative number, sometimes the sum is positive and sometimes it is negative. This depends on which of the numbers has the larger absolute value. This leads us to the second part of our addition rule.

ADDING SIGNED NUMBERS CASE 2: DIFFERENT SIGNS

If two numbers have different signs, subtract their absolute values, the smaller from the larger. Give the sum the sign of the number with the larger absolute value.

Example 5

(a) $7 + (-19) = -12$

Since the two numbers have different signs, subtract the absolute values ($19 - 7 = 12$). The sum has the sign ($-$) of the number with the larger absolute value, -19.

(b) $-13 + 7 = -6$

Subtract the absolute values ($13 - 7 = 6$). The sum has the sign ($-$) of the number with the larger absolute value, -13.

Remember, signed numbers can be fractions and decimals as well as integers.

(c) $-8.2 + (4.5) = -3.7$

Subtract the absolute values ($8.2 - 4.5 = 3.7$). The sum has the sign ($-$) of the number with the larger absolute value, -8.2.

CHECK YOURSELF 5

Add mentally.

1. $5 + (-14)$ **2.** $-7 + (-8)$ **3.** $-8 + 15$

4. $7 + (-8)$ **5.** $-\dfrac{2}{3} + \left(-\dfrac{7}{3}\right)$ **6.** $5.3 + (-2.3)$

There are two other properties of addition that we should mention before concluding this section. First, the sum of any number and 0 is always that number. In symbols,

> **ADDITIVE IDENTITY PROPERTY**
>
> For any number *a*,
>
> $a + 0 = 0 + a = a$

No number loses its identity after addition with 0. Zero is called the *additive identity*.

Example 6

Add.

(*a*) $9 + 0 = 9$
(*b*) $0 + (-8) = -8$
(*c*) $(-25) + 0 = -25$

CHECK YOURSELF 6

Add.

1. $8 + 0$ **2.** $0 + (-7)$ **3.** $(-36) + 0$

We'll need one further definition to state our second property. Every number has an *opposite*. It corresponds to a point the same distance from the origin as the given integer, but in the opposite direction.

The opposite of a number is also called the *additive inverse* of that number.

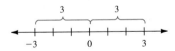

3 and −3 are opposites.

The opposite of 9 is -9.

The opposite of -15 is 15.

Our second property states that the sum of any number and its opposite is 0.

> **ADDITIVE INVERSE PROPERTY**
>
> For any number *a*, there exists a number $-a$ such that
>
> $a + (-a) = (-a) + a = 0$

Here $-a$ represents the opposite of the number *a*. The sum of any number and its opposite, or additive inverse, is 0.

Example 7

(a) $9 + (-9) = 0$

(b) $-15 + 15 = 0$

(c) $(-2.3) + 2.3 = 0$

(d) $\dfrac{4}{5} + \left(-\dfrac{4}{5}\right) = 0$

CHECK YOURSELF 7

Add.

1. $(-17) + 17$ **2.** $12 + (-12)$

3. $\dfrac{1}{3} + \left(-\dfrac{1}{3}\right)$ **4.** $(-1.6) + 1.6$

All properties of addition from Section 1.3 apply when negative numbers are involved.

We can now use the associative and commutative properties of addition, introduced in Section 1.3, to find the sum when more than two signed numbers are involved. Our final example illustrates.

Example 8

We use the commutative property to reverse the order of addition for -3 and 5. We then group -5 and 5. Do you see why?

$(-5) + (-3) + 5$
$= (-5) + 5 + (-3)$
$= [(-5) + 5] + (-3)$
$= 0 + (-3) = -3$

CHECK YOURSELF 8

Add.

1. $(-4) + 5 + (-3)$ **2.** $(-8) + 4 + 8$

CHECK YOURSELF ANSWERS

1. (1) 11; (2) 3.   **2.** (1) -9; (2) -10; (3) -20; (4) -4.

3. (1) 16; (2) -16; (3) -9; (4) -10.

4. (1) 2; (2) -4; (3) 5; (4) -4.

5. (1) -9; (2) -15; (3) 7; (4) -1; (5) -3; (6) 3.

6. (1) 8; (2) -7; (3) -36.

7. (1) 0; (2) 0; (3) 0; (4) 0.

8. (1) -2; (2) 4.

Name _____

Date _____

Build Your Skills

Add.

1. $4 + 5$

2. $6 + 8$

3. $9 + 7$

4. $8 + 7$

5. $\dfrac{2}{3} + \dfrac{4}{3}$

6. $\dfrac{3}{4} + \dfrac{9}{4}$

7. $\dfrac{1}{2} + \dfrac{4}{5}$

8. $\dfrac{2}{3} + \dfrac{5}{9}$

9. $(-2) + (-3)$

10. $(-1) + (-9)$

11. $\left(-\dfrac{3}{4}\right) + \left(-\dfrac{5}{4}\right)$

12. $\left(-\dfrac{3}{5}\right) + \left(\dfrac{12}{5}\right)$

13. $\left(-\dfrac{1}{2}\right) + \left(-\dfrac{3}{8}\right)$

14. $\left(-\dfrac{3}{7}\right) + \left(-\dfrac{5}{14}\right)$

15. $(-1.6) + (-2.3)$

16. $(-3.5) + (-2.6)$

17. $9 + (-3)$

18. $10 + (-4)$

19. $6 + (-12)$

20. $5 + (-9)$

21. $\left(\dfrac{3}{4}\right) + \left(-\dfrac{1}{2}\right)$

22. $\left(\dfrac{2}{3}\right) + \left(-\dfrac{1}{6}\right)$

ANSWERS

1. _____
2. _____
3. _____
4. _____
5. _____
6. _____
7. _____
8. _____
9. _____
10. _____
11. _____
12. _____
13. _____
14. _____
15. _____
16. _____
17. _____
18. _____
19. _____
20. _____
21. _____

23. _____

24. _____

25. _____

26. _____

27. _____

28. _____

29. _____

30. _____

31. _____

32. _____

33. _____

34. _____

35. _____

36. _____

37. _____

38. _____

39. _____

40. _____

41. _____

42. _____

43. _____

44. _____

45. _____

46. _____

47. _____

48. _____

49. _____

50. _____

23. $\left(-\dfrac{4}{5}\right) + \left(\dfrac{9}{20}\right)$

24. $\left(-\dfrac{11}{6}\right) + \left(\dfrac{5}{12}\right)$

25. $-6.5 + 8.5$

26. $-3.4 + 7.4$

27. $-9.6 + 13.6$

28. $-1.2 + 3.5$

29. $-9 + 0$

30. $-15 + 0$

31. $18 + 0$

32. $14 + 0$

33. $7 + (-7)$

34. $12 + (-12)$

35. $-14 + 14$

36. $-5 + 5$

37. $(-3 + 1) + (-4)$

38. $(-8 + 7) + (-5)$

39. $-2 + (-3) + (-8)$

40. $7 + (-9) + (-10)$

41. $-9 + (-17) + 9$

42. $15 + (-3) + (-15)$

43. $5 + 3 + (-9) + 1$

44. $6 + (-8) + (-4) + 5$

45. $(-3) + 0 + (-6) + 8$

46. $5 + (-6) + 7 + (-8)$

47. $1 + (-2) + 3 + (-4)$

48. $(-9) + 0 + (-2) + 12$

49. $\dfrac{5}{3} + \left(-\dfrac{4}{3}\right) + \dfrac{5}{3}$

50. $-\dfrac{6}{5} + \left(-\dfrac{13}{5}\right) + \dfrac{4}{5}$

78

51. $-\dfrac{3}{2} + \left(-\dfrac{7}{4}\right) + \dfrac{1}{4}$

52. $\dfrac{1}{3} + \left(-\dfrac{5}{6}\right) + \left(-\dfrac{1}{2}\right)$

53. $2.3 + (-5.4) + (-2.9)$

54. $-(5.4) + (-2.1) + (-3.5)$

Evaluate each of the following expressions.

55. $|2 + (-3)|$

56. $|(-7) + 5|$

57. $|-15 + 6|$

58. $|-19 + 6|$

59. $|-3 + 2 + (-4)|$

60. $|-2 + 7 + (-5)|$

61. $|2 + (-3)| + |(-3) + 2|$

62. $|8 + (-10)| + |-12 + 14|$

Think About These

Label each of the following statements as true or false.

63. $-10 + 6 = 6 - 10$

64. $5 + (-9) = -9 + 5$

65. $|-3| + |2| = |-3 + (+2)|$

66. $|-8| + |3| = |-8 + 3|$

Place absolute value bars in the proper location on the left side of the expression so that the statement is true.

67. $-3 + 7 = 10$

68. $-5 + 9 = 14$

69. $-6 + 7 + (-4) = 3$

70. $-10 + 15 + (-9) = 4$

In Chapter 1 we saw that the words "increased by" generally mean to add two or more quantities. If we increase soil formation, we will be adding positive numbers. If we "increase" soil erosion, we will be adding negative numbers. This last idea becomes clearer when we realize that increased erosion results in greater loss of soil.

71. If a wheat field has 15 centimeters (cm) of topsoil but erosion is increased by 3 cm, how much topsoil will be left after the erosion occurs?

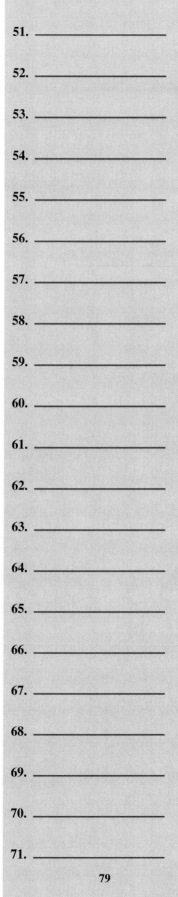

ANSWERS

51. _____

52. _____

53. _____

54. _____

55. _____

56. _____

57. _____

58. _____

59. _____

60. _____

61. _____

62. _____

63. _____

64. _____

65. _____

66. _____

67. _____

68. _____

69. _____

70. _____

71. _____

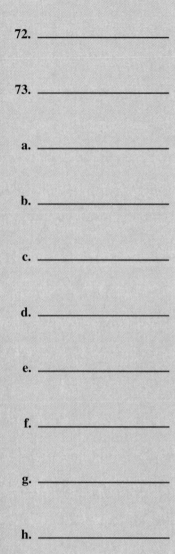

72. Over a period of 10 years a cotton field lost 4 cm of topsoil to erosion. Over the next 10 years erosion increased by 7 cm. How much soil was lost over the 20-year period?

73. A northwest forest lost 15 cm of topsoil over a 50-year period. After soil conservation practices were instituted, the same forest increased its soil formation by 5.5 cm over the next 25-year period. What was the total gain or loss of soil over this 75-year period?

Skillscan (Appendix 1)

Subtract as indicated.

a. $\dfrac{5}{2} - \dfrac{3}{2}$

b. $\dfrac{7}{2} - 2$

c. $\dfrac{4}{3} - \dfrac{1}{6}$

d. $\dfrac{9}{4} - \dfrac{3}{2}$

e. $5 - \dfrac{2}{3}$

f. $\dfrac{9}{10} - \dfrac{1}{4}$

g. $\dfrac{1}{5} - \dfrac{1}{7}$

h. $\dfrac{12}{5} - \dfrac{2}{3}$

ANSWERS

1. 9   **3.** 16   **5.** 2   **7.** $\dfrac{13}{10}$ **9.** -5 **11.** -2 **13.** $-\dfrac{7}{8}$ **15.** -3.9 **17.** 6

19. -6 **21.** $\dfrac{1}{4}$ **23.** $-\dfrac{7}{20}$ **25.** 2   **27.** 4   **29.** -9 **31.** 18   **33.** 0   **35.** 0

37. -6 **39.** -13 **41.** -17 **43.** 0   **45.** -1 **47.** -2 **49.** 2   **51.** -3

53. -6 **55.** 1   **57.** 9   **59.** 5   **61.** 2   **63.** True   **65.** False

67. $|-3| + |7| = 10$ **69.** $|-6 + 7 + (-4)| = 3$ **71.** 12 cm   **73.** 9.5 cm lost   **a.** 1

b. $\dfrac{3}{2}$ **c.** $\dfrac{7}{6}$ **d.** $\dfrac{3}{4}$ **e.** $\dfrac{13}{3}$ **f.** $\dfrac{13}{20}$ **g.** $\dfrac{2}{35}$ **h.** $\dfrac{26}{15}$

2.3 Subtracting Signed Numbers

OBJECTIVE
To find the difference of two signed numbers

To begin our discussion of subtraction when signed numbers are involved, we can look back at a problem using natural numbers. Of course, we know that

$$8 - 5 = 3 \tag{1}$$

From our work in adding signed numbers in the last section, we know that it is also true that

$$8 + (-5) = 3 \tag{2}$$

Comparing Equations (1) and (2), we see that the results are the same. This leads us to an important pattern. Any subtraction problem can be written as a problem in addition. Subtracting 5 is the same as adding the opposite of 5, or -5. We can write this fact as follows:

$$8 - 5 = 8 + (-5) = 3$$

This leads us to the following rule for subtracting signed numbers.

SUBTRACTING SIGNED NUMBERS

STEP 1   Rewrite the subtraction problem as an addition problem by
    a Changing the minus sign to a plus sign
    b Replacing the number being subtracted with its opposite

STEP 2   Add the resulting signed numbers as before.

In symbols,

This is the *definition* of subtraction.

$$a - b = a + (-b)$$

The following example illustrates the use of this definition in performing subtraction.

Example 1

Subtraction　　　　　　　　　　　　　　　*Addition*

Change the subtraction symbol $(-)$ to an addition symbol $(+)$.

(*a*) $15 - 7 \quad = 15 + (-7)$

Replace 7 with its opposite, -7.

$$= 8$$

(b) $9 - 12$ $= 9 + (-12)$
$= -3$

(c) $-6 - 7$ $= -6 + (-7)$
$= -13$

(d) $-\dfrac{3}{5} - \dfrac{7}{5} = -\dfrac{3}{5} + \left(-\dfrac{7}{5}\right)$
$= \dfrac{-10}{5} = -2$

(e) $2.1 - 3.4 = 2.1 + (-3.4) = -1.3$

(f) Subtract 5 from -2. We write the statement as

$-2 - 5$

and proceed as before:

$-2 - 5 = -2 + (-5)$
$= -7$

CHECK YOURSELF 1

Subtract.

1. $18 - 7$

2. $5 - 13$

3. $-7 - 9$

4. $-\dfrac{5}{6} - \dfrac{7}{6}$

5. $-2 - 7$

6. $5.6 - 7.8$

The subtraction rule is used in the same way when the number being subtracted is negative. Again change the subtraction to addition. Replace the negative number being subtracted with its opposite—that will now be positive. The following example illustrates.

Example 2

Subtraction   *Addition*

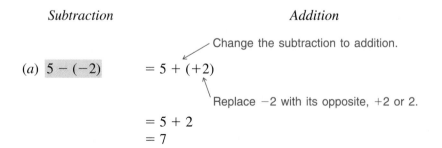

(a) $5 - (-2)$ $= 5 + (+2)$

Change the subtraction to addition.

Replace -2 with its opposite, $+2$ or 2.

$= 5 + 2$
$= 7$

(b) 7 − (−8) $= 7 + (+8)$
$= 7 + 8$
$= 15$

(c) −9 − (−5) $= -9 + 5$
$= -4$

(d) −12.7 − (−3.7) $= -12.7 + 3.7 = -9$

(e) $-\dfrac{3}{4} - \left(-\dfrac{7}{4}\right)$ $= -\dfrac{3}{4} + \left(+\dfrac{7}{4}\right) = \dfrac{4}{4} = 1$

(f) Subtract −4 from −5. We write

$$-5 - (-4) = -5 + 4 = -1$$

CHECK YOURSELF 2

Subtract.

1. $8 - (-2)$ **2.** $3 - (-10)$
3. $-7 - (-2)$ **4.** $-9.8 - (-5.8)$
5. $7 - (-7)$

As before, when parentheses are involved in an expression, you should do any operations inside parentheses as the first step. Our next example illustrates.

Example 3

Evaluate.

$-5 - 4 = -5 + (-4)$
$\qquad\quad\; = -9$

$5 - (-5 - 4)$

Evaluate as the first step.

$= 5 - (-9)$
$= 5 + 9$
$= 14$

CHECK YOURSELF 3

Evaluate.

$7 - (-9 + 6)$

Your scientific calculator can be used to do arithmetic with signed numbers. Before we look at an example, there are some keys on your calculator with which you should become familiar.

There are two buttons you must find on the calculator. The first is used for subtraction ($\boxed{-}$) and is usually found in the right column of calculator buttons. The second will "change the sign" of a number. It is usually a $\boxed{+/-}$ and is found on the bottom row.

We will now use these buttons in our next example.

Some graphing calculators have a negative sign $\boxed{(-)}$ that goes in front of a negative number.

Example 4

Using your calculator, find the difference.

(*a*) $-12.43 - 3.516$

Enter the 12.43 and push the $\boxed{+/-}$ to make it negative. Then push $\boxed{-}$ 3.516 $\boxed{=}$. The result should be -15.946.

If you have a graphing calculator, the key sequence will be

$\boxed{(-)}$ 12.43 $\boxed{-}$ 3.516 $\boxed{=}$

(*b*) $23.56 - (-4.7)$

The key sequence will probably be

23.56 $\boxed{-}$ 4.7 $\boxed{+/-}$ $\boxed{=}$

The answer should be 28.26.

CHECK YOURSELF 4

Use your calculator to find the difference

1. $-13.46 - 5.71$ **2.** $-3.575 - (-6.825)$

CHECK YOURSELF ANSWERS

1. (1) 11; (2) -8; (3) -16; (4) -2; (5) -9; (6) -2.2.
2. (1) 10; (2) 13; (3) -5; (4) -4; (5) 14.
3. 10.
4. (1) -19.17; (2) 3.25.

Name

Date

Build Your Skills

Subtract.

1. $17 - 9$

2. $25 - 11$

3. $76 - 39$

4. $125 - 78$

5. $\dfrac{15}{7} - \dfrac{8}{7}$

6. $\dfrac{17}{8} - \dfrac{9}{8}$

7. $6.8 - 4.3$

8. $9.7 - 2.5$

9. $5 - 7$

10. $7 - 12$

11. $24 - 45$

12. $136 - 352$

13. $\dfrac{5}{4} - \dfrac{13}{4}$

14. $\dfrac{10}{7} - \dfrac{31}{7}$

15. $6.5 - 10.3$

16. $17.7 - 28.9$

17. $-5 - 3$

18. $-15 - 8$

19. $-9 - 14$

20. $-8 - 12$

21. $-\dfrac{2}{5} - \dfrac{7}{10}$

22. $-\dfrac{5}{9} - \dfrac{7}{18}$

ANSWERS

1. _____
2. _____
3. _____
4. _____
5. _____
6. _____
7. _____
8. _____
9. _____
10. _____
11. _____
12. _____
13. _____
14. _____
15. _____
16. _____
17. _____
18. _____
19. _____
20. _____
21. _____
22. _____

23. _____

24. _____

25. _____

26. _____

27. _____

28. _____

29. _____

30. _____

31. _____

32. _____

33. _____

34. _____

35. _____

36. _____

37. _____

38. _____

39. _____

40. _____

41. _____

42. _____

43. _____

44. _____

45. _____

46. _____

23. $-3.4 - 4.7$

24. $-8.1 - 7.6$

25. $5 - (-2)$

26. $9 - (-5)$

27. $7 - (-9)$

28. $4 - (-8)$

29. $7 - (-12)$

30. $3 - (-10)$

31. $\dfrac{3}{4} - \left(-\dfrac{3}{2}\right)$

32. $\dfrac{5}{6} - \left(-\dfrac{7}{6}\right)$

33. $\dfrac{6}{7} - \left(-\dfrac{5}{14}\right)$

34. $\dfrac{11}{16} - \left(-\dfrac{7}{8}\right)$

35. $8.3 - (-5.7)$

36. $6.5 - (-4.3)$

37. $8.9 - (-11.7)$

38. $14.5 - (-24.6)$

39. $-36 - (-24)$

40. $-28 - (-11)$

41. $-16 - (-24)$

42. $-8 - (-13)$

43. $-\dfrac{2}{3} - \left(-\dfrac{8}{3}\right)$

44. $-\dfrac{5}{8} - \left(-\dfrac{3}{4}\right)$

45. $-12.7 - (-5.7)$

46. $-5.6 - (-2.6)$

47. $-6.9 - (-10.1)$

48. $-3.4 - (-7.6)$

49. $-11 - (-11)$

50. $-15 - (-15)$

51. $0 - (-8)$

52. $0 - (-11)$

Do the indicated operations.

53. $2 - 5 - 7$

54. $3 - 8 - 4$

55. $5 + 9 - 12$

56. $7 + 11 - 5$

57. $-9 + 6 - 12$

58. $-11 + 7 - 6$

59. $-7 - (-5) - 2$

60. $-8 - (-2) - 3$

61. $-8 - (-3) + 7$

62. $-7 - (-5) + 9$

63. $-3 - 8 + 4$

64. $-12 - 6 + 14$

65. $-2 - (6 + 3)$

66. $-11 - (5 + 13)$

67. $-3 - 8 - 10$

68. $-2 - 9 - 7$

69. $10 - 6 - 3$

70. $15 - 6 - 4$

71. $-4 - (6 - 9)$

72. $-5 - (8 - 12)$

73. $11 - (9 - 8)$

74. $15 - (6 - 3)$

ANSWERS

47. _____

48. _____

49. _____

50. _____

51. _____

52. _____

53. _____

54. _____

55. _____

56. _____

57. _____

58. _____

59. _____

60. _____

61. _____

62. _____

63. _____

64. _____

65. _____

66. _____

67. _____

68. _____

69. _____

70. _____

71. _____

72. _____

73. _____

74. _____

75. _____

76. _____

77. _____

78. _____

79. _____

80. _____

81. _____

82. _____

83. _____

84. _____

85. _____

86. _____

87. _____

88. _____

89. _____

90. _____

91. _____

92. _____

93. _____

94. _____

95. _____

96. _____

75. $7 - (3 - 5) + 2$

76. $4 - (5 - 7) + 3$

77. $8 - (6 - 2) - (-3)$

78. $12 - (8 - 3) - (-2)$

79. $-3 - (-2 - 5) - 8$

80. $-6 - (-1 - 4) - 9$

81. Subtract -8 from 5

82. Subtract -7 from 3

83. Subtract -7 from -2

84. Subtract -9 from -2

85. Subtract 7 from -2

86. Subtract 15 from -9

87. Subtract 12 from the sum of -6 and 3

88. Subtract 18 from the sum of -3 and 8

89. Subtract the sum of 8 and -14 from 6

90. Subtract the sum of -9 and -3 from -2

Evaluate each expression.

91. $|6 - (-9)|$

92. $|4 - (-3)|$

93. $|-22 - (-2)|$

94. $|23 - (-6)|$

95. $|18| - |-2|$

96. $|23| - |-6|$

97. $|-28 - (-34)|$

98. $|-64 - (-81)|$

99. $|-28| - |-34|$

100. $|-64| - |-81|$

Think About These

Label the following as true or false.

101. $8 - 5 = 5 - 8$

102. $-16 - (-12) = -12 - (-16)$

103. $|-9| - |-10| = 9 - 10$

104. $|-12| - |-5| = 12 - 5$

Insert parentheses in the following expressions so that the statement is true.

105. $-9 - 5 - 6 = -8$

106. $-9 - 5 - 6 = -20$

In Chapter 1 we saw that the words "decreased by" generally mean to subtract one number from another. If we decrease erosion, we will be subtracting negative numbers. From our work in this section we have seen that subtracting a negative number is equivalent to adding a positive number. This is consistent with erosional processes because a decrease in erosion means less soil loss.

107. The current erosion rate on some California cropland is 0.4 cm/yr. If this erosion rate were decreased by 0.25 cm/yr, what would the new erosion rate be?

108. The erosion rate on some African croplands has averaged 0.35 cm per year recently. If soil conservation measures could decrease the average erosion rate by .43 cm per year, would the croplands be losing soil or gaining soil? In either case, what is the new rate?

ANSWERS

97. _____

98. _____

99. _____

100. _____

101. _____

102. _____

103. _____

104. _____

105. _____

106. _____

107. _____

108. _____

Skillscan (Section 2.2)

Add.

a. $(-1) + (-1) + (-1) + (-1)$ **b.** $3 + 3 + 3 + 3 + 3$

c. $9 + 9 + 9$ **d.** $(-10) + (-10) + (-10)$

e. $(-5) + (-5) + (-5) + (-5) + (-5)$ **f.** $(-8) + (-8) + (-8) + (-8)$

ANSWERS

1. 8   **3.** 37   **5.** 1   **7.** 2.5   **9.** -2 **11.** -21 **13.** -2 **15.** -3.8 **17.** -8
19. -23 **21.** $-\dfrac{11}{10}$ **23.** -8.1 **25.** 7   **27.** 16   **29.** 19   **31.** $\dfrac{9}{4}$ **33.** $\dfrac{17}{14}$
35. 14   **37.** 20.6   **39.** -12 **41.** 8   **43.** 2   **45.** -7 **47.** 3.2   **49.** 0
51. 8   **53.** -10 **55.** 2   **57.** -15 **59.** -4 **61.** 2   **63.** -7 **65.** -11
67. -21 **69.** 1   **71.** -1 **73.** 10   **75.** 11   **77.** 7   **79.** -4 **81.** 13
83. 5   **85.** -9 **87.** -15 **89.** 12   **91.** 15   **93.** 20   **95.** 16   **97.** 6
99. -6 **101.** False   **103.** True   **105.** $-9 - (5 - 6) = -8$
107. 0.15 cm/yr of erosion   **a.** -4 **b.** 15   **c.** 27   **d.** -30 **e.** -25 **f.** -32

Multiplying Signed Numbers

OBJECTIVE
To find the product of two or more signed numbers

When you first considered multiplication in arithmetic, it was thought of as repeated addition. Let's see what our work with the addition of signed numbers can tell us about multiplication when signed numbers are involved. For example,

$$3 \cdot 4 = \underbrace{4 + 4 + 4} = 12$$

We interpret multiplication as repeated addition to find the product, 12.

Now, consider the product $(3)(-4)$:

$$(3)(-4) = (-4) + (-4) + (-4) = -12$$

Looking at this product suggests the first portion of our rule for multiplying signed numbers. The product of a positive number and a negative number is negative.

MULTIPLYING SIGNED NUMBERS CASE 1: DIFFERENT SIGNS

The product of two numbers with different signs is negative.

To use this rule in multiplying two numbers with different signs, multiply their absolute values and attach a negative sign.

Example 1

Multiply.

(a) $(5)(-6) = -30$

The product is negative.

(b) $(-10)(10) = -100$

(c) $(8)(-12) = -96$

(d) $\left(-\dfrac{3}{4}\right)\left(\dfrac{2}{5}\right) = -\dfrac{3}{10}$

Remember to multiply together numerators and then denominators and reduce.

Multiply.

1. $(-7)(5)$ **2.** $(-12)(9)$ **3.** $(-15)(8)$ **4.** $\left(-\dfrac{4}{7}\right)\left(\dfrac{14}{5}\right)$

The product of two negative numbers is harder to visualize. The following pattern may help you see how we can determine the sign of the product.

This number is decreasing by 1.

$$(3)(-2) = -6$$

$$(2)(-2) = -4$$

$$(1)(-2) = -2$$

$$(0)(-2) = 0$$

$$(-1)(-2) = 2$$

$$(-2)(-2) = 4$$

Do you see that the product is *increasing* by 2 each time?

What should the product $(-3)(-2)$ be? Continuing the pattern shown, we see that

$$(-3)(-2) = 6$$

This suggests that the product of two negative numbers is positive, and that is the case. We can extend our multiplication rule.

If you would like a more detailed explanation, see the discussion at the end of this section.

MULTIPLYING SIGNED NUMBERS CASE 2: SAME SIGNS

The product of two numbers with the same sign is positive.

Example 2

Multiply.

(a) $9 \cdot 7 = 63$ The product of two positive numbers (same sign, $+$) is positive.

(b) $(-8)(-5) = 40$ The product of two negative numbers (same sign, $-$) is positive.

(c) $\left(-\dfrac{1}{2}\right)\left(-\dfrac{1}{3}\right) = \dfrac{1}{6}$

CHECK YOURSELF 2

Multiply.

1. $10 \cdot 12$ **2.** $(-8)(-9)$ **3.** $\left(-\dfrac{2}{3}\right)\left(-\dfrac{6}{7}\right)$

Two numbers, 0 and 1, have special properties in multiplication.

The number 1 is called the *multiplicative identity* for this reason.

MULTIPLICATIVE IDENTITY PROPERTY

The product of 1 and any number is that number. In symbols,

$a \cdot 1 = 1 \cdot a = a$

MULTIPLICATIVE PROPERTY OF ZERO

The product of 0 and any number is 0. In symbols,

$a \cdot 0 = 0 \cdot a = 0$

Example 3

(*a*) $(1)(-7) = -7$
(*b*) $(15)(1) = 15$
(*c*) $(-7)(0) = 0$
(*d*) $0 \cdot 12 = 0$
(*e*) $\left(-\dfrac{4}{5}\right)(0) = 0$

CHECK YOURSELF 3

Multiply.

1. $(-10)(1)$ **2.** $(0)(-17)$ **3.** $\left(\dfrac{5}{7}\right)(1)$ **4.** $(0)\left(\dfrac{3}{4}\right)$

To complete our discussion of the properties of multiplication, we state the following.

$\dfrac{1}{a}$ is called the *multiplicative inverse,* or the *reciprocal,* of *a.* The product of any nonzero number and its reciprocal is 1.

MULTIPLICATIVE INVERSE PROPERTY

For any number *a,* where $a \neq 0$, there is a number $\dfrac{1}{a}$ such that

$$a \cdot \dfrac{1}{a} = 1$$

The following example illustrates.

Example 4

(a) $3 \cdot \dfrac{1}{3} = 1$ The reciprocal of 3 is $\dfrac{1}{3}$.

(b) $-5\left(-\dfrac{1}{5}\right) = 1$ The reciprocal of -5 is $\dfrac{1}{-5}$ or $-\dfrac{1}{5}$.

(c) $\dfrac{2}{3} \cdot \dfrac{3}{2} = 1$ The reciprocal of $\dfrac{2}{3}$ is $\dfrac{1}{\frac{2}{3}}$, or $\dfrac{3}{2}$.

CHECK YOURSELF 4

Find the multiplicative inverse (or the reciprocal) of each of the following numbers.

1. 6   **2.** -4 **3.** $\dfrac{1}{4}$ **4.** $-\dfrac{3}{5}$

In addition to the properties just mentioned, we can extend the commutative and associative properties for multiplication to signed numbers. Our next example is an application of the associative property of multiplication.

Example 5

Find the following product:

$$(-3)(2)(-7)$$

Applying the associative property, we can group the first two factors to write

Once again, this "grouping" can be done mentally.

$[(-3)(2)](-7)$

$= (-6)(-7)$ Evaluate first.

$= 42$

CHECK YOURSELF 5

Find the product.

$(-5)(-8)(-2)$

When symbols of grouping, or more than one operator, are involved in an expression, we must again follow our rules for the order of operations. Consider the following example.

Example 6

Evaluate each expression.

(a) $7(-9 + 12)$ Evaluate inside the parentheses first.
    $= 7(3) = 21$

(b) $(-8)(-7) - 40$ Multiply first, then subtract.
    $= 56 - 40$
    $= 16$

(c) $(-5)^2 - 3$ Evaluate the power first.
    $= (-5)(-5) - 3$ Note that $(-5)^2 = (-5)(-5)$
    $= 25 - 3$ $= 25$
    $= 22$

(d) $-5^2 - 3$ Note that
    $= -25 - 3$ $-5^2 = -25$
    $= -28$ The power applies *only* to the 5.

CHECK YOURSELF 6

Evaluate each expression.

1. $8(-9 + 7)$ **2.** $(-3)(-5) + 7$
3. $(-4)^2 - (-4)$ **4.** $-4^2 - (-4)$

Here is a more detailed explanation of why the product of two negative numbers is positive.

THE PRODUCT OF TWO NEGATIVE NUMBERS

From our earlier work, we know that the sum of a number and its opposite is 0:

$$5 + (-5) = 0$$

Multiply both sides of the equation by -3:

$$(-3)[5 + (-5)] = (-3)(0)$$

Since the product of 0 and any number is 0, on the right we have 0.

$$(-3)[5 + (-5)] = 0$$

We use the distributive law on the left.

$$(-3)(5) + (-3)(-5) = 0$$

We know that $(-3)(5) = -15$, so the equation becomes

$$-15 + (-3)(-5) = 0$$

We now have a statement of the form

$$-15 + \square = 0$$

where $\square$ is the value of $(-3)(-5)$. We also know that $\square$ is the number that must be added to -15 to get 0, so $\square$ is the opposite of -15, or 15. This means that

$$(-3)(-5) = 15 \qquad \text{The product is positive!}$$

It doesn't matter what numbers we use in this argument. The resulting product of two negative numbers will always be positive.

CHECK YOURSELF ANSWERS

1. (1) -35; (2) -108; (3) -120; (4) $-\dfrac{8}{5}$.   **2.** (1) 120; (2) 72; (3) $\dfrac{4}{7}$.

3. (1) -10; (2) 0; (3) $\dfrac{5}{7}$; (4) 0.   **4.** (1) $\dfrac{1}{6}$; (2) $-\dfrac{1}{4}$; (3) 4; (4) $-\dfrac{5}{3}$.

5. -80.   **6.** (1) -16; (2) 22; (3) 20; (4) -12.

Name _____

Date _____

Build Your Skills

Multiply.

1. $8 \cdot 5$

2. $7 \cdot 6$

3. $(6)(-10)$

4. $(5)(-4)$

5. $(-8)(9)$

6. $(-12)(3)$

7. $(4)\left(-\dfrac{3}{2}\right)$

8. $(9)\left(-\dfrac{2}{3}\right)$

9. $\left(-\dfrac{1}{4}\right)(8)$

10. $\left(-\dfrac{3}{2}\right)(4)$

11. $(3.25)(-4)$

12. $(5.4)(-5)$

13. $(-8)(-7)$

14. $(-9)(-8)$

15. $(-10)(-6)$

16. $(-7)(-3)$

17. $(-9)\left(-\dfrac{2}{3}\right)$

18. $(-6)\left(-\dfrac{3}{2}\right)$

19. $(-2.5)(-6)$

20. $(-3.75)(-8)$

1. _____

2. _____

3. _____

4. _____

5. _____

6. _____

7. _____

8. _____

9. _____

10. _____

11. _____

12. _____

13. _____

14. _____

15. _____

16. _____

17. _____

18. _____

19. _____

20. _____

21. _____

22. _____

23. _____

24. _____

25. _____

26. _____

27. _____

28. _____

29. _____

30. _____

31. _____

32. _____

33. _____

34. _____

35. _____

36. _____

37. _____

38. _____

39. _____

40. _____

41. _____

42. _____

21. $(0)(-18)$

22. $(-17)(0)$

23. $(15)(0)$

24. $(0)(25)$

25. $\left(-\dfrac{5}{6}\right)(0)$

26. $\left(-\dfrac{7}{8}\right)(0)$

27. $(-4.63)(0)$

28. $(-1.75)(0)$

29. $\left(-\dfrac{3}{2}\right)\left(-\dfrac{2}{3}\right)$

30. $\left(-\dfrac{4}{5}\right)\left(-\dfrac{5}{4}\right)$

31. $\left(\dfrac{3}{7}\right)\left(-\dfrac{7}{3}\right)$

32. $\left(\dfrac{6}{7}\right)\left(-\dfrac{7}{6}\right)$

33. $(-5)(3)(-2)$

34. $(-4)(2)(-3)$

35. $(8)(-3)(7)$

36. $(13)(-2)(6)$

37. $(-2)(-4)(-2)$

38. $(-7)(5)(-2)$

39. $(-7)(-5)(2)$

40. $(-2)(-3)(8)$

41. $\left(-\dfrac{1}{3}\right)\left(\dfrac{6}{5}\right)(-10)$

42. $\left(-\dfrac{1}{2}\right)\left(\dfrac{4}{3}\right)(-6)$

43. $(-9)(-12)(0)$ **44.** $(-13)(-0)(-7)$

Do the indicated operations. Remember the rules for the order of operations.

45. $5(7 - 2)$ **46.** $7(8 - 5)$

47. $3(8 - 10)$ **48.** $2(11 - 17)$

49. $-2(8 - 5)$ **50.** $-3(14 - 8)$

51. $-3(-2 - 5)$ **52.** $-2(-7 - 3)$

53. $(-2)(3) - 5$ **54.** $(-6)(8) - 27$

55. $4(-7) - 5$ **56.** $(-3)(-9) - 11$

57. $(-5)(-2) - 12$ **58.** $(-7)(-3) - 25$

59. $(3)(-7) + 20$ **60.** $(2)(-6) + 8$

61. $-4 + (-3)(6)$ **62.** $-5 + (-2)(3)$

63. $5 - (-2)(-3)$ **64.** $7 - (-4)(-3)$

65. $-8 - (-2)(-5)$ **66.** $-9 - (-3)(-2)$

ANSWERS

43. _____

44. _____

45. _____

46. _____

47. _____

48. _____

49. _____

50. _____

51. _____

52. _____

53. _____

54. _____

55. _____

56. _____

57. _____

58. _____

59. _____

60. _____

61. _____

62. _____

63. _____

64. _____

65. _____

66. _____

ANSWERS

67. _____

68. _____

69. _____

70. _____

71. _____

72. _____

73. _____

74. _____

75. _____

76. _____

77. _____

78. _____

79. _____

80. _____

81. _____

82. _____

83. _____

84. _____

85. _____

86. _____

87. _____

88. _____

89. _____

90. _____

91. _____

92. _____

67. $-12 - (-6)(5)$

68. $-10 - (-2)(7)$

69. $(-2)(-7) + (2)(-3)$

70. $(-3)(-6) + (4)(-2)$

71. $(-7)(3) - (-2)(-8)$

72. $(-5)(2) - (-3)(-4)$

73. $(-6)^2 - 4$

74. $(-5)^2 - 9$

75. $(-6)^2 + 7$

76. $(-3)^2 + 5$

77. $-6^2 - 4$

78. $-5^2 - 3$

79. $(-4)^2 - (-2)(-5)$

80. $(-3)^3 - (-8)(-2)$

81. $(-8)^2 - 5^2$

82. $(-6)^2 - 4^2$

83. $(-6)^2 - (-3)^2$

84. $(-8)^2 - (-4)^2$

85. $-8^2 - 5^2$

86. $-6^2 - 3^2$

87. $-8^2 - (-5)^2$

88. $-9^2 - (-6)^2$

89. $-11^2 - (-11)^2$

90. $-5^2 - (-5)^2$

91. $(-7)^2 - 7^2$

92. $(-6)^2 - 6^2$

Think About These

In each of the following, place parentheses in the proper location so that the statement is true.

93. $-8 - 4 = -4$

94. $-6 - 3 \cdot 2 = -18$

95. $-3 \cdot 4 - 2 = -6$

96. $-7 \cdot 4 - 2 \cdot 3 = -42$

Write an expression for each of the following and evaluate.

97. The product of -2 and 7, subtracted from -3

98. The product of -5 and the sum of 2 and 6

In this section we will need to assign positive and negative values to time. We will consider time measured forward from a given point as being positive and time measured backward from a given point as negative.

Example: Five years from now in the future can be represented by the number 5. Ten years ago can be represented by the number -10.

 99. If an annual erosion rate of 0.42 cm occurs for 15 consecutive years, how much soil loss will occur over that time?

100. Assume a Kansas farmer's wheat field is losing 0.52 cm of topsoil every year to erosion. How much soil could she have saved if she had eliminated soil loss from her land 5 years ago?

ANSWERS

93. _____

94. _____

95. _____

96. _____

97. _____

98. _____

99. _____

100. _____

ANSWERS

a. _____

b. _____

c. _____

d. _____

e. _____

f. _____

g. _____

h. _____

Skillscan

Simplify each of the following.

a. $\dfrac{25}{5}$ b. $\dfrac{49}{7}$

c. $\dfrac{66}{11}$ d. $\dfrac{84}{12}$

e. $\dfrac{90}{15}$ f. $\dfrac{144}{24}$

g. $\dfrac{81}{18}$ h. $\dfrac{80}{15}$

ANSWERS

1. 40   **3.** −60   **5.** −72   **7.** −6   **9.** −2   **11.** −13   **13.** 56   **15.** 60
17. 6   **19.** 15   **21.** 0   **23.** 0   **25.** 0   **27.** 0   **29.** 1   **31.** −1   **33.** 30
35. −168   **37.** −16   **39.** 70   **41.** 4   **43.** 0   **45.** 25   **47.** −6   **49.** −6
51. 21   **53.** −11   **55.** −33   **57.** −2   **59.** −1   **61.** −22   **63.** −1   **65.** −18
67. 18   **69.** 8   **71.** −37   **73.** 32   **75.** 43   **77.** −40   **79.** 6   **81.** 39
83. 27   **85.** −89   **87.** −89   **89.** −242   **91.** 0   **93.** −(8 − 4) = −4
95. −3 · (4 − 2) = −6   **97.** −3 − (−2)(7); 11   **99.** 6.3 cm of soil loss

a. 5   **b.** 7   **c.** 6   **d.** 7   **e.** 6   **f.** 6   **g.** $\dfrac{9}{2}$ **h.** $\dfrac{16}{3}$

Dividing Signed Numbers

OBJECTIVE

To find the quotient of two signed numbers

You know from your work in arithmetic that multiplication and division are related operations. We can use that fact, and our work of the last section, to determine rules for the division of signed numbers. Every division problem can be stated as an equivalent multiplication problem. For instance,

$$\frac{15}{5} = 3 \qquad \text{since} \qquad 15 = 5 \cdot 3$$

$$\frac{-24}{6} = -4 \qquad \text{since} \qquad -24 = (6)(-4)$$

$$\frac{-30}{-5} = 6 \qquad \text{since} \qquad -30 = (-5)(6)$$

The examples above illustrate that because the two operations are related, the rule of signs that we stated in the last section for multiplication is also true for division.

DIVIDING SIGNED NUMBERS

1. The quotient of two numbers with different signs is negative.
2. The quotient of two numbers with the same sign is positive.

Again, the rule is easy to use. To divide two signed numbers, divide their absolute values. Then attach the proper sign according to the rule above.

Example 1

Divide.

(a) Positive $\longrightarrow$ $\dfrac{28}{7}$ $\longleftarrow$ Positive $= 4 \longleftarrow$ Positive

(b) Negative $\longrightarrow$ $\dfrac{-36}{-4}$ $\longleftarrow$ Negative $= 9 \longleftarrow$ Positive

(c) Negative $\longrightarrow$ $\dfrac{-42}{7}$ $\longleftarrow$ Positive $= -6 \longleftarrow$ Negative

(d) Positive $\longrightarrow$ $\dfrac{75}{-3}$ $\longleftarrow$ Negative $= -25 \longleftarrow$ Negative

(e) Positive $\longrightarrow$ $\dfrac{15.2}{-3.8}$ $\longleftarrow$ Negative $= -4 \longleftarrow$ Negative

CHECK YOURSELF 1

Divide.

1. $\dfrac{-55}{11}$ **2.** $\dfrac{80}{20}$ **3.** $\dfrac{-48}{-8}$ **4.** $\dfrac{144}{-12}$ **5.** $\dfrac{-13.5}{-2.7}$

You should be very careful when 0 is involved in a division problem. Remember that 0 divided by any nonzero number is just 0. Recall that

$$\dfrac{0}{-7} = 0 \qquad \text{because} \qquad 0 = (-7)(0)$$

However, if zero is the *divisor,* we have a special problem. Consider

$$\dfrac{9}{0} = ?$$

This means that $9 = 0 \cdot ?$.

Can 0 times a number ever be 9? No, so there is no solution.

Since $\dfrac{9}{0}$ cannot be replaced by any number, we agree that *division by 0 is not allowed.* We say that

Division by 0 is undefined.

Example 2

Divide, if possible.

(a) $\dfrac{7}{0}$ is undefined.

(b) $\dfrac{-9}{0}$ is undefined.

(c) $\dfrac{0}{5} = 0$

(d) $\dfrac{0}{-8} = 0$

Note: The expression $\dfrac{0}{0}$ is called an *indeterminate form*. You will learn more about this in later mathematics classes.

CHECK YOURSELF 2

Divide if possible.

1. $\dfrac{0}{3}$ **2.** $\dfrac{5}{0}$ **3.** $\dfrac{-7}{0}$ **4.** $\dfrac{0}{-9}$

Recall that the fraction bar serves as a grouping symbol. This means that all operations in the numerator and denominator should be performed separately. Then the division is done as the last step. Our next example illustrates.

Example 3

Evaluate each expression.

(a) $\dfrac{(-6)(-7)}{3} = \dfrac{42}{3} = 14$ Multiply in the numerator, then divide.

(b) $\dfrac{3 + (-12)}{3} = \dfrac{-9}{3} = -3$ Add in the numerator, then divide.

(c) $\dfrac{-4 + (2)(-6)}{-6 - 2} = \dfrac{-4 + (-12)}{-6 - 2}$ Multiply in the numerator. Then add in the numerator and subtract in the denominator.

$\qquad\qquad = \dfrac{-16}{-8} = 2$ Divide as the last step.

CHECK YOURSELF 3

Evaluate each expression.

1. $\dfrac{-4 + (-8)}{6}$ **2.** $\dfrac{3 - (2)(-6)}{-5}$ **3.** $\dfrac{(-2)(-4) - (-6)(-5)}{(-4)(11)}$

Evaluating fractions with a calculator poses a special problem. Our next example illustrates.

Example 4

Use your scientific calculator to evaluate each fraction.

(a) $\dfrac{4}{2-3}$

As you can see, the correct answer should be -4. To get this answer with your calculator, you must place the denominator in parentheses. The key stroke sequence will be

4 ÷ (2 − 3) =

(b) $\dfrac{-7-7}{3-10}$

In this problem, the correct answer is 2. This can be found on your calculator by placing the numerator in parentheses and then placing the denominator in parentheses. The key stroke sequence will be

(7 +/− − 7) ÷ (3 − 10) =

When evaluating a fraction with a calculator, it is safest to use parentheses in both the numerator and the denominator.

CHECK YOURSELF 4

Evaluate using your calculator

1. $\dfrac{-8}{5-7}$ **2.** $\dfrac{-3-2}{-13+23}$

CHECK YOURSELF ANSWERS

1. (1) -5; (2) 4; (3) 6; (4) -12; (5) 5.
2. (1) 0; (2) undefined; (3) undefined; (4) 0.
3. (1) -2; (2) -3; (3) $\dfrac{1}{2}$.
4. (1) 4; (2) -0.5

Name

Date

Build Your Skills

Divide.

1. $\dfrac{-15}{-3}$

2. $\dfrac{60}{12}$

3. $\dfrac{72}{9}$

4. $\dfrac{-27}{9}$

5. $\dfrac{50}{-5}$

6. $\dfrac{-32}{-8}$

7. $\dfrac{-52}{4}$

8. $\dfrac{56}{-7}$

9. $\dfrac{-75}{-3}$

10. $\dfrac{-60}{15}$

11. $\dfrac{0}{-8}$

12. $\dfrac{-125}{-25}$

13. $\dfrac{-9}{-1}$

14. $\dfrac{-10}{0}$

15. $\dfrac{-84}{-7}$

16. $\dfrac{-10}{1}$

17. $\dfrac{15}{0}$

18. $\dfrac{0}{-15}$

1. _____

2. _____

3. _____

4. _____

5. _____

6. _____

7. _____

8. _____

9. _____

10. _____

11. _____

12. _____

13. _____

14. _____

15. _____

16. _____

17. _____

18. _____

ANSWERS

19. _____

20. _____

21. _____

22. _____

23. _____

24. _____

25. _____

26. _____

27. _____

28. _____

29. _____

30. _____

31. _____

32. _____

33. _____

34. _____

35. _____

36. _____

37. _____

38. _____

19. $\dfrac{-17}{1}$

20. $\dfrac{-27}{-1}$

21. $\dfrac{-144}{-16}$

22. $\dfrac{-150}{6}$

23. $\dfrac{-22.2}{3.7}$

24. $\dfrac{-16.8}{-2.4}$

25. $\dfrac{-5}{20}$

26. $\dfrac{-7}{-35}$

27. $\dfrac{24}{-16}$

28. $\dfrac{-25}{10}$

29. $\dfrac{-28}{-42}$

30. $\dfrac{-125}{-75}$

Perform the indicated operations.

31. $\dfrac{(-6)(-3)}{2}$

32. $\dfrac{(-9)(5)}{-3}$

33. $\dfrac{(-8)(2)}{-4}$

34. $\dfrac{(7)(-8)}{-14}$

35. $\dfrac{18}{-3-6}$

36. $\dfrac{27}{-10+7}$

37. $\dfrac{-8-8}{-2}$

38. $\dfrac{15-24}{-3}$

39. $\dfrac{43 - 15}{-10 - 4}$

40. $\dfrac{-7 - 2}{-12 + 9}$

41. $\dfrac{7 - 5}{2 - 2}$

42. $\dfrac{10 - 6}{4 - 4}$

43. $\dfrac{-15 - (-3)}{3 - (-1)}$

44. $\dfrac{21 - (-4)}{-3 - 2}$

45. $\dfrac{(-5)(-8) - 4}{15 - (-3)}$

46. $\dfrac{6 - 2(-10)}{-18 - (-5)}$

47. $\dfrac{(-3)(-6) - (-4)(8)}{6 - (-4)}$

48. $\dfrac{(5)(-2) - (-4)(-5)}{-4 - 2}$

49. $\dfrac{2(-5) + 4(6 - 8)}{3(-4 + 2)}$

50. $\dfrac{(-3)(-5) - 3(5 - 8)}{4(-8 + 6)}$

51. $\dfrac{(-5)^2 - (-4)(5)}{3(5 - 8)}$

52. $\dfrac{(-3)^2 - (-9)(-5)}{4 - (-2)}$

Think About These

Insert parentheses in each of the following so that the statement is true.

53. $8 \div 4 \cdot 2 = 1$

54. $6^2 \cdot \dfrac{1}{6} - 5 \cdot 3 = 3$

55. $|-8| - (-8) \div (-4) = -4$

56. $-5^2 \cdot (-2) \div (-10) = 5$

Label each of the following as true or false.

57. $12 \div 3^2 = 12 \div 3 \cdot 3$

58. $64 \div 4^3 = 64 \div 4^2 \cdot 4$

ANSWERS

39. _____

40. _____

41. _____

42. _____

43. _____

44. _____

45. _____

46. _____

47. _____

48. _____

49. _____

50. _____

51. _____

52. _____

53. _____

54. _____

55. _____

56. _____

57. _____

58. _____

59. $8 \div 4 \cdot 2 = 8 \div 2 \cdot 4$

60. $(-18) \div (-3) \cdot (-2) = (-18) \cdot (-3) \div (-2)$

The following is a list of erosion rates in metric ton per hectare per year for selected regions of the world.

Midwest United States	-35.6
Southern plains United States	-51.5
China	-43.0
East Germany	-13.0
Nigeria	-14.4
El Salvador	-55.0

61. What is the average rate of erosion for the listed regions?

62. What percentage is the midwest U.S. erosion rate of the southern plains U.S. erosion rate?

63. Express the southern plains regional erosion rate as a percentage of the average rate for the regions.

Skillscan (Section 1.6)

Evaluate each expression if $x = 2$, $y = 3$, and $z = 5$.

a. $5x + 3z$

b. $y^2 - 4x$

c. $3x^2 - 2z$

d. $x^3 + 2z^2$

e. $\dfrac{2x + 4z}{2y}$

f. $\dfrac{x(z - y)}{y + z}$

g. $\dfrac{2(x + y + 2z)}{y^2 - 2x}$

h. $\dfrac{3x - 2y + 5z}{3y - 2x}$

ANSWERS

1. 5   **3.** 8   **5.** -10 **7.** -13 **9.** 25   **11.** 0   **13.** 9   **15.** 12

17. Undefined   **19.** -17 **21.** 9   **23.** -6 **25.** $-\dfrac{1}{4}$ **27.** $-\dfrac{3}{2}$ **29.** $\dfrac{2}{3}$ **31.** 9

33. 4   **35.** -2 **37.** 8   **39.** -2 **41.** Undefined   **43.** -3 **45.** 2   **47.** 5

49. 3   **51.** -5 **53.** $8 \div (4 \cdot 2) = 1$ **55.** $(|-8| - (-8)) \div (-4) = -4$ **57.** False

59. False   **61.** -35.4 cm   **63.** 145.5 percent   **a.** 25   **b.** 1   **c.** 2   **d.** 58

e. 4   **f.** $\dfrac{1}{2}$ **g.** 6   **h.** 5

Name

Date

Build Your Skills

Evaluate each of the expressions if $a = -2$, $b = 5$, $c = -4$, and $d = 6$.

1. $3a + 4c$ **2.** $3b + 5c$

3. $4b - 3c$ **4.** $5c - 7a$

5. $-b^2 + b$ **6.** $(-b)^2 + b$

7. $3a^2$ **8.** $6c^2$

9. $c^2 - 2d$ **10.** $3a^2 + 4c$

11. $2a^2 + 3b^2$ **12.** $4b^2 - 2c^2$

13. $2(a + b)$ **14.** $5(b - c)$

15. $4(2a - d)$ **16.** $6(3c - d)$

17. $a(b + 3c)$ **18.** $c(3a - d)$

19. $\dfrac{6d}{c}$ **20.** $\dfrac{8b}{5c}$

21. $\dfrac{2b - a}{d}$ **22.** $\dfrac{3d + 2b}{c}$

ANSWERS

1. _____
2. _____
3. _____
4. _____
5. _____
6. _____
7. _____
8. _____
9. _____
10. _____
11. _____
12. _____
13. _____
14. _____
15. _____
16. _____
17. _____
18. _____
19. _____
20. _____
21. _____
22. _____

23. _____

24. _____

25. _____

26. _____

27. _____

28. _____

29. _____

30. _____

31. _____

32. _____

33. _____

34. _____

35. _____

36. _____

37. _____

38. _____

39. _____

40. _____

41. _____

42. _____

43. _____

44. _____

45. _____

46. _____

23. $\dfrac{2b - 3a}{c + 2d}$

24. $\dfrac{3d - 2b}{5a + d}$

25. $d^2 - b^2$

26. $c^2 - a^2$

27. $(d - b)^2$

28. $(c - a)^2$

29. $(d - b)(d + b)$

30. $(c - a)(c + a)$

31. $d^3 - b^3$

32. $c^3 + a^3$

33. $(d - b)^3$

34. $(c + a)^3$

35. $(d - b)(d^2 + db + b^2)$

36. $(c + a)(c^2 - ac + a^2)$

37. $a^2 + d^2$

38. $b^2 - c^2$

39. $(a + d)^2$

40. $(b - c)^2$

41. $a^2 + 2ad + d^2$

42. $b^2 - 2bc + c^2$

Evaluate each expression if $X = -2$, $Y = -5$, and $Z = 3$.

43. $X + Y * Z$

44. $Y - 2 * Z$

45. $X^2 - Z^2$

46. $X^2 + Y^2$

116

47. $(X * Y)/(Z - X)$

48. $Y\char`^2/(Z * Y)$

49. $(2 * X + Y)/(2 * X + Z)$

50. $(X\char`^2 * Y\char`^2)/(X * Y)$

Think About These

Solve the following problems.

51. The perimeter of a rectangle of length L and width W is given by the formula $P = 2L + 2W$. Find the perimeter when L is 10 inches (in) and W is 5 in.

52. The simple interest I on a principal of P dollars at interest rate r for time t, in years, is given by $I = Prt$. Find the simple interest on a principal of \$6000 at 8 percent for 3 years.
(Note: $8\% = 0.08$.)

In each of the following problems, decide if the given number makes the statement true or false.

53. $x - 7 = 2y + 5$; $x = 22$, $y = 5$

54. $3(x - y) = 6$; $x = 5$, $y = -3$

55. $2(x + y) = 2x + y$; $x = -4$, $y = -2$

56. $x^2 - y^2 = x - y$; $x = 4$, $y = -3$

The equation $D = S + T(E + A)$ represents the total soil depth on an experimental soil plot.

$D =$ soil depth after some given time
$S =$ original depth of soil
$T =$ time elapsed in years
$E =$ yearly erosion rate
$A =$ yearly accumulation rate

47. _____

48. _____

49. _____

50. _____

51. _____

52. _____

53. _____

54. _____

55. _____

56. _____

57. _____

58. _____

 57. Find the soil depth after 25 years if the original depth was 15 cm, the erosion rate was −0.35 cm/yr, and the accumulation rate was 0.17 cm/yr.

 58. Find the soil depth after 200 years if the accumulation rate is predicted to be 0.25 cm/yr, the erosion rate is predicted to be −0.21 cm/yr, and the original soil depth is 7 cm.

ANSWERS

1. −22   **3.** 32   **5.** −20   **7.** 12   **9.** 4   **11.** 83   **13.** 6   **15.** −40   **17.** 14
19. −9   **21.** 2   **23.** 2   **25.** 11   **27.** 1   **29.** 11   **31.** 91   **33.** 1   **35.** 91
37. 40   **39.** 16   **41.** 16   **43.** −17   **45.** −5   **47.** 2   **49.** 9   **51.** 30 in
53. True   **55.** False   **57.** 10.5 cm

Summary

Signed Numbers—The Terms [2.1]

Negative numbers / Positive numbers

Positive Numbers   Numbers used to name points to the right of the origin on the number line.

Negative Numbers   Numbers used to name points to the left of the origin on the number line.

Signed Numbers   The positive and negative numbers.

Integers   The natural (or counting) numbers, their negatives, and zero.

The integers are

$$\{ \ldots, -3, -2, -1, 0, 1, 2, 3, \ldots \}$$

$|7| = 7$
$|-10| = 10$

Absolute Value   The distance (on the number line) between the point named by a signed number and the origin.

The absolute value of x is written $|x|$

Adding Signed Numbers [2.2]

To Add Signed Numbers

$9 + 7 = 16$
$(-9) + (-7) = -16$
$15 + (-10) = 5$
$(-12) + 9 = -3$

1. If two numbers have the same sign, add their absolute values. Give the sum the sign of the original numbers.
2. If two numbers have different signs, subtract their absolute values, the smaller from the larger. Give the sum the sign of the number with the larger absolute value.

Subtracting Signed Numbers [2.3]

To Subtract Signed Numbers

$16 - 8 = 16 + (-8)$
$\qquad = 8$
$8 - 15 = 8 + (-15)$
$\qquad = -7$

$-9 - (-7) = -9 + 7$
$\qquad = -2$

1. Rewrite the subtraction problem as an addition problem by
   (*a*) Changing the subtraction symbol to an addition symbol
   (*b*) Replacing the number being subtracted with its opposite
2. Add the resulting signed numbers as before.

Multiplying Signed Numbers [2.4]

$5(-7) = -35$

$(-10)(9) = -90$
$8 \cdot 7 = 56$

$(-9)(-8) = 72$

To Multiply Signed Numbers   Multiply the absolute values of the two numbers.

1. If the numbers have different signs, the product is negative.
2. If the numbers have the same sign, the product is positive.

$$\frac{-32}{4} = -8$$

$$\frac{75}{-5} = -15$$

$$\frac{20}{5} = 4$$

$$\frac{-18}{-9} = 2$$

Dividing Signed Numbers [2.5]

To Divide Signed Numbers   Divide the absolute values of the two numbers.

1. If the numbers have different signs, the quotient is negative.
2. If the numbers have the same sign, the quotient is positive.

Evaluate

$$\frac{4a - b}{2c}$$

if $a = -6$, $b = 8$, and $c = -4$.

$$\frac{4a - b}{2c} = \frac{4(-6) - 8}{2(-4)}$$

$$= \frac{-24 - 8}{-8}$$

$$= \frac{-32}{-8} = 4$$

Evaluating Algebraic Expressions [2.6]

To Evaluate an Expression

1. Replace each variable by the given number value.
2. Do the necessary arithmetic operations. (Be sure to follow the rules for the order of operations.)

Summary Exercises Chapter 2

This supplementary exercise set will give you practice with each of the objectives of the chapter. Each exercise is keyed to the appropriate chapter section. The answers are provided in the instructor's manual. Your instructor will give you guidelines on how to best use these exercises in your instructional setting.

[2.1] Represent the integers on the number line shown.

1. 6, −18, −3, 2, 15, −9

[2.1] Evaluate.

2. $|9|$ **3.** $|-9|$ **4.** $-|9|$ **5.** $-|-9|$

6. $|12 - 8|$ **7.** $|8| - |12|$ **8.** $-|8 - 12|$ **9.** $|-8| - |-12|$

[2.2] Add.

10. $-3 + (-8)$ **11.** $10 + (-4)$ **12.** $6 + (-6)$ **13.** $-16 + (-16)$

14. $-18 + 0$ **15.** $\dfrac{3}{8} + \left(-\dfrac{11}{8}\right)$ **16.** $5.7 + (-9.7)$ **17.** $-18 + 7 + (-3)$

[2.3] Subtract.

18. $8 - 13$ **19.** $-7 - 10$ **20.** $10 - (-7)$ **21.** $-5 - (-1)$

22. $-9 - (-9)$ **23.** $0 - (-2)$ **24.** $-\dfrac{5}{4} - \left(-\dfrac{17}{4}\right)$ **25.** $7.9 - (-8.1)$

[2.3] Perform the indicated operations.

26. $|8 - 12|$ **27.** $|8| - |12|$ **28.** $|-8 - 12|$ **29.** $|-8| - |-12|$

30. $-6 - (-2) + 3$ **31.** $-5 - (5 - 8)$ **32.** $7 - (3 - 7) + 4$

33. Subtract -7 from -8.   **34.** Subtract -9 from the sum of 6 and -2.

[2.4] Multiply.

35. $(10)(-7)$ **36.** $(-8)(-5)$ **37.** $(-3)(-15)$ **38.** $(1)(-15)$

39. $(0)(-8)$ **40.** $\left(\dfrac{2}{3}\right)\left(-\dfrac{3}{2}\right)$ **41.** $(-4)\left(\dfrac{3}{8}\right)$ **42.** $\left(-\dfrac{5}{4}\right)(-1)$

43. $(-8)(-2)(5)$ **44.** $(-4)(-3)(2)$ **45.** $\left(\dfrac{2}{5}\right)(-10)\left(-\dfrac{5}{2}\right)$ **46.** $\left(\dfrac{4}{3}\right)(-6)\left(-\dfrac{3}{4}\right)$

[2.4] Perform the indicated operations.

47. $2(-4 + 3)$ **48.** $(2)(-3) - (-5)(-3)$ **49.** $(2 - 8)(2 + 8)$

[2.5] Divide.

50. $\dfrac{80}{16}$ **51.** $\dfrac{-63}{7}$ **52.** $\dfrac{-81}{-9}$

53. $\dfrac{0}{-5}$ **54.** $\dfrac{32}{-8}$ **55.** $\dfrac{-7}{0}$

[2.5] Perform the indicated operations.

56. $\dfrac{-8 + 6}{-8 - (-10)}$ **57.** $\dfrac{2(-3) - 1}{5 - (-2)}$ **58.** $\dfrac{(-5)^2 - (-2)^2}{-5 - (-2)}$

[2.6] Evaluate the expressions if $x = -3$, $y = 6$, $z = -4$, and $w = 2$.

59. $3x + w$ **60.** $5y - 4z$ **61.** $x + y - 3z$ **62.** $5z^2$

63. $3x^2 - 2w^2$ **64.** $3x^3$ **65.** $5(x^2 - w^2)$ **66.** $\dfrac{6z}{2w}$

67. $\dfrac{2x - 4z}{y - z}$ **68.** $\dfrac{3x - y}{w - x}$ **69.** $\dfrac{x(y^2 - z^2)}{(y + z)(y - z)}$ **70.** $\dfrac{y(x - w)^2}{x^2 - 2xw + w^2}$

Self-Test for CHAPTER 2

Name

Section

Date

The purpose of this self-test is to help you check your progress and to review for a chapter test in class. Allow yourself about an hour to take the test. When you are done, check your answers in the back of the book. If you missed any problems, go back and review the appropriate sections in the chapter and the exercises provided.

ANSWERS

Represent the integers on the number line shown.

$$\xleftarrow{\quad\;|\quad\;|\quad\;|\quad\;|\quad\;|\quad\;}\xrightarrow{}$$
$$\quad\;-20\quad\;-10\quad\;\;0\quad\;\;10\quad\;\;20$$

1. 5, −12, 4, −7, 18, −17

Evaluate.

2. $|7|$

3. $|-7|$

4. $|18 - 7|$

5. $|18| - |-7|$

Add.

6. $-8 + (-5)$

7. $6 + (-9)$

8. $(-9) + (-12)$

9. $-\dfrac{5}{3} + \dfrac{8}{3}$

Subtract.

10. $9 - 15$

11. $-9 - 15$

12. $5 - (-4)$

13. $-7 - (-7)$

Multiply.

14. $(-8)(5)$

15. $(-9)(-7)$

16. $(4.5)(-6)$

17. $(-2)(-3)(-4)$

1. _____

2. _____

3. _____

4. _____

5. _____

6. _____

7. _____

8. _____

9. _____

10. _____

11. _____

12. _____

13. _____

14. _____

15. _____

16. _____

17. _____

18. _____

19. _____

20. _____

21. _____

22. _____

23. _____

24. _____

25. _____

Divide.

18. $\dfrac{75}{-3}$

19. $\dfrac{-27}{-9}$

20. $\dfrac{-45}{9}$

21. $\dfrac{9}{0}$

Evaluate if $a = -2$, $b = 6$, and $c = -4$.

22. $4a - c$

23. $5c^2$

24. $6(2b - 3c)$

25. $\dfrac{3a - 4b}{a + c}$

CHAPTER 3

Equations and Inequalities

Habitat Loss and Extinction

Human activities have invaded the far reaches of the planet. The effects have been noted in places as diverse as the Pacific northwest, the high Himalayas, Antarctica, the coral reefs in the warm oceans of the world, and the desert regions of the Middle East. Our impact on wildlife habitat is now global, and we must learn to consider our actions carefully if we are to protect wildlife from the negative impacts of our presence.

Today, one of the greatest threats to wildlife is the loss of habitat. Plant and animal habitats are destroyed or altered in a variety of ways by human activities. Clear-cutting forests, fouling lakes and streams, and dumping toxic chemicals are seen by nearly everyone as being harmful to wildlife. Not so easily recognized are individual activities such as hiking and camping or building vacation homes in woodland settings. Although necessary for human needs, dam building for power generation or expanding agriculture to provide food for a growing world population takes a toll on wildlife habitat. Because of the rapid growth of human population and expansion around the globe, many plants and animals are being forced to adapt to changes in their environment. Those unable to adapt are becoming extinct.

The loss of tropical moist forests around the world is becoming a major concern of the scientific community. Although tropical forests cover only 6 percent of the land surface of the earth, they contain over 50 percent of the plant and animal species. The loss of tropical forest habitat therefore has a proportionally greater impact on the loss of plants and animals than other parts of the earth do. In the Amazonia region of Brazil there were 1250 square kilometers of cleared forestland in 1975. By 1987 the amount of forest cut down had increased to 60,000 square kilometers. Three times that amount of forest had been adversely impacted by the expansion of settlements into the area. It is estimated that we may be losing over 1000 species of plants and animals per year to extinction because of this type of forest loss in the tropical regions.

Current research is beginning to show our true lack of knowledge about the variety of life on earth. There are 1.7 million documented (scientifically named) species. It is generally accepted that there are at least 5 million species of plants and animals, with some researchers claiming there may be as many as 30 million insects alone in the tropical forests. Investigators have studied only 10 percent of all plant species and have thoroughly investigated only 1 percent of them. With so little known about the web of life on earth, we should be very careful about which species we allow to become extinct because of our needs for space and resources. Many of these plants and animals may prove useful, and all play some role in the earth's ecosystems.

Equations—An Introduction

OBJECTIVE

To determine whether a given number is a solution for an equation

In this chapter you will begin working with one of the most important tools of mathematics, the equation. The ability to recognize and solve various types of equations is probably the most useful algebraic skill you will learn. We will continue to build upon the methods of this chapter throughout the remainder of the text. To start, let's describe what we mean by an equation.

> An *equation* is a mathematical statement that two expressions are equal.

Some examples are

$$3 + 4 = 7$$
$$x + 3 = 5$$
$$P = 2L + 2W$$

As you can see, an equals sign $(=)$ separates the two equal expressions. These expressions are usually called the *left side* and the *right side* of the equation.

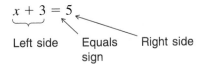

Left side   Equals sign   Right side

An equation such as $x + 3 = 5$ is called a *conditional equation* because it can be either true or false depending on the value given to the variable.

An equation may be either true or false. For instance, $3 + 4 = 7$ is true because both sides name the same number. What about an equation such as $x + 3 = 5$ that has a letter or variable on one side? Any numbers can replace x in the equation. However, only one number will make this equation a true statement.

$$\text{If } x = \begin{cases} 1 & 1 + 3 = 5 \text{ is false} \\ 2 & 2 + 3 = 5 \text{ is true} \\ 3 & 3 + 3 = 5 \text{ is false} \end{cases}$$

The number 2 is called the *solution* (or *root*) of the equation $x + 3 = 5$ because substituting 2 for x gives a true statement.

A *solution* for an equation is any value for the variable that makes the equation a true statement.

Example 1

(*a*) Is 3 a solution for the equation $2x + 4 = 10$?

To find out, replace x with 3 and evaluate $2x + 4$ on the left.

Left side	Right side
$2 \cdot 3 + 4 \overset{?}{=}$	10
$6 + 4 \overset{?}{=}$	10
$10 =$	10

Since $10 = 10$ is a true statement, 3 is a solution of the equation.

(*b*) Is 5 a solution of the equation $3x - 2 = 2x + 1$?

To find out, replace x with 5 and evaluate each side separately.

Remember the rules for the order of operation. Multiply first; then add or subtract.

Left side	Right side
$3 \cdot 5 - 2 \overset{?}{=}$	$2 \cdot 5 + 1$
$15 - 2 \overset{?}{=}$	$10 + 1$
$13 \neq$	11

Since the two sides do not name the same number, we do not have a true statement and 5 is not a solution.

CHECK YOURSELF 1

For the equation

$$2x - 1 = x + 5$$

1. Is 4 a solution?
2. Is 6 a solution?

You may be wondering whether an equation can have more than one solution. It certainly can. For instance,

This is an example of a *quadratic equation.* We will consider methods of solution in Chapter 5 and then again in Chapter 10.

$$x^2 = 9$$

has two solutions. They are 3 and -3 because

$$3^2 = 9 \qquad \text{and} \qquad (-3)^2 = 9$$

In this chapter, however, we will always be working with *linear equations in one variable*. These are equations that can be put into the form

$$ax + b = 0$$

where the variable is x, where a and b are any numbers, and a is not equal to 0. In a linear equation, the variable can appear only to the first power. No other power (x^2, x^3, etc.) can appear. Linear equations are also called *first-degree equations*. The degree of an equation in one variable is the highest degree to which the variable appears.

> Linear equations in one variable that can be written in the form
>
> $$ax + b = 0 \qquad a \neq 0$$
>
> will have exactly one solution.

Example 2

Label each of the following as an expression, a linear equation, or an equation that is not linear.

(a) $4x + 5$ is an expression
(b) $2x + 8 = 0$ is a linear equation
(c) $3x^2 - 9 = 0$ is not a linear equation
(d) $5x = 15$ is a linear equation

CHECK YOURSELF 2

Label each of the following as an expression, a linear equation, or an equation that is not linear.

1. $2x^2 = 8$
2. $2x - 3 = 0$
3. $5x - 10$
4. $2x + 1 = 7$

It is not difficult to find the solution for an equation such as $x + 3 = 8$ by guessing the answer to the following question:

What plus 3 is 8?

Here the answer to the question is 5, and that is also the solution for the equation. But for more complicated equations you are going to need something more than guesswork. So in Section 3.2 we will begin looking at a set of rules that will let you solve any linear equation in one variable.

CHECK YOURSELF ANSWERS

1. (1) 4 is not a solution; (2) 6 is a solution.
2. (1) Nonlinear equation; (2) linear equation; (3) expression; (4) linear equation.

Name

Date

ANSWERS

Build Your Skills

Is the number shown in parentheses a solution for the given equation?

1. $x + 7 = 8$ (1)

2. $x + 5 = 9$ (3)

3. $x - 7 = 4$ (10)

4. $x - 5 = 3$ (8)

5. $5 - x = 2$ (4)

6. $10 - x = 7$ (3)

7. $4 - x = 6$ (-2)

8. $5 - x = 6$ (-3)

9. $2x + 3 = 9$ (5)

10. $4x + 5 = 17$ (3)

11. $2x - 1 = 3$ (4)

12. $3x - 2 = 4$ (2)

13. $3x - 5 = 16$ (7)

14. $5x - 3 = 21$ (5)

15. $3x - 8 = -12$ (-2)

16. $7x + 21 = 0$ (-3)

17. $5 - 2x = 7$ (-1)

18. $4 - 5x = 9$ (-2)

19. $12 + 2x = 0$ (-6)

20. $4x - 3 = -23$ (-5)

21. $4x - 5 = 2x + 3$ (4)

22. $5x + 4 = 2x + 10$ (4)

23. $3x + 2 = x - 6$ (-5)

24. $6x + 3 = 2x - 9$ (-3)

25. $x + 3 + 2x = 5 + x + 8$ (5)

26. $5x - 3 + 2x = 3 + x - 12$ (-2)

1. _____
2. _____
3. _____
4. _____
5. _____
6. _____
7. _____
8. _____
9. _____
10. _____
11. _____
12. _____
13. _____
14. _____
15. _____
16. _____
17. _____
18. _____
19. _____
20. _____
21. _____
22. _____
23. _____
24. _____
25. _____
26. _____

27. _____

28. _____

29. _____

30. _____

31. _____

32. _____

33. _____

34. _____

35. _____

36. _____

37. _____

38. _____

a. _____

b. _____

c. _____

d. _____

e. _____

f. _____

g. _____

h. _____

27. $\dfrac{2}{3}x = 12$ (36)

28. $\dfrac{3}{4}x = 18$ (24)

29. $\dfrac{3}{5}x + 5 = 11$ (10)

30. $\dfrac{2}{3}x + 8 = -12$ (−6)

Label each of the following as an expression or a linear equation.

31. $2x + 1 = 9$

32. $7x + 14$

33. $2x - 8$

34. $5x - 3 = 12$

35. $7x + 2x + 8 - 3$

36. $x + 5 = 13$

37. $2x - 8 = 3$

38. $12x - 5x + 2 + 5$

Skillscan (Section 1.4)

Simplify by combining like terms.

a. $x + 3x$

b. $5x - 7x$

c. $5a - 3a$

d. $3b + 7b$

e. $-7x + 5x$

f. $9w - 9w$

g. $2y + 1 + 3y$

h. $5p - 3 + 2p$

ANSWERS

1. Yes   **3.** No   **5.** No   **7.** Yes   **9.** No   **11.** No   **13.** Yes   **15.** No
17. Yes   **19.** Yes   **21.** Yes   **23.** No   **25.** Yes   **27.** No   **29.** Yes   **31.** Linear
equation   **33.** Expression   **35.** Expression   **37.** Linear equation   **a.** $4x$ **b.** $-2x$
c. $2a$ **d.** $10b$ **e.** $-2x$ **f.** 0 **g.** $5y + 1$ **h.** $7p - 3$

 **3.2**

Solving Equations by Adding or Subtracting

OBJECTIVE
To use addition and/or subtraction to solve equations

As we said in Section 3.1, guesswork is not a very good approach to solving complicated equations. A better method is to transform the given equation to an *equivalent equation* whose solution can be found by inspection. Let's make a definition.

> Equations that have the same solution are called *equivalent equations.*

As examples,

$$2x + 3 = 5$$
$$2x = 2$$
$$x = 1$$

are all equivalent equations. They all have the same solution, 1. We say that a linear equation is *solved* when it is transformed to an equivalent equation of the form

Note: In some cases we'll write the equation in the form

$$\square = x$$

The number will be our solution when the equation has the variable isolated on the left or on the right.

$$x = \square$$

↑
The variable is alone on the left side.

The right side is some number, the solution.

Here is the first property you will need to transform an equation to an equivalent form.

Remember: An equation is a statement that the two sides are equal. Adding the same quantity to both sides does not change the equality or "balance."

> **THE ADDITION PROPERTY OF EQUALITY**
>
> If $a = b$
> then $a + c = b + c$
>
> In words, adding the same quantity to both sides of an equation gives an equivalent equation.

Let's look at an example of applying this property to solve an equation.

Example 1

Solve

$$x - 3 = 9$$

Remember that our goal is to isolate x on one side of the equation. Since 3 is being subtracted from x, we can add 3 to remove it. We must use the addition property to add 3 to both sides of the equation.

$$
\begin{array}{rr}
x - 3 = & 9 \\
+3 & +3 \\
\hline
x \quad = & 12
\end{array}
$$

Adding 3 "undoes" the subtraction and leaves x alone on the left.

Since 12 is the solution for the equivalent equation $x = 12$, it is the solution for our original equation.

To check, replace x with 12 in the original equation:

$x - 3 = 9$	The given equation
$12 - 3 = 9$	Substitute 12 for x
$9 = 9$	A true statement

Since we have a true statement, 12 is the solution.

CHECK YOURSELF 1

Solve and check.

$x - 5 = 4$

The addition property also allows us to add a negative number to both sides of an equation. This is really the same as subtracting the same quantity from both sides.

Example 2

Solve

$x + 5 = 9$

In this case 5 is *added* to x on the left. We can use the addition property to subtract 5 from both sides. This will "undo" the addition and leave the variable x alone on one side of the equation.

$$
\begin{array}{rr}
x + 5 = & 9 \\
-5 & -5 \\
\hline
x \quad = & 4
\end{array}
$$

The solution is 4. To check, replace x with 4:

$4 + 5 = 9$ \quad (true)

CHECK YOURSELF 2

Solve and check.

$x + 6 = 13$

What if the equation has a variable term on both sides? You will have to use the addition property to add or subtract a term involving the variable to get the desired result.

Example 3

Solve

$5x = 4x + 7$

We will start by subtracting $4x$ from both sides of the equation. Do you see why? Remember that an equation is solved when we have an equivalent equation of the form $x = \square$.

$$
\begin{array}{rl}
5x = & 4x + 7 \\
\underline{-4x} & \underline{-4x} \\
x = & 7
\end{array}
\left.\begin{array}{l}
\\
\\
\\
\end{array}\right\{
\begin{array}{l}
\text{Subtracting } 4x \text{ from} \\
\text{both sides } \textit{removes} \\
4x \text{ from the right.}
\end{array}
$$

To check: Since 7 is a solution for the equivalent equation $x = 7$, it should be a solution for the original equation. To find out, replace x with 7:

$$5 \cdot 7 \stackrel{?}{=} 4 \cdot 7 + 7$$
$$35 \stackrel{?}{=} 28 + 7$$
$$35 = 35 \qquad \text{(true)}$$

CHECK YOURSELF 3

Solve and check.

$7x = 6x + 3$

You may have to apply the addition property more than once to solve an equation. Look at the following example.

Example 4

Solve

$7x - 8 = 6x$

We want all variables on *one* side of the equation. If we choose the left, we can begin by subtracting $6x$ from both sides of the equation. This will remove $6x$ from the right:

$$\begin{array}{rcl} 7x - 8 = & & 6x \\ -6x & & -6x \\ \hline x - 8 = & & 0 \end{array}$$

Now we want the variable alone, so we add 8 to both sides. This will isolate x on the left side.

$$\begin{array}{rcl} x - 8 = & & 0 \\ +8 & & +8 \\ \hline x & = & 8 \end{array}$$

The solution is 8. We'll leave it to you to check this result.

CHECK YOURSELF 4

Solve and check.

$9x + 3 = 8x$

Often an equation will have more than one variable term *and* more than one number. You will have to apply the addition property twice in solving these equations.

Example 5

Solve

$5x - 7 = 4x + 3$

We would like the variable terms on the left, so we start by subtracting $4x$ to remove that term from the right side of the equation:

$$\begin{array}{rcl} 5x - 7 = & & 4x + 3 \\ -4x & & -4x \\ \hline x - 7 = & & 3 \end{array}$$

Now, to isolate the variable, we add 7 to both sides to undo the subtraction on the left:

$$\begin{array}{rcl} x - 7 = & & 3 \\ +7 & & +7 \\ \hline x & = & 10 \end{array}$$

You could just as easily have added 7 to both sides and *then* subtracted 4x. The result would be the same. In fact, some students prefer to combine the two steps.

The solution is 10. To check, replace x with 10 in the original equation:

$$5 \cdot 10 - 7 = 4 \cdot 10 + 3$$
$$43 = 43 \qquad \text{(true)}$$

CHECK YOURSELF 5

Solve and check.

1. $4x - 5 = 3x + 2$ **2.** $6x + 2 = 5x - 4$

Remember, by *simplify* we mean to combine any like terms that may appear on one side of the equation.

In solving an equation, you should always simplify each side as much as possible before using the addition property.

Example 6

Solve

Like terms   Like terms

$$5 + 8x - 2 = 2x - 3 + 5x$$

Since like terms appear on each side of the equation, we start by combining the numbers on the left (5 and -2). Then we combine the like terms ($2x$ and $5x$) on the right. We have

$$3 + 8x = 7x - 3$$

Now we can apply the addition property, as before:

$$
\begin{array}{rl}
3 + 8x = 7x - 3 & \\
\underline{ -7x = -7x} & \text{Subtract } 7x. \\
3 + x = - 3 & \\
\underline{-3 - 3} & \text{Subtract 3.} \\
x = - 6 & \text{Isolate } x.
\end{array}
$$

The solution is -6. To check, always return to the original equation. That will catch any possible errors in simplifying. Replacing x with -6 gives

$$5 + 8(-6) - 2 \stackrel{?}{=} 2(-6) - 3 + 5(-6)$$
$$5 - 48 - 2 \stackrel{?}{=} -12 - 3 - 30$$
$$-45 = -45 \qquad \text{(true)}$$

Solve and check.

1. $3 + 6x + 4 = 8x - 3 - 3x$
2. $5x + 21 + 3x = 20 + 7x - 2$

We may have to apply some of the properties discussed in Section 1.3 in solving equations. Our final example illustrates the use of the distributive property to clear an equation of parentheses.

Example 7

Solve

$$2(3x + 4) = 5x - 6$$

Applying the distributive property on the left, we have

$$6x + 8 = 5x - 6$$

Note:

$2(3x + 4)$
$= 2(3x) + 2(4)$
$= 6x + 8$

We can then proceed as before:

$$
\begin{array}{rcl}
6x + 8 & = & 5x - 6 \\
-5x & & -5x \\
\hline
x + 8 & = & -6 \\
-8 & & -8 \\
\hline
x & = & -14
\end{array}
$$

Subtract $5x$.

Subtract 8.

The solution is -14. We will leave the checking of this result to the reader. **Remember:** Always return to the original equation to check.

Recall our comment that we could write an equation in the equivalent forms $x = \square$ or $\square = x$, where $\square$ represents some number. Suppose we have an equation like

$12 = x + 7$

Subtracting 7 will isolate x *on the right*:

$$
\begin{array}{rcl}
12 & = & x + 7 \\
-7 & & -7 \\
\hline
5 & = & x
\end{array}
$$

and the solution is 5.

Solve and check each of the following equations.

1. $4(5x - 2) = 19x + 4$
2. $3(5x + 1) = 2(7x - 3) - 4$

1. 9.   **2.** 7.   **3.** 3.   **4.** -3.
5. (1) 7; (2) -6.   **6.** (1) -10; (2) -3.   **7.** (1) 12; (2) -13.

Name _____

Date _____

Build Your Skills

Solve and check the following equations.

1. $x + 3 = 5$

2. $x - 8 = 2$

3. $x - 6 = 5$

4. $x + 6 = 10$

5. $x - 8 = -10$

6. $x + 5 = 2$

7. $x + 4 = -3$

8. $x - 5 = -4$

9. $11 = x + 5$

10. $x + 7 = 0$

11. $6 + x = 9$

12. $5 = x - 8$

13. $x - 5 = 0$

14. $9 + x = 13$

15. $3x = 2x + 4$

16. $5x = 4x - 8$

17. $8x = 7x - 10$

18. $7x = 6x + 5$

19. $6x + 3 = 5x$

20. $12x - 6 = 11x$

21. $8x - 4 = 7x$

22. $9x - 7 = 8x$

23. $2x + 3 = x + 5$

24. $3x - 2 = 2x + 1$

25. $5x - 7 = 4x - 3$

26. $8x + 5 = 7x - 2$

27. $7x - 2 = 6x + 4$

28. $10x - 3 = 9x - 6$

ANSWERS

1. _____
2. _____
3. _____
4. _____
5. _____
6. _____
7. _____
8. _____
9. _____
10. _____
11. _____
12. _____
13. _____
14. _____
15. _____
16. _____
17. _____
18. _____
19. _____
20. _____
21. _____
22. _____
23. _____
24. _____
25. _____
26. _____
27. _____
28. _____

29. _____

30. _____

31. _____

32. _____

33. _____

34. _____

35. _____

36. _____

37. _____

38. _____

39. _____

40. _____

41. _____

42. _____

43. _____

44. _____

45. _____

46. _____

47. _____

48. _____

49. _____

50. _____

51. _____

29. $2 + 5x = 5 + 4x$

30. $3 + 6x = 2 + 5x$

31. $3 + 7x = 6x - 8$

32. $2 + 10x = 9x - 7$

33. $2 + 4x + 1 = x + 9 + 2x$

34. $5x - 2 + x = 3 + 5x + 6$

35. $4x + 7 + 3x = 5x + 13 + x$

36. $5x + 9 + 4x = 9 + 8x - 7$

37. $3x - 5 + 2x - 7 + x = 5x + 2$

38. $5x + 8 + 3x - x + 5 = 6x - 3$

39. $3(2x + 5) = 5x - 3$

40. $4(3x - 2) = 11x + 5$

41. $3(7x + 2) = 5(4x + 1) + 17$

42. $5(5x + 3) = 3(8x - 2) + 4$

43. $6(6x - 1) = 5(7x + 3) - 3$

44. $9(5x - 8) = 4(11x - 16) - 2$

45. $\dfrac{5}{4}x - 1 = \dfrac{1}{4}x + 7$

46. $\dfrac{7}{5}x + 3 = \dfrac{2}{5}x - 8$

47. $\dfrac{9}{2}x - \dfrac{3}{4} = \dfrac{7}{2}x + \dfrac{5}{4}$

48. $\dfrac{11}{3}x + \dfrac{1}{6} = \dfrac{8}{3}x + \dfrac{19}{6}$

Translate each statement to an algebraic equation. Let x represent the number in each case.

49. 3 more than a number is 7.

50. 5 less than a number is 12.

51. 7 less than 3 times a number is twice that same number.

52. 4 more than 5 times a number is 6 times that same number.

53. 2 times the sum of a number and 5 is 18 more than that same number.

54. 3 times the sum of a number and 7 is 4 times that same number.

Think About These

55. Which of the following is equivalent to the equation $8x + 5 = 9x - 4$?
  a. $17x = -9$
  b. $x = -9$
  c. $8x + 9 = 9x$
  d. $9 = 17x$

56. Which of the following is equivalent to the equation $5x - 7 = 4x - 12$?
  a. $9x = 19$
  b. $9x - 7 = -12$
  c. $x = -18$
  d. $x - 7 = -12$

57. Which of the following is equivalent to the equation $12x - 6 = 8x + 14$?
  a. $4x - 6 = 14$
  b. $x = 20$
  c. $20x = 20$
  d. $4x = 8$

58. Which of the following is equivalent to the equation $7x + 5 = 12x - 10$?
  a. $5x = -15$
  b. $7x - 5 = 12x$
  c. $-5 = 5x$
  d. $7x + 15 = 12x$

True or false?

59. Every linear equation with one variable has exactly one solution.

ANSWERS

52. _____

53. _____

54. _____

55. _____

56. _____

57. _____

58. _____

59. _____

60. _____

61. _____

62. _____

a. _____

b. _____

c. _____

d. _____

e. _____

f. _____

g. _____

h. _____

142

60. Isolating the variable on the right side of the equation will result in a negative solution.

61. In 1975, 1250 square kilometers (km^2) of Amazonian forest had been cleared in Brazil. By 1987, the area of cleared forest had increased to 60,000 km^2. If we let x be the amount of forest cut between 1975 and 1987, we get the equation

$$1250 + x = 60,000$$

Find the amount of forest cut from 1975 to 1987.

62. The whooping crane is a highly endangered species of bird which lives in North America. In 1941, only 16 of these birds were known to exist. Today, there are about 217 whooping cranes with at least 170 of them living in the wild. Use the equation

Captives + wild = total

to determine the number of whooping cranes held in captivity.

Skillscan (Section 2.4)

Multiply.

a. $\left(\dfrac{1}{3}\right)(3)$

b. $(-6)\left(-\dfrac{1}{6}\right)$

c. $(7)\left(\dfrac{1}{7}\right)$

d. $\left(-\dfrac{1}{4}\right)(-4)$

e. $\left(\dfrac{3}{5}\right)\left(\dfrac{5}{3}\right)$

f. $\left(\dfrac{7}{8}\right)\left(\dfrac{8}{7}\right)$

g. $\left(-\dfrac{4}{7}\right)\left(-\dfrac{7}{4}\right)$

h. $\left(-\dfrac{6}{11}\right)\left(-\dfrac{11}{6}\right)$

ANSWERS

1. 2   **3.** 11   **5.** −2   **7.** −7   **9.** 6   **11.** 3   **13.** 5   **15.** 4   **17.** −10
19. −3   **21.** 4   **23.** 2   **25.** 4   **27.** 6   **29.** 3   **31.** −11   **33.** 6   **35.** 6
37. 14   **39.** −18   **41.** 16   **43.** 18   **45.** 8   **47.** 2   **49.** $x + 3 = 7$
51. $3x − 7 = 2x$ **53.** $2(x + 5) = x + 18$ **55.** c   **57.** a   **59.** True
61. 58,750 km^2 **a.** 1   **b.** 1   **c.** 1   **d.** 1   **e.** 1   **f.** 1   **g.** 1   **h.** 1

3.3

Solving Equations by Multiplying or Dividing

OBJECTIVE

To use multiplication and/or division to solve equations

Let's look at a different type of equation. For instance, what if we want to solve an equation like the following?

$$6x = 18$$

Adding or subtracting by the property of the last section won't help. We will need a second property for solving equations.

> **THE MULTIPLICATION PROPERTY OF EQUALITY**
>
> If $a = b$ then $ac = bc$ where $c \neq 0$
> In words, multiplying both sides of an equation by the same nonzero number gives an equivalent equation.

Again, as long as you do the *same* thing to *both* sides of the equation, the "balance" is maintained.

Do you see why the number cannot be 0? Multiplying by 0 gives $0 = 0$. We have lost the variable!

Let's work through some examples, using this second rule.

Example 1

Solve

$$6x = 18$$

Here the variable x is multiplied by 6. So we apply the multiplication property and multiply both sides by $\frac{1}{6}$. Keep in mind that we want an equation of the form

$$x = \square$$

$$\frac{1}{6}(6x) = \left(\frac{1}{6}\right)18$$

$$\frac{1}{6}(6x) = \left(\frac{1}{6} \cdot 6\right)x$$
$$= 1 \cdot x, \text{ or } x$$

We then have x alone on the left, which is what we want.

We can now simplify.

$$1 \cdot x = 3 \quad \text{or} \quad x = 3$$

The solution is 3. To check, replace x with 3:

$$6 \cdot 3 \overset{?}{=} 18$$
$$18 = 18 \quad \text{(true)}$$

CHECK YOURSELF 1

Solve and check.

$8x = 32$

In Example 1 we solved the equation by multiplying both sides by the reciprocal of the coefficient of the variable.

Our next example illustrates a slightly different approach to solving an equation by using the multiplication property.

Example 2

Since division is defined in terms of multiplication, we can also divide both sides of an equation by the same nonzero number.

Solve

$5x = -35$

The variable x is multiplied by 5. We *divide* both sides by 5 to "undo" that multiplication:

$$\frac{5x}{5} = \frac{-35}{5}$$
$$x = -7$$

Note that the right side reduces to -7. Be careful with the rules for signs.

We will leave it to you to check the solution.

CHECK YOURSELF 2

Solve and check.

$7x = -42$

Example 3

Solve

$-9x = 54$

In this case x is multiplied by -9. So we divide both sides by -9 to isolate x on the left:

$$\frac{-9x}{-9} = \frac{54}{-9}$$
$$x = -6$$

The solution is -6. To check:

$$(-9)(-6) \overset{?}{=} 54$$
$$54 = 54 \qquad \text{(true)}$$

CHECK YOURSELF 3

Solve and check.

$$-10x = -60$$

The following examples illustrate the use of the multiplication property when fractions appear in an equation.

Example 4

(a) Solve

$$\frac{x}{3} = 6$$

Here x is *divided* by 3. We will use multiplication to isolate x.

$$3\left(\frac{x}{3}\right) = 3 \cdot 6$$
$$x = 18$$

$\left\{ \begin{array}{l} \text{This leaves } x \text{ alone on the} \\ \text{left because} \end{array} \right.$

$$3\left(\frac{x}{3}\right) = \frac{3}{1} \cdot \frac{x}{3} = \frac{x}{1} = x$$

To check:

$$\frac{18}{3} \overset{?}{=} 6$$
$$6 = 6 \qquad \text{(true)}$$

(b) Solve

$$\frac{x}{5} = -9$$

Since x is divided by 5, we multiply both sides by 5:

$$5\left(\frac{x}{5}\right) = 5(-9)$$
$$x = -45$$

The solution is -45. To check, we replace x with -45:

$$\frac{-45}{5} \stackrel{?}{=} -9$$
$$-9 = -9 \qquad \text{(true)}$$

The solution is verified.

CHECK YOURSELF 4

Solve and check.

1. $\dfrac{x}{7} = 3$ **2.** $\dfrac{x}{4} = -8$

When the variable is multiplied by a fraction that has a numerator other than 1, there are two approaches to finding the solution.

Example 5

Solve

$$\frac{3}{5}x = 9$$

One approach is to multiply by 5 as the first step.

$$5\left(\frac{3}{5}x\right) = 5 \cdot 9$$
$$3x = 45$$

Now we divide by 3.

$$\frac{3x}{3} = \frac{45}{3}$$
$$x = 15$$

To check:

$$\frac{3}{5} \cdot 15 \stackrel{?}{=} 9$$
$$9 = 9 \qquad \text{(true)}$$

Recall that $\dfrac{5}{3}$ is the *reciprocal* of $\dfrac{3}{5}$, and the product of a number and its reciprocal is just 1! So

$$\left(\dfrac{5}{3}\right)\left(\dfrac{3}{5}\right) = 1$$

A second method combines the multiplication and division steps and is generally a bit more efficient. We multiply by $\dfrac{5}{3}$.

$$\dfrac{5}{3}\left(\dfrac{3}{5}x\right) = \dfrac{5}{3} \cdot 9$$

So $x = 15$, as before.

CHECK YOURSELF 5

Solve and check.

$$\dfrac{2}{3}x = 18$$

You may sometimes have to simplify an equation before applying the methods of this section. Our final example illustrates.

Example 6

Solve and check:

$$3x + 5x = 40$$

Using the distributive property, we can combine the like terms on the left to write

$$8x = 40$$

We can now proceed as before.

$$\dfrac{8x}{8} = \dfrac{40}{8} \qquad \text{Divide by 8.}$$
$$x = 5$$

The solution is 5. To check, we return to the original equation. Substituting 5 for x yields

$$3 \cdot 5 + 5 \cdot 5 \stackrel{?}{=} 40$$
$$15 + 25 \stackrel{?}{=} 40$$
$$40 = 40 \qquad \text{(true)}$$

The solution is verified.

Solve and check.

$$7x + 4x = -66$$

CHECK YOURSELF ANSWERS

1. 4.
2. -6.
3. 6.
4. (1) 21; (2) -32.
5. 27.
6. -6.

Name _____

Date _____

Build Your Skills

Solve for x and check your result.

1. $3x = 12$ **2.** $4x = 20$

3. $8x = 48$ **4.** $5x = -35$

5. $63 = 9x$ **6.** $66 = 6x$

7. $4x = -16$ **8.** $-3x = 27$

9. $-9x = 72$ **10.** $10x = -100$

11. $6x = -54$ **12.** $-7x = 49$

13. $-4x = -12$ **14.** $52 = -4x$

15. $-35 = 5x$ **16.** $-5x = -25$

17. $-8x = -72$ **18.** $-10x = -60$

19. $\dfrac{x}{2} = 4$ **20.** $\dfrac{x}{3} = 2$

21. $\dfrac{x}{5} = 3$ **22.** $\dfrac{x}{8} = 5$

ANSWERS

1. _____
2. _____
3. _____
4. _____
5. _____
6. _____
7. _____
8. _____
9. _____
10. _____
11. _____
12. _____
13. _____
14. _____
15. _____
16. _____
17. _____
18. _____
19. _____
20. _____
21. _____
22. _____

23. _____

24. _____

25. _____

26. _____

27. _____

28. _____

29. _____

30. _____

31. _____

32. _____

33. _____

34. _____

35. _____

36. _____

37. _____

38. _____

39. _____

40. _____

41. _____

42. _____

43. _____

44. _____

23. $6 = \dfrac{x}{7}$

24. $6 = \dfrac{x}{3}$

25. $\dfrac{x}{4} = -5$

26. $\dfrac{x}{5} = -7$

27. $-\dfrac{x}{8} = 3$

28. $-\dfrac{x}{3} = -4$

29. $\dfrac{2}{3}x = 6$

30. $\dfrac{4}{5}x = 8$

31. $\dfrac{3}{4}x = -15$

32. $\dfrac{7}{8}x = -21$

33. $-\dfrac{2}{5}x = 10$

34. $-\dfrac{5}{6}x = -15$

35. $5x + 4x = 36$

36. $8x - 3x = -50$

37. $13x - 6x = -42$

38. $9x + 3x = 60$

39. $7x - 3x + 5x = 36$

40. $9x + 3x - 4x = -48$

Once again, certain equations involving decimal fractions can be solved by the methods of this section. For instance, to solve $2.3x = 6.9$ we simply use our multiplication property to divide both sides of the equation by 2.3. This will isolate x on the left as desired. Use this idea to solve each of the following equations for x.

41. $3.2x = 12.8$

42. $5.1x = -15.3$

43. $-4.5x = 13.5$

44. $-8.2x = -32.8$

45. $1.3x + 2.8x = 12.3$

46. $2.7x + 5.4x = -16.2$

47. $9.3x - 6.2x = 12.4$

48. $12.5x - 7.2x = -21.2$

Translate each of the following statements to an equation. Let x represent the number in each case.

49. 5 times a number is 40.

50. Twice a number is 36.

51. A number divided by 7 is equal to 6.

52. A number divided by 5 is equal to -4.

53. $\dfrac{1}{3}$ of a number is 8.

54. $\dfrac{1}{5}$ of a number is 10.

55. $\dfrac{3}{4}$ of a number is 18.

56. $\dfrac{2}{7}$ of a number is 8.

57. Twice a number, divided by 5, is 12.

58. 3 times a number, divided by 4, is 36.

Although tropical forests contain the greatest number of plant and animal species, Pacific northwest forests contain a greater amount of living material per hectare. The richest tropical forests contain approximately 45 percent of the plant material per hectare compared to the average Pacific northwest forest. (1 hectare = 2.471 acres.)

59. If tropical forests contain 457 metric tons (t) of plants per hectare, we can use the equation

$$0.45X = 457$$

to describe the amount of plant material in the average Pacific northwest forest. How many metric tons of plant material per hectare does a typical northwest forest contain?

60. The typical tropical forest may contain only one-tenth of the plant material of some redwood forests of the northern California coast. Use an equation similar to that of problem 59 to determine the amount of plant material per hectare in some redwood forests.

45. _____

46. _____

47. _____

48. _____

49. _____

50. _____

51. _____

52. _____

53. _____

54. _____

55. _____

56. _____

57. _____

58. _____

59. _____

60. _____

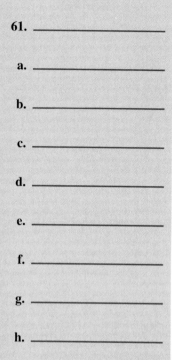

61. In the mid-1980s, Canada had 1.5 times as much forest as the United States. If Canada had 4.4 million square kilometers (km^2) of forest in 1986, approximately how much forestland was in the United States?

Skillscan (Section 1.3)

Use the distributive property to remove the parentheses in the following expressions.

a. $2(x - 3)$ **b.** $3(a + 4)$

c. $5(2b + 1)$ **d.** $3(3p - 4)$

e. $7(3x - 4)$ **f.** $-4(5x + 4)$

g. $-3(4x - 3)$ **h.** $-5(3y - 2)$

ANSWERS

1. 4   **3.** 6   **5.** 7   **7.** -4 **9.** -8 **11.** -9 **13.** 3   **15.** -7 **17.** 9
19. 8   **21.** 15   **23.** 42   **25.** -20 **27.** -24 **29.** 9   **31.** -20 **33.** -25
35. 4   **37.** -6 **39.** 4   **41.** 4   **43.** -3 **45.** 3   **47.** 4   **49.** $5x = 40$
51. $\dfrac{x}{7} = 6$ **53.** $\dfrac{x}{3} = 8$ **55.** $\dfrac{3}{4}x = 18$ **57.** $\dfrac{2x}{5} = 12$ **59.** 1016 t
61. 2.9 million km^2 **a.** $2x - 6$ **b.** $3a + 12$ **c.** $10b + 5$ **d.** $9p - 12$
e. $21x - 28$ **f.** $-20x - 16$ **g.** $-12x + 9$ **h.** $-15y + 10$

Combining the Rules to Solve Equations

OBJECTIVE

To use both the addition/subtraction and the multiplication/division properties to solve equations

In all our examples thus far, either the addition property or the multiplication property was used in solving an equation. Often, finding a solution will require the use of both properties.

Example 1

(*a*) Solve

$$4x - 5 = 7$$

Here *x* is *multiplied* by 4. The result, $4x$, then has 5 subtracted from it on the left side of the equation. These two operations mean that both properties must be applied in solving the equation.

Since the variable term is already on the left, we start by adding 5 to both sides:

Note: From now on we'll show the steps of the solution in a *horizontal form,* which is what you will probably want to use in practice.

$$4x - 5 + 5 = 7 + 5 \qquad \text{or} \qquad 4x = 12$$

We now divide both sides by 4:

$$\frac{4x}{4} = \frac{12}{4}$$

$$x = 3$$

The solution is 3. To check, replace *x* with 3 in the original equation. Be careful to follow the rules for the order of operations.

$$4 \cdot 3 - 5 \stackrel{?}{=} 7$$
$$12 - 5 \stackrel{?}{=} 7$$
$$7 = 7 \qquad \text{(true)}$$

(*b*) Solve

$$3x + 8 = -4$$

Again, we want the numbers on the right, so in this case we want to subtract 8 from both sides. Do you see why?

$$3x + 8 = -4$$
$$3x + 8 - 8 = -4 - 8$$
$$3x = -12$$

Now divide both sides by 3 to isolate x on the left.

$$\frac{3x}{3} = \frac{-12}{3}$$
$$x = -4$$

The solution is -4. We'll leave the check of this result to you.

CHECK YOURSELF 1

Solve and check.

1. $6x + 9 = -15$

2. $5x - 8 = 7$

The variable may appear in any position in an equation. Just apply the rules carefully as you try to write an equivalent equation, and you will find the solution. The following example illustrates.

Example 2

Solve

$$3 - 2x = 9$$

First subtract 3 from both sides.

$$3 - 2x = 9$$
$$3 - 3 - 2x = 9 - 3$$
$$-2x = 6$$

Note: $\frac{-2}{-2} = 1$, so we divide by -2 to isolate x on the left.

Now divide both sides by -2. This will leave x alone on the left.

$$\frac{-2x}{-2} = \frac{6}{-2}$$
$$x = -3$$

The solution is -3. To check:

$$3 - 2(-3) \stackrel{?}{=} 9$$
$$3 + 6 \stackrel{?}{=} 9$$
$$9 = 9 \quad \text{(true)}$$

CHECK YOURSELF 2

Solve and check.

$$10 - 3x = 1$$

You may also have to combine multiplication with addition or subtraction to solve an equation. Consider the following example.

Example 3

(a) Solve

$$\frac{x}{5} - 3 = 4$$

To get the x term alone, we first add 3 to both sides.

$$\frac{x}{5} - 3 + 3 = 4 + 3$$

$$\frac{x}{5} = 7$$

Now how can we isolate x on the left? We can undo the division by multiplying both sides of the equation by 5.

$$5\left(\frac{x}{5}\right) = 5 \cdot 7$$

$$x = 35$$

The solution is 35. The check is the same as before. Just return to the original equation.

$$\frac{35}{5} - 3 = 4$$

$$7 - 3 = 4$$

$$4 = 4 \qquad \text{(true)}$$

(b) Solve

$$\frac{2}{3}x + 5 = 13$$

First subtract 5 from both sides.

$$\frac{2}{3}x + 5 \boxed{-5} = 13 \boxed{-5}$$

$$\frac{2}{3}x = 8$$

Now multiply both sides by $\frac{3}{2}$, the reciprocal of $\frac{2}{3}$.

$$\left(\frac{3}{2}\right)\left(\frac{2}{3}x\right) = \left(\frac{3}{2}\right)8$$

or

$$x = 12$$

The solution is 12. We'll leave it to you to check this result.

CHECK YOURSELF 3

Solve and check.

1. $\dfrac{x}{6} + 5 = 3$ **2.** $\dfrac{3}{4}x - 8 = 10$

In Section 3.2 you learned how to solve certain equations when the variable appeared on both sides. The following examples will show you how to extend that work by using the multiplication property of equality.

Example 4

Solve

$$6x - 4 = 3x - 2$$

First add 4 to both sides. This will undo the subtraction on the left.

$$6x - 4 = 3x - 2$$
$$6x - 4 \boxed{+4} = 3x - 2 \boxed{+4}$$
$$6x = 3x + 2$$

Now subtract $3x$ so that the terms in x will be on the left only.

$$6x = 3x \qquad + 2$$
$$6x \boxed{-3x} = 3x \boxed{-3x} + 2$$
$$3x = \qquad\quad 2$$

Finally divide by 3.

$$\frac{3x}{3} = \frac{2}{3}$$

$$x = \frac{2}{3}$$

Check:

$$6\left(\frac{2}{3}\right) - 4 \stackrel{?}{=} 3\left(\frac{2}{3}\right) - 2$$

$$4 - 4 \stackrel{?}{=} 2 - 2$$

$$0 = 0 \qquad \text{(true)}$$

Note: As you know, the basic idea is to use our two properties to form an equivalent equation with the x isolated. Here we added 4 and then subtracted $3x$. You can do these steps in either order. Try it for yourself the other way. In either case, the multiplication property is then used as the *last step* in finding the solution.

CHECK YOURSELF 4

Solve and check.

$$7x - 5 = 3x + 5$$

Let's look at two approaches to solving equations in which the coefficient on the right side is greater than the coefficient on the left.

Example 5

Solve $4x - 8 = 7x + 7$.

$$4x - 8 = 7x + 7$$
$$4x - 8 + 8 = 7x + 7 + 8 \qquad \text{Adding 8 will leave}$$
$$\text{the } x \text{ term alone on the left.}$$
$$4x = 7x + 15$$
$$4x - 7x = 7x - 7x + 15 \qquad \text{Subtracting } 7x \text{ will get}$$
$$\text{the variable terms on the left.}$$
$$-3x = 15$$
$$\frac{-3x}{-3} = \frac{15}{-3} \qquad \text{Dividing by } -3 \text{ will isolate } x \text{ on the left.}$$
$$x = -5$$

We'll let you check this result.

To avoid the negative coefficient (-3) in Example 5, some students prefer a different approach. Let's return to our example.

Example 5 (An Alternative Method)

This time we'll work toward having the number on the *left* and the x term on the *right,* or

$$\square = x$$

It is usually easier to isolate the variable term on the side that will result in a positive coefficient.

$$4x - 8 = 7x + 7$$
$$4x - 8\,\boxed{-\,7} = 7x + 7\,\boxed{-\,7} \qquad \text{Subtract 7.}$$
$$4x - 15 = 7x$$
$$4x\,\boxed{-\,4x} - 15 = 7x\,\boxed{-\,4x} \qquad \text{Subtract } 4x \text{ to get the variables}$$
$$-15 = 3x \qquad\qquad\qquad\quad \text{on the right.}$$
$$\frac{-15}{3} = \frac{3x}{3} \qquad\qquad\quad \text{Divide by 3 to isolate } x \text{ on the right.}$$
$$-5 = x$$

Since $-5 = x$ and $x = -5$ are equivalent equations, it really makes no difference; the solution is still -5! You can use whichever approach you prefer.

CHECK YOURSELF 5

Solve $5x + 3 = 9x - 21$ by finding equivalent equations of the form $x = \square$ and $\square = x$ to compare the two methods of finding the solution.

Where possible, we start by combining like terms on each side of the equation.

Example 6

Solve

$$7x - 3 + 5x + 4 = 6x + 25$$

We start the solution process by combining like terms on the left. This gives

$$12x + 1 = 6x + 25$$
$$12x + 1\,\boxed{-\,1} = 6x + 25\,\boxed{-\,1} \qquad \text{Subtract 1.}$$
$$12x = 6x + 24$$
$$12x\,\boxed{-\,6x} = 6x\,\boxed{-\,6x} + 24 \qquad \text{Subtract } 6x.$$
$$6x = 24$$
$$\frac{6x}{6} = \frac{24}{6} \qquad\qquad\qquad \text{Divide by 6.}$$
$$\text{or} \qquad x = 4$$

The solution is 4. We leave the checking of this result to the reader.

CHECK YOURSELF 6

Solve and check.

$$9x - 6 - 3x + 1 = 2x + 15$$

It may also be necessary to remove grouping symbols in solving an equation. Our final example illustrates.

Example 7

Solve and check.

$$5(x - 3) - 2x = x + 7$$

As our first step we apply the distributive property to simplify on the left.

$$5x - 15 - 2x = x + 7$$

Combining like terms, we have

$$3x - 15 = x + 7$$

We then proceed as in previous examples.

Note: $5(x - 3)$
$= 5(x) - 5(3)$
$= 5x - 15$

$$
\begin{aligned}
3x - 15 + 15 &= x + 7 + 15 && \text{Add 15.} \\
3x &= x + 22 \\
3x - x &= x - x + 22 && \text{Subtract } x. \\
2x &= 22 && \text{Divide by 2.} \\
x &= 11
\end{aligned}
$$

The solution is 11. To check, substitute 11 for x in the original equation. Again note the use of our rules for the order of operations.

$$
\begin{aligned}
5(11 - 3) - 2 \cdot 11 &\overset{?}{=} 11 + 7 && \text{Simplify terms in parentheses.} \\
5 \cdot 8 - 2 \cdot 11 &\overset{?}{=} 11 + 7 && \text{Multiply.} \\
40 - 22 &\overset{?}{=} 11 + 7 && \text{Add and subtract.} \\
18 &= 18 && \text{A true statement.}
\end{aligned}
$$

CHECK YOURSELF 7

Solve and check.

$$7(x + 5) - 3x = x - 7$$

Such an outline of steps is sometimes called an *algorithm* for the process.

Let's summarize our work with an outline of the steps involved in solving linear equations.

SOLVING LINEAR EQUATIONS

STEP 1   Use the distributive property to remove any grouping symbols. Then simplify by combining like terms on each side of the equation.

STEP 2   Add or subtract the same term on each side of the equation until the term involving the variable is on one side and a number is on the other.

STEP 3   Multiply or divide both sides of the equation by the same nonzero number so that the variable is alone on one side of the equation.

STEP 4   Check the solution in the original equation.

CHECK YOURSELF ANSWERS

1. (1) -4; (2) 3.
2. 3.
3. (1) -12; (2) 24.
4. $\dfrac{5}{2}$.
5. 6.
6. 5.
7. -14.

Name

Date

Build Your Skills

Solve for x and check your result.

1. $2x + 1 = 9$ **2.** $3x - 1 = 17$

3. $3x - 2 = 7$ **4.** $5x + 3 = 23$

5. $4x + 7 = 35$ **6.** $7x - 8 = 13$

7. $2x + 9 = 5$ **8.** $6x + 25 = -5$

9. $3x - 5 = -1$ **10.** $2x + 3 = -2$

11. $3x - 2 = -26$ **12.** $4x - 3 = -19$

13. $2 - 5x = 12$ **14.** $5 - 3x = -4$

15. $7 - 6x = -11$ **16.** $4 - 7x = 39$

17. $\dfrac{x}{2} + 1 = 5$ **18.** $\dfrac{x}{3} - 2 = 3$

19. $\dfrac{x}{4} - 5 = 3$ **20.** $\dfrac{x}{5} + 3 = 8$

ANSWERS

1. _____

2. _____

3. _____

4. _____

5. _____

6. _____

7. _____

8. _____

9. _____

10. _____

11. _____

12. _____

13. _____

14. _____

15. _____

16. _____

17. _____

18. _____

19. _____

20. _____

21. _____

22. _____

23. _____

24. _____

25. _____

26. _____

27. _____

28. _____

29. _____

30. _____

31. _____

32. _____

33. _____

34. _____

35. _____

36. _____

37. _____

38. _____

39. _____

40. _____

41. _____

42. _____

21. $\dfrac{2}{3}x + 5 = 17$

22. $\dfrac{3}{4}x - 5 = 4$

23. $\dfrac{4}{5}x - 3 = 13$

24. $\dfrac{5}{7}x + 4 = 14$

25. $4x = x + 9$

26. $3x = 8 - x$

27. $5x = 30 - x$

28. $8x = 3x + 20$

29. $8x = 4x - 3$

30. $3x = 4 - 2x$

31. $5x - 2 = 2x + 19$

32. $7x + 3 = 2x + 18$

33. $9x + 2 = 3x + 38$

34. $8x - 3 = 4x + 17$

35. $4x - 8 = x - 14$

36. $6x - 5 = 3x - 29$

37. $5x + 7 = 2x - 3$

38. $9x + 7 = 5x - 3$

39. $7x - 3 = 9x + 5$

40. $5x - 2 = 8x - 11$

41. $5x + 4 = 7x - 8$

42. $2x + 23 = 6x - 5$

43. $3x - 5 = 6x - 10$

44. $2x + 9 = 4x - 2$

45. $2x - 3 + 5x = 7 + 4x + 2$

46. $8x - 7 - 2x = 2 + 4x - 5$

47. $6x + 7 - 4x = 8 + 7x - 26$

48. $7x - 2 - 3x = 5 + 8x + 13$

49. $9x - 2 + 7x + 13 = 10x - 13$

50. $5x + 3 + 6x - 11 = 8x + 25$

51. $8x - 7 + 5x - 10 = 10x - 12$

52. $10x - 9 + 2x - 3 = 8x - 18$

53. $7(2x - 1) - 5x = x + 25$

54. $9(3x + 2) - 10x = 12x - 7$

55. $3x + 2(4x - 3) = 6x - 9$

56. $7x + 3(2x + 5) = 10x + 17$

57. $\dfrac{8}{3}x - 3 = \dfrac{2}{3}x + 15$

58. $\dfrac{12}{5}x + 7 = 31 - \dfrac{3}{5}x$

59. $\dfrac{2}{5}x - 5 = \dfrac{12}{5}x + 8$

60. $\dfrac{3}{7}x - 5 = \dfrac{24}{7}x + 7$

61. $5.3x - 7 = 2.3x + 5$

62. $9.8x + 2 = 3.8x + 20$

Translate each of the following statements to an equation. Let x represent the number in each case.

63. 3 more than twice a number is 7.

43. _____

44. _____

45. _____

46. _____

47. _____

48. _____

49. _____

50. _____

51. _____

52. _____

53. _____

54. _____

55. _____

56. _____

57. _____

58. _____

59. _____

60. _____

61. _____

62. _____

63. _____

64. _____

65. _____

66. _____

67. _____

68. _____

69. _____

70. _____

71. _____

72. _____

64. 5 less than 3 times a number is 25.

65. 7 less than 4 times a number is 41.

66. 10 more than twice a number is 44.

67. 5 more than two-thirds of a number is 21.

68. 3 less than three-fourths of a number is 24.

69. 3 times a number is 12 more than that number.

70. 5 times a number is 8 less than that number.

Think About These

Each of the following can be translated into either an expression or an equation. Do that translation, then identify which it is.

71. 6 more than twice a number

72. 4 less than 3 times a number is 11.

73. 4 times a number is 9 more than that number.

74. The product of 4 less than a number and 4 more than the same number.

75. Remember that a sentence must have a subject and a verb. In the preceding four exercises, which were sentences, the expressions, or the equations?

76. The following equation can be read as a sentence:

$2x + 3 = 7$

When you do so, which symbol is the verb?

 77. North and Central America contain approximately 802 million hectares of forest. The Central American countries except Mexico contain 21.6 million hectares of forest. If the United States contains 6.1 times as much forest as Mexico and if Canada contains 9 times as much forest as Mexico, how many hectares of forest are located in Mexico?
   This problem can be translated to the equation

$x + 9x + 6.1x + 21.6 = 802$

where x represents the amount of Mexican forest in millions of hectares. Now solve for x.

78. There are approximately 750 protected natural areas in Canada, the United States, and Mexico. If Canada has three-fourths as many as the United States and if Mexico has 50 protected natural areas, how many such areas does the United States have?
   The following equation might help you:

Areas in U.S. + areas in Canada + areas in Mexico = total

ANSWERS

73. _____

74. _____

75. _____

76. _____

77. _____

78. _____

ANSWERS

a. _____

b. _____

c. _____

d. _____

e. _____

f. _____

g. _____

h. _____

Skillscan (Section 1.5)

Divide.

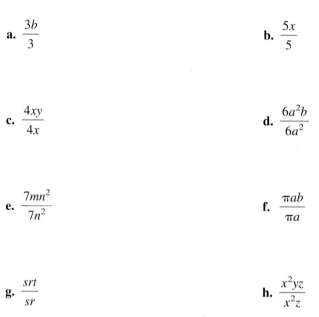

a. $\dfrac{3b}{3}$

b. $\dfrac{5x}{5}$

c. $\dfrac{4xy}{4x}$

d. $\dfrac{6a^2b}{6a^2}$

e. $\dfrac{7mn^2}{7n^2}$

f. $\dfrac{\pi ab}{\pi a}$

g. $\dfrac{srt}{sr}$

h. $\dfrac{x^2yz}{x^2z}$

ANSWERS

1. 4   **3.** 3   **5.** 7   **7.** -2 **9.** $\dfrac{4}{3}$ **11.** -8 **13.** -2 **15.** 3   **17.** 8

19. 32   **21.** 18   **23.** 20   **25.** 3   **27.** 5   **29.** $-\dfrac{3}{4}$ **31.** 7   **33.** 6   **35.** -2

37. $-\dfrac{10}{3}$ **39.** -4 **41.** 6   **43.** $\dfrac{5}{3}$ **45.** 4   **47.** 5   **49.** -4 **51.** $\dfrac{5}{3}$ **53.** 4

55. $-\dfrac{3}{5}$ **57.** 9   **59.** $-\dfrac{13}{2}$ **61.** 4   **63.** $2x + 3 = 7$ **65.** $4x - 7 = 41$

67. $\dfrac{2}{3}x + 5 = 21$ **69.** $3x = x + 12$ **71.** $2x + 6$, expression   **73.** $4x = x + 9$, equation

75. Equations   **77.** $16.1x = 780.4$, $x = 48.5$ million hectares   **a.** b **b.** x **c.** y
d. b **e.** m **f.** b **g.** t **h.** y

166

Solving Literal Equations

OBJECTIVE

To be able to solve a literal equation for any one of its variables

Formulas are extremely useful tools in any field in which mathematics is applied. Formulas are simply equations that express a relationship between more than one letter or variable. You are no doubt familiar with all kinds of formulas, such as

$$A = \frac{1}{2}bh \qquad \text{The area of a triangle.}$$

$$I = Prt \qquad \text{Interest.}$$

$$V = \pi r^2 h \qquad \text{The volume of a cylinder.}$$

Actually a formula is also called a *literal* equation because it involves several letters or variables. For instance, our first formula or literal equation, $A = \frac{1}{2}bh$, involves the three letters A (for area), b (for base), and h (for height).

Unfortunately, formulas are not always given in the form needed to solve a particular problem. Then algebra is needed to change the formula to a more useful equivalent equation, which is solved for a particular letter or variable. The steps used in the process are very similar to those you used in solving linear equations. Let's consider an example.

Example 1

Suppose that we know the area A and the base b of a triangle and want to find its height h.

We are given

$$A = \frac{1}{2}b \cdot h$$

Our job is to find an equivalent equation with h, the unknown, by itself on one side. We call $\frac{1}{2}b$ the *coefficient* of h. We can remove the two *factors* of that coefficient, $\frac{1}{2}$ and b, separately.

Note:

$$2\left(\frac{1}{2}bh\right) = \left(2 \cdot \frac{1}{2}\right)(bh)$$
$$= 1 \cdot bh$$
$$= bh$$

$$2A = 2\left(\frac{1}{2}bh\right) \qquad \text{Multiply both sides by 2 to clear the equation of fractions.}$$

or

$$2A = bh$$

$$\frac{2A}{b} = \frac{bh}{b} \qquad \text{Divide by } b \text{ to isolate } h.$$

$$\frac{2A}{b} = h$$

or

$$h = \frac{2A}{b}$$ Reverse the sides to write h on the left.

We now have the height h in terms of the area A and the base b. This is called *solving the equation for h* and means that we are rewriting the formula as an equivalent equation of the form

Here $\square$ means an expression containing all the numbers or letters *other than* h.

$$h = \square$$

You have already learned the methods needed to solve most literal equations or formulas for some specified variable. As Example 1 illustrates, the rules of Sections 3.2 and 3.3 are applied in exactly the same way as they were applied to equations with one variable.

CHECK YOURSELF 1

Solve $V = \frac{1}{3}B \cdot h$ for h.

You may have to apply both the addition and the multiplication properties when solving a formula for a specified variable. The following example illustrates this.

Example 2

This is a linear equation in two variables. You will see this again in Chapter 7.

Solve $y = mx + b$ for x.

Remember that we want to end up with x alone on one side of the equation. Let's start by subtracting b from both sides to undo the addition on the right.

$$y = mx + b$$
$$y - b = mx + b - b$$
$$y - b = mx$$

If we now divide both sides by m, then x will be alone on the right-hand side.

$$\frac{y - b}{m} = \frac{mx}{m}$$

$$\frac{y - b}{m} = x$$

or

$$x = \frac{y - b}{m}$$

CHECK YOURSELF 2

Solve $v = v_0 + gt$ for t.

Let's summarize the steps illustrated by our examples.

SOLVING A FORMULA OR LITERAL EQUATION

STEP 1   If necessary, multiply both sides of the equation by the same term to clear of fractions.

STEP 2   Add or subtract the same term on both sides of the equation so that all terms involving the variable that you are solving for are on one side of the equation and all other terms are on the other side.

STEP 3   Divide both sides of the equation by the coefficient of the variable that you are solving for.

Let's look at one more example, using the above steps.

Example 3

This is a formula for the *amount* of money in an account after interest has been earned.

Solve $A = P + Prt$ for r.

$$A = P + Prt$$

$$A - P = P - P + Prt$$

Subtracting P from both sides will leave the term involving r alone on the right.

$$A - P = Prt$$

$$\frac{A - P}{Pt} = \frac{Prt}{Pt}$$

Dividing both sides by Pt will isolate r on the right.

$$\frac{A - P}{Pt} = r$$

or

$$r = \frac{A - P}{Pt}$$

CHECK YOURSELF 3

Solve $2x + 3y = 6$ for y.

Now let's look at an application of solving a literal equation.

Example 4

Suppose that the amount in an account, 3 years after a principal of $5000 was invested, is $6050. What was the interest rate?

From our previous example,

$$A = P + Prt \tag{1}$$

where A is the amount in the account, P is the principal, r is the interest rate, and t is the time that the money has been invested. By the result of Example 4 we have

$$r = \frac{A - P}{Pt} \tag{2}$$

Do you see the advantage of having our equation solved for the desired variable?

and we can substitute the known values in Equation (2):

$$r = \frac{6050 - 5000}{(5000)(3)}$$

$$= \frac{1050}{15,000} = 0.07 = 7\%$$

The interest rate is 7 percent.

CHECK YOURSELF 4

Suppose that the amount in an account, 4 years after a principal of $3000 was invested, is $3720. What was the interest rate?

CHECK YOURSELF ANSWERS

1. $h = \dfrac{3V}{B}$.

2. $t = \dfrac{v - v_0}{g}$.

3. $y = \dfrac{6 - 2x}{3}$.

4. 6 percent.

Name _____

Date _____

Build Your Skills

Solve each literal equation for the indicated variable.

1. $p = 4s$ (for s) Perimeter of a square

2. $V = Bh$ (for B) Volume of a prism

3. $E = IR$ (for R) Voltage in an electric circuit

4. $I = Prt$ (for r) Simple interest

5. $V = LWH$ (for H) Volume of a rectangular solid

6. $V = \pi r^2 h$ (for h) Volume of a cylinder

7. $A + B + C = 180$ (for B) Measure of angles in a triangle

8. $V - E + F = 2$ (for F) Euler's formula

9. $ax + b = 0$ (for x) Linear equation in one variable

10. $y = mx + b$ (for m) Point-slope form for a line

11. $s = \dfrac{1}{2}gt^2$ (for g) Distance

12. $K = \dfrac{1}{2}mv^2$ (for m) Energy

13. $x + 3y = 6$ (for y) Linear equation

ANSWERS

1. _____

2. _____

3. _____

4. _____

5. _____

6. _____

7. _____

8. _____

9. _____

10. _____

11. _____

12. _____

13. _____

ANSWERS

14. _____

15. _____

16. _____

17. _____

18. _____

19. _____

20. _____

21. _____

22. _____

23. _____

24. _____

25. _____

14. $3x + 4y = 12$ (for x)   Linear equation

15. $P = 2L + 2W$ (for L)   Perimeter of a rectangle

16. $ax + by = c$ (for y)   Linear equation in two variables

17. $V = \dfrac{KT}{P}$ (for T)   Volume of a gas

18. $V = \dfrac{1}{3}\pi r^2 h$ (for h)   Volume of a cone

19. $x = \dfrac{a + b}{2}$ (for b)   Average of two numbers

20. $D = \dfrac{C - s}{n}$ (for s)   Depreciation

21. $F = \dfrac{9}{5}C + 32$ (for C)   Celsius/Fahrenheit

22. $A = P + Prt$ (for t)   Amount at simple interest

23. $S = 2\pi r^2 + 2\pi rh$ (for h)   Total surface area of a cylinder

24. $A = \dfrac{1}{2}h\,(B + b)$ (for b)   Area of a trapezoid

25. A rectangular solid has a base with length 6 cm and width 4 cm. If the volume of the solid is 72 cm^3, find the height of the solid. See Exercise 5.

26. A cylinder has a radius of 4 in. If the volume of the cylinder is 144π in^3, what is the height of the cylinder? See Exercise 6.

26. _____

27. A principal of $2000 was invested in a savings account for 4 years. If the interest earned for that period was $480, what was the interest rate? See Exercise 4.

27. _____

28. If the perimeter of a rectangle is 60 ft and its width is 12 ft, find its length. See Exercise 15.

28. _____

29. The high temperature in New York for a particular day was reported at 77°F. How would that same temperature have been given in degrees Celsius? See Exercise 21.

29. _____

30. The area of a trapezoid is 36 in^2. If its height is 4 in and the length of one of the bases is 11 in, find the length of the other base. See Exercise 24.

Think About These

30. _____

We considered the notation for the four arithmetic operations in the programming language BASIC in Section 2.6. To review, those operations are indicated by the following:

ALGEBRAIC EXPRESSION	EXPRESSION IN BASIC
$a + b$	A + B
$a - b$	A − B
ab	A * B
$\dfrac{a}{b}$	A/B

31. _____

Using the above information, write a BASIC expression for the given formula solved for the specified letter. *Hint:* In BASIC, the operations of multiplication and division are done *in order* from left to right. Then the operations of addition and subtraction are performed in the same manner. You may have to insert parentheses in the following answers to achieve your desired result.

31. $V = Bh$ (for h) **32.** $I = Prt$ (for t)

32. _____

33. $A = P + Prt$ (for t) **34.** $A = \dfrac{1}{2}h(B + b)$ (for b)

33. _____

34. _____

Skillscan (Section 2.1)

Locate each of the following numbers on the number line.

a. 4

b. −5

c. −3

d. 2

e. $-\dfrac{7}{2}$

f. $\dfrac{2}{3}$

g. 2.5

h. −1.1

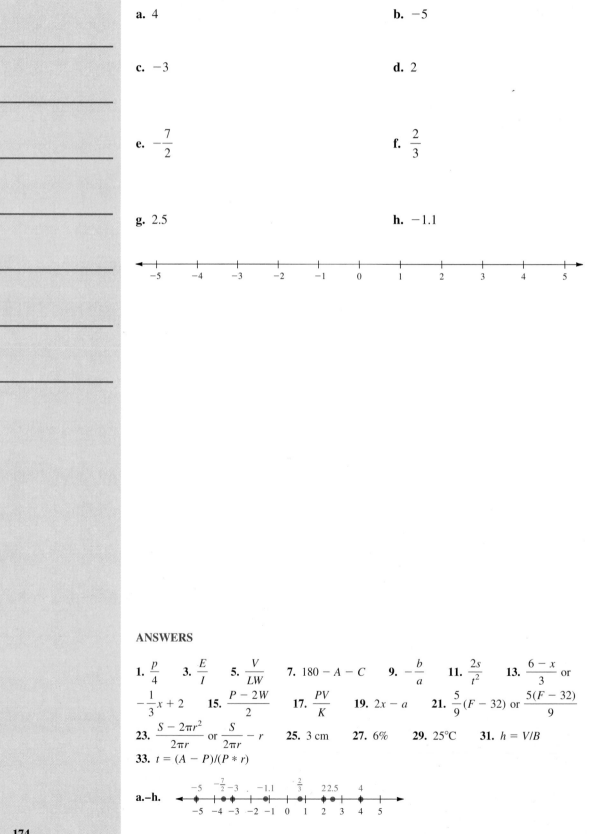

ANSWERS

1. $\dfrac{p}{4}$ **3.** $\dfrac{E}{I}$ **5.** $\dfrac{V}{LW}$ **7.** $180 - A - C$ **9.** $-\dfrac{b}{a}$ **11.** $\dfrac{2s}{t^2}$ **13.** $\dfrac{6 - x}{3}$ or

$-\dfrac{1}{3}x + 2$ **15.** $\dfrac{P - 2W}{2}$ **17.** $\dfrac{PV}{K}$ **19.** $2x - a$ **21.** $\dfrac{5}{9}(F - 32)$ or $\dfrac{5(F - 32)}{9}$

23. $\dfrac{S - 2\pi r^2}{2\pi r}$ or $\dfrac{S}{2\pi r} - r$ **25.** 3 cm **27.** 6% **29.** 25°C **31.** $h = V/B$

33. $t = (A - P)/(P * r)$

3.6 Inequalities—An Introduction

OBJECTIVES
1. To understand the notation of inequalities
2. To graph the solution sets of inequalities

As pointed out in this chapter's introduction, an equation is just a statement that two expressions are equal. In algebra, an *inequality* is a statement that one expression is less than or greater than another. Four new symbols are used in writing inequalities. The use of two of them is illustrated in the following example.

Example 1

The strict inequality symbols are less than ($<$) and greater than ($>$).

$5 < 8$ is an inequality read ''5 is less than 8.''

To help you remember, the "arrowhead" always points toward the smaller quantity.

$9 > 6$ is an inequality read ''9 is greater than 6.''

CHECK YOURSELF 1

Fill in the blanks, using the symbols $<$ and $>$.

1. 12_____8   **2.** 20_____25

Just as was the case with equations, inequalities that involve variables may be either true or false depending on the value that we give to the variable. For instance, consider the inequality

$x < 6$

If $x = \begin{cases} 3 & 3 < 6 \text{ is true} \\ 5 & 5 < 6 \text{ is true} \\ -10 & -10 < 6 \text{ is true} \\ 8 & 8 < 6 \text{ is false} \end{cases}$

Therefore 3, 5, and -10 are some *solutions* for the inequality $x < 6$; they make the inequality a true statement. You should see that 8 is *not* a solution. We call the set of all solutions the *solution set* for the inequality. Of course, there are many possible solutions.

Since there are so many solutions (an infinite number, in fact), we certainly do not want to try to list them all! A convenient way to show the solution set of an inequality is with the use of a number line.

Example 2

To graph the solution set for the inequality $x < 6$, we want to include all real numbers that are "less than" 6. This means all numbers *to the left* of 6 on the number line. We then start at 6 and draw an arrow extending left, as shown:

The colored arrow indicated the direction of the *solution*.

Note: The open circle at 6 means that we do not include 6 in the solution set (6 is not less than itself). The colored arrow shows all the numbers in the solution set, with the arrowhead indicating that the solution set continues indefinitely to the left.

CHECK YOURSELF 2

Graph the solution set of $x < -2$.

Two other symbols are used in writing inequalities. They are used with inequalities such as

$$x \geq 5 \qquad \text{and} \qquad x \leq 2$$

Here $x \geq 5$ is really a combination of the two statements $x > 5$ and $x = 5$. It is read "x is greater than or equal to 5." The solution set includes 5 in this case.

The inequality $x \leq 2$ combines the statements $x < 2$ and $x = 2$. It is read "x is less than or equal to 2."

Example 3

The solution set for $x \geq 5$ is graphed as follows.

Here the filled-in circle means that we want to include 5 in the solution set. This is often called a "closed" circle.

CHECK YOURSELF 3

Graph the solution sets.

1. $x \leq -4$ **2.** $x \geq 3$

CHECK YOURSELF ANSWERS

1. (1) $12 > 8$; (2) $20 < 25$.
2. $x < -2$.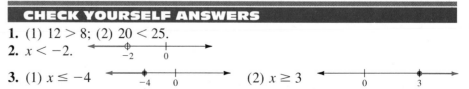
3. (1) $x \leq -4$ (2) $x \geq 3$

Name

Date

Build Your Skills

Complete the statements, using the symbol $<$ or $>$.

1. 5 _____ 10

2. 9 _____ 8

3. 7 _____ -2

4. 0 _____ -5

5. 0 _____ 4

6. -10 _____ -5

7. -3 _____ -8

8. -8 _____ -12

9. -7 _____ 0

10. -8 _____ -9

Write each of the inequalities in words.

11. $x < 3$

12. $x \leq -5$

13. $x \geq -4$

14. $x < -2$

15. $-5 \leq x$

16. $2 < x$

17. $x \geq 0$

18. $x \geq -7$

ANSWERS

1. _____

2. _____

3. _____

4. _____

5. _____

6. _____

7. _____

8. _____

9. _____

10. _____

11. _____

12. _____

13. _____

14. _____

15. _____

16. _____

17. _____

18. _____

Graph the solution set of each of the following inequalities.

19. _____

19. $x > 1$

20. $x < -2$

20. _____

21. $x < 8$

22. $x > 3$

21. _____

23. $x > -5$

24. $x < -4$

22. _____

23. _____

25. $x \geq 9$

26. $x \geq 0$

24. _____

27. $x < 0$

28. $x \leq -3$

25. _____

26. _____

29. $x \leq -10$

30. $x \geq -8$

27. _____

28. _____

29. _____

30. _____

Think About These

Match each inequality on the right with a statement on the left.

31. x is nonnegative

    a. $x \geq 0$

32. x is negative

    b. $x \geq 5$

33. x is no more than 5

    c. $x \leq 5$

34. x is positive

    d. $x > 0$

35. x is at least 5

    e. $x < 5$

36. x is less than 5

    f. $x < 0$

37. There may be less than 1000 wild giant pandas left in the bamboo forests of China. Write an inequality expressing this relationship.

38. Let C represent the amount of Canadian forest and M represent the amount of Mexican forest. Write an inequality showing the relationship of the forests of Canada and Mexico if Canada contains at least 9 times as much forest as Mexico.

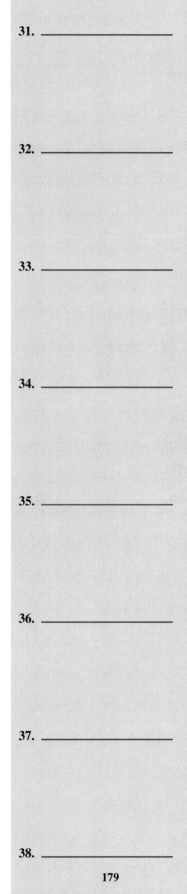

ANSWERS

31. _____

32. _____

33. _____

34. _____

35. _____

36. _____

37. _____

38. _____

ANSWERS

a. _____

b. _____

c. _____

d. _____

e. _____

f. _____

g. _____

h. _____

Skillscan (Section 3.4)

Solve the following equations.

a. $3x + 4 = 19$

b. $2w - 5 = -13$

c. $5a - 8 = 3a$

d. $9w + 14 = 7w$

e. $8x - 5 = 5x - 23$

f. $4y + 7 = 7y + 5$

g. $3(x - 8) = x - 2$

h. $\dfrac{2}{3}x - 4 = 8$

ANSWERS

1. $5 < 10$ **3.** $7 > -2$ **5.** $0 < 4$ **7.** $-3 > -8$ **9.** $-7 < 0$ **11.** x is less than 3
13. x is greater than or equal to -4 **15.** -5 is less than or equal to x **17.** x is greater than or equal to 0

19. 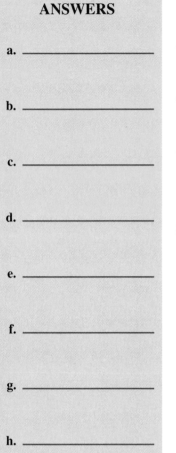 1   **21.** 8   **23.** -5

25. 9   **27.** 0   **29.** -10

31. a   **33.** c   **35.** b   **37.** $P < 1000$ **a.** 5   **b.** -4 **c.** 4   **d.** -7 **e.** -6
f. $\dfrac{2}{3}$ **g.** 11   **h.** 18

Solving Linear Inequalities

OBJECTIVE

To solve and graph the solution sets for linear inequalities in one variable

You learned how to graph the solution sets of some simple inequalities, such as $x < 8$ or $x \geq 10$, in the last section. Now we will look at more complicated inequalities, such as

$$2x - 3 < x + 4$$

This is called a *linear inequality in one variable*. Only one variable is involved in the inequality, and it appears only to the first power. Fortunately, the methods used to solve this type of inequality are very similar to those we used earlier in this chapter to solve linear equations in one variable. Here is our first property for inequalities.

> **THE ADDITION PROPERTY OF INEQUALITY**
>
> If $a < b$ then $a + c < b + c$
>
> In words, adding the same quantity to both sides of an inequality gives an *equivalent inequality*.

Equivalent inequalities have exactly the same solution sets.

The following example illustrates the use of this property.

Example 1

Solve and graph the solution set for $x - 8 < 7$.

The inequality is solved when an equivalent inequality has the form

$$x < \square \quad \text{or} \quad x > \square$$

That should look familiar!

To solve $x - 8 < 7$, we will add 8 to both sides of the inequality by the addition property.

$$x - 8 < 7$$
$$x - 8 + 8 < 7 + 8$$
$$x < 15 \quad \text{(the solution)}$$

The graph of the solution set is

181

CHECK YOURSELF 1

Solve and graph the solution set for

$x - 9 > -3$

Example 2

Solve and graph the solution set for $4x - 2 \geq 3x + 5$.

First, we subtract $3x$ from both sides of the inequality.

As with equations, the addition property allows us to *subtract* the same quantity from both sides of an inequality.

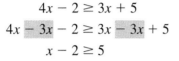

$$x - 2 \geq 5$$

Now we add 2 to both sides.

$$x - 2 \boxed{+ 2} \geq 5 \boxed{+ 2}$$
$$x \geq 7$$

We subtracted $3x$ and then added 2 to both sides. If these steps are done in the other order, the resulting inequality will be the same. You might want to verify that yourself.

The graph of the solution set is

```
◄──┼────┼────●──┼──────┼──►
   0         7  10     15
```

CHECK YOURSELF 2

Solve and graph the solution set.

$7x - 8 \leq 6x + 2$

You will also need a rule for multiplying on both sides of an inequality. Here you'll have to be a bit careful. There is a difference between the multiplication property for inequalities and that for equations. Look at the following:

$2 < 7$ (a true inequality)

Let's multiply both sides by 3.

$$2 < 7$$
$$3 \cdot 2 < 3 \cdot 7$$
$$6 < 21 \quad \text{(a true inequality)}$$

Now we multiply both sides by -3.

$$2 < 7$$
$$(-3)(2) < (-3)(7)$$
$$-6 < -21 \quad (\textit{not} \text{ a true inequality})$$

 CAUTION

Let's try something different.

$$2 < 7$$ ———— Change the "sense" of
$$(-3)(2) > (-3)(7)$$ the inequality: $<$ becomes $>$.
$$-6 > -21$$ (this is now a true inequality)

This suggests that multiplying both sides of an inequality by a negative number changes the ''sense'' of the inequality.

We can state the following general property.

THE MULTIPLICATION PROPERTY OF INEQUALITY

If $a < b$ then $ac < bc$ where $c > 0$
and $ac > bc$ where $c < 0$

In words, multiplying both sides of an inequality by the same *positive* number gives an equivalent inequality.

When both sides of an inequality are multiplied by the same *negative* number, it is necessary to *reverse the sense* of the inequality to give an equivalent inequality.

Example 3

(a) Solve and graph the solution set for $5x < 30$.

Multiplying both sides of the inequality by $\dfrac{1}{5}$ gives

$$\frac{1}{5}(5x) = \frac{1}{5}(30)$$

Simplifying, we have

$$x < 6$$

The graph of the solution set is

(b) Solve and graph the solution set for $-4x \geq 28$.

In this case we want to multiply both sides of the inequality by $-\dfrac{1}{4}$ to leave x alone on the left.

$$\left(-\frac{1}{4}\right)(-4x) \leq \left(-\frac{1}{4}\right)(28)$$ Reverse the sense of the inequality because you are multiplying by a negative number!

or $x \leq -7$

The graph of the solution set is

Solve and graph the solution sets:

1. $7x > 35$ **2.** $-8x \le 48$

The following example illustrates the use of the multiplication property when fractions are involved in an inequality.

Example 4

(a) Solve and graph the solution set for

$$\frac{x}{4} > 3$$

Here we multiply both sides of the inequality by 4. This will isolate x on the left.

$$4\left(\frac{x}{4}\right) > 4(3)$$
$$x > 12$$

The graph of the solution set is

(b) Solve and graph the solution set for

$$-\frac{x}{6} \ge -3$$

In this case, we multiply both sides of the inequality by -6:

Note that we reverse the sense of the inequality since we are multiplying by a negative number.

$$(-6)\left(-\frac{x}{6}\right) \le (-6)(-3)$$
$$x \le 18$$

The graph of the solution set is

CHECK YOURSELF 4

Solve and graph the solution sets for the following inequalities.

1. $\dfrac{x}{5} \leq 4$ **2.** $-\dfrac{x}{3} < -7$

Example 5

(a) Solve and graph the solution set for $5x - 3 < 2x$.

First, add 3 to both sides to undo the subtraction on the left.

$$5x - 3 < 2x$$
$$5x - 3 + 3 < 2x + 3$$
$$5x < 2x + 3$$

Now subtract $2x$, so that only the number remains on the right.

$$5x < 2x + 3$$
$$5x - 2x < 2x - 2x + 3$$
$$3x < 3$$

Note that the multiplication property also allows us to divide both sides by a nonzero number.

Next *divide* both sides by 3.

$$\frac{3x}{3} < \frac{3}{3}$$
$$x < 1$$

The graph of the solution set is

(b) Solve and graph the solution set for $2 - 5x < 7$.

$$2 - 5x < 7$$
$$2 - 2 - 5x < 7 - 2 \qquad \text{Subtract 2.}$$
$$-5x < 5$$
$$\frac{-5x}{-5} > \frac{5}{-5} \qquad \begin{array}{l}\text{Divide by } -5. \text{ Be sure to reverse the}\\ \text{sense of the inequality.}\end{array}$$
or $\qquad x > -1$

The graph is

Solve and graph the solution sets.

1. $4x + 9 \geq x$ **2.** $5 - 6x < 41$

As with equations, we will collect all variable terms on one side and all constant terms on the other.

Example 6

Solve and graph the solution set for $5x - 5 \geq 3x + 4$.

$$5x - 5 \geq 3x + 4$$
$$5x - 5 + 5 \geq 3x + 4 + 5 \qquad \text{Add 5.}$$
$$5x \geq 3x + 9$$
$$5x - 3x \geq 3x - 3x + 9 \qquad \text{Subtract } 3x.$$
$$2x \geq 9$$
$$\frac{2x}{2} \geq \frac{9}{2} \qquad \text{Divide by 2.}$$
$$x \geq \frac{9}{2}$$

The graph of the solution set is

Solve and graph the solution set for

$8x + 3 < 4x - 13$

Again, you should be especially careful when negative coefficients occur in the solution process. Consider the following example.

Example 7

Solve and graph the solution set for $2x + 4 < 5x - 2$.

$$2x + 4 < 5x - 2$$
$$2x + 4 \boxed{- 4} < 5x - 2 \boxed{- 4} \qquad \text{Subtract 4.}$$
$$2x < 5x - 6$$
$$2x \boxed{- 5x} < 5x \boxed{- 5x} - 6 \qquad \text{Subtract } 5x.$$
$$-3x < -6$$
$$\frac{-3x}{-3} \boxed{>} \frac{-6}{-3} \qquad \text{Divide by } -3, \text{ and reverse the sense of the inequality.}$$
$$x > 2$$

The graph of the solution set is

CHECK YOURSELF 7

Solve and graph the solution set.

$$5x + 12 \geq 10x - 8$$

The solution of inequalities may also require the use of the distributive property to clear an inequality of parentheses. Our final example illustrates.

Example 8

Solve and graph the solution set for

$$5(x - 2) \geq -8$$

Applying the distributive property on the left yields

$$5x - 10 \geq -8$$

Solving as before yields

$$5x - 10 \boxed{+ 10} \geq -8 \boxed{+ 10} \qquad \text{Add 10.}$$
$$5x \geq 2$$
or
$$x \geq \frac{2}{5} \qquad \text{Divide by 5.}$$

The graph of the solution set is

CHECK YOURSELF 8

Solve and graph the solution set.

$4(x + 3) < 9$

The following outline (or algorithm) summarizes our work in this section.

SOLVING LINEAR INEQUALITIES

STEP 1   Remove any grouping symbols and combine any like terms appearing on either side of the inequality.

STEP 2   Apply the addition property to write an equivalent inequality with the variable term on one side of the inequality and the number on the other.

STEP 3   Apply the multiplication property to write an equivalent inequality with the variable isolated on one side of the inequality. Be sure to reverse the sense of the inequality if you multiply or divide by a negative number. The solution derived in step 3 can then be graphed on a number line.

CHECK YOURSELF ANSWERS

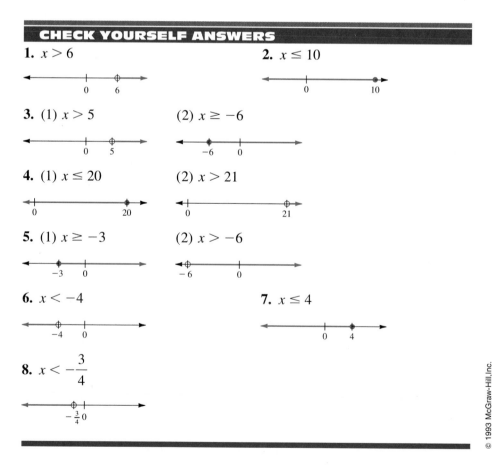

1. $x > 6$

2. $x \leq 10$

3. (1) $x > 5$ (2) $x \geq -6$

4. (1) $x \leq 20$ (2) $x > 21$

5. (1) $x \geq -3$ (2) $x > -6$

6. $x < -4$ **7.** $x \leq 4$

8. $x < -\dfrac{3}{4}$

Name _____

Date _____

Build Your Skills

Solve and graph the solution set of each of the following inequalities.

1. $x - 5 < 8$

$\longleftarrow \qquad \longrightarrow$

2. $x + 3 \leq 2$

$\longleftarrow \qquad \longrightarrow$

3. $x + 9 \geq 11$

$\longleftarrow \qquad \longrightarrow$

4. $x - 5 > -8$

$\longleftarrow \qquad \longrightarrow$

5. $5x < 4x + 7$

$\longleftarrow \qquad \longrightarrow$

6. $3x \geq 2x - 4$

$\longleftarrow \qquad \longrightarrow$

7. $6x - 8 \leq 5x$

$\longleftarrow \qquad \longrightarrow$

8. $3x + 2 > 2x$

$\longleftarrow \qquad \longrightarrow$

9. $4x - 3 \geq 3x + 5$

$\longleftarrow \qquad \longrightarrow$

10. $5x + 2 \leq 4x - 6$

$\longleftarrow \qquad \longrightarrow$

11. $7x + 5 < 6x - 4$

$\longleftarrow \qquad \longrightarrow$

12. $8x - 7 > 7x + 3$

$\longleftarrow \qquad \longrightarrow$

13. $3x \leq 9$

$\longleftarrow \qquad \longrightarrow$

14. $5x > 20$

$\longleftarrow \qquad \longrightarrow$

15. $5x > -35$

$\longleftarrow \qquad \longrightarrow$

16. $7x \leq -21$

$\longleftarrow \qquad \longrightarrow$

17. $-6x \geq 18$

$\longleftarrow \qquad \longrightarrow$

18. $-9x < 45$

$\longleftarrow \qquad \longrightarrow$

19. $-10x < -60$

$\longleftarrow \qquad \longrightarrow$

20. $-12x \geq -48$

$\longleftarrow \qquad \longrightarrow$

21. $\dfrac{x}{4} > 5$

$\longleftarrow \qquad \longrightarrow$

22. $\dfrac{x}{3} \leq -3$

$\longleftarrow \qquad \longrightarrow$

23. $-\dfrac{x}{2} \geq -3$

$\longleftarrow \qquad \longrightarrow$

24. $-\dfrac{x}{5} < 4$

$\longleftarrow \qquad \longrightarrow$

1. _____
2. _____
3. _____
4. _____
5. _____
6. _____
7. _____
8. _____
9. _____
10. _____
11. _____
12. _____
13. _____
14. _____
15. _____
16. _____
17. _____
18. _____
19. _____
20. _____
21. _____
22. _____
23. _____
24. _____

25. _____

26. _____

27. _____

28. _____

29. _____

30. _____

31. _____

32. _____

33. _____

34. _____

35. _____

36. _____

37. _____

38. _____

39. _____

40. _____

41. _____

42. _____

43. _____

44. _____

45. _____

46. _____

47. _____

48. _____

49. _____

50. _____

25. $\dfrac{2x}{3} < 6$

$\longleftrightarrow$

26. $\dfrac{3x}{4} \geq -9$

$\longleftrightarrow$

27. $5x > 3x + 8$

$\longleftrightarrow$

28. $4x \leq x - 9$

$\longleftrightarrow$

29. $8x - 25 < 3x$

$\longleftrightarrow$

30. $7x + 12 \geq x$

$\longleftrightarrow$

31. $5x - 2 > 3x$

$\longleftrightarrow$

32. $7x + 3 \geq 2x$

$\longleftrightarrow$

33. $3 - 2x > 5$

$\longleftrightarrow$

34. $5 - 3x \leq 17$

$\longleftrightarrow$

35. $2x \geq 5x + 18$

$\longleftrightarrow$

36. $3x < 7x - 28$

$\longleftrightarrow$

37. $4x < 8x - 3$

$\longleftrightarrow$

38. $5x < 10x + 4$

$\longleftrightarrow$

39. $5x - 3 \leq 3x + 15$

$\longleftrightarrow$

40. $8x + 7 > 5x + 34$

$\longleftrightarrow$

41. $9x + 7 > 2x - 28$

$\longleftrightarrow$

42. $10x - 5 \leq 8x - 25$

$\longleftrightarrow$

43. $7x - 5 < 3x + 2$

$\longleftrightarrow$

44. $5x - 2 \geq 2x - 7$

$\longleftrightarrow$

45. $3x - 8 \leq 9x + 16$

$\longleftrightarrow$

46. $2x + 7 > 7x - 18$

$\longleftrightarrow$

47. $5x + 7 > 8x - 17$

$\longleftrightarrow$

48. $4x - 3 \leq 9x + 27$

$\longleftrightarrow$

49. $3x - 2 \leq 5x + 3$

$\longleftrightarrow$

50. $2x + 3 > 8x - 2$

$\longleftrightarrow$

51. $3(x - 2) > 9$

52. $5(x + 4) \le -15$

53. $4(x + 7) \le 2x + 31$

54. $6(x - 5) > 3x - 26$

55. $2(x - 7) > 5x - 12$

56. $3(x + 4) \le 7x + 7$

Translate the following statements into inequalities. Let x represent the number in each case.

57. 5 more than a number is greater than 3.

58. 3 less than a number is less than or equal to 5.

59. 4 less than twice a number is less than or equal to 7.

60. 10 more than a number is greater than negative 2.

61. 4 times a number, decreased by 15, is greater than that number.

62. 2 times a number, increased by 28, is less than or equal to 6 times that number.

63. More than three-fifths of the wetlands of southeast Asia and India have been lost. If there are now approximately 150,000 km^2 of wetland habitat in this part of the world, use the following inequality to determine how much wetland habitat there was originally.

$$\frac{3}{5}x > 150,000$$

64. The continental United States (all states except Alaska and Hawaii) contains approximately 385,000 km^2 of wetlands. If the United States has lost at least 55 percent of its original wetlands, how much wetland habitat did the United States originally have?

$0.55 \times$ original wetlands $\le$ current wetlands

ANSWERS

51. _____

52. _____

53. _____

54. _____

55. _____

56. _____

57. _____

58. _____

59. _____

60. _____

61. _____

62. _____

63. _____

64. _____

Skillscan (Section 1.1)

Write each statement, using symbols.

a. 7 more than x

b. 5 less than a

c. 4 times p

d. 3 times w

e. 3 more than twice s

f. 2 less than 5 times x

g. The sum of twice x and $x + 1$

h. The sum of 3 times x and $x + 2$

© 1993 McGraw-Hill, Inc.

a. _____

b. _____

c. _____

d. _____

e. _____

f. _____

g. _____

h. _____

ANSWERS

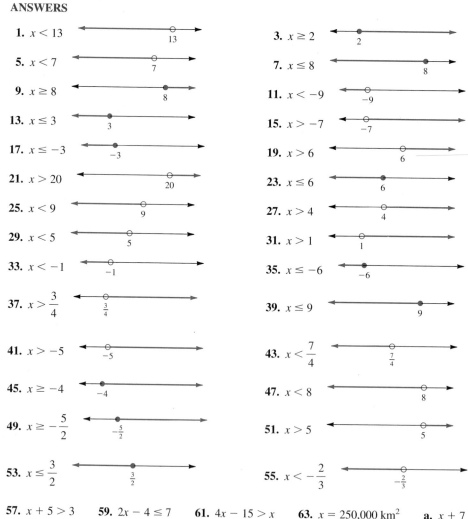

1. $x < 13$

3. $x \geq 2$

5. $x < 7$

7. $x \leq 8$

9. $x \geq 8$

11. $x < -9$

13. $x \leq 3$

15. $x > -7$

17. $x \leq -3$

19. $x > 6$

21. $x > 20$

23. $x \leq 6$

25. $x < 9$

27. $x > 4$

29. $x < 5$

31. $x > 1$

33. $x < -1$

35. $x \leq -6$

37. $x > \dfrac{3}{4}$

39. $x \leq 9$

41. $x > -5$

43. $x < \dfrac{7}{4}$

45. $x \geq -4$

47. $x < 8$

49. $x \geq -\dfrac{5}{2}$

51. $x > 5$

53. $x \leq \dfrac{3}{2}$

55. $x < -\dfrac{2}{3}$

57. $x + 5 > 3$ **59.** $2x - 4 \leq 7$ **61.** $4x - 15 > x$ **63.** $x = 250,000 \text{ km}^2$ **a.** $x + 7$
b. $a - 5$ **c.** $4p$ **d.** $3w$ **e.** $2s + 3$ **f.** $5x - 2$ **g.** $2x + x + 1$ or $3x + 1$
h. $3x + x + 2$ or $4x + 2$

192

Applying Equations

OBJECTIVE

To use equations for the solution of word problems

Earlier in this chapter you learned how to solve linear equations in one variable. The next step is to use this work in the solution of word problems. If you feel a bit uneasy about this subject, don't be too nervous. You have lots of company! To help you feel more comfortable when solving word problems, we are going to present a step-by-step approach that will, *with practice,* allow you to organize your work. Organization is the key to the solution of these problems.

TO SOLVE WORD PROBLEMS

STEP 1   Read the problem carefully. Then reread it to decide what you are asked to find.

STEP 2   Choose a letter to represent one of the unknowns in the problem. Then represent all other unknowns of the problem with expressions that use the same letter.

STEP 3   Translate the problem to the language of algebra to form an equation.

STEP 4   Solve the equation and answer the question of the original problem.

STEP 5   Verify your solution by returning to the original problem.

We discussed these translations in Section 1.1. You might find it helpful to review that section before going on.

The third step is usually the hardest part. We must translate words to the language of algebra. Before we look at a complete example, the table on the next page may help you review that translation step.

TRANSLATING WORDS TO ALGEBRA

WORDS	ALGEBRA
The sum of x and y	$x + y$
3 plus a	$3 + a$ or $a + 3$
5 more than m	$m + 5$
b increased by 7	$b + 7$
The difference of x and y	$x - y$
4 less than a	$a - 4$
s decreased by 8	$s - 8$
The product of x and y	$x \cdot y$ or xy
5 times a	$5 \cdot a$ or $5a$
Twice m	$2m$
The quotient of x and y	$\dfrac{x}{y}$
a divided by 6	$\dfrac{a}{6}$
One-half of b	$\dfrac{b}{2}$ or $\dfrac{1}{2}b$

Now let's look at some typical examples of translating phrases to algebra.

Example 1

Translate each statement to an algebraic expression.

(*a*) The sum of a and 2 times b

$a + 2b$

Sum   2 times b

(*b*) 5 times m increased by 1

$5m + 1$

5 times m Increased by 1

(*c*) 5 less than 3 times x

$3x - 5$

3 times x 5 less than

(*d*) The product of *x* and *y*, divided by 3

The product of *x* and *y*

$$\frac{xy}{3}$$

Divided by 3

Translate to algebra.

1. 2 more than twice *x*
2. 4 less than 5 times *n*
3. The product of twice *a* and *b*
4. The sum of *s* and *t*, divided by 5

Now let's work through a complete example, using our five-step approach.

Example 2

The sum of twice a number and 5 is 17. What is the number?

Step 1 *Read carefully*. You must find the unknown number.

Step 2 *Choose letters or variables*. Let *x* represent the unknown number. There are no other unknowns.

Step 3 *Translate*.

The sum of

$$2x + 5 = 17$$

Twice *x*   is

Step 4 *Solve*.

$$2x + 5 = 17$$
$$2x + 5 - 5 = 17 - 5 \qquad \text{Subtract 5.}$$
$$2x = 12$$
$$\frac{2x}{2} = \frac{12}{2} \qquad \text{Divide by 2.}$$
$$x = 6$$

So the number is 6.

Always return to the *original problem* to check your result and *not* to the equation of step 3. This will prevent possible errors!

Step 5 *Check*. Is the sum of twice 6 and 5 equal to 17? Yes (12 + 5 = 17). We have checked our solution.

CHECK YOURSELF 2

The sum of 3 times a number and 8 is 35. What is the number?

Consecutive integers are integers that follow one another, like 10, 11, and 12. To represent them in algebra:

If x is an integer, then $x + 1$ is the next consecutive integer, $x + 2$ is the next, and so on.

We'll need this idea in the following example.

Example 3

The sum of two consecutive integers is 41. What are the two integers?

REMEMBER THE STEPS!

Read the problem carefully. What do you need to find?

Step 1 We want to find the two consecutive integers.

Assign letters to the unknown or unknowns.

Step 2 Let x be the first integer. Then $x + 1$ must be the next.

Write an equation.

Step 3

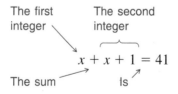

Solve the equation.

Step 4

$$x + x + 1 = 41$$
$$2x + 1 = 41$$
$$2x = 40$$
$$x = 20$$

The first integer (x) is 20, and the next integer ($x + 1$) is 21.

Check.

Step 5 The sum of the two integers 20 and 21 is 41.

CHECK YOURSELF 3

The sum of three consecutive integers is 51. What are the three integers?

Example 4

There were 55 more yes votes than no votes on an election measure. If 735 votes were cast in all, how many yes votes were there? How many no votes?

What do you need to find?

Step 1   We want to find the number of yes votes and the number of no votes.

Assign letters to the unknowns.

Step 2   Let x be the number of no votes. Then

$$\underbrace{x + 55}$$

55 more than x

is the number of yes votes.

Write an equation.

Step 3

$$x + \underbrace{x + 55} = 735$$

No votes   Yes votes

Solve the equation.

Step 4

$$x + x + 55 = 735$$
$$2x + 55 = 735$$
$$2x = 680$$
$$x = 340$$
$$\text{No votes } (x) = 340$$
$$\text{Yes votes } (x + 55) = 395$$

Check.

Step 5   Thus 340 no votes plus 395 yes votes equals 735 total votes. The solution checks.

CHECK YOURSELF 4

Francine earns $120 per month more than Rob. If they earn a total of $2680 per month, what are their monthly salaries?

Similar methods will allow you to solve a variety of word problems. Look at the following example.

Example 5

Juan worked twice as many hours as Jerry. Marcia worked 3 more hours than Jerry. If they worked a total of 31 hours, find out how many hours each worked.

Step 1 We want to find the hours each worked, so there are three unknowns.

There are other choices for *x*, but choosing the smallest quantity will usually give the easiest equation to write and solve.

Step 2 Let x be the hours that Jerry worked.

Twice Jerry's hours

Then $2x$ is Juan's hours worked

3 more hours than Jerry worked

and $x + 3$ is Marcia's hours.

Step 3

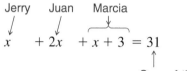

$$x \quad + 2x \quad + x + 3 = 31$$

Sum of their hours

Step 4

$$x + 2x + x + 3 = 31$$
$$4x + 3 = 31$$
$$4x = 28$$
$$x = 7$$

Jerry's hours $(x) = 7$

Juan's hours $(2x) = 14$

Marcia's hours $(x + 3) = 10$

Step 5 The sum of their hours $(7 + 14 + 10)$ is 31, and the solution is verified.

CHECK YOURSELF 5

Lucy jogged twice as many miles as Paul but 3 miles (mi) less than Isaac. If the three ran a total of 23 mi, how far did each person run?

Many word problems involve geometric figures and measurements. Let's look at an example.

Example 6

The perimeter of a rectangle is 46 cm. If the length is 3 cm more than the width, what are the dimensions of the rectangle?

Step 1 We want to find the dimensions (length and width) of the rectangle.

Step 2 Let x be the width. Then $x + 3$ is the length.

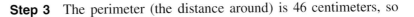

3 more than x

Always draw a sketch at this point when you can do so. It will help you form an equation in step 3.

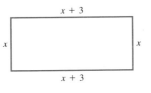

Step 3 The perimeter (the distance around) is 46 centimeters, so

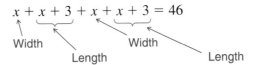

$$x + x + 3 + x + x + 3 = 46$$

Width   Length   Width   Length

Step 4

$$x + x + 3 + x + x + 3 = 46$$
$$4x + 6 = 46$$
$$4x = 40$$
$$x = 10$$
$$\text{Width } (x) = 10 \text{ cm}$$
$$\text{Length } (x + 3) = 13 \text{ cm}$$

Step 5 The perimeter is $10 + 13 + 10 + 13$, or 46 centimeter. The solution is verified.

CHECK YOURSELF 6

The perimeter of a triangle is 55 inches. If the length of the base is 5 inches less than the length of the two equal legs, find the lengths of the legs and the base.

That is all we propose to do with word problems for the moment. Remember to use the five-step approach as you work on the problems for this section. We hope that you find it helpful.

CHECK YOURSELF ANSWERS

1. (1) $2x + 2$; (2) $5n - 4$; (3) $2ab$; (4) $\dfrac{s + t}{5}$.

2. The equation is $3x + 8 = 35$. The number is 9.

3. The equation is $x + x + 1 + x + 2 = 51$. The integers are 16, 17, and 18.

4. The equation is $x + x + 120 = 2680$. Rob's salary is $1280, and Francine's is $1400.

5. Paul: 4 mi; Lucy: 8 mi; Isaac: 11 mi.

6. Legs: 20 in; base: 15 in.

Name _____

Date _____

Build Your Skills

Solve the following word problems. Be sure to label the unknowns and to show the equation you use for the solution.

1. The sum of twice a number and 7 is 33. What is the number?

2. 3 times a number, increased by 8, is 50. Find the number.

3. 5 times a number, minus 12, is 78. Find the number.

4. 4 times a number, decreased by 20, is 44. What is the number?

5. The sum of two consecutive integers is 71. Find the two integers.

6. The sum of two consecutive integers is 145. Find the two integers.

7. The sum of three consecutive integers is 63. What are the three integers?

8. If the sum of three consecutive integers is 93, find the three integers.

9. The sum of two consecutive even integers is 66. What are the two integers? (*Hint:* Consecutive even integers such as 10, 12, and 14 can be represented by x, $x + 2$, $x + 4$, and so on.)

10. If the sum of two consecutive even integers is 86, find the two integers.

1. _____

2. _____

3. _____

4. _____

5. _____

6. _____

7. _____

8. _____

9. _____

10. _____

11. _____

12. _____

13. _____

14. _____

15. _____

16. _____

17. _____

18. _____

19. _____

20. _____

21. _____

11. If the sum of two consecutive odd integers is 52, what are the two integers? (*Hint:* Consecutive odd integers such as 21, 23, and 25 can be represented by x, $x + 2$, $x + 4$, and so on.)

12. The sum of two consecutive odd integers is 88. Find the two integers.

13. The sum of three consecutive odd integers is 105. What are the three integers?

14. The sum of three consecutive even integers is 126. What are the three integers?

15. If the sum of four consecutive integers is 86, what are the four integers?

16. The sum of four consecutive integers is 62. What are the four integers?

17. 4 times an integer is 9 more than 3 times the next consecutive integer. What are the two integers?

18. 4 times an integer is 30 less than 5 times the next consecutive even integer. Find the two integers.

19. In an election, the winning candidate had 160 more votes than the loser. If the total number of votes cast was 3260, how many votes did each candidate receive?

20. Jody earns $140 more per month than Frank. If their monthly salaries total $2760, what amount does each earn?

21. A washer-dryer combination costs $650. If the washer costs $70 more than the dryer, what does each appliance cost?

22. Morgan has a board that is 98 inches long. He wishes to cut the board into two pieces so that one piece will be 10 inches longer than the other. What should be the length of each piece?

23. Ken is 1 year less than twice as old as his sister. If the sum of their ages is 14 years, how old is Ken?

24. Diane is twice as old as her brother Dan. If the sum of their ages is 27 years, how old are Diane and her brother?

25. José is 3 years less than 4 times as old as his daughter. If the sum of their ages is 37 years, how old is José?

26. Mrs. Jackson is 2 years more than 3 times as old as her son. If the difference between their ages is 22 years, how old is Mrs. Jackson?

27. On her vacation in Europe, Jovita's expenses for food and lodging were $60 less than twice as much as her airfare. If she spent $2400 in all, what was her airfare?

28. Rachel earns $6000 less than twice as much as Tom. If their two incomes total $48,000, how much does each earn?

29. There are 99 students registered in three sections of algebra. There are twice as many students in the 10 o'clock section as in the 8 o'clock section and 7 more students at 12 o'clock than at 8 o'clock. How many students are in each section?

30. The Randolphs used 12 more gallons (gal) of fuel oil in October than in September and twice as much oil in November as in September. If they used 132 gal for the 3 months, how much was used during each month?

ANSWERS

22. _____

23. _____

24. _____

25. _____

26. _____

27. _____

28. _____

29. _____

30. _____

31. _____

31. The length of a rectangle is 5 cm more than its width. The perimeter is 98 cm. What are the dimensions? (*Hint:* Remember to draw a sketch whenever geometric figures are involved in a word problem.)

32. The length of a rectangle is 3 times its width. If the perimeter of the rectangle is 48 in, find the length and width of the rectangle.

32. _____

33. The length of a rectangle is 2 ft more than 3 times its width. If the perimeter is 68 ft, what are the length and width of the rectangle?

34. The length of a rectangle is 3 cm more than twice its width. What are the dimensions of the rectangle if its perimeter is 48 cm?

33. _____

35. One side of a triangle is 4 meters (m) longer than the shortest side. The third side is twice the length of the shortest side. If the perimeter of the triangle is 44 m, find the lengths of the three sides of the triangle.

36. The equal legs of an isosceles triangle are each 4 ft more than twice the length of the base. If the perimeter of the triangle is 68 ft, find the lengths of the three sides of the triangle.

34. _____

35. _____

ANSWERS

1. 13   **3.** 18   **5.** 35, 36   **7.** 20, 21, 22   **9.** 32, 34   **11.** 25, 27   **13.** 33, 35, 37
15. 20, 21, 22, 23   **17.** 12, 13   **19.** 1710 votes, 1550 votes   **21.** Washer $360, dryer
$290   **23.** 9 years old   **25.** 29 years old   **27.** $820   **29.** 8 o'clock, 23; 10 o'clock, 46;
12 o'clock, 30   **31.** 22 cm, 27 cm   **33.** 26 ft, 8 ft   **35.** 10 m, 14 m, 20 m

36. _____

Summary

Algebraic Equations [3.1 to 3.4]

Equation   A statement that two expressions are equal.

$$2x - 7 = x + 3$$

Left side   Right side
Equals sign

4 is a solution for the equation
$3x - 5 = 7$
because
$3 \cdot 4 - 5 = 7$
$12 - 5 = 7$
$7 = 7$ (true)

$4x - 3 = 5$
$4x = 8$
$x = 2$

are all equivalent equations that have same solution, 2.

$5x = 20$ and $x = 4$ are equivalent equations.

Solution   A value for the variable that will make an equation a true statement.

Equivalent Equations   Equations that have exactly the same solutions.

Writing Equivalent Equations   There are two basic properties that will yield equivalent equations.

1. If $a = b$, then $a + c = b + c$.
   Adding (or subtracting) the same quantity on each side of an equation gives an equivalent equation.
2. If $a = b$, then $ac = bc$, $c \neq 0$.
   Multiplying (or dividing) both sides of an equation by the same number gives an equivalent equation.

Solving Linear Equations   We say that an equation is "solved" when we have an equivalent equation of the form

$x = \boxed{}$ Some number

The steps of solving a linear equation are as follows:

1. Use the distributive property to remove any grouping symbols that appear. Then simplify by combining any like terms.
2. Add or subtract the same term on both sides of the equation until the term involving the variable is on one side and a number is on the other.
3. Multiply or divide both sides of the equation by the same nonzero number so that the variable is alone on one side of the equation.
4. Check the solution in the original equation.

Solve:
$$
\begin{aligned}
3(x - 2) + 4x &= 3x + 14 \\
3x - 6 + 4x &= 3x + 14 \\
7x - 6 &= 3x + 14 \\
+ 6 & \qquad + 6 \\
\hline
7x &= 3x + 20 \\
-3x & \qquad -3x \\
\hline
4x &= 20 \\
\frac{4x}{4} &= \frac{20}{4} \\
x &= 5
\end{aligned}
$$

Solve for b:
$$
\begin{aligned}
a &= \frac{2b + c}{3} \\
3 \cdot a &= \left(\frac{2b + c}{3}\right)3 \\
3a &= 2b + c \\
- c &= \qquad - c \\
\hline
3a - c &= 2b \\
\frac{3a - c}{2} &= b
\end{aligned}
$$

Literal Equations [3.5]

Literal Equation   An equation that involves more than one letter or variable.

Solving Literal Equations

1. Multiply both sides of the equation by the LCD to clear of fractions.
2. Add or subtract the same term on both sides of the equation so that all terms involving the variable you are solving for are on one side and all other terms are on the other side.
3. Divide both sides by any numbers or letters multiplying the variable that you are solving for.

Inequalities [3.6 and 3.7]

Inequality   A statement that one quantity is less than (or greater than) another. Four symbols are used:

$$a < b \qquad a > b \qquad a \le b \qquad a \ge b$$

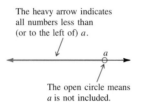

a is less than *b*   *a* is greater than *b*   *a* is less than
or equal to *b*   *a* is greater than
or equal to *b*

Graphing Inequalities   To graph $x < a$, we use an open circle and an arrow pointing left.

The heavy arrow indicates
all numbers less than
(or to the left of) *a*.

a

The open circle means
a is not included.

To graph $x \ge b$, we use a closed circle and an arrow pointing right.

b

The closed circle means
that in this case *b*
is included.

Solving Inequalities   An inequality is ''solved'' when it is in the form $x < \square$ or $x > \square$.

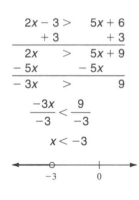

$$2x - 3 > 5x + 6$$
$$\underline{+\ 3 \qquad\quad +\ 3}$$
$$2x > 5x + 9$$
$$\underline{-\ 5x \qquad -\ 5x}$$
$$-3x > 9$$

$$\frac{-3x}{-3} < \frac{9}{-3}$$

$$x < -3$$

−3   0

Proceed as in solving equations by using the following properties.

1. If $a < b$, then $a + c < b + c$.

Adding (or subtracting) the same quantity to both sides of an inequality gives an equivalent inequality.

2. If $a < b$, then $ac < bc$, when $c > 0$; and $ac > bc$, when $c < 0$.

Multiplying both sides of an inequality by the same *positive number* gives an equivalent inequality. When both sides of an inequality are multiplied by the same *negative number, you must reverse the sense* of the inequality to give an equivalent inequality.

Applying Equations [3.8]

Using Equations to Solve Word Problems   Follow these steps.

1. Read the problem carefully. Then reread it to decide what you are asked to find.
2. Choose a letter to represent one of the unknowns in the problem. Then represent each of the unknowns of the problem with an expression that uses the same letter.
3. Translate the problem to the language of algebra to form an equation.
4. Solve the equation and answer the question of the original problem.
5. Verify your solution by returning to the original problem.

Summary Exercises Chapter 3

This summary exercise set is provided to give you practice with each of the objectives of the chapter. Each exercise is keyed to the appropriate chapter section. The answers are provided in the instructor's manual. Your instructor will give you guidelines on how to best use these exercises in your instructional setting.

[3.1] Tell whether the number shown in parentheses is a solution for the given equation.

1. $7x + 2 = 16$ (2)

2. $5x - 8 = 3x + 2$ (4)

3. $7x - 2 = 2x + 8$ (2)

4. $4x + 3 = 2x - 11$ (-7)

5. $x + 5 + 3x = 2 + x + 23$ (6)

6. $\dfrac{2}{3}x - 2 = 10$ (21)

[3.2] Solve the following equations and check your results.

7. $x + 5 = 7$

8. $x - 9 = 3$

9. $5x = 4x - 5$

10. $3x - 9 = 2x$

11. $5x - 3 = 4x + 2$

12. $9x + 2 = 8x - 7$

13. $7x - 5 = 6x - 4$

14. $3 + 4x - 1 = x - 7 + 2x$

15. $4(2x + 3) = 7x + 5$

16. $5(5x - 3) = 6(4x + 1)$

[3.3] Solve the following equations and check your results.

17. $5x = 35$

18. $7x = -28$

19. $-6x = 24$

20. $-9x = -63$

21. $\dfrac{x}{4} = 8$

22. $-\dfrac{x}{5} = -3$

23. $\dfrac{2}{3}x = 18$

24. $\dfrac{3}{4}x = 24$

[3.4] Solve the following equations and check your results.

25. $5x - 3 = 12$

26. $4x + 3 = -13$

27. $7x + 8 = 3x$

28. $3 - 5x = -17$

29. $3x - 7 = x$

30. $2 - 4x = 5$

31. $\dfrac{x}{3} - 5 = 1$

32. $\dfrac{3}{4}x - 2 = 7$

33. $6x - 5 = 3x + 13$

34. $3x + 7 = x - 9$

35. $7x + 4 = 2x + 6$

36. $9x - 8 = 7x - 3$

37. $2x + 7 = 4x - 5$

38. $3x - 15 = 7x - 10$

39. $3x - 2 + 5 = 7 + 2x + 21$

40. $8x + 3 - 2x + 5 = 3 - 4x$

41. $5(3x - 1) - 6x = 3x - 2$

42. $5x + 2(3x - 4) = 14x + 7$

43. $\dfrac{10}{3}x - 5 = \dfrac{4}{3}x + 7$

44. $\dfrac{11}{4}x - 15 = 5 - \dfrac{5}{4}x$

45. $3.7x + 8 = 1.7x + 16$

46. $5.4x - 3 = 8.4x + 9$

[3.5] Solve for the indicated variable.

47. $V = LWH$ (for L)

48. $P = 2L + 2W$ (for L)

49. $ax + by = c$ (for y)

50. $A = \dfrac{1}{2}bh$ (for h)

51. $A = P + Prt$ (for t)

52. $m = \dfrac{n - p}{q}$ (for n)

[3.6] Graph the solution sets.

53. $x > 5$

54. $x < -3$

55. $x \leq -4$

56. $x \geq 9$

57. $x \geq -6$

58. $x < 0$

[3.7] Solve and graph the solution sets for the following inequalities.

59. $x - 4 \leq 7$

60. $x + 3 > -2$

61. $5x > 4x - 3$

62. $4x \geq -12$

63. $-12x < 36$

64. $-\dfrac{x}{5} \geq 3$

65. $2x \leq 8x - 3$

66. $2x + 3 \geq 9$

67. $4 - 3x > 8$

68. $5x - 2 \leq 4x + 5$

69. $7x + 13 \geq 3x + 19$

70. $4x - 2 < 7x + 16$

71. $5(x - 3) < 2x + 12$

72. $4(x + 3) \geq x + 7$

[3.8] Solve the following word problems. Be sure to label the unknowns and to show the equation you used for the solution.

73. The sum of 3 times a number and 7 is 25. What is the number?

74. 5 times a number, decreased by 8, is 32. Find the number.

75. If the sum of two consecutive integers is 85, find the two integers.

76. The sum of three consecutive odd integers is 57. What are the three integers?

77. Rafael earns $35 more per week than Andrew. If their weekly salaries total $715, what amount does each earn?

78. Larry is 2 years older than Susan, while Nathan is twice as old as Susan. If the sum of their ages is 30 years, find each of their ages.

79. The perimeter of a rectangle is 28 in. If the length is 1 in less than twice the width, what are the dimensions of the rectangle?

80. One side of a triangle is 3 cm longer than the shortest side. The third side is twice the length of the shortest side. If the perimeter of the triangle is 35 cm, find the lengths of the three sides of the triangle.

Name

Section

Date

The purpose of this self-test is to help you check your progress and to review for a chapter test in class. Allow yourself about an hour to take the test. When you are done, check your answers in the back of the book. If you missed any problems, be sure to go back and review the appropriate sections in the chapter and the exercises that are provided.

Tell whether the number shown in parentheses is a solution for the given equation.

1. $7x - 3 = 25$ (5)

2. $8x - 3 = 5x + 9$ (4)

Solve the following equations and check your results.

3. $x - 7 = 4$

4. $7x - 12 = 6x$

5. $9x - 2 = 8x + 5$

Solve the following equations and check your results.

6. $7x = 49$

7. $\dfrac{1}{4}x = -3$

8. $\dfrac{4}{5}x = 20$

Solve the following equations and check your results.

9. $7x - 5 = 16$

10. $10 - 3x = -2$

11. $7x - 3 = 4x - 5$

12. $2x - 7 = 5x + 8$

Solve for the indicated variable.

13. $C = 2\pi r$ (for r)

14. $V = \dfrac{1}{3}Bh$ (for h)

ANSWERS

1. _____

2. _____

3. _____

4. _____

5. _____

6. _____

7. _____

8. _____

9. _____

10. _____

11. _____

12. _____

13. _____

14. _____

15. _____

15. $3x + 2y = 6$ (for y)

Graph the solution sets.

16. _____

16. $x \geq 9$ **17.** $x < -3$

Solve and graph the solution sets for the following inequalities.

17. _____

18. $x - 5 \leq 9$ **19.** $5 - 3x > 17$

20. $5x + 13 \geq 2x + 17$ **21.** $2x - 3 < 7x + 2$

18. _____

Solve the following word problems. Be sure to show the equation you used for the solution.

22. 5 times a number, decreased by 7, is 28. What is the number?

19. _____

23. The sum of three consecutive integers is 66. Find the three integers.

20. _____

24. Jan is twice as old as Steve, while Rick is 5 years older than Jan. If the sum of their ages is 35 years, find each of their ages.

21. _____

25. The perimeter of a rectangle is 62 in. If the length of the rectangle is 1 in more than twice its width, what are the dimensions of the rectangle?

22. _____

23. _____

24. _____

25. _____

This test is provided to help you in the process of reviewing the previous chapters. Answers are provided in the back of the book. If you missed any problems, be sure to go back and review the appropriate chapter sections.

Write, using symbols.

1. 3 times the sum of r and s

2. The quotient when 5 less than x is divided by 3

Write in exponential form.

3. $5 \cdot 5 \cdot 5 \cdot 5 \cdot 5$

4. $8 \cdot x \cdot x \cdot x \cdot y \cdot y$

Identify the property that is illustrated by each of the following statements.

5. $7 + (5 + 2) = (7 + 5) + 2$

6. $2(3 + 5) = 2 \cdot 3 + 2 \cdot 5$

Simplify each of the following expressions.

7. $7a^2b - 2a^2b$

8. $10a^2 + 5a + 2a^2 - 2a$

9. $3m^2n \cdot 5m^3n^2$

10. $\dfrac{25x^3y^4}{5x^2y^3}$

Evaluate each of the following expressions.

11. $2 \cdot 3^2 - 8 \cdot 2$

12. $5(7 - 3)^2$

13. $|12 - 5|$

14. $|12| - |5|$

15. $(-7) + (-9)$

16. $\dfrac{17}{3} + \left(-\dfrac{5}{3}\right)$

17. $(-7)(-9)$

18. $(-3.2)(5)$

19. $\dfrac{0}{-13}$

Evaluate each of the following expressions if $x = -2$, $y = 3$, and $z = 5$.

20. $3x - y$

21. $4x^2 - y$

22. $\dfrac{5z - 4x}{2y + z}$

ANSWERS

1. _____

2. _____

3. _____

4. _____

5. _____

6. _____

7. _____

8. _____

9. _____

10. _____

11. _____

12. _____

13. _____

14. _____

15. _____

16. _____

17. _____

18. _____

19. _____

20. _____

21. _____

22. _____

Solve the following equations and check your results.

23. $9x - 5 = 8x$

24. $-\dfrac{3}{4}x = 18$

25. $6x - 8 = 2x - 3$

26. $2x + 3 = 7x + 5$

27. $\dfrac{4}{3}x - 6 = 4 - \dfrac{2}{3}x$

Solve the following equations for the indicated variable.

28. $I = Prt$ (for r)

29. $A = \dfrac{1}{2}bh$ (for h)

30. $ax + by = c$ (for y)

Solve and graph the solution sets for the following inequalities.

31. $3x - 5 < 4$

32. $7 - 2x \geq 10$

33. $7x - 2 > 4x + 10$

34. $2x + 5 \leq 8x - 3$

Solve the following word problems. Be sure to show the equation used for the solution.

35. If 4 times a number, decreased by 7, is 45, find that number.

36. The sum of two consecutive integers is 85. What are those two integers?

37. If 3 times an integer is 12 more than the next consecutive odd integer, what is that integer?

38. Michelle earns $120 more per week than David. If their weekly salaries total $720, how much does Michelle earn?

39. The length of a rectangle is 2 cm more than 3 times its width. If the perimeter of the rectangle is 44 cm, what are the dimensions of the rectangle?

40. One side of a triangle is 5 in longer than the shortest side. The third side is twice the length of the shortest side. If the triangle perimeter is 37 in, find the length of each leg.

CHAPTER 4

Polynomials

Old-Growth Forests

When the first European settlers landed on the east coast of the United States, they surely thought the forests that lay before them were inexhaustible. As large as these forests appeared to the early immigrants, they have proved to be no match for modern technology. As recently as 40 years ago, most of the forests of the Pacific northwest were largely uncut. Today, less than 10 percent of the original forest is left.

The rain forest of the Pacific northwest stretches from the southern Alaska panhandle to just north of San Francisco. It ranges inland from the coast to the mountains of the Coast Range in northern California and the Cascade Range in Oregon, Washington, and southern Canada.

This region is characterized by a mild year-round climate and high levels of rainfall, enabling extremely favorable tree growing conditions. As a result, this forest contains some of the largest trees on earth. The trees in this "old-growth" forest can reach heights of over 250 feet and be over 25 feet in diameter. Many of these trees are over 200 years old, and some live as long as 500 years or more.

This forest directly affects the stability of the regional climate and is thought to influence much of the world's climate. It causes as much as one-third of the local rainfall by stripping moisture from clouds as they move inland from the Pacific Ocean. It stores more carbon than any other ecosystem by taking up carbon dioxide from the atmosphere in the growth and maintenance of these huge trees. This, in turn, affects the level of carbon dioxide in the atmosphere which is linked to the greenhouse effect. As this forest is cut down and consumed, the carbon dioxide level in the atmosphere is predicted to rise and increase global warming.

The forest of the Pacific northwest is a temperate rainforest. This forest is one of the most productive biological regions on earth. In recent years it has been discovered that there is more living material, and possibly more individual types of living things, in these forests than in well-known, abundant tropical forest regions. Many of the plants and animals found in these old-growth forests are unique to this region. They are not found anywhere else, and they cannot survive anywhere else. Scientists are just now discovering how these plants and animals work together to make these forests unique.

These forests not only play a major role in the natural systems of the region and planet, but also are a major component of the regional and global economic systems. Timber production from these forests is the primary source of employment for much of the region's population. Over 250,000 people in the northwest rely directly or indirectly on these forests for their livelihood. Billions of dollars are generated annually by the forest products provided by old-growth forests.

It has recently become apparent the forest can no longer maintain this dual identity. If its role in the environment is to be maintained, its role as a primary source of wood products will have to be diminished. If its role as a resource base is to be continued, its role in the environment will undoubtedly be altered. This need to change how we view these forests has caused a major controversy.

Environmentalists want to stop all cutting of these forests to preserve the remaining trees. They claim that old-growth forests are necessary to maintain the health of the environment and that wood products should be obtained from the second- and third-growth forests that are returning in previously harvested regions.

The timber industry maintains that continued harvesting of old-growth trees is necessary to avoid catastrophic upheaval in the regional economy. If the old-growth forests of the northwest are placed off limits to harvesting, it is predicted that as many as 40,000 people could be put out of work. Entire towns could be destroyed by this reduction in workforce. The economic impact could be felt clearly throughout the nation.

The decisions to be made about the future role of old-growth forests are not simple. As with all environmental decisions, they are complex and tend to generate heated debate. The actions taken to protect or harvest the remaining old-growth forests will have environmental and economic impacts for many years. These actions need to be taken with the utmost caution.

Polynomials—An Introduction

OBJECTIVES
1. To identify polynomials, monomials, binomials, and trinomials
2. To find the degree of a polynomial
3. To write polynomials in descending-exponent form

Our work in this chapter deals with the most common kind of algebraic expression, a *polynomial*. To define a polynomial let's recall our earlier definition of the word "term."

> A *term* is a number or the product of a number and one or more variables.

For example, x^5, $3x$, $-4xy^2$, 8, $\dfrac{5}{x}$, and $-14\sqrt{x}$ are terms. A polynomial consists of one or more terms in which the only allowable exponents are the whole numbers, 0, 1, 2, 3, . . . and so on. These terms are connected by addition or subtraction signs.

> In each term of a polynomial, the number is called the *numerical coefficient,* or more simply the *coefficient,* of that term.

Example 1

Note: Each sign (+ or −) is attached to the term that follows that sign.

(*a*) $x + 3$ is a polynomial. The terms are x and 3. The coefficients are 1 and 3.
(*b*) $3x^2 - 2x + 5$ is also a polynomial. Its terms are $3x^2$, $-2x$, and 5. The coefficients are 3, −2, and 5.
(*c*) $5x^3 + 2 - \dfrac{3}{x}$ is *not* a polynomial because of the division by x in the third term.

CHECK YOURSELF 1

Which of the following are polynomials?

1. $5x^2$ **2.** $3y^3 - 2y + \dfrac{5}{y}$ **3.** $4x^2 - 2x + 3$

Certain polynomials are given special names because of the number of terms that they have.

The prefix "mono" means 1.

The prefix "bi" means 2.

The prefix "tri" means 3.

A polynomial with one term is called a *monomial.*

A polynomial with two terms is called a *binomial.*

A polynomial with three terms is called a *trinomial.*

Example 2

In a polynomial, terms are separated by plus and minus signs.

(*a*) $3x^2y$ is a monomial. It has one term.
(*b*) $2x^3 + 5x$ is a binomial. It has two terms, $2x^3$ and $5x$.
(*c*) $5x^2 - 4x + 3$ is a trinomial. Its three terms are $5x^2$, $-4x$, and 3.

Note: There are no special names for polynomials with more than three terms.

CHECK YOURSELF 2

Classify each of these as a monomial, binomial, or trinomial.

1. $5x^4 - 2x^3$ **2.** $4x^7$ **3.** $2x^2 + 5x - 3$

Remember, in a polynomial the allowable exponents are the whole numbers 0, 1, 2, 3, and so on. The degree will be a whole number.

We also classify polynomials by their *degree.* The degree of a polynomial that has only one variable is the highest power appearing in any one term.

Example 3

The highest power

(*a*) $5x^3 - 3x^2 + 4x$ has degree 3.

The highest power

(*b*) $4x - 5x^4 + 3x^3 + 2$ has degree 4.
(*c*) $8x$ has degree 1.

(Because $8x = 8x^1$)

$x^0 = 1$ because of a definition in algebra. We will discuss this in Chapter 9.

(*d*) 7 has degree 0.

(Because $7 = 7 \cdot 1 = 7x^0$)

Note: Polynomials can have more than one variable, such as $4x^2y^3 + 5xy^2$. The degree is then the sum of the highest powers in any single term (here $2 + 3$, or 5). In general, we will be working with polynomials in a single variable, such as x.

CHECK YOURSELF 3

Find the degree of each polynomial.

1. $6x^5 - 3x^3 - 2$ **2.** $5x$ **3.** $3x^3 + 2x^6 - 1$ **4.** 9

Working with polynomials is much easier if you get used to writing them in *descending-exponent,* or *-power form.* This simply means that the term with the highest exponent is written first, then the term with the next highest exponent, and so on.

Example 4

The exponents get smaller from left to right.

(a) $5x^7 - 3x^4 + 2x^2$ is in descending-exponent form.
(b) $4x^4 + 5x^6 - 3x^5$ is *not* in descending-exponent form. The polynomial should be written as

$$5x^6 - 3x^5 + 4x^4$$

Notice that the degree of the polynomial is the power of the *first,* or *leading,* term once the polynomial is arranged in descending-exponent form.

CHECK YOURSELF 4

Write the following polynomials in descending-exponent form.

1. $5x^4 - 4x^5 + 7$ **2.** $4x^3 + 9x^4 + 6x^8$

A polynomial may represent different numbers depending on the value given to the variable. Our final example illustrates.

Example 5

Given the polynomial

$$3x^3 - 2x^2 - 4x + 1$$

(a) Find the value of the polynomial when $x = 2$.

Substituting 2 for x, we have

Again note how the rules for the order of operations are applied. See Section 1.6 for a review if you would like.

$$3(2)^3 - 2(2)^2 - 4(2) + 1$$
$$= 3(8) - 2(4) - 4(2) + 1$$
$$= 24 - 8 - 8 + 1$$
$$= 9$$

(b) Find the value of the polynomial when $x = -2$.

Now we substitute -2 for x.

Be particularly careful when dealing with powers of negative numbers!

$3(-2)^3 - 2(-2)^2 - 4(-2) + 1$

$= 3(-8) - 2(4) - 4(-2) + 1$

$= -24 - 8 + 8 + 1$

$= -23$

CHECK YOURSELF 5

For the polynomial

$$4x^3 - 3x^2 + 2x - 1$$

find its value when

1. $x = 3$ **2.** $x = -3$

CHECK YOURSELF ANSWERS

1. (1) and (3) are polynomials.
2. (1) Binomial; (2) monomial; (3) trinomial.
3. (1) 5; (2) 1; (3) 6; (4) 0.
4. (1) $-4x^5 + 5x^4 + 7$; (2) $6x^8 + 9x^4 + 4x^3$.
5. (1) 86; (2) -142.

Name _____

Date _____

Build Your Skills

Which of the following expressions are polynomials?

1. $5x^2$

2. $4x^2 - \dfrac{2}{x}$

3. $5x^2y - 2xy^2$

4. 3

5. -7

6. $4x^3 + x$

7. $\dfrac{3 + x}{x^2}$

8. $5a^2 - 2a + 7$

For each of the following polynomials, list the terms and the coefficients.

9. $2x^2 - 3x$

10. $5x^3 + x$

11. $4x^3 - 3x + 2$

12. $7x^2$

Classify each of the following as a monomial, binomial, or trinomial where possible.

13. $4x^2 - 2x$

14. $3x^5$

15. $6y^2 + 3y + 3$

16. $x^2 + 2xy + y^2$

17. $2x^4 - 3x^2 + 5x - 2$

18. $x^4 + \dfrac{5}{x} + 7$

19. $5y^7$

20. $3x^3 - 4x^2 + 5x - 3$

ANSWERS

1. _____

2. _____

3. _____

4. _____

5. _____

6. _____

7. _____

8. _____

9. _____

10. _____

11. _____

12. _____

13. _____

14. _____

15. _____

16. _____

17. _____

18. _____

19. _____

20. _____

21. _____

22. _____

23. _____

24. _____

25. _____

26. _____

27. _____

28. _____

29. _____

30. _____

31. _____

32. _____

33. _____

34. _____

35. _____

36. _____

37. _____

38. _____

39. _____

40. _____

41. _____

42. _____

43. _____

44. _____

21. $x^5 - \dfrac{3}{x^2}$

22. $4x^2 - 9$

Arrange in descending-exponent form if necessary, and give the degree of each polynomial.

23. $4x^5 - 3x^2$

24. $5x^2 + 3x^3 + 4$

25. $7x^7 - 5x^9 + 4x^3$

26. $2 + x$

27. $4x$

28. $x^{17} - 3x^4$

29. $5x^2 - 3x^5 + x^6 - 7$

30. 5

Find the values of each of the following polynomials for the given values of the variable.

31. $5x + 2$, $x = 1$ and $x = -1$

32. $3x - 3$, $x = 2$ and $x = -2$

33. $x^2 - x$, $x = 2$ and $x = -2$

34. $2x^2 + 5$, $x = 3$ and $x = -3$

35. $3x^2 + 4x - 2$, $x = 4$ and $x = -4$

36. $2x^2 - 5x + 1$, $x = 2$ and $x = -2$

37. $-x^2 - 2x + 3$, $x = 1$ and $x = -3$

38. $-x^2 - 5x - 6$, $x = -3$ and $x = -2$

Indicate whether each of the following statements is always, sometimes, or never true.

39. A monomial is a polynomial.

40. A binomial is a trinomial.

41. The degree of a trinomial is 3.

42. A trinomial has three terms.

43. A polynomial has four or more terms.

44. A binomial must have two coefficients.

45. If x equals 0, the value of a polynomial in x equals 0.

46. The coefficient of the leading term in a polynomial is the largest coefficient of the polynomial.

Think About These

Capital italic letters such as P or Q are often used to name polynomials. For example, we might write $P(x) = 3x^3 - 5x^2 + 2$ where $P(x)$ is read "P of x." The notation permits a convenient shorthand. We write $P(2)$, read "P of 2," to indicate the value of the polynomial when $x = 2$. Here

$$P(2) = 3(2)^3 - 5(2)^2 + 2$$
$$= 3 \cdot 8 - 5 \cdot 4 + 2$$
$$= 6$$

Use the information above in the following problems.

If $P(x) = x^3 - 2x^2 + 5$ and $Q(x) = 2x^2 + 3$, find:

47. $P(1)$

48. $P(-1)$

49. $Q(2)$

50. $Q(-2)$

51. $P(3)$

52. $Q(-3)$

53. $P(0)$

54. $Q(0)$

55. $P(2) + Q(-1)$

56. $P(-2) + Q(3)$

57. $P(3) - Q(-3) \div Q(0)$

58. $Q(-2) \div Q(2) \cdot P(0)$

59. $|Q(4)| - |P(4)|$

60. $\dfrac{P(-1) + Q(0)}{P(0)}$

ANSWERS

45. _____

46. _____

47. _____

48. _____

49. _____

50. _____

51. _____

52. _____

53. _____

54. _____

55. _____

56. _____

57. _____

58. _____

59. _____

60. _____

Skillscan (Section 1.4)

Combine like terms where possible.

a. $8m + 7m$　　　　　　　　　**b.** $9x - 5x$

c. $9m^2 - 8m$　　　　　　　　**d.** $8x^2 - 7x^2$

e. $5c^3 + 15c^3$　　　　　　　**f.** $9s^3 + 8s^3$

g. $8c^2 - 6c + 2c^2$　　　　　**h.** $8r^3 - 7r^2 + 5r^3$

ANSWERS

1. Polynomial　**3.** Polynomial　**5.** Polynomial　**7.** Not a polynomial　**9.** $2x^2, -3x$; 2, -3　**11.** $4x^3, -3x, 2$; 4, -3, 2　**13.** Binomial　**15.** Trinomial　**17.** Not classified　**19.** Monomial　**21.** Not a polynomial　**23.** $4x^5 - 3x^2$; 5　**25.** $-5x^9 + 7x^7 + 4x^3$; 9　**27.** $4x$; 1　**29.** $x^6 - 3x^5 + 5x^2 - 7$; 6　**31.** 7, -3　**33.** 2, 6　**35.** 62, 30　**37.** 0, 0　**39.** Always　**41.** Sometimes　**43.** Sometimes　**45.** Sometimes　**47.** 4　**49.** 11　**51.** 14　**53.** 5　**55.** 10　**57.** 7　**59.** -2　**a.** $15m$　**b.** $4x$　**c.** $9m^2 - 8m$　**d.** x^2　**e.** $20c^3$　**f.** $17s^3$　**g.** $10c^2 - 6c$　**h.** $13r^3 - 7r^2$

4.2

Adding and Subtracting Polynomials

OBJECTIVES
1. To add polynomials
2. To subtract polynomials

Addition is always a matter of combining like quantities (two apples plus three apples, four books plus five books, and so on). If you keep that basic idea in mind, adding polynomials will be easy. It is just a matter of combining like terms. Suppose that you want to add

$$5x^2 + 3x + 4 \qquad \text{and} \qquad 4x^2 + 5x - 6$$

Parentheses are sometimes used in adding, so for the sum of these polynomials, we can write

The plus sign between the parentheses indicates the addition.

$$(5x^2 + 3x + 4) + (4x^2 + 5x - 6)$$

Now what about the parentheses? You can use the following rule.

REMOVING SIGNS OF GROUPING CASE 1

If a plus sign (+) or nothing at all appears in front of parentheses, just remove the parentheses. No other changes are necessary.

Now let's return to the addition.

Just remove the parentheses. No other changes are necessary.

$$(5x^2 + 3x + 4) + (4x^2 + 5x - 6)$$
$$= 5x^2 + 3x + 4 + 4x^2 + 5x - 6$$

Like terms   Like terms

Like terms

Note the use of the associative and commutative properties in reordering and regrouping.

Collect like terms. (*Remember,* like terms have the same variables raised to the same power.)

$$= (5x^2 + 4x^2) + (3x + 5x) + (4 - 6)$$

Combine like terms for the result:

Here we use the distributive property. For example,
$$5x^2 + 4x^2 = 9x^2$$

$$= 9x^2 + 8x - 2$$

As should be clear, much of this work can be done mentally. You can then write the sum directly by locating like terms and combining. The following examples illustrate.

Example 1

Add $3x - 5$ and $2x + 3$.

Write the sum.

$(3x - 5) + (2x + 3)$
$= 3x - 5 + 2x + 3 = 5x - 2$

Like terms

Like terms

CHECK YOURSELF 1

Add $6x^2 + 2x$ and $4x^2 - 7x$.

The same technique is used to find the sum of two trinomials.

Example 2

Add $4a^2 - 7a + 5$ and $3a^2 + 3a - 4$.

Write the sum.

$(4a^2 - 7a + 5) + (3a^2 + 3a - 4)$
$= 4a^2 - 7a + 5 + 3a^2 + 3a - 4 = 7a^2 - 4a + 1$

Like terms

Like terms

Like terms

CHECK YOURSELF 2

Add $5y^2 - 3y + 7$ and $3y^2 - 5y - 7$.

Remember that it is only the like terms that are combined in the sum.

Example 3

Add $2x^2 + 7x$ and $4x - 6$.

Write the sum.

$(2x^2 + 7x) + (4x - 6)$
$= 2x^2 + 7x + 4x - 6$

These are the only like terms;
$2x^2$ and -6 cannot be combined.

$= 2x^2 + 11x - 6$

CHECK YOURSELF 3

Add $5m^2 + 8$ and $8m^2 - 3m$.

As we mentioned in Section 4.1, writing polynomials in descending-exponent form usually makes the work easier. Look at the following example.

Example 4

Add $3x - 2x^2 + 7$ and $5 + 4x^2 - 3x$.

Write the polynomials in descending-exponent form, then add.

$(-2x^2 + 3x + 7) + (4x^2 - 3x + 5)$
$= 2x^2 + 12$

CHECK YOURSELF 4

Add $8 - 5x^2 + 4x$ and $7x - 8 + 8x^2$.

Subtracting polynomials requires another rule for removing signs of grouping.

REMOVING SIGNS OF GROUPING CASE 2

If a minus sign (−) appears in front of a set of parentheses, the parentheses can be removed by changing the sign of each term inside the parentheses.

The use of this rule is illustrated in our next example.

Example 5

In each of the following, remove the parentheses.

(a) $-(2x + 3y) = -2x - 3y$ Change each sign to remove the parentheses.

(b) $m - (5n - 3p) = m \underbrace{- 5n + 3p}$
  Sign changes.

(c) $2x - (-3y + z) = 2x \underbrace{+ 3y - z}$
  Sign changes.

Note: This uses the distributive property, since

$-(2x + 3y) = (-1)(2x + 3y)$
$= -2x - 3y$

Remove the parentheses.

1. $-(3m + 5n)$ **2.** $-(5w - 7z)$
3. $3r - (2s - 5t)$ **4.** $5a - (-3b - 2c)$

    Subtracting polynomials is now a matter of using the previous rule to remove the parentheses and then combining the like terms. Consider the following example.

Example 6

(*a*) Subtract $5x - 3$ from $8x + 2$.

    Write

Note: The expression after "from" is written first in the problem.

$$(8x + 2) - (5x - 3)$$
$$= 8x + 2 \underbrace{- 5x + 3}_{\text{Sign changes.}}$$
$$= 3x + 5$$

(*b*) Subtract $4x^2 - 8x + 3$ from $8x^2 + 5x - 3$.

    Write

$$(8x^2 + 5x - 3) - (4x^2 - 8x + 3)$$
$$= 8x^2 + 5x - 3 \underbrace{- 4x^2 + 8x - 3}_{\text{Sign changes.}}$$
$$= 4x^2 + 13x - 6$$

1. Subtract $7x + 3$ from $10x - 7$.
2. Subtract $5x^2 - 3x + 2$ from $8x^2 - 3x - 6$.

    Again, writing all polynomials in descending-exponent form will make locating and combining like terms much easier. Look at the following example.

Example 7

(a) Subtract $4x^2 - 3x^3 + 5x$ from $8x^3 - 7x + 2x^2$.

Write

$$(8x^3 + 2x^2 - 7x) - (-3x^3 + 4x^2 + 5x)$$
$$= 8x^3 + 2x^2 \underbrace{- 7x + 3x^3 - 4x^2 - 5x}_{\text{Sign changes.}}$$

$$= 11x^3 - 2x^2 - 12x$$

(b) Subtract $8x - 5$ from $-5x + 3x^2$.

Write

$$(3x^2 - 5x) - (8x - 5)$$
$$= 3x^2 \underbrace{- 5x - 8x}_{} + 5$$

Only the like terms can be combined.

$$= 3x^2 - 13x + 5$$

CHECK YOURSELF 7

1. Subtract $7x - 3x^2 + 5$ from $5 - 3x + 4x^2$.
2. Subtract $3a - 2$ from $5a + 4a^2$.

If you think back to addition and subtraction in arithmetic, you'll remember that the work was arranged vertically. That is, the numbers being added or subtracted were placed under one another so that each column represented the same place value. This meant that in adding or subtracting columns you were always dealing with "like quantities."

It is also possible to use a vertical method for adding or subtracting polynomials. First rewrite the polynomials in descending-exponent form, then arrange them one under another, so that each column contains like terms. Then add or subtract in each column.

Example 8

Add $2x^2 - 5x$, $3x^2 + 2$, and $6x - 3$.

Like terms

$$
\begin{array}{r}
2x^2 - 5x \\
3x^2 \qquad + 2 \\
6x - 3 \\
\hline
5x^2 + x - 1
\end{array}
$$

CHECK YOURSELF 8

Add $3x^2 + 5$, $x^2 - 4x$, and $6x + 7$.

The following example illustrates subtraction by the vertical method.

Example 9

(a) Subtract $5x - 3$ from $8x - 7$.

Write

$$
\begin{array}{r}
8x - 7 \\
(-)\ \underline{5x - 3} \\
3x - 4
\end{array}
\qquad
\begin{array}{l}
\text{To subtract, change} \\
\text{each sign of } 5x - 3 \text{ to get} \\
-5x + 3, \text{ then add.}
\end{array}
$$

$$
=\ \begin{array}{r}
8x - 7 \\
\underline{-5x + 3} \\
3x - 4
\end{array}
$$

(b) Subtract $5x^2 - 3x + 4$ from $8x^2 + 5x - 3$.

Write

$$
\begin{array}{r}
8x^2 + 5x - 3 \\
(-)\ \underline{5x^2 - 3x + 4} \\
3x^2 + 8x - 7
\end{array}
\qquad
\begin{array}{l}
\text{To subtract, change each} \\
\text{sign of } 5x^2 - 3x + 4 \text{ to get} \\
-5x^2 + 3x - 4. \text{ Then add.}
\end{array}
$$

$$
=\ \begin{array}{r}
8x^2 + 5x - 3 \\
\underline{-5x^2 + 3x - 4} \\
3x^2 + 8x - 7
\end{array}
$$

Subtracting in this form takes some practice. Take time to study the method carefully. You'll be using it in long division in Section 4.4.

CHECK YOURSELF 9

Subtract, using the vertical method.

1. $4x^2 - 3x$ from $8x^2 + 2x$ **2.** $8x^2 + 4x - 3$ from $9x^2 - 5x + 7$

CHECK YOURSELF ANSWERS

1. $10x^2 - 5x$.   **2.** $8y^2 - 8y$.   **3.** $13m^2 - 3m + 8$.   **4.** $3x^2 + 11x$.
5. (1) $-3m - 5n$; (2) $-5w + 7z$; (3) $3r - 2s + 5t$; (4) $5a + 3b + 2c$.
6. (1) $3x - 10$; (2) $3x^2 - 8$.   **7.** (1) $7x^2 - 10x$; (2) $4a^2 + 2a + 2$.
8. $4x^2 + 2x + 12$.   **9.** (1) $4x^2 + 5x$; (2) $x^2 - 9x + 10$.

Name _____

Date _____

Build Your Skills

Add.

1. $5a - 3$ and $4a + 7$

2. $7x + 8$ and $5x - 9$

3. $9b^2 - 12b$ and $4b^2 - 6b$

4. $3m^2 + 2m$ and $5m^2 - 7m$

5. $3x^2 - 2x$ and $-5x^2 + 2x$

6. $3p^2 + 5p$ and $-7p^2 - 5p$

7. $2x^2 + 5x - 3$ and $3x^2 - 7x + 4$

8. $4d^2 - 8d + 7$ and $5d^2 - 6d - 9$

9. $2b^2 + 8$ and $5b + 8$

10. $4x - 3$ and $3x^2 - 9x$

11. $8y^3 - 5y^2$ and $5y^2 - 2y$

12. $9x^4 - 2x^2$ and $2x^2 + 3$

13. $a^2 - 3a^3$ and $2a^3 + 3a^2$

14. $7m^3 - 3m$ and $-5m - 2m^3$

15. $x^2 - 5 + 3x$ and $8 - 4x - 3x^2$

16. $b^3 - 5b + b^2$ and $4b^2 + 2b - 3b^3$

Remove the parentheses in each of the following expressions, and simplify where possible.

17. $-(2a + 3b)$

18. $-(7x - 4y)$

19. $5a - (2b - 3c)$

20. $7x - (4y + 3z)$

ANSWERS

1. _____
2. _____
3. _____
4. _____
5. _____
6. _____
7. _____
8. _____
9. _____
10. _____
11. _____
12. _____
13. _____
14. _____
15. _____
16. _____
17. _____
18. _____
19. _____
20. _____

21. _____

22. _____

23. _____

24. _____

25. _____

26. _____

27. _____

28. _____

29. _____

30. _____

31. _____

32. _____

33. _____

34. _____

35. _____

36. _____

37. _____

38. _____

39. _____

40. _____

21. $9r - (3r + 5s)$

22. $10m - (3m - 2n)$

23. $5p - (-3p + 2q)$

24. $8d - (-7c - 2d)$

Subtract.

25. $x + 4$ from $2x - 3$

26. $x - 2$ from $3x + 5$

27. $2m^2 - m$ from $3m^2 - 4m$

28. $5a^2 + 2a$ from $7a^2 - 3a$

29. $4y^2 + 3y$ from $2y^2 + 3y$

30. $7n^2 - 3n$ from $5n^2 - 3n$

31. $x^2 - 4x - 3$ from $3x^2 - 5x - 2$

32. $3x^2 - 2x + 4$ from $5x^2 - 8x - 3$

33. $3a + 7$ from $8a^2 - 9a$

34. $3x^3 + x^2$ from $4x^3 - 5x$

35. $4b^2 - 3b$ from $5b - 2b^2$

36. $7y - 3y^2$ from $3y^2 - 2y$

37. $x^2 - 5 - 8x$ from $3x^2 - 8x + 7$

38. $4x - 2x^2 + 4x^3$ from $4x^3 + x - 3x^2$

Perform the indicated operations.

39. Subtract $3b + 2$ from the sum of $4b - 2$ and $5b + 3$.

40. Subtract $5m - 7$ from the sum of $2m - 8$ and $9m - 2$.

41. Subtract $3x^2 + 2x - 1$ from the sum of $x^2 + 5x - 2$ and $2x^2 + 7x - 8$.

42. Subtract $4x^2 - 5x - 3$ from the sum of $x^2 - 3x - 7$ and $2x^2 - 2x + 9$.

43. Subtract $2x^2 - 3x$ from the sum of $4x^2 - 5$ and $2x - 7$.

44. Subtract $5a^2 - 3a$ from the sum of $3a - 3$ and $5a^2 + 5$.

45. Subtract the sum of $3y^2 - 3y$ and $5y^2 + 3y$ from $2y^2 - 8y$.

46. Subtract the sum of $7r^3 - 4r^2$ and $-3r^3 + 4r^2$ from $2r^3 + 3r^2$.

Add, using the vertical method.

47. $w^2 + 5$, $2w - 3$, and $5w^2 - 4w$

48. $x^2 - 3x - 2$, $5x - 2$, and $4x^2 + 7$

49. $2x^2 + x - 3$, $3x^2 - x - 5$, and $4x^2 - x + 8$

ANSWERS

41. _____

42. _____

43. _____

44. _____

45. _____

46. _____

47. _____

48. _____

49. _____

50. _____

51. _____

52. _____

53. _____

54. _____

55. _____

56. _____

57. _____

58. _____

59. _____

60. _____

234

50. $5x^2 + 2x - 4$, $x^2 - 2x - 3$, and $2x^2 - 4x - 3$

Subtract, using the vertical method.

51. $3a^2 - 2a$ from $5a^2 + 3a$ $5A^2 +3A$
$-3A^2 +2A$
$\overline{}$
$2A^2 +5A$

52. $6r^3 + 4r^2$ from $4r^3 - 2r^2$

53. $5x^2 - 6x + 7$ from $8x^2 - 5x + 7$

54. $8x^2 - 4x + 2$ from $9x^2 - 8x + 6$

55. $5x^2 - 3x$ from $8x^2 - 9$

56. $7x^2 + 6x$ from $9x^2 - 3$

Think About These

Perform the indicated operations.

57. $[(9x^2 - 3x + 5) - (3x^2 + 2x - 1)] - (x^2 - 2x - 3)$

58. $[(5x^2 + 2x - 3) - (-2x^2 + x - 2)] - (2x^2 + 3x - 5)$

Find values for a, b, c, and d so that the following equations are true.

59. $3ax^4 - 5x^3 + x^2 - cx + 2 = 9x^4 - bx^3 + x^2 - 2d$

60. $(4ax^3 - 3bx^2 - 10) - 3(x^3 + 4x^2 - cx - d) = x^2 - 6x + 8$

Northern spotted owls require large amounts of old-growth forest in which to breed and forage for food. Assume the typical breeding pair of spotted owls requires x^2 acres of old growth for food gathering and an additional $3x$ acres of old growth for breeding and raising young in Oregon. Additional research shows that Washington owls typically require $2x^2$ acres for food gathering and $5x$ acres for breeding and raising young.

61. If there are 350 pairs of owls in Oregon and 275 pairs of owls in Washington, we can represent the total acreage of old growth necessary to support these owls by the following expression:

$$350(x^2 + 3x) + 275(2x^2 + 5x)$$

Simplify the above expression to find the total acreage needed to support the given numbers of owls.

62. If there are 275 pairs of owls in Oregon and 400 pairs of owls in Washington, we can represent the total acreage of old growth necessary to support these owls by the following expression:

$$275(x^2 + 3x) + 400(2x^2 + 5x)$$

Simplify the above expression to find the total acreage needed to support the given numbers of owls.

63. Hemlock, Douglas fir, and alder are three species of trees associated with old-growth forests.

Let the expression $nD + mH - pA$ represent the profit from harvesting a given section of forest where D, H, and A are the profit or loss from cutting Douglas fir, hemlock, and alder, respectively, and n, m, and p are the number of trees of each type. If one section of forest yields $12D + 15H - 6A$ and another section of forest yields $6D + 9H - 10A$, the total profit for the two sections is found by adding the two expressions. What is the total profit for harvesting these sections of forest?

64. The volume of wood from one clear-cut can be approximated by the expression $100R^2 + 16R + 15$ and in a second clear-cut by $80R^2 - 17R - 21$. Find the difference in volumes by subtracting the second expression from the first.

Skillscan (Section 1.5)

Multiply.

a. $x^5 \cdot x^7$ **b.** $y^8 \cdot y^{12}$

c. $2a^3 \cdot a^4$ **d.** $3m^5 \cdot m^2$

e. $4r^5 \cdot 3r$ **f.** $6w^2 \cdot 5w^3$

g. $(-2x^2)(8x^7)$ **h.** $(-10a)(-3a^5)$

ANSWERS

1. $9a + 4$ **3.** $13b^2 - 18b$ **5.** $-2x^2$ **7.** $5x^2 - 2x + 1$ **9.** $2b^2 + 5b + 16$
11. $8y^3 - 2y$ **13.** $-a^3 + 4a^2$ **15.** $-2x^2 - x + 3$ **17.** $-2a - 3b$ **19.** $5a - 2b + 3c$
21. $6r - 5s$ **23.** $8p - 2q$ **25.** $x - 7$ **27.** $m^2 - 3m$ **29.** $-2y^2$ **31.** $2x^2 - x + 1$
33. $8a^2 - 12a - 7$ **35.** $-6b^2 + 8b$ **37.** $2x^2 + 12$ **39.** $6b - 1$ **41.** $10x - 9$
43. $2x^2 + 5x - 12$ **45.** $-6y^2 - 8y$ **47.** $6w^2 - 2w + 2$ **49.** $9x^2 - x$ **51.** $2a^2 + 5a$
53. $3x^2 + x$ **55.** $3x^2 + 3x - 9$ **57.** $5x^2 - 3x + 9$ **59.** $a = 3, b = 5, c = 0, d = -1$
61. $900x^2 + 2425x$ **63.** $18D + 24H - 16A$ **a.** x^{12} **b.** y^{20} **c.** $2a^7$ **d.** $3m^7$
e. $12r^6$ **f.** $30w^5$ **g.** $-16x^9$ **h.** $30a^6$

4.3 Multiplying Polynomials

OBJECTIVES

1. To find the product of a monomial and any polynomial
2. To find the product of two polynomials

You have already had some experience in multiplying polynomials. In Section 1.5 we stated the first property of exponents and used that property to find the product of two monomials. Let's review briefly.

TO FIND THE PRODUCT OF MONOMIALS

STEP 1   Multiply the coefficients.

STEP 2   Use the first property of exponents to combine the variables.

The first property of exponents:

$$x^m \cdot x^n = x^{m+n}$$

Example 1

Multiply $3x^2y$ and $2x^3y^5$.

Write

Once again we have used the commutative and associative properties to rewrite the problems.

$$(3x^2y)(2x^3y^5)$$
$$= (3 \cdot 2)(x^2 \cdot x^3)(y \cdot y^5)$$

Multiply   Add the exponents.
the coefficients.

$$= 6x^5y^6$$

CHECK YOURSELF 1

Multiply.

1. $(5a^2b)(3a^2b^4)$ **2.** $(-3xy)(4x^3y^5)$

Our next task is to find the product of a monomial and a polynomial. Here we use the distributive property, which we introduced in Section 1.3. That property leads us to the following rule for multiplication.

You might want to review Section 1.3 now before going on.

TO MULTIPLY A POLYNOMIAL BY A MONOMIAL

Use the distributive property to multiply each term of the polynomial by the monomial.

Distributive property:
$a(b + c) = ab + ac$

Example 2

(a) Multiply $2x + 3$ by x.

Write

Note: With practice you will do this step mentally.

$x(2x + 3)$

$= \boxed{x} \cdot 2x + \boxed{x} \cdot 3$

$= 2x^2 + 3x$

Multiply x by $2x$ and then by 3, the terms of the polynomial. That is, "distribute" the multiplication over the sum.

(b) Multiply $2a^3 + 4a$ by $3a^2$.

Write

$3a^2(2a^3 + 4a)$

$= \boxed{3a^2} \cdot 2a^3 + \boxed{3a^2} \cdot 4a$

$= 6a^5 + 12a^3$

CHECK YOURSELF 2

Multiply.

1. $2y(y^2 + 3y)$ **2.** $3w^2(2w^3 + 5w)$

The distributive property lets us extend the patterns of Example 2 to a polynomial with *any* number of terms.

Example 3

Multiply the following.

(a) $3x(4x^3 + 5x^2 + 2)$

$= \boxed{3x} \cdot 4x^3 + \boxed{3x} \cdot 5x^2 + \boxed{3x} \cdot 2$

$= 12x^4 + 15x^3 + 6x$

(b) $5y^2(2y^3 - 4)$

$= \boxed{5y^2} \cdot 2y^3 - \boxed{(5y^2)}(4)$

$= 10y^5 - 20y^2$

(c) $-5c(4c^2 - 8c)$

$= \boxed{(-5c)}(4c^2) - \boxed{(-5c)}(8c)$

$= -20c^3 + 40c^2$

(d) $3c^2d^2(7cd^2 - 5c^2d^3)$

$= \boxed{3c^2d^2} \cdot 7cd^2 - \boxed{(3c^2d^2)}(5c^2d^3)$

$= 21c^3d^4 - 15c^4d^5$

Again we have shown all the steps of the process. With practice you can write the product directly, and you should try to do so.

CHECK YOURSELF 3

Multiply.

1. $3(5a^2 + 2a + 7)$ **2.** $4x^2(8x^3 - 6)$

3. $-5m(8m^2 - 5m)$ **4.** $9a^2b(3a^3b - 6a^2b^4)$

We also use the distributive property in writing the product of two binomials. Consider the following example.

Example 4

(*a*) Multiply $x + 2$ by $x + 3$.

We can think of $x + 2$ as a single quantity and apply the distributive property.

Note that this ensures that each term, *x* and 2, of the first binomial is multiplied by each term, *x* and 3, of the second binomial.

$$(x + 2)(x + 3)$$

    Multiply $x + 2$ by x and then by 3.

$$= (x + 2)x + (x + 2)3$$
$$= x \cdot x + 2 \cdot x + x \cdot 3 + 2 \cdot 3$$
$$= x^2 + 2x + 3x + 6$$
$$= x^2 + 5x + 6$$

(*b*) Multiply $a - 3$ by $a - 4$. (Think of $a - 3$ as a single quantity and distribute.)

$$(a - 3)(a - 4)$$
$$= (a - 3)a - (a - 3)(4)$$
$$= a \cdot a - 3 \cdot a - [(a \cdot 4) - (3 \cdot 4)]$$
$$= a^2 - 3a - (4a - 12)$$
$$= a^2 - 3a - 4a + 12$$
$$= a^2 - 7a + 12$$

Note that the parentheses are needed here because a *minus sign* precedes the binomial.

CHECK YOURSELF 4

Multiply.

1. $(x + 4)(x + 5)$ **2.** $(y + 5)(y - 6)$

Fortunately, there is a pattern to this kind of multiplication that allows you to write the product of the two binomials directly without going through all these steps. We call it the *FOIL method* of multiplying. The reason for this name will be clear as we look at the process in more detail.

To multiply $(x + 2)(x + 3)$:

1. $(x + 2)(x + 3)$

Remember this by F!
$x \cdot x$

Find the product of the *first* terms of the factors.

2. $(x + 2)(x + 3)$

Remember this by O!
$x \cdot 3$

Find the product of the *outer* terms.

3. $(x + 2)(x + 3)$

Remember this by I!
$2 \cdot x$

Find the product of the *inner* terms.

4. $(x + 2)(x + 3)$

Remember this by L!
$2 \cdot 3$

Find the product of the *last* terms.

Combining the four steps, we have

Of course these are the same four terms found in Example 4a.

$(x + 2)(x + 3)$
$= x^2 + 3x + 2x + 6$
$= x^2 + 5x + 6$

It's called FOIL to give you an easy way of remembering the steps: *F*irst, *O*uter, *I*nner, and *L*ast.

With practice, the FOIL method will let you write the products quickly and easily. Consider the following examples that illustrate this approach.

Example 5

Find the following products, using the FOIL method.

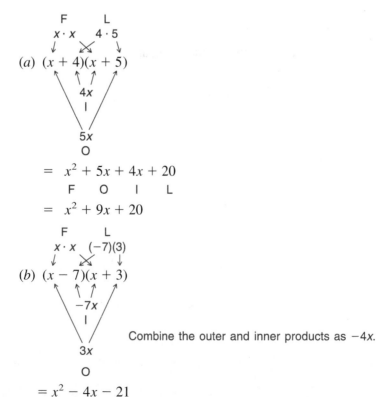

When possible, you should combine the outer and inner products mentally and write just the final product.

$= x^2 + 5x + 4x + 20$
   F   O   I   L
$= x^2 + 9x + 20$

Combine the outer and inner products as $-4x$.

$= x^2 - 4x - 21$

Multiply.

1. $(x + 6)(x + 7)$ **2.** $(x + 3)(x - 5)$ **3.** $(x - 2)(x - 8)$

Using the FOIL method, you can also find the product of binomials with coefficients other than 1 or with more than one variable.

Example 6

Find the following products, using the FOIL method.

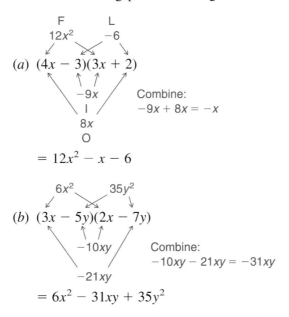

(a) $(4x - 3)(3x + 2)$

$$= 12x^2 - x - 6$$

(b) $(3x - 5y)(2x - 7y)$

$$= 6x^2 - 31xy + 35y^2$$

The following rule summarizes our work in multiplying binomials.

TO MULTIPLY TWO BINOMIALS

STEP 1   Find the first term of the product of the binomials by multiplying the first terms of the binomials (F).

STEP 2   Find the middle term of the product as the sum of the outer and inner products (O + I).

STEP 3   Find the last term of the product by multiplying the last terms of the binomials (L).

Multiply.

1. $(5x + 2)(3x - 7)$ **2.** $(4a - 3b)(5a - 4b)$
3. $(3m + 5n)(2m + 3n)$

You should now be able to multiply any two binomials, using the FOIL method. But what if one of the factors has three or more terms? The vertical format, shown in the following example, works for factors with any number of terms.

Example 7

Multiply $x^2 - 5x + 8$ by $x + 3$.

Step 1

$$
\begin{array}{r}
x^2 - 5x + 8 \\
x + 3 \\
\hline
3x^2 - 15x + 24
\end{array}
$$

Multiply each term of $x^2 - 5x + 8$ by 3.

Step 2

$$
\begin{array}{r}
x^2 - 5x + 8 \\
x + 3 \\
\hline
3x^2 - 15x + 24 \\
x^3 - 5x^2 + 8x
\end{array}
$$

Now multiply each term by x.

Note that this line is shifted over so that like terms are in the same columns.

Step 3

$$
\begin{array}{r}
x^2 - 5x + 8 \\
x + 3 \\
\hline
3x^2 - 15x + 24 \\
x^3 - 5x^2 + 8x \\
\hline
x^3 - 2x^2 - 7x + 24
\end{array}
$$

Now add to combine like terms to write the product.

Note: Using this vertical method ensures that each term of one factor multiplies each term of the other. That's why it works!

CHECK YOURSELF 7

Multiply $2x^2 - 5x + 3$ by $3x + 4$.

CHECK YOURSELF ANSWERS

1. (1) $15a^4b^5$; (2) $-12x^4y^6$.
2. (1) $2y^3 + 6y^2$; (2) $6w^5 + 15w^3$.
3. (1) $15a^2 + 6a + 21$; (2) $32x^5 - 24x^2$; (3) $-40m^3 + 25m^2$; (4) $27a^5b^2 - 54a^4b^5$.
4. (1) $x^2 + 9x + 20$; (2) $y^2 - y - 30$.
5. (1) $x^2 + 13x + 42$; (2) $x^2 - 2x - 15$; (3) $x^2 - 10x + 16$.
6. (1) $15x^2 - 29x - 14$; (2) $20a^2 - 31ab + 12b^2$; (3) $6m^2 + 19mn + 15n^2$.
7. $6x^3 - 7x^2 - 11x + 12$.

4.3 Exercises

Name _____

Date _____

Build Your Skills

Multiply.

1. $(5x^2)(3x^3)$

2. $(7a^5)(4a^6)$

3. $(-7b^4)(4b^6)$

4. $(7y^5)(-8y^5)$

5. $(-5p^5)(-8p^8)$

6. $(-9m^9)(6m^6)$

7. $(4m^5)(-3m)$

8. $(-5r^7)(-3r)$

9. $(4x^3y^2)(8x^2y)$

10. $(-3r^4s^2)(-7r^2s^5)$

11. $(-3m^5n^2)(2m^4n)$

12. $(7a^3b^5)(-6a^4b)$

Use the distributive property to multiply.

13. $4(3x + 7)$

14. $5(8b - 6)$

15. $5a(2a + 3)$

16. $4x(3x - 5)$

17. $3s^2(4s^2 - 7s)$

18. $9a^2(3a^3 + 5a)$

19. $2x(4x^2 - 2x + 1)$

20. $5m(4m^3 - 3m^2 + 2)$

ANSWERS

1. _____
2. _____
3. _____
4. _____
5. _____
6. _____
7. _____
8. _____
9. _____
10. _____
11. _____
12. _____
13. _____
14. _____
15. _____
16. _____
17. _____
18. _____
19. _____
20. _____

243

21. _____

22. _____

23. _____

24. _____

25. _____

26. _____

27. _____

28. _____

29. _____

30. _____

31. _____

32. _____

33. _____

34. _____

35. _____

36. _____

37. _____

38. _____

39. _____

40. _____

41. _____

42. _____

244

21. $3xy(2x^2y + xy^2 + 5xy)$

22. $5ab^2(ab - 3a + 5b)$

23. $6m^2n(3m^2n - 2mn + mn^2)$

24. $8pq^2(2pq - 3p + 5q)$

Multiply.

25. $(x + 3)(x + 2)$

26. $(a - 3)(a - 7)$

27. $(m - 5)(m - 9)$

28. $(b + 7)(b + 5)$

29. $(p - 8)(p + 7)$

30. $(x - 10)(x + 9)$

31. $(w + 10)(w + 20)$

32. $(s - 12)(s - 8)$

33. $(3x - 5)(x - 8)$

34. $(w + 5)(4w - 7)$

35. $(2x - 3)(3x + 4)$

36. $(5a + 1)(3a + 7)$

37. $(3a - b)(4a - 9b)$

38. $(7s - 3t)(3s + 8t)$

39. $(4p - 3q)(5p + 7q)$

40. $(4x - 5y)(3x - 2y)$

41. $(5x + 4y)(6x + 5y)$

42. $(6x - 7y)(6x + 3y)$

43. $(x + 5)^2$

44. $(y + 8)^2$

45. $(y - 9)^2$

46. $(2a + 3)^2$

47. $(6m + n)^2$

48. $(7b - c)^2$

49. $(a - 5)(a + 5)$

50. $(x - 7)(x + 7)$

51. $(p - 3q)(p + 3q)$

52. $(5x + y)(5x - y)$

53. $(7m + 4n)(7m - 4n)$

54. $(8a - 3b)(8a + 3b)$

Multiply, using the vertical method.

55. $(x + 2)(x^2 + 2x - 3)$

56. $(a - 3)(a^2 + 4a + 1)$

57. $(2m - 5)(2m^2 + 3m - 2)$

58. $(5p + 3)(p^2 + 4p + 1)$

59. $(3x + 4y)(x^2 + xy + 5y^2)$

60. $(7a - 2b)(2a^2 - ab + 4b^2)$

61. $(a^2 + 3ab - b^2)(a^2 - 5ab + b^2)$

62. $(m^2 - 5mn + 3n^2)(m^2 + 4mn - 2n^2)$

63. $(x - 2y)(x^2 + 2xy + 4y^2)$

64. $(m + 3n)(m^2 - 3mn + 9n^2)$

43. _____

44. _____

45. _____

46. _____

47. _____

48. _____

49. _____

50. _____

51. _____

52. _____

53. _____

54. _____

55. _____

56. _____

57. _____

58. _____

59. _____

60. _____

61. _____

62. _____

63. _____

64. _____

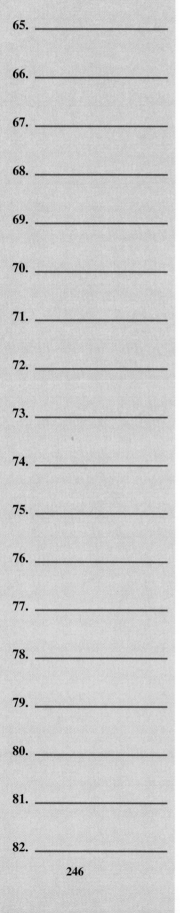

65. $(3a + b)(9a^2 - 3ab + b^2)$

66. $(4r - s)(16r^2 + 4rs + s^2)$

Multiply.

67. $2x(3x - 2)(4x + 1)$

68. $3x(2x + 1)(2x - 1)$

69. $5a(4a - 3)(4a + 3)$

70. $6m(3m - 2)(3m - 7)$

71. $3s(5s - 2)(4s - 1)$

72. $7w(2w - 3)(2w + 3)$

73. $(x - 2)(x + 1)(x - 3)$

74. $(y + 3)(y - 2)(y - 4)$

75. $(a - 1)^3$

76. $(x + 1)^3$

Think About These

Multiply the following.

77. $\left(\dfrac{x}{2} + \dfrac{2}{3}\right)\left(\dfrac{2x}{3} - \dfrac{2}{5}\right)$

78. $\left(\dfrac{x}{3} + \dfrac{3}{4}\right)\left(\dfrac{3x}{4} - \dfrac{3}{5}\right)$

79. $[x + (y - 2)][x - (y - 2)]$

80. $[x + (3 - y)][x - (3 - y)]$

Label the following as true or false.

81. $(x + y)^2 = x^2 + y^2$

82. $(x - y)^2 = x^2 - y^2$

83. $(x + y)^2 = x^2 + 2xy + y^2$ **84.** $(x - y)^2 = x^2 - 2xy + y^2$

85. An area of forest restricted from logging is currently $2K \times 3K$ square kilometers (km^2). An environmental group wants to expand the protected area by 3 km on each side. The area of this new protected area would be $(2K + 3) \times (3K + 3)$. Simplify the expression for the new area that will be protected.

86. An area of forest restricted from logging is currently $4K \times 2K$ km^2. An environmental group wants to expand the protected area by 2 km on each side. The area of this new protected area would be $(4K + 2) \times (2K + 2)$. Simplify the expression for the new area that will be protected.

87. An environmental group has proposed a new wilderness in the shape of a circle. The local logging community has said it will accept the proposal if the wilderness is reduced in radius by 150 meters (m). Using the formula for area of a circle, we get the expression πR^2 for the area of the environmental group proposal and $\pi(R - 150)^2$ for the community proposal.
  a. To express the difference in area of the two proposals, we get $\pi R^2 - \pi(R - 150)^2$. Write this expression in simplest form.
  b. What is the difference in area for the two proposals if the environmental group wants a wilderness with a radius of 1200 m?

88. An environmental group has proposed a new wilderness in the shape of a circle. The local logging community has said it will accept the proposal if the wilderness is reduced in radius by 175 m. Using the formula for area of a circle, we get the expression πR^2 for the area of the environmental group proposal and $\pi(R - 175)^2$ for the community proposal.
  a. To express the difference in area of the two proposals, we get $\pi R^2 - \pi(R - 175)^2$. Write this expression in simplest form.
  b. What is the difference in areas for the two proposals if the environmental group wants a wilderness with a radius of 1500 m?

ANSWERS

83. _____

84. _____

85. _____

86. _____

87. **a.** _____

b. _____

88. **a.** _____

b. _____

a. _____

b. _____

c. _____

d. _____

e. _____

f. _____

g. _____

h. _____

Skillscan (Section 1.5)

a. $(3a)(3a)$

b. $(3a)^2$

c. $(5x)(5x)$

d. $(5x)^2$

e. $(-2w)(-2w)$

f. $(-2w)^2$

g. $(-4r)(-4r)$

h. $(-4r)^2$

ANSWERS

1. $15x^5$ **3.** $-28b^{10}$ **5.** $40p^{13}$ **7.** $-12m^6$ **9.** $32x^5y^3$ **11.** $-6m^9n^3$
13. $12x + 28$ **15.** $10a^2 + 15a$ **17.** $12s^4 - 21s^3$ **19.** $8x^3 - 4x^2 + 2x$
21. $6x^3y^2 + 3x^2y^3 + 15x^2y^2$ **23.** $18m^4n^2 - 12m^3n^2 + 6m^3n^3$ **25.** $x^2 + 5x + 6$
27. $m^2 - 14m + 45$ **29.** $p^2 - p - 56$ **31.** $w^2 + 30w + 200$ **33.** $3x^2 - 29x + 40$
35. $6x^2 - x - 12$ **37.** $12a^2 - 31ab + 9b^2$ **39.** $20p^2 + 13pq - 21q^2$
41. $30x^2 + 49xy + 20y^2$ **43.** $x^2 + 10x + 25$ **45.** $y^2 - 18y + 81$
47. $36m^2 + 12mn + n^2$ **49.** $a^2 - 25$ **51.** $p^2 - 9q^2$ **53.** $49m^2 - 16n^2$
55. $x^3 + 4x^2 + x - 6$ **57.** $4m^3 - 4m^2 - 19m + 10$ **59.** $3x^3 + 7x^2y + 19xy^2 + 20y^3$
61. $a^4 - 2a^3b - 15a^2b^2 + 8ab^3 - b^4$ **63.** $x^3 - 8y^3$ **65.** $27a^3 + b^3$
67. $24x^3 - 10x^2 - 4x$ **69.** $80a^3 - 45a$ **71.** $60s^3 - 39s^2 + 6s$

73. $x^3 - 4x^2 + x + 6$ **75.** $a^3 - 3a^2 + 3a - 1$ **77.** $\dfrac{x^2}{3} + \dfrac{11x}{45} - \dfrac{4}{15}$

79. $x^2 - y^2 + 4y - 4$ **81.** False **83.** True **85.** $6K^2 + 15K + 9$
87. *a.* $300\pi R - 22{,}500\pi$, *b.* $1.06 \times 10^6 \, \text{m}^2$ **a.** $9a^2$ **b.** $9a^2$ **c.** $25x^2$ **d.** $25x^2$
e. $4w^2$ **f.** $4w^2$ **g.** $16r^2$ **h.** $16r^2$

Special Products

OBJECTIVES

1. To square a binomial
2. To find the product of two binomials that differ only in sign

Certain products occur frequently enough in algebra that it is worth learning special formulas for dealing with them. First, let's look at the product of two equal binomial factors. This is called the *square of a binomial.*

$$(x + y)^2 = (x + y)(x + y)$$
$$= x^2 + 2xy + y^2$$

$$(x - y)^2 = (x - y)(x - y)$$
$$= x^2 - 2xy + y^2$$

The patterns above lead us to the following rule.

TO SQUARE A BINOMIAL

STEP 1   Find the first term of the square by squaring the first term of the binomial.

STEP 2   Find the middle term of the square as twice the product of the two terms of the binomial.

STEP 3   Find the last term of the square by squaring the last term of the binomial.

Example 1

(a) $(x + 3)^2 = x^2 + 2 \cdot x \cdot 3 + 3^2$

Square of first term — Twice the product of the two terms — Square of the last term

$$= x^2 + 6x + 9$$

(b) $(3a + 4b)^2 = (3a)^2 + 2(3a)(4b) + (4b)^2$
$$= 9a^2 + 24ab + 16b^2$$

(c) $(y - 5)^2 = y^2 + 2 \cdot y \cdot (-5) + (-5)^2$
$$= y^2 - 10y + 25$$

(d) $(5c - 3d)^2 = (5c)^2 + 2(5c)(-3d) + (-3d)^2$
$$= 25c^2 - 30cd + 9d^2$$

Again we have shown all the steps. With practice you can write just the square.

Multiply.

1. $(2x + 1)^2$ **2.** $(4x - 3y)^2$

⟨CAUTION⟩

Be careful! A very common mistake in squaring binomials is to forget the middle term.

Example 2

Find $(y + 4)^2$.

$(y + 4)^2$ is *not* equal to $y^2 + 4^2$ or $y^2 + 16$

The correct square is

You should note that
$(2 + 3)^2 \neq 2^2 + 3^2$ because
$5^2 \neq 4 + 9$

$$(y + 4)^2 = y^2 + 8y + 16$$

The middle term is twice the product of y and 4.

Multiply.

1. $(x + 5)^2$ **2.** $(3a + 2)^2$ **3.** $(y - 7)^2$ **4.** $(5x - 2y)^2$

A second special product will be very important in the next chapter, which deals with factoring. Suppose the form of a product is

$$(x + y)(x - y)$$

The two terms differ
only in sign.

Let's see what happens when we multiply.

$$(x + y)(x - y)$$
$$= x^2 - \underbrace{xy + xy} - y^2$$
$$= 0$$
$$= x^2 - y^2$$

Since the middle term becomes 0, we have the following rule.

The product of two binomials that differ only in the sign between the terms is the square of the first term minus the square of the second term.

Let's look at the application of this rule in our final example.

Example 3

Multiply each pair of factors.

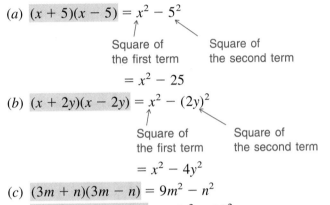

(a) $(x + 5)(x - 5) = x^2 - 5^2$

Square of the first term

Square of the second term

$= x^2 - 25$

(b) $(x + 2y)(x - 2y) = x^2 - (2y)^2$

Square of the first term

Square of the second term

$= x^2 - 4y^2$

Note:

$(2y)^2 = (2y)(2y)$
$\qquad = 4y^2$

(c) $(3m + n)(3m - n) = 9m^2 - n^2$

(d) $(4a - 3b)(4a + 3b) = 16a^2 - 9b^2$

CHECK YOURSELF 3

Find the products.

1. $(a - 6)(a + 6)$ **2.** $(x - 3y)(x + 3y)$
3. $(5n + 2p)(5n - 2p)$ **4.** $(7b - 3c)(7b + 3c)$

When finding the product of three or more factors, it is useful to look for this pattern first.

Example 4

(a) $x(x - 3)(x + 3)$ These binomials differ only in the sign.
$\quad = x(x^2 - 9)$
$\quad = x^3 - 9x$

(b) $(x + 1)(x - 5)(x + 5)$ These binomials differ only in the sign.
$\quad = (x + 1)(x^2 - 25)$ With two binomials, use the FOIL method.
$\quad = x^3 + x^2 - 25x - 25$

(c) $(2x - 1)(x + 3)(2x + 1)$ These two binomials differ only in the sign of the second term. We can use the commutative property to rearrange the terms.

$\quad = (x + 3)(2x - 1)(2x + 1)$
$\quad = (x + 3)(4x^2 - 1)$
$\quad = 4x^3 + 12x^2 - x - 3$

CHECK YOURSELF 4

Multiply.

1. $3x(x - 5)(x + 5)$

2. $(x - 4)(2x + 3)(2x - 3)$

3. $(x - 7)(3x - 1)(x + 7)$

CHECK YOURSELF ANSWERS

1. (1) $4x^2 + 4x + 1$; (2) $16x^2 - 24xy + 9y^2$.

2. (1) $x^2 + 10x + 25$; (2) $9a^2 + 12a + 4$; (3) $y^2 - 14y + 49$;
(4) $25x^2 - 20xy + 4y^2$.

3. (1) $a^2 - 36$; (2) $x^2 - 9y^2$; (3) $25n^2 - 4p^2$; (4) $49b^2 - 9c^2$.

4. (1) $3x^3 - 75x$; (2) $4x^3 - 16x^2 - 9x + 36$; (3) $3x^3 - x^2 - 147x + 49$.

Name

Date

Build Your Skills

Find each of the following squares.

1. $(x + 4)^2$

2. $(y + 8)^2$

3. $(w - 7)^2$

4. $(a - 9)^2$

5. $(z + 12)^2$

6. $(p - 20)^2$

7. $(2a - 1)^2$

8. $(3x - 2)^2$

9. $(6m + 1)^2$

10. $(7b - 2)^2$

11. $(3x - y)^2$

12. $(5m + n)^2$

13. $(2r + 5s)^2$

14. $(3a - 4b)^2$

15. $(6a - 7b)^2$

16. $(5p + 4q)^2$

17. $\left(x + \dfrac{1}{2}\right)^2$

18. $\left(w - \dfrac{1}{4}\right)^2$

Find each of the following products.

19. $(x - 5)(x + 5)$

20. $(y + 7)(y - 7)$

1. _____

2. _____

3. _____

4. _____

5. _____

6. _____

7. _____

8. _____

9. _____

10. _____

11. _____

12. _____

13. _____

14. _____

15. _____

16. _____

17. _____

18. _____

19. _____

20. _____

21. _____

22. _____

23. _____

24. _____

25. _____

26. _____

27. _____

28. _____

29. _____

30. _____

31. _____

32. _____

33. _____

34. _____

35. _____

36. _____

37. _____

38. _____

39. _____

40. _____

254

21. $(m + 12)(m - 12)$

22. $(w - 10)(w + 10)$

23. $\left(x - \dfrac{1}{2}\right)\left(x + \dfrac{1}{2}\right)$

24. $\left(x + \dfrac{2}{3}\right)\left(x - \dfrac{2}{3}\right)$

25. $(p - 0.4)(p + 0.4)$

26. $(m - 0.6)(m + 0.6)$

27. $(a - 3b)(a + 3b)$

28. $(p + 4q)(p - 4q)$

29. $(4r - s)(4r + s)$

30. $(7x - y)(7x + y)$

31. $(7w + 4z)(7w - 4z)$

32. $(8c + 3d)(8c - 3d)$

33. $(5x - 9y)(5x + 9y)$

34. $(6s - 5t)(6s + 5t)$

35. $x(x - 2)(x + 2)$

36. $a(a + 5)(a - 5)$

37. $2s(s - 3r)(s + 3r)$

38. $5w(2w - z)(2w + z)$

39. $4r(r + 5)^2$

40. $6x(x - 3)^2$

Think About These

For each of the following problems, let x represent the number, then write an expression for product.

41. The product of 6 more than a number and 6 less than that number

42. The square of 5 more than a number

43. The square of 4 less than a number

44. The product of 5 less than a number and 5 more than that number

Note that $(28)(32) = (30 - 2)(30 + 2) = 900 - 4 = 896$. Use this pattern to find each of the following products.

45. $(49)(51)$ **46.** $(27)(33)$

47. $(34)(26)$ **48.** $(98)(102)$

49. $(55)(65)$ **50.** $(64)(56)$

51. Recent legislation has mandated that all clear-cuts in national forest be enclosed by a buffer zone of uncut trees. If a clear-cut is in the shape of a square, the expression for the area of the buffer zone can be written as $(X + 2B)^2 - X^2$, where X is the length of one side of the clear-cut and B is the width of the buffer zone.
    Simplify the expression for the area of the clearcut.

52. Use the simplified expression in problem 51 to find the area of a buffer strip 10 meters (m) wide around a clearcut 115 m on a side.

ANSWERS

41. _____

42. _____

43. _____

44. _____

45. _____

46. _____

47. _____

48. _____

49. _____

50. _____

51. _____

52. _____

Skillscan (Section 1.5)

Divide.

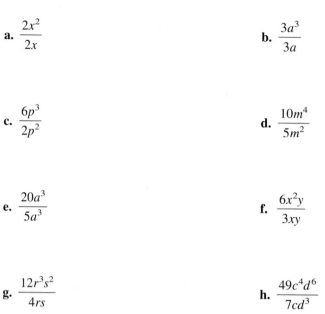

a. $\dfrac{2x^2}{2x}$

b. $\dfrac{3a^3}{3a}$

c. $\dfrac{6p^3}{2p^2}$

d. $\dfrac{10m^4}{5m^2}$

e. $\dfrac{20a^3}{5a^3}$

f. $\dfrac{6x^2y}{3xy}$

g. $\dfrac{12r^3s^2}{4rs}$

h. $\dfrac{49c^4d^6}{7cd^3}$

ANSWERS

1. $x^2 + 8x + 16$ **3.** $w^2 - 14w + 49$ **5.** $z^2 + 24z + 144$ **7.** $4a^2 - 4a + 1$
9. $36m^2 + 12m + 1$ **11.** $9x^2 - 6xy + y^2$ **13.** $4r^2 + 20rs + 25s^2$

15. $36a^2 - 84ab + 49b^2$ **17.** $x^2 + x + \dfrac{1}{4}$ **19.** $x^2 - 25$ **21.** $m^2 - 144$ **23.** $x^2 - \dfrac{1}{4}$

25. $p^2 - 0.16$ **27.** $a^2 - 9b^2$ **29.** $16r^2 - s^2$ **31.** $49w^2 - 16z^2$ **33.** $25x^2 - 81y^2$
35. $x^3 - 4x$ **37.** $2s^3 - 18r^2s$ **39.** $4r^3 + 40r^2 + 100r$ **41.** $x^2 - 36$
43. $x^2 - 8x + 16$ **45.** 2499 **47.** 884 **49.** 3575 **51.** $4XB + 4B^2$ **a.** x **b.** a^2
c. $3p$ **d.** $2m^2$ **e.** 4 **f.** $2x$ **g.** $3r^2s$ **h.** $7c^3d^3$

Dividing Polynomials

OBJECTIVES

1. To divide a polynomial by a monomial
2. To divide a polynomial by another polynomial

In Section 1.5, we introduced the second property of exponents, which was used to divide one monomial by another monomial. Let's review that process.

The second property says:
If x is not zero and $m > n$,

$$\frac{x^m}{x^n} = x^{m-n}$$

TO DIVIDE A MONOMIAL BY A MONOMIAL

STEP 1 Divide the coefficients.

STEP 2 Use the second property of exponents to combine the variables.

Example 1

$$\text{Divide: } \frac{8}{2} = 4$$

(a) $\dfrac{8x^4}{2x^2} = 4x^{4-2}$

Subtract the exponents.

$$= 4x^2$$

(b) $\dfrac{45a^5b^3}{9a^2b} = 5a^3b^2$

CHECK YOURSELF 1

Divide.

1. $\dfrac{16a^5}{8a^3}$
2. $\dfrac{28m^4n^3}{7m^3n}$

Now let's look at how this can be extended to divide any polynomial by a monomial. Suppose that you want to divide $12a^3 + 8a^2$ by $4a$. Write

$$\frac{12a^3 + 8a^2}{4a}$$

This is the same as

Technically this step depends on the distributive law and the definition of division.

$$\frac{12a^3}{4a} + \frac{8a^2}{4a}$$

Now do each division.

$$= 3a^2 + 2a$$

The work above leads us to the following rule.

> **TO DIVIDE A POLYNOMIAL BY A MONOMIAL**
>
> Divide each term of the polynomial by the monomial. Then combine the results.

Example 2

Divide each term by 2.

$(a)\quad \dfrac{4a^2 + 8}{2} = \dfrac{4a^2}{2} + \dfrac{8}{2}$

$\qquad\qquad = 2a^2 + 4$

Divide each term by 6y.

$(b)\quad \dfrac{24y^3 - 18y^2}{6y} = \dfrac{24y^3}{6y} - \dfrac{18y^2}{6y}$

$\qquad\qquad = 4y^2 - 3y$

Remember the rules for signs in division.

$(c)\quad \dfrac{15x^2 + 10x}{-5x} = \dfrac{15x^2}{-5x} + \dfrac{10x}{-5x}$

$\qquad\qquad = -3x - 2$

With practice you can write just the quotient.

$(d)\quad \dfrac{14x^4 + 28x^3 - 21x^2}{7x^2} = \dfrac{14x^4}{7x^2} + \dfrac{28x^3}{7x^2} - \dfrac{21x^2}{7x^2}$

$\qquad\qquad = 2x^2 + 4x - 3$

$(e)\quad \dfrac{9a^3b^4 - 6a^2b^3 + 12ab^4}{3ab} = \dfrac{9a^3b^4}{3ab} - \dfrac{6a^2b^3}{3ab} + \dfrac{12ab^4}{3ab}$

$\qquad\qquad = 3a^2b^3 - 2ab^2 + 4b^3$

CHECK YOURSELF 2

Divide.

1. $\dfrac{20y^3 - 15y^2}{5y}$
　　　　　　　　　　　　　2. $\dfrac{8a^3 - 12a^2 + 4a}{-4a}$

3. $\dfrac{16m^4n^3 - 12m^3n^2 + 8mn}{4mn}$

We are now ready to look at dividing one polynomial by another polynomial (with more than one term). The process is very much like long division in arithmetic. The following example will illustrate.

Example 3

Divide $x^2 + 7x + 10$ by $x + 2$.

Divide x^2 by x to get x.

The first term in the dividend, x^2, is divided by the first term in the divisor, x.

Step 1
$$x + 2 \overline{)x^2 + 7x + 10}$$
with x on top.

Step 2
$$\begin{array}{r} x \\ x + 2 \overline{)x^2 + 7x + 10} \\ x^2 + 2x \end{array}$$

Multiply the divisor, $x + 2$, by x.

Remember: To subtract $x^2 + 2x$, mentally change each sign to $-x^2 - 2x$, and add. Take your time and be careful here. It's where most errors are made.

Step 3
$$\begin{array}{r} x \\ x + 2 \overline{)x^2 + 7x + 10} \\ x^2 + 2x \\ \hline 5x + 10 \end{array}$$

Subtract and bring down 10.

Step 4
$$\begin{array}{r} x + 5 \\ x + 2 \overline{)x^2 + 7x + 10} \\ x^2 + 2x \\ \hline 5x + 10 \end{array}$$

Divide $5x$ by x to get 5.

Note that we repeat the process until the degree of the remainder is less than that of the divisor or until there is no remainder.

Step 5
$$\begin{array}{r} x + 5 \\ x + 2 \overline{)x^2 + 7x + 10} \\ x^2 + 2x \\ \hline 5x + 10 \\ 5x + 10 \\ \hline 0 \end{array}$$

Multiply $x + 2$ by 5 and then subtract.

The quotient is $x + 5$.

CHECK YOURSELF 3

Divide $x^2 + 9x + 20$ by $x + 4$.

In the above example, we have shown all the steps separately to help you see the process. In practice, the work can be shortened.

Example 4

Divide $x^2 + x - 12$ by $x - 3$.

You might want to write out a problem like $408 \div 17$, to compare the steps.

$$
\begin{array}{r}
x + 4 \\
x - 3 \overline{) x^2 + x - 12} \\
\underline{x^2 - 3x} \\
4x - 12 \\
\underline{4x - 12} \\
0
\end{array}
$$

THE STEPS

1. Divide x^2 by x to get x, the first term of the quotient.
2. Multiply $x - 3$ by x.
3. Subtract and bring down -12. Remember to mentally change the signs to $-x^2 + 3x$ and add.
4. Divide $4x$ by x to get 4, the second term of the quotient.
5. Multiply $x - 3$ by 4 and subtract.

The quotient is $x + 4$.

CHECK YOURSELF 4

Divide.

$(x^2 + 2x - 24) \div (x - 4)$

You may have a remainder in algebraic long division just as in arithmetic. Consider the following example.

Example 5

Divide $4x^2 - 8x + 11$ by $2x - 3$.

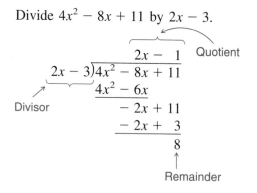

This result can be written as

$$
\frac{4x^2 - 8x + 11}{2x - 3}
$$

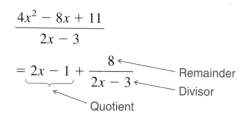

CHECK YOURSELF 5

Divide.

$(6x^2 - 7x + 15) \div (3x - 5)$

The division process shown in our previous examples can be extended to dividends of a higher degree. The steps involved in the division process are exactly the same. The following example illustrates.

Example 6

Divide $6x^3 + x^2 - 4x - 5$ by $3x - 1$.

$$
\begin{array}{r}
2x^2 + x - 1 \\
3x - 1 \overline{)6x^3 + x^2 - 4x - 5} \\
\underline{6x^3 - 2x^2} \\
3x^2 - 4x \\
\underline{3x^2 - x} \\
-3x - 5 \\
\underline{-3x + 1} \\
-6
\end{array}
$$

This result can be written as

$$
\frac{6x^3 + x^2 - 4x - 5}{3x - 1} = 2x^2 + x - 1 + \frac{-6}{3x - 1}
$$

CHECK YOURSELF 6

Divide $4x^3 - 2x^2 + 2x + 15$ by $2x + 3$.

Suppose that the dividend is "missing" a term in some power of the variable. You can use 0 as the coefficient for the missing term. Consider the following example.

Example 7

Divide $x^3 - 2x^2 + 5$ by $x + 3$.

$$
\begin{array}{r}
x^2 - 5x + 15 \\
x + 3 \overline{)x^3 - 2x^2 + 0x + 5} \\
\underline{x^3 + 3x^2} \\
-5x^2 + 0x \\
\underline{-5x^2 - 15x} \\
15x + 5 \\
\underline{15x + 45} \\
-40
\end{array}
$$

Write $0x$ for the "missing" term in x.

This result can be written as

$$\frac{x^3 - 2x^2 + 5}{x + 3} = x^2 - 5x + 15 + \frac{-40}{x + 3}$$

CHECK YOURSELF 7

Divide.

$(4x^3 + x + 10) \div (2x - 1)$

You should always arrange the terms of the divisor and dividend in descending-exponent form before starting the long division process. This is illustrated in our final example.

Example 8

Divide $5x^2 - x + x^3 - 5$ by $-1 + x^2$.
   Write the divisor as $x^2 - 1$ and the dividend as $x^3 + 5x^2 - x - 5$.

Write $x^3 - x$, the product of x and $x^2 - 1$, so that like terms fall in the same columns.

$$
\begin{array}{r}
x + 5 \\
x^2 - 1 \overline{)x^3 + 5x^2 - x - 5} \\
\underline{x^3 \qquad - x} \\
5x^2 \qquad - 5 \\
\underline{5x^2 \qquad - 5} \\
0
\end{array}
$$

CHECK YOURSELF 8

Divide:

$(5x^2 + 10 + 2x^3 + 4x) \div (2 + x^2)$

CHECK YOURSELF ANSWERS

1. (1) $2a^2$; (2) $4mn^2$.
2. (1) $4y^2 - 3y$; (2) $-2a^2 + 3a - 1$; (3) $4m^3n^2 - 3m^2n + 2$.
3. $x + 5$.   4. $x + 6$.
5. $2x + 1 + \dfrac{20}{3x - 5}$.   6. $2x^2 - 4x + 7 + \dfrac{-6}{2x + 3}$.
7. $2x^2 + x + 1 + \dfrac{11}{2x - 1}$.   8. $2x + 5$.

Name

Date

Build Your Skills

Divide.

1. $\dfrac{18x^6}{9x^2}$

2. $\dfrac{20a^7}{5a^5}$

3. $\dfrac{35m^3n^2}{7mn^2}$

4. $\dfrac{42x^5y^2}{6x^3y}$

5. $\dfrac{3a + 6}{3}$

6. $\dfrac{4x - 8}{4}$

7. $\dfrac{9b^2 - 12}{3}$

8. $\dfrac{10m^2 + 5m}{5}$

9. $\dfrac{16a^3 - 24a^2}{4a}$

10. $\dfrac{9x^3 + 12x^2}{3x}$

11. $\dfrac{12m^2 + 6m}{-3m}$

12. $\dfrac{20b^3 - 25b^2}{-5b}$

13. $\dfrac{18a^4 + 12a^3 - 6a^2}{6a}$

14. $\dfrac{21x^5 - 28x^4 + 14x^3}{7x}$

15. $\dfrac{20x^4y^2 - 15x^2y^3 + 10x^3y}{5x^2y}$

16. $\dfrac{16m^3n^3 + 24m^2n^2 - 40mn^3}{8mn^2}$

ANSWERS

1. _____

2. _____

3. _____

4. _____

5. _____

6. _____

7. _____

8. _____

9. _____

10. _____

11. _____

12. _____

13. _____

14. _____

15. _____

16. _____

Perform the indicated divisions.

17. $\dfrac{x^2 + 5x + 6}{x + 2}$

18. $\dfrac{x^2 + 8x + 15}{x + 3}$

19. $\dfrac{x^2 - x - 20}{x + 4}$

20. $\dfrac{x^2 - 2x - 35}{x + 5}$

21. $\dfrac{2x^2 + 5x - 3}{2x - 1}$

22. $\dfrac{3x^2 + 20x - 32}{3x - 4}$

23. $\dfrac{2x^2 - 3x - 5}{x - 3}$

24. $\dfrac{3x^2 + 17x - 12}{x + 6}$

25. $\dfrac{4x^2 - 18x - 15}{x - 5}$

26. $\dfrac{3x^2 - 18x - 32}{x - 8}$

27. $\dfrac{6x^2 - x - 10}{3x - 5}$

28. $\dfrac{4x^2 + 6x - 25}{2x + 7}$

29. $\dfrac{x^3 + x^2 - 4x - 4}{x + 2}$

30. $\dfrac{x^3 - 2x^2 + 4x - 21}{x - 3}$

31. $\dfrac{4x^3 + 7x^2 + 10x + 5}{4x - 1}$

32. $\dfrac{2x^3 - 3x^2 + 4x + 4}{2x + 1}$

33. $\dfrac{x^3 - x^2 + 5}{x - 2}$

34. $\dfrac{x^3 + 4x - 3}{x + 3}$

35. $\dfrac{25x^3 + x}{5x - 2}$

36. $\dfrac{8x^3 - 6x^2 + 2x}{4x + 1}$

37. $\dfrac{2x^2 - 8 - 3x + x^3}{x - 2}$

38. $\dfrac{x^2 - 18x + 2x^3 + 32}{x + 4}$

39. $\dfrac{x^4 - 1}{x - 1}$

40. $\dfrac{x^4 + x^2 - 16}{x + 2}$

41. $\dfrac{x^3 - 3x^2 - x + 3}{x^2 - 1}$

42. $\dfrac{x^3 + 2x^2 + 3x + 6}{x^2 + 3}$

43. $\dfrac{x^4 + 2x^2 - 2}{x^2 + 3}$

44. $\dfrac{x^4 + x^2 - 5}{x^2 - 2}$

Think About These

45. $\dfrac{y^3 + 1}{y + 1}$

46. $\dfrac{y^3 - 8}{y - 2}$

47. $\dfrac{x^4 - 1}{x^2 - 1}$

48. $\dfrac{x^6 - 1}{x^3 - 1}$

33. _____

34. _____

35. _____

36. _____

37. _____

38. _____

39. _____

40. _____

41. _____

42. _____

43. _____

44. _____

45. _____

46. _____

47. _____

48. _____

49. Find the value of c so that $\dfrac{y^2 - y + c}{y + 1} = y - 2$

50. Find the value of c so that $\dfrac{x^3 + x^2 + x + c}{x^2 + 1} = x + 1$

Skillscan (Section 1.2)

Use the distributive property to simplify each expression.

a. $3(4a - 3)$ **b.** $5(3m + 7)$

c. $7(6x + 8)$ **d.** $6(4b - 7)$

e. $-3(2w + 5)$ **f.** $-7(3x - 8)$

g. $-5(-3y - 2)$ **h.** $-6(-4a + 2)$

ANSWERS

1. $2x^4$ **3.** $5m^2$ **5.** $a + 2$ **7.** $3b^2 - 4$ **9.** $4a^2 - 6a$ **11.** $-4m - 2$

13. $3a^3 + 2a^2 - a$ **15.** $4x^2y - 3y^2 + 2x$ **17.** $x + 3$ **19.** $x - 5$ **21.** $x + 3$

23. $2x + 3 + \dfrac{4}{x - 3}$ **25.** $4x + 2 + \dfrac{-5}{x - 5}$ **27.** $2x + 3 + \dfrac{5}{3x - 5}$ **29.** $x^2 - x - 2$

31. $x^2 + 2x + 3 + \dfrac{8}{4x - 1}$ **33.** $x^2 + x + 2 + \dfrac{9}{x - 2}$ **35.** $5x^2 + 2x + 1 + \dfrac{2}{5x - 2}$

37. $x^2 + 4x + 5 + \dfrac{2}{x - 2}$ **39.** $x^3 + x^2 + x + 1$ **41.** $x - 3$ **43.** $x^2 - 1 + \dfrac{1}{x^2 + 3}$

45. $y^2 - y + 1$ **47.** $x^2 + 1$ **49.** $c = -2$ **a.** $12a - 9$ **b.** $15m + 35$ **c.** $42x + 56$

d. $24b - 42$ **e.** $-6w - 15$ **f.** $-21x + 56$ **g.** $15y + 10$ **h.** $24a - 12$

More on Linear Equations

OBJECTIVE
To solve linear equations when signs of grouping are involved

In Section 4.2, we removed signs of grouping from polynomials. The ideas of that section are useful in solving equations where parentheses are involved.

Let's start by reviewing an example similar to those we considered in Chapter 3. We will then solve other equations involving grouping symbols.

Example 1

Solve $5(2x - 1) = 25$ for x.

First, multiply on the left to remove the parentheses, then solve as before.

$10x - 5 = 25$

$10x - 5 + 5 = 25 + 5$ Add 5.

$10x = 30$

$\dfrac{10x}{10} = \dfrac{30}{10}$ Divide by 10.

$x = 3$

Again, returning to the *original equation* will catch any possible errors in the removal of the parentheses.

Left side	Right side
$5(2 \cdot 3 - 1) \overset{?}{=} 25$	
$5(6 - 1) \overset{?}{=} 25$	
$5 \cdot 5 \overset{?}{=} 25$	
$25 = 25$ (true)	

The answer is 3. To check, return to the *original equation*. Substitute 3 for x on the left and right. Then evaluate separately.

CHECK YOURSELF 1

Solve for x.

$8(3x + 5) = 16$

Be especially careful if a minus sign precedes a grouping symbol. The sign of each term inside the grouping symbol must be changed.

Example 2

Solve $8 - (3x + 1) = -8$.

First, remove the parentheses. The original equation then becomes

Remember,

$-(3x + 1) = -3x - 1$
$\qquad\quad\uparrow\quad\uparrow$
Change *both* signs.

$8 - 3x - 1 = -8$

$-3x + 7 = -8$ Combine like terms.

$-3x + 7 - 7 = -8 - 7$ Subtract 7.

$-3x = -15$

$x = 5$ Divide by -3.

The solution is 5. You should verify this result.

Solve for x.

$7 - (4x - 3) = 22$

Our next example illustrates the solution process when more than one grouping symbol is involved in an equation.

Example 3

Solve $2(3x - 1) - 3(x + 5) = 4$.

$2(3x - 1) - 3(x + 5) = 4$

$6x - 2 - 3x - 15 = 4$ Use the distributive property to remove the parentheses.

$3x - 17 = 4$ Combine like terms on the left.

$3x = 21$ Add 17.

$x = 7$ Divide by 3.

The solution is 7.

To check, return to the original equation to replace x with 7.

Note how the rules for the order of operations are applied.

$2(3 \cdot 7 - 1) - 3(7 + 5) \stackrel{?}{=} 4$

$2(21 - 1) - 3(7 + 5) \stackrel{?}{=} 4$

$2 \cdot 20 - 3 \cdot 12 \stackrel{?}{=} 4$

$40 - 36 \stackrel{?}{=} 4$

$4 = 4$ (a true statement)

The solution is verified.

Solve for x.

$5(2x + 4) = 7 - 3(1 - 2x)$

1. -1.   **2.** -3.   **3.** -4.

Build Your Skills

Solve each of the following equations for x, and check your results.

1. $2(x - 3) = 8$ **2.** $3(x + 4) = -6$

3. $4(3x + 1) = 28$ **4.** $5(2x - 3) = 65$

5. $7(5x + 8) = -84$ **6.** $6(3x + 2) = -60$

7. $10 - (x - 2) = 15$ **8.** $12 - (x + 3) = 3$

9. $5 - (2x + 1) = 12$ **10.** $9 - (3x - 2) = 2$

11. $7 - (3x - 5) = 13$ **12.** $5 - (4x + 3) = 4$

13. $5x = 3(x - 6)$ **14.** $5x = 2(x + 12)$

15. $7(2x + 1) = 12x$ **16.** $5(3x - 2) = 12x$

17. $4(7 - x) = 3x$ **18.** $5(8 - x) = 3x$

1. _____

2. _____

3. _____

4. _____

5. _____

6. _____

7. _____

8. _____

9. _____

10. _____

11. _____

12. _____

13. _____

14. _____

15. _____

16. _____

17. _____

18. _____

19. _____

20. _____

21. _____

22. _____

23. _____

24. _____

25. _____

26. _____

27. _____

28. _____

29. _____

30. _____

31. _____

32. _____

33. _____

19. $2(2x - 1) = 3(x + 1)$

20. $3(3x - 1) = 4(2x + 1)$

21. $5(4x + 2) = 6(3x + 4)$

22. $4(6x - 1) = 7(3x + 2)$

23. $9(8x - 1) = 5(4x + 6)$

24. $7(3x + 11) = 9(3 - 6x)$

25. $-4(2x - 1) + 3(3x + 1) = 9$

26. $7(3x + 4) = 8(2x + 5) + 13$

27. $5(2x - 1) - 3(x - 4) = 4(x + 4)$

28. $2(x - 3) - 3(x + 5) = 3(x - 2) - 7$

29. $3(3 - 4x) + 30 = 5x - 2(6x - 7)$

30. $3x - 5(3x - 7) = 2(x + 9) + 45$

Translate each of the following statements to equations. Let x represent the number in each case.

31. Twice the sum of a number and 4 is 20.

32. The sum of twice a number and 4 is 20.

33. 3 times the difference of a number and 5 is 21.

34. The difference of 3 times a number and 5 is 21.

35. The sum of twice an integer and 3 times the next consecutive integer is 48.

36. The sum of 4 times an odd integer and twice the next consecutive odd integer is 46.

Think About These

Solve each of the following equations.

37. $-2x + [3x - (-2x + 5)] = -(15 + 2x)$

38. $-3x + [5x - (-x + 4)] = -2(x - 3)$

39. $3x^2 - 2(x^2 + 2) = (x + 2)(x - 2)$

40. $5x^2 - [2(2x^2 + 3)] - 3 = (x + 3)(x - 3)$

ANSWERS

34. _____

35. _____

36. _____

37. _____

38. _____

39. _____

40. _____

a. _____

b. _____

c. _____

d. _____

e. _____

f. _____

g. _____

h. _____

Skillscan (Section 1.1)

Translate each of the following phrases to symbols.

a. 3 more than a **b.** 2 less than x

c. 5 more than twice m **d.** 7 less than 4 times p

e. 3 times r, decreased by 9   **f.** 5 times w, increased by 10

g. 12 less than 4 times q **h.** 9 more than twice c

ANSWERS

1. 7   **3.** 2   **5.** -4 **7.** -3 **9.** -4 **11.** $-\dfrac{1}{3}$ **13.** -9 **15.** $-\dfrac{7}{2}$ **17.** 4

19. 5   **21.** 7   **23.** $\dfrac{3}{4}$ **25.** 2   **27.** 3   **29.** 5   **31.** $2(x + 4) = 20$

33. $3(x - 5) = 21$ **35.** $2x + 3(x + 1) = 48$ **37.** -2 **39.** All numbers   **a.** $a + 3$
b. $x - 2$ **c.** $2m + 5$ **d.** $4p - 7$ **e.** $3r - 9$ **f.** $5w + 10$ **g.** $4q - 12$ **h.** $2c + 9$

More Applications

OBJECTIVES

1. To solve word problems involving numbers
2. To solve word problems involving geometric figures
3. To solve mixture problems
4. To solve motion problems

Many applications lead to equations involving parentheses. That means the methods of Section 4.6 will have to be applied during the solution process. Before we look at examples, let's review the five-step process for solving word problems.

TO SOLVE WORD PROBLEMS

STEP 1 Read the problem carefully. Then reread it to decide what you are asked to find.

STEP 2 Choose a letter to represent the unknown or unknowns. Then represent all other unknowns of the problem with expressions using the same letter.

STEP 3 Translate the problem to the language of algebra to form an equation.

STEP 4 Solve the equation, and answer the question of the original problem.

STEP 5 Verify your solution by returning to the original problem.

These steps are illustrated in the following example.

Example 1

One number is 5 more than a second number. If 3 times the smaller number plus 4 times the larger is 104, find the two numbers.

Step 1 What are you asked to find? You must find the two numbers.

Step 2 Represent the unknowns. Let x be the smaller number. Then

$x + 5$ is the larger.

  "5 more than" x

Step 3 Write an equation.

Note that the parentheses are *essential* in writing the correct equation.

$3x + 4(x + 5) = 104$

3 times   Plus   4 times
the smaller   the larger

Step 4   Solve the equation.

$$3x + 4(x + 5) = 104$$
$$3x + 4x + 20 = 104$$
$$7x + 20 = 104$$
$$7x = 84$$
$$x = 12$$

The smaller number (x) is 12, and the larger number ($x + 5$) is 17.

Step 5   Check the solution: 12 is the smaller number, and 17 is the larger number.

$$3 \cdot 12 + 4 \cdot 17 = 104 \qquad \text{(true)}$$

CHECK YOURSELF 1

One number is 4 more than another. If 6 times the smaller minus 4 times the larger is 4, what are the two numbers?

The solutions for many problems from geometry will also yield equations involving parentheses. Consider the following.

Example 2

The length of a rectangle is 1 cm less than 3 times the width. If the perimeter is 54 cm, find the dimensions of the rectangle.

Step 1   You want to find the dimensions (the width and length).

Step 2   Let x be the width.

Then $3x - 1$ is the length.

3 times
the width   1 less than

Whenever you are working on an application involving geometric figures, you should draw a sketch of the problem including the labels assigned in this step. In this example we have

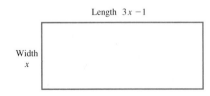

Length $3x - 1$

Width
x

Step 3 To write an equation, we'll use this formula for the perimeter of a rectangle:

$$P = 2W + 2L$$

So

$$2x + 2(3x - 1) = 54$$

Twice the width   Twice the length   Perimeter

Step 4 Solve the equation.

$$2x + 2(3x - 1) = 54$$
$$2x + 6x - 2 = 54$$
$$8x = 56$$
$$x = 7$$

The width x is 7 cm, and the length, $3x - 1$, is 20 cm. We leave step 5, the check, to you.

Be sure to return to the original statement of the problem when checking your result.

CHECK YOURSELF 2

The length of a rectangle is 5 in more than twice the width. If the perimeter of the rectangle is 76 in, what are the dimensions of the rectangle?

Let's look at another group of applications, called *coin problems*. There is one key idea in solving this type of word problem.

Suppose that you have 8 nickels. How much money do you have? To find the value of the coins, you must *multiply the number of coins by the value of one coin* (in this case 5¢). So if you have 8 nickels, you have $8 \cdot 5¢$, or 40¢. Let's see how this is used in the following example.

Example 3

Jacob has 4 more dimes than nickels. If the value of the coins is $3.10, how many nickels and how many dimes does he have?

Step 1 You want to find the number of nickels and dimes.

Step 2 Let x be the number of nickels.

Then $x + 4$ is the number of dimes.

4 more than x

Step 3   Write an equation (in cents).

Note that we have chosen to write our expressions in terms of *cents* to avoid decimals.

The value of the nickels is $5x$.

          ↑

         5¢ times the
number of nickels

The value of the dimes is $10(x + 4)$.

          ↑

         10¢ times the
number of dimes

So

$$5x + 10(x + 4) = 310$$

          ↑

         The total value is
310¢, or $3.10.

Step 4   Solve the equation.

$$5x + 10(x + 4) = 310$$
$$5x + 10x + 40 = 310$$
$$15x + 40 = 310$$
$$15x = 270$$
$$x = 18 \qquad \text{(nickels)}$$
$$\text{and} \qquad x + 4 = 22 \qquad \text{(dimes)}$$

Jacob has 18 nickels and 22 dimes.

Step 5   To check our result,

18 nickels have value	90¢
22 dimes have value	220¢
Total value	310¢ or $3.10

The solution is verified.

CHECK YOURSELF 3

Carlos has 3 more quarters than dimes. If the total value of the coins is $3.90, how many of each type does he have?

Coin problems are actually a part of a larger group called *mixture problems*. Mixture problems give a relationship between groups of different kinds of objects. Look at the following example.

Example 4

Four hundred tickets were sold for a school play. General admission tickets were $4, while student tickets were $3. If the total ticket sales were $1350, how many of each type of ticket were sold?

Step 1   You want to find the number of each type of ticket sold.

Step 2   Let x be the number of general admission tickets.

We subtract x, the number of general admission tickets, from 400, the total number of tickets, to find the number of student tickets.

Then $\underbrace{400 - x}$ student tickets were sold.

400 tickets were
sold in all.

Step 3   The value of each kind of ticket is found in exactly the same way as in the coin problem of Example 3.

General admission tickets: $4x$ $4 for each of the x tickets
Student tickets: $3(400 - x)$ $3 for each of the $400 - x$ tickets

So to form an equation, we have

$$4x + 3(400 - x) = 1350$$

Value of   Value of   Total
general   student   value
admission   tickets
tickets

Step 4   Solve the equation.

$$4x + 3(400 - x) = 1350$$
$$4x + 1200 - 3x = 1350$$
$$x + 1200 = 1350$$
$$x = 150$$

So 150 general admission and 250 student tickets were sold. We leave the check to you.

CHECK YOURSELF 4

Beth bought 25¢ stamps and 15¢ stamps at the post office. If she purchased 60 stamps at a cost of $13, how many of each kind did she buy?

The last group of applications we will look at in this section involves *motion problems*. They involve a distance traveled, a rate or speed, and time. To solve motion problems, we need a relationship among these three quantities.

Suppose you travel at a rate of 50 miles per hour (mi/h) on a highway for 6 hours (h). How far (the distance) will you have gone? To find the distance, you multiply:

$$(50 \text{ mi/h})(6 \text{ h}) = 300 \text{ mi}$$

Speed or   Time   Distance
rate

In general, if r is a rate, t is the time, and d is the distance traveled,

> $$d = r \cdot t$$

This is the key relationship, and it will be used in all motion problems. Let's see how it is applied in the following example.

Be careful to make your units consistent. If a rate is given in miles per hour, *then the time must be given in* hours *and the distance in* miles.

Example 5

On Friday morning Ricardo drove from his house to the beach in 4 h. In coming back on Sunday afternoon, heavy traffic slowed his speed by 10 mi/h, and the trip took 5 h. What was his average speed (rate) in each direction?

Step 1   We want the speed or rate in each direction.

Step 2   Let x be Ricardo's speed to the beach. Then $x - 10$ is his return speed.
It is always a good idea to sketch the given information in a motion problem. Here we would have

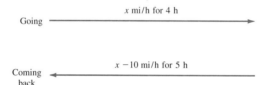

Going   *x mi/h for 4 h* →

Coming back   ← *x − 10 mi/h for 5 h*

Step 3   Since we know that the distance is the same each way, we can write an equation, using the fact that the product of the rate and the time each way must be the same.
So

$$4x = 5(x - 10)$$

Time · rate   Time · rate
(going)   (coming back)

Distance (going)
= distance (coming back)

or

Time · rate (going) = time · rate (coming back)

A chart can also be used to summarize the given information. We begin by filling in the information given in the problem.

	DISTANCE	RATE	TIME
Going		x	4
Coming		$x - 10$	5

Now we will fill in the missing information. Here we use the fact that $d = rt$ to complete it.

	DISTANCE	RATE	TIME
Going	$4x$	x	4
Coming	$5(x - 10)$	$x - 10$	5

From here we set the two distances equal to each other and solve as before.

Step 4 Solve.

$$4x = 5(x - 10)$$
$$4x = 5x - 50$$
$$-x = -50$$
$$x = 50 \text{ mi/h}$$

x was his rate going, $x - 10$ her rate coming back.

So Ricardo's rate going to the beach was 50 mi/h, and his rate coming back was 40 mi/h.

Step 5 To check, you should verify that the product of the time and the rate is the same in each direction.

CHECK YOURSELF 5

A plane made a flight (with the wind) between two towns in 2 h. Returning against the wind, the plane's speed was 60 mi/h slower, and the flight took 3 h. What was the plane's speed in each direction?

The next example illustrates another way of using the distance relationship.

Example 6

Katy leaves Las Vegas for Los Angeles at 10 A.M., driving at 50 mi/h. At 11 A.M. Jensen leaves Los Angeles for Las Vegas, driving at 55 mi/h along the same route. If the cities are 260 mi apart, at what time will they meet?

Step 1 Let's find the time that Katy's travels until they meet.

Step 2 Let x be Katy's time.

Then $x - 1$ is Jensen's time.

Jensen left 1 h later!

Again, you should draw a sketch of the given information.

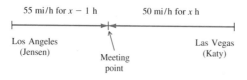

Los Angeles (Jensen) — Meeting point — Las Vegas (Katy)

Step 3 To write an equation, we will again need the relationship $d = r \cdot t$. From this equation, we can write

Katy's distance $= 50x$

Jensen's distance $= 55(x - 1)$

As before, we can use a table to solve.

	DISTANCE	RATE	TIME
Katy	$50x$	50	x
Jensen	$55(x - 1)$	55	$x - 1$

From the original problem, the sum of those distances is 260 mi, so

$$50x + 55(x - 1) = 260$$

Step 4

$$50x + 55(x - 1) = 260$$
$$50x + 55x - 55 = 260$$
$$105x - 55 = 260$$
$$105x = 315$$
$$x = 3 \text{ h}$$

Be sure to answer the question asked in the problem.

Finally, since Katy left at 10 A.M., the two will meet at 1 P.M. We leave the check of this result to you.

CHECK YOURSELF 6

At noon a jogger leaves one point, running at 8 mi/h. One hour later a bicyclist leaves the same point, traveling at 20 mi/h in the opposite direction. At what time will they be 36 mi apart?

CHECK YOURSELF ANSWERS

1. The numbers are 10 and 14.
2. The width is 11; the length is 27.
3. 9 dimes and 12 quarters.
4. 40 at 25¢ and 20 at 15¢.
5. 180 mi/h with the wind and 120 mi/h against the wind.
6. At 2 P.M.

Name

Date

ANSWERS

Build Your Skills

Solve the following word problems. Be sure to show the equation you use for the solution.

1. One number is 8 more than another. If the sum of the smaller number and twice the larger number is 46, find the two numbers.

2. One number is 3 less than another. If 4 times the smaller number minus 3 times the larger number is 4, find the two numbers.

3. One number is 7 less than another. If 4 times the smaller number plus 2 times the larger number is 62, find the two numbers.

4. One number is 10 more than another. If the sum of twice the smaller number and 3 times the larger number is 55, find the two numbers.

5. Find two consecutive integers such that the sum of twice the first integer and 3 times the second integer is 28. (*Hint:* If x represents the first integer, $x + 1$ represents the next consecutive integer.)

6. Find two consecutive odd integers such that 3 times the first integer is 5 more than twice the second. (*Hint:* If x represents the first integer, $x + 2$ represents the next consecutive odd integer.)

7. The length of a rectangle is 1 in more than twice its width. If the perimeter of the rectangle is 74 in, find the dimensions of the rectangle.

8. The length of a rectangle is 5 cm less than 3 times its width. If the perimeter of the rectangle is 46 cm, find the dimensions of the rectangle.

1. _____

2. _____

3. _____

4. _____

5. _____

6. _____

7. _____

8. _____

9. _____

10. _____

11. _____

12. _____

13. _____

14. _____

15. _____

16. _____

17. _____

9. The length of a rectangular garden is 4 m more than 3 times its width. The perimeter of the garden is 56 m. What are the dimensions of the garden?

10. The length of a rectangular playing field is 5 ft less than twice its width. If the perimeter of the field is 230 ft, find the length and width of the field.

11. The base of an isosceles triangle is 3 cm less than the length of the equal sides. If the perimeter of the triangle is 36 cm, find the length of each of the sides.

12. The length of one of the equal legs of an isosceles triangle is 3 in less than twice the length of the base. If the perimeter is 29 in, find the length of each of the sides.

13. Tickets for a play cost $8 for the main floor and $6 in the balcony. If the total receipts from 500 tickets were $3600, how many of each type of ticket were sold?

14. Tickets for a basketball tournament were $6 for students and $9 for nonstudents. Total sales were $10,500, and 250 more student tickets were sold than nonstudent tickets. How many of each type of ticket were sold?

15. Maria bought 80 stamps at the post office in 25¢ and 20¢ denominations. If she paid $18.50 for the stamps, how many of each denomination did she buy?

16. A bank teller has a total of 125 $10 bills and $20 bills to start a day. If the value of the bills was $1650, how many of each denomination did he have?

17. Tickets for a train excursion were $120 for a sleeping room, $80 for a berth, and $50 for a coach seat. The total ticket sales were $8600. If there were 20 more berth tickets sold than sleeping room tickets, and 3 times as many coach tickets as sleeping room tickets, how many of each type of ticket were sold?

18. Admission for a baseball game is $6 for box seats, $5 for the grandstand, and $3 for the bleachers. The total receipts for one evening were $9000. There were 100 more grandstand tickets sold than box seat tickets. Twice as many bleacher tickets were sold as box seat tickets. How many tickets of each type were sold?

19. Patrick drove for 3 h to attend a meeting. On the return trip, his speed was 10 mi/h less, and the trip took 4 h. What was his speed each way?

20. A bicyclist rode into the country for 5 h. In returning, her speed was 5 mi/h faster and the trip took 4 h. What was her speed each way?

21. A car leaves a city and goes north at a rate of 50 mi/h at 2 P.M. One hour later a second car leaves, traveling south at a rate of 40 mi/h. At what time will the two cars be 320 mi apart?

22. A passenger bus leaves a station at 1 P.M., traveling west at an average rate of 44 mi/h. One hour later a second bus leaves the same station, traveling east at a rate of 48 mi/h. At what time will the two buses be 274 mi apart?

23. At 8:00 A.M., Catherine leaves on a trip at 45 mi/h. One hour later, Max decides to join her and leaves along the same route, traveling at 54 mi/h. When will Max catch up with Catherine?

24. Martina leaves home at 9 A.M., bicycling at a rate of 24 mi/h. Two hours later, John leaves, driving at a rate of 48 mi/h. At what time will John catch up with Martina?

25. Jean leaves Boston for Baltimore at 10 A.M., traveling at 45 mi/h. One hour later, Bill leaves Baltimore for Boston on the same route, traveling at 50 mi/h. If the two cities are 425 mi apart, when will Jean and Bill meet?

18. _____

19. _____

20. _____

21. _____

22. _____

23. _____

24. _____

25. _____

26. _____

26. A train leaves town A for town B, traveling at a rate of 35 mi/h. At the same time, a second train leaves town B for town A at 45 mi/h. If the two towns are 320 mi apart, how long will it take for the trains to meet?

27. _____

27. There are 500 Douglas fir and hemlock trees in a section of forest bought by Hoodoo Logging Co. The company paid an average of $250 for each Douglas fir and $300 for each hemlock. If the company paid $132,000 for the trees, how many of each kind of tree did the company buy?

28. _____

28. There are 850 Douglas fir and ponderosa pine trees in a section of forest bought by Skookum Logging Co. The company paid an average of $300 for each Douglas fir and $225 for each ponderosa pine. If the company paid $217,500 for the trees, how many of each kind of tree did the company buy?

ANSWERS

1. 10, 18   **3.** 8, 15   **5.** 5, 6   **7.** 12 in, 25 in   **9.** 6 m, 22 m   **11.** Legs, 13 cm; base, 10 cm   **13.** 200 $6 tickets, 300 $8 tickets   **15.** 30 20¢ stamps, 50 25¢ stamps   **17.** 60 coach, 40 berth, and 20 sleeping room   **19.** 40 mi/h, 30 mi/h   **21.** 6 P.M.   **23.** 2 P.M.   **25.** 3 P.M.   **27.** 140 hemlock, 360 Douglas fir

Summary

Polynomials [4.1]

$4x^3 - 3x^2 + 5x$ is a polynomial. The terms of $4x^3 - 3x^2 + 5x$ are $4x^3$, $-3x^2$, and $5x$.

Polynomial   An algebraic expression made up of terms in which the exponents are whole numbers. These terms are connected by a plus or minus signs. Each sign ($+$ or $-$) is attached to the term following that sign.

Term   A number, or the product of a number and one or more variables, raised to a power.

The coefficients of $4x^3 - 3x^2$ are 4 and -3.

Coefficient   In each term of a polynomial, the number is called the *numerical coefficient* or, more simply, the *coefficient* of that term.

Types of Polynomials   A polynomial can be classified according to the number of terms it has.

$2x^3$ is a monomial.

A *mono*mial has one term.

$3x^2 - 7x$ is a binomial.

A *bi*nomial has two terms.

$5x^5 - 5x^3 + 2$ is a trinomial.

A *tri*nomial has three terms.

The degree of $4x^5 - 5x^3 + 3x$ is 5.

Degree   The highest power of the variable appearing in any one term.

$4x^5 - 5x^3 + 3x$ is written in descending-exponent form.
$5x^3 - 3x^4$ is *not* in descending-exponent form.

Descending-Exponent Form   The form of a polynomial when it is written with the highest-degree term first, the next highest-degree term second, and so on.

Adding and Subtracting Polynomials [4.2]

Removing Signs of Grouping

$(2x + 3) + (3x - 5)$
$= 2x + 3 + 3x - 5$

1. If a plus sign ($+$) or no sign at all appears in front of parentheses, just remove the parentheses. No other changes are necessary.

$-(2x^2 - 4x + 5)$
$= -2x^2 + 4x - 5$
Sign changes

2. If a minus sign ($-$) appears in front of parentheses, the parentheses can be removed by changing the sign of each term inside the parentheses.

$(2x^2 + 3x) + (4x^2 - 5x)$
$= 2x^2 + 3x + 4x^2 - 5x$
$= 6x^2 - 2x$

Adding Polynomials   Remove the signs of grouping. Then collect and combine any like terms.

$(4x^2 - 5x) - (3x^2 - 2x)$
$= 4x^2 - 5x - 3x^2 + 2x$
$= x^2 - 3x$

Subtracting Polynomials   Remove the signs of grouping by changing the sign of each term in the polynomial being subtracted. Then combine any like terms.

Multiplying Polynomials [4.3]

$3x^3(2x^2 - 5x)$
$= 3x^3 \cdot 2x^2 - 3x^3 \cdot 5x$
$= 6x^5 - 15x^4$

To Multiply a Polynomial by a Monomial   Multiply each term of the polynomial by the monomial, and add the results.

To Multiply a Binomial by a Binomial   Use the FOIL method:

$(2x - 3)(3x + 5)$

$= 6x^2 + 10x - 9x - 15$

$\quad$ F $\quad$ O $\quad$ I $\quad$ L

$= 6x^2 + x - 15$

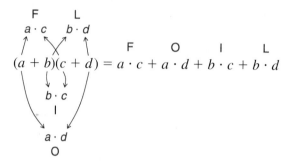

$$
\begin{array}{r}
x^2 - 3x + 5 \\
2x - 3 \\
\hline
-3x^2 + 9x - 15 \\
2x^3 - 6x^2 + 10x \\
\hline
2x^3 - 9x^2 + 19x - 15
\end{array}
$$

To Multiply a Polynomial by a Polynomial   Arrange the polynomials vertically. Multiply each term of the upper polynomial by each term of the lower polynomial, and add the results.

Special Products [4.4]

The Square of a Binomial

$$(a + b)^2 = a^2 + 2ab + b^2$$

$(2x - 5)^2$

$= 4x^2 + 2 \cdot 2x \cdot (-5) + 25$

$= 4x^2 - 20x + 25$

1. The first term of the square is the square of the first term of the binomial.
2. The middle term is twice the product of the two terms of the binomial.
3. The last term is the square of the last term of the binomial.

$(2x - 5y)(2x + 5y)$

$= (2x)^2 - (5y)^2$

$= 4x^2 - 25y^2$

The Product of Binomials That Differ Only in Sign   Subtract the square of the second term from the square of the first term.

$$(a + b)(a - b) = a^2 - b^2$$

Dividing Polynomials [4.5]

$$\dfrac{9x^4 + 6x^3 - 15x^2}{3x}$$

$$= \dfrac{9x^4}{3x} + \dfrac{6x^3}{3x} - \dfrac{15x^2}{3x}$$

$$= 3x^3 + 2x^2 - 5x$$

To Divide a Polynomial by a Monomial   Divide each term of the polynomial by the monomial. Then combine the results.

$$
\begin{array}{r}
x + 5 \\
x - 3 \overline{)x^2 + 2x - 7} \\
\underline{x^2 - 3x} \\
5x - 7 \\
\underline{5x - 15} \\
8
\end{array}
$$

To Divide a Polynomial by a Polynomial   Use the long division method.

Linear Equations and Applications [4.6 and 4.7]

Solve:

$5 - 2(3x - 1) = 31$

$5 - 6x + 2 = 31$

$-6x = 24$

$\dfrac{-6x}{-6} = \dfrac{24}{-6}$

$x = -4$

Solving Linear Equations   If parentheses (or other signs of grouping) are in an equation, use the distributive law and remove the signs of grouping. Then solve the resulting equation by the methods of Chapter 3.

Applications Involving Linear Equations   Use the five-step process introduced in Section 3.8 and repeated in Section 4.7.

Summary Exercises Chapter 4

This summary exercise set is provided to give you practice with each of the objectives of the chapter. Each exercise is keyed to the appropriate chapter section. The answers are provided in the instructor's manual. Your instructor will give you guidelines on how to best use these exercises in your instructional setting.

[4.1] Classify each of the following polynomials as monomials, binomials, or trinomials, where possible.

1. $5x^3 - 2x^2$ **2.** $7x^5$ **3.** $4x^5 - 8x^3 + 5$

4. $x^3 + 2x^2 - 5x + 3$ **5.** $9a - 18a^2$

[4.1] Arrange in descending-exponent form, if necessary, and give the degree of each polynomial.

6. $5x^5 + 3x^2$ **7.** $9x$ **8.** $6x^2 + 4x^4 + 6$

9. $5 + x$ **10.** -8 **11.** $9x^4 - 3x + 7x^6$

[4.2] Add.

12. $9a^2 - 5a$ and $12a^2 + 3a$ **13.** $5x^2 + 3x - 5$ and $4x^2 - 6x - 2$

14. $5y^3 - 3y^2$ and $4y + 3y^2$

[4.2] Subtract.

15. $4x^2 - 3x$ from $8x^2 + 5x$ **16.** $2x^2 - 5x - 7$ from $7x^2 - 2x + 3$

17. $5x^2 + 3$ from $9x^2 - 4x$

[4.2] Perform the indicated operations.

18. Subtract $5x - 3$ from the sum of $9x + 2$ and $-3x - 7$.

19. Subtract $5a^2 - 3a$ from the sum of $5a^2 + 2$ and $7a - 7$.

20. Subtract the sum of $16w^2 - 3w$ and $8w + 2$ from $7w^2 - 5w + 2$.

[4.2] Add, using the vertical method.

21. $x^2 + 5x - 3$ and $2x^2 + 4x - 3$ **22.** $9b^2 - 7$ and $8b + 5$

23. $x^2 + 7$, $3x - 2$, and $4x^2 - 8x$

[4.2] Subtract, using the vertical method.

24. $5x^2 - 3x + 2$ from $7x^2 - 5x - 7$ **25.** $8m - 7$ from $9m^2 - 7$

[4.3] Multiply.

26. $(5a^3)(a^2)$ **27.** $(2x^2)(3x^5)$ **28.** $(-9p^3)(-6p^2)$

29. $(3a^2b^3)(-7a^3b^4)$ **30.** $5(3x - 8)$ **31.** $4a(3a + 7)$

32. $(-5rs)(2r^2s - 5rs)$ **33.** $7mn(3m^2n - 2mn^2 + 5mn)$ **34.** $(x + 5)(x + 4)$

35. $(w - 9)(w - 10)$ **36.** $(a - 7b)(a + 7b)$ **37.** $(p - 3q)^2$

38. $(a + 4b)(a + 3b)$ **39.** $(b - 8)(2b + 3)$ **40.** $(3x - 5y)(2x - 3y)$

41. $(5r + 7s)(3r - 9s)$ **42.** $(y + 2)(y^2 - 2y + 3)$ **43.** $(b + 3)(b^2 - 5b - 7)$

44. $(x - 2)(x^2 + 2x + 4)$ **45.** $(m^2 - 3)(m^2 + 7)$ **46.** $2x(x + 5)(x - 6)$

47. $3a(2a - 5b)(2a - 7b)$

Find the following products.

48. $(x + 7)^2$ **49.** $(a - 8)^2$ **50.** $(2w - 5)^2$

51. $(3p + 4)^2$ **52.** $(a + 7b)^2$ **53.** $(8x - 3y)^2$

54. $(x - 5)(x + 5)$ **55.** $(y + 9)(y - 9)$ **56.** $(2m + 3)(2m - 3)$

57. $(3r - 7)(3r + 7)$ **58.** $(5r - 2s)(5r + 2s)$ **59.** $(7a + 3b)(7a - 3b)$

60. $2x(x - 5)^2$ **61.** $3c(c + 5d)(c - 5d)$

[4.5] Divide.

62. $\dfrac{9a^5}{3a^2}$ **63.** $\dfrac{24m^4n^2}{6m^2n}$

64. $\dfrac{15a - 10}{5}$ **65.** $\dfrac{32a^3 + 24a}{8a}$

66. $\dfrac{9r^2s^3 - 18r^3s^2}{-3rs^2}$ **67.** $\dfrac{35x^3y^2 - 21x^2y^3 + 14x^3y}{7x^2y}$

[4.5] Perform the indicated long division.

68. $\dfrac{x^2 - 2x - 15}{x + 3}$ **69.** $\dfrac{2x^2 + 9x - 35}{2x - 5}$

70. $\dfrac{x^2 - 8x + 17}{x - 5}$ **71.** $\dfrac{6x^2 - x - 10}{3x + 4}$

72. $\dfrac{6x^3 + 14x^2 - 2x - 6}{6x + 2}$ **73.** $\dfrac{4x^3 + x + 3}{2x - 1}$

74. $\dfrac{3x^2 + x^3 + 5 + 4x}{x + 2}$ **75.** $\dfrac{2x^4 - 2x^2 - 10}{x^2 - 3}$

[4.6] Solve the following equations for x.

76. $4(3x - 2) = 52$ **77.** $8 - (4x + 1) = 15$

78. $7 - (3x - 5) = 14$ **79.** $9 - 3(x - 2) = 1 - x$

80. $5(4x - 3) = 3(5x - 2) + 16$ **81.** $5(8x + 4) - 2(4x - 2) = 48$

[4.7] Solve the following word problems.

82. One number is 6 more than another. If the sum of 3 times the smaller number and twice the larger number is 47, find the two numbers.

83. Find two consecutive integers such that twice the first integer is 18 less than 3 times the second integer.

84. Find three consecutive even integers such that the sum of the first integer and 3 times the second integer is 3 times the third integer.

85. The length of a rectangle is 5 in more than its width. If the perimeter of the rectangle is 50 in, find the dimensions of the rectangle.

86. The length of a doubles tennis court is 6 ft more than twice its width. If the perimeter of the court is 228 ft, find the dimensions of the court.

87. Jovita has 33 coins, all nickels and dimes, with a value of $2.70. How many of each type of coin does she have?

88. A cashier starts the day with $1690 in $5, $10, and $20 bills. If he has 4 more $10 bills than $5 bills and twice as many $20 bills as $5 bills, how many of each denomination does he have?

89. A boat makes a trip upriver against the current in 6 h. Coming back down the river, the boat can travel 6 mi/h faster and makes the trip in 4 h. What is the speed of the boat in each direction?

90. At 9 A.M. David left New Orleans for Tallahassee, averaging 47 mi/h. Two hours later, Gloria left Tallahassee for New Orleans along the same route, driving 5 mi/h faster than David. If the two cities are 391 mi apart, at what time did David and Gloria meet?

Name

Section

Date

The purpose of this self-test is to help you check your progress and to review for a chapter test in class. Allow yourself about an hour to take the test. When you are done, check your answers in the back of the book. If you missed any problems, be sure to go back and review the appropriate sections in the chapter and do the exercises that are provided.

Classify each of the following polynomials as monomials, binomials, or trinomials.

1. $6x^2 + 7x$

2. $5x^2 + 8x - 8$

Arrange in descending-exponent form, and give the coefficients and degree of the polynomial.

3. $-3x^2 + 8x^4 - 7$

Add.

4. $3x^2 - 7x + 2$ and $7x^2 - 5x - 9$

5. $7a^2 - 3a$ and $7a^3 + 4a^2$

Subtract.

6. $5x^2 - 2x + 5$ from $8x^2 + 9x - 7$

7. $2b^2 + 5$ from $3b^2 - 7b$

8. $5a^2 + a$ from the sum of $3a^2 - 5a$ and $9a^2 - 4a$

Add, using the vertical method.

9. $x^2 + 3$, $5x - 7$, and $3x^2 - 2$

Subtract, using the vertical method.

10. $3x^2 - 5$ from $5x^2 - 7x$

Multiply.

11. $5ab(3a^2b - 2ab + 4ab^2)$

12. $(x - 2)(3x + 7)$

13. $(a - 7b)(a + 7b)$

ANSWERS

1. _____

2. _____

3. _____

4. _____

5. _____

6. _____

7. _____

8. _____

9. _____

10. _____

11. _____

12. _____

13. _____

14. _____

15. _____

16. _____

17. _____

18. _____

19. _____

20. _____

21. _____

22. _____

23. _____

24. _____

25. _____

292

Multiply.

14. $(3m + 2n)^2$

15. $(2x + y)(x^2 + 3xy - 2y^2)$

Divide.

16. $\dfrac{14x^3y - 21xy^2}{7xy}$

17. $\dfrac{20c^3d - 30cd + 45c^2d^2}{5cd}$

18. $(x^2 - 2x - 24) \div (x + 4)$

19. $(2x^2 + x + 4) \div (2x - 3)$

20. $(6x^3 - 7x^2 + 3x + 9) \div (3x + 1)$

Solve the following equations for x.

21. $3(5x - 3) = 51$

22. $7 - 2(3x - 4) = 5 + 11(6 + 2x)$

Solve the following word problems.

23. One number is 5 more than another. If the sum of 4 times the smaller number and 3 times the larger is 57, find the two numbers.

24. Sydney has 45 coins, all dimes and quarters, with a value of $8.25. How many dimes and quarters does she have?

25. A plane makes a trip against a headwind in 9 h. Returning with the wind, the plane can travel 50 mi/h faster and makes the trip in 7 h. What is the speed of the plane in each direction?

CHAPTER 5

Factoring

The ENVIRONMENT

Acid Rain

Although not the hot topic it once was, acid rain continues to be a problem today. The high levels of concern expressed in the 1970s about air pollution and acid rain brought about many changes in air pollution regulations. These regulations have led to a significant reduction in carbon monoxide (CO), nitrogen dioxide (NO_2), sulfur dioxide (SO_2), ground-level ozone (O_3), and particulates, the principal air pollutants and components of acid rain. Even with these reduced levels of pollutants, however, evidence shows a continuing problem of acid rain in some areas.

Acid rain, or more correctly acid deposition, is caused mainly by the burning of fossil fuels. When the fuel is burned, nitrogen and sulfur compounds in these fuels combine with oxygen to form NO_2 and SO_2. These air pollutants then either fall to the earth in solid form or combine with water vapor in the atmosphere and fall to the earth as rain or snow. Although rainwater is naturally acidic to a small degree, these additional acids can be harmful to the environment.

Acidified lakes have been found in North America and northern Europe. In 1988 and 1989 it was discovered that over 150,000 lakes in Canada had suffered some biological damage from acidification. Over 3000 lakes in the United States showed signs of damage. More than 2000 lifeless lakes have been found in Sweden; fish have been eliminated from 13,000 square kilometers of surface water in Norway; and Scotland, Wales, and the Lake District of Great Britain all contain acidified lakes.

Acid deposition is suspected of causing forest decline in Germany, Scandinavia, and Canada. Although it is difficult to claim that acid rain is the direct cause of large-scale forest death, many researchers claim forests, after being weakened by acid conditions in the environment, are dying from other causes.

The natural environment is not the only arena where the effects of acid rain have been detected. Ancient Greek monuments in Athens have deteriorated more in the last two decades than in the last 20 centuries. Trains in southern Poland must slow down in some places because the rails have been corroded by acid pollution. In the United States, Independence Hall in Philadelphia and the Gettysburg Civil War battlefield are just a small part of our historical heritage that is being degraded by air pollution.

The developing world is also suffering the impact of these air pollutants. Tropical forests are beginning to show some of the same stresses as their northern counterparts. The Taj Mahal in India appears to be eroding from the emissions of a nearby oil refinery. The Mayan temples of southern Mexico and the Yucatan peninsula are being destroyed much more quickly than natural erosive forces would indicate. Oil wells and smokestacks near the Gulf of Mexico are believed to be responsible. Even the exhaust from tour buses and cars may be impacting these ancient treasures.

Methods of combatting acid rain include the installation of scrubbers to remove SO_2 from flue gases, electrostatic precipitators and baghouse filters to control particulate emissions, and "clean" coal technologies to lower both SO_2 and NO_2 emissions. These technologies provide some reductions in acid-causing emissions, but they can create environmental problems by creating toxic wastes of their own which cause a whole new set of concerns.

Nontechnological approaches to the acid rain problem need to be considered as well. Conservation and energy efficiency improvements could lessen emissions from power plants and automobiles. Materials recycling could become very important as a pollution prevention alternative. Newsprint made from waste paper instead of raw fiber reduces air pollution by 75 percent. Aluminum made from recycled cans reduces the emissions of NO_2 by 95 percent and SO_2 by 99 percent.

Engine modifications, catalytic converters, and improved mileage efficiencies have reduced the emissions from automobiles in many countries, but these improvements have been negated by the increased numbers of cars and trucks on the world's highways. It may become necessary to make some major changes in our transportation systems away from automobiles toward mass transit before we see any lasting change in the air quality in major urban areas.

Acid rain has been recognized as a problem for more than 20 years. Successful technologies have been developed to reduce acid-causing emissions, yet the problem persists. The next step is to make the necessary cultural changes to reduce the demand for the pollution-causing industries and machines. This cultural change will be difficult, but it can be done if we make the commitment.

Factoring—An Introduction

OBJECTIVE

To factor a monomial from a polynomial

In Chapter 4 you were given factors and asked to find a product. We are now going to reverse the process. You will be given a polynomial and asked to find its factors. This is called *factoring*.

Let's start with an example from arithmetic. To *multiply* $5 \cdot 7$, you write

You find the product, 35, by multiplying.

$5 \cdot 7 = 35$

To *factor* 35, you would write

$35 = 5 \cdot 7$

Factoring is just the *reverse* of multiplication.

Now let's look at factoring in algebra. Up to now you have used the distributive property as

$$a(b + c) = ab + ac$$

For instance,

$3(x + 5) = 3x + 15$

3 and $x + 5$ are the factors of $3x + 15$.

To use the distributive property in factoring, we apply that property in the opposite fashion, as

$$ab + ac = a(b + c)$$

The property allows us to remove the common monomial factor a from the terms of $ab + ac$.

To use this in factoring, the first step is to see whether each term of the polynomial has a common monomial factor.

In our earlier example,

$3x + 15 = 3 \cdot x + 3 \cdot 5$

Common factor

So, by the distributive property,

$$3x + 15 = 3(x + 5)$$

Again, factoring is just the reverse of multiplication.

To check this, multiply $3(x + 5)$.

Here is a diagram that will relate the ideas of multiplication and factoring.

Multiplying

$$3(x + 5) = 3x + 15$$

Factoring

The first step in factoring is to identify the *greatest common factor* (GCF) of a set of terms. This is the monomial with the largest common numerical coefficient and the largest power of each common variable.

Example 1

Find the GCF for each set of terms.

(*a*) 9 and 12

The largest number that is a factor of both is 3.

(*b*) 10, 25, 150

The GCF is 5.

(*c*) x^4 and x^7

The largest power common to the two variables is x^4.

(*d*) $12a^3$ and $18a^2$

The GCF is $6a^2$.

CHECK YOURSELF 1

Find the GCF for each set of terms.

1. 14, 24   **2.** 9, 27, 81
3. a^9, a^5 **4.** $10x^5$, $35x^4$

To factor a monomial from a polynomial, you can use these steps:

Checking your answer is always important and perhaps is never easier than after you have factored.

TO FACTOR A MONOMIAL FROM A POLYNOMIAL

STEP 1   Find the *greatest common factor* (GCF) for all the terms.

STEP 2   Factor the GCF from each term, then apply the distributive law.

STEP 3   Mentally check your factoring by multiplication.

The following examples illustrate the use of this rule for factoring polynomials.

Example 2

(a) Factor $8x^2 + 12x$.

The largest common numerical factor of 8 and 12 is 4, and x is the variable factor with the largest common power. So $4x$ is the GCF. Write

$$8x^2 + 12x = 4x \cdot 2x + 4x \cdot 3$$
GCF

Now, by the distributive property, we have

$$8x^2 + 12x = 4x(2x + 3)$$

It is always a good idea to check your answer by multiplying to make sure that you get the original polynomial. Try it here. Multiply $4x$ by $2x + 3$.

(b) Factor $6a^4 - 18a^2$.

The GCF in this case is $6a^2$. Write

$$6a^4 - 18a^2 = 6a^2 \cdot a^2 - 6a^2 \cdot 3$$
GCF

Again, using the distributive property yields

$$6a^4 - 18a^2 = 6a^2(a^2 - 3)$$

You should check this by multiplying.

Note: In part b of the previous example, it is also true that

$$6a^4 - 18a^2 = 3a(2a^3 - 6a)$$

However, this is *not completely factored.* Do you see why? You want to find the common monomial factor with the *largest* possible coefficient and the *largest* exponent, in this case $6a^2$.

CHECK YOURSELF 2

Factor each of the following polynomials.

1. $5x + 20$ **2.** $6x^2 - 24x$ **3.** $10a^3 - 15a^2$

The process is exactly the same for polynomials with more than two terms. Consider the following examples.

Example 3

(*a*) Factor $5x^2 - 10x + 15$.

The GCF is 5.

$$5x^2 - 10x + 15 = \underline{5} \cdot x^2 - \underline{5} \cdot 2x + \underline{5} \cdot 3$$

$$\text{GCF}$$

$$= 5(x^2 - 2x + 3)$$

(*b*) Factor $6ab + 9ab^2 - 15a^2$.

The GCF is $3a$.

$$6ab + 9ab^2 - 15a^2 = \underline{3a} \cdot 2b + \underline{3a} \cdot 3b^2 - \underline{3a} \cdot 5a$$

$$\text{GCF}$$

$$= 3a(2b + 3b^2 - 5a)$$

(*c*) Factor $4a^4 + 12a^3 - 20a^2$.

The GCF is $4a^2$.

$$4a^4 + 12a^3 - 20a^2 = \underline{4a^2} \cdot a^2 + \underline{4a^2} \cdot 3a - \underline{4a^2} \cdot 5$$

$$\text{GCF}$$

$$= 4a^2(a^2 + 3a - 5)$$

CHECK YOURSELF 3

Factor each of the following polynomials.

1. $8b^2 + 16b - 32$ **2.** $4xy - 8x^2y + 12x^3$
3. $7x^4 - 14x^3 + 21x^2$

With practice you should be able to factor polynomials like the ones we have been considering without having to write out the factors in each term. Try it in the following example.

Example 4

(*a*) Factor $\underbrace{6a^2b + 9ab^2 + 3ab}$.

Mentally note that 3, *a*, and *b* are factors of each term, so

In each of these examples, you will want to practice checking the result by multiplying the factors.

$$6a^2b + 9ab^2 + 3ab = 3ab(2a + 3b + 1)$$

(*b*) Factor $\underbrace{9x^4 - 27x^3 - 18x^2}$.

Mentally note that 9 and x^2 are factors of each term.

Thus

$$9x^4 - 27x^3 - 18x^2 = 9x^2(x^2 - 3x - 2)$$

(c) Factor $\underbrace{3m^4 + 9m^3 - 12m^2 + 15m}$.
Here 3 and m are factors of each term.

$$3m^4 + 9m^3 - 12m^2 + 15m = 3m(m^3 + 3m^2 - 4m + 5)$$

CHECK YOURSELF 4

Factor each of the following polynomials.

1. $5x^2y^2 - 10xy^2 + 15x^2y$ **2.** $8p^4 - 16p^3 - 40p^2$
3. $8a^4 - 16a^3 + 20a^2 - 24a$

Sometimes the GCF of an expression will be a binomial. Our final example illustrates.

Example 5

(a) Factor $x(x + y) + 3(x + y)$.

Note that $x + y$ is a common binomial factor for each term. Removing that factor gives

$$x(x + y) + 3(x + y) = (x + y)(x + 3)$$

(b) Factor $a(a - b) - 2(a - b)$.

Here $a - b$ is the common factor, so

$$a(a - b) - 2(a - b) = (a - b)(a - 2)$$

CHECK YOURSELF 5

Factor each of the following.

1. $y(y - 1) - 7(y - 1)$ **2.** $a(a + b) + 3b(a + b)$

CHECK YOURSELF ANSWERS

1. (1) 2; (2) 9; (3) a^5; (4) $5x^4$.
2. (1) $5(x + 4)$; (2) $6x(x - 4)$; (3) $5a^2(2a - 3)$.
3. (1) $8(b^2 + 2b - 4)$; (2) $4x(y - 2xy + 3x^2)$; (3) $7x^2(x^2 - 2x + 3)$.
4. (1) $5xy(xy - 2y + 3x)$; (2) $8p^2(p^2 - 2p - 5)$;
   (3) $4a(2a^3 - 4a^2 + 5a - 6)$.
5. (1) $(y - 1)(y - 7)$; (2) $(a + b)(a + 3b)$.

Name

Date

Build Your Skills

Find the greatest common factor for the following sets of terms.

1. 6, 8

2. 10, 125

3. 24, 16, 80

4. 121, 33, 66

5. x^2, x^5

6. y^7, y^9

7. a^3, a^6, a^9

8. b^4, b^6, b^8

9. $5x^4$, $10x^5$

10. $8y^9$, $24y^3$

11. $8a^4$, $6a^6$, $10a^{10}$

12. $9b^3$, $6b^5$, $12b^4$

13. $6xy^3$, $9x^3y$, $21x^2y^2$

14. $24a^2b^2$, $12a^4b^4$, $18a^5b^3$

15. $15a^2b$, $5b^2c$, $10b$

16. $27x^2$, $18x^2y^2$, $6y^2$

17. $15a^2bc^2$, $9ab^2c^2$, $6a^2b^2c^2$

18. $18x^3y^2z^3$, $27x^4y^2z^3$, $81xy^2z$

19. $(x + y)^2$, $(x + y)^3$

20. $12(a + b)^4$, $4(a + b)^3$

Factor each of the following polynomials.

21. $8a + 4$

22. $5x - 15$

ANSWERS

1. _____

2. _____

3. _____

4. _____

5. _____

6. _____

7. _____

8. _____

9. _____

10. _____

11. _____

12. _____

13. _____

14. _____

15. _____

16. _____

17. _____

18. _____

19. _____

20. _____

21. _____

22. _____

23. _____

24. _____

25. _____

26. _____

27. _____

28. _____

29. _____

30. _____

31. _____

32. _____

33. _____

34. _____

35. _____

36. _____

37. _____

38. _____

39. _____

40. _____

41. _____

42. _____

43. _____

44. _____

45. _____

46. _____

47. _____

48. _____

49. _____

50. _____

23. $14m - 21n$

24. $6p + 12q$

25. $6m^2 - 9m$

26. $18n^2 + 27n$

27. $10s^2 + 5s$

28. $12y^2 - 6y$

29. $12x^2 + 24x$

30. $14b^2 - 28b$

31. $15a^3 - 25a^2$

32. $36b^4 + 24b^2$

33. $6pq + 18p^2q$

34. $8ab - 24ab^2$

35. $7m^3n - 21mn^3$

36. $36p^2q^2 - 9pq$

37. $6x^2 - 18x + 30$

38. $7a^2 + 21a - 42$

39. $3a^3 + 6a^2 - 12a$

40. $5x^3 - 15x^2 + 25x$

41. $4m + 8mn - 16mn^2$

42. $9s - 12st + 15st^2$

43. $7x^2y - 14xy + 28xy^2$

44. $6a^2b - 18ab + 24ab^2$

45. $10r^3s^2 + 25r^2s^2 - 15r^2s^3$

46. $28x^2y^3 - 35x^2y^2 + 42x^3y$

47. $9a^5 - 15a^4 + 21a^3 - 27a$

48. $8p^6 - 40p^4 + 24p^3 - 16p^2$

49. $15m^3n^2 - 20m^2n + 35mn^3 - 10mn$

50. $14ab^4 + 21a^2b^3 - 35a^3b^2 + 28ab^2$

51. $x(x - 2) + 3(x - 2)$

52. $y(y + 5) - 3(y + 5)$

53. $p(p - 2q) - q(p - 2q)$

54. $2c(c + d) + 3d(c + d)$

Think About These

55. $3(x + y)^2 + 9(x + y)$

56. $6(a - b)^2 - 12(a - b)$

57. The GCF of $2x - 6$ is 2. The GCF of $5x + 10$ is 5. Find the greatest common factor of the product $(2x - 6)(5x + 10)$.

58. The GCF of $3z + 12$ is 3. The GCF of $4z + 8$ is 4. Find the GCF of the following product: $(3z + 12)(4z + 8)$.

59. The GCF of $2x^3 - 4x$ is $2x$. The GCF of $3x + 6$ is 3. Find the GCF of the product $(2x^3 - 4x)(3x + 6)$.

60. State, in a sentence, the rule that the previous three problems illustrated.

Find the GCF for each product.

61. $(2a + 8)(3a - 6)$

62. $(5b - 10)(2b + 4)$

63. $(2x^2 + 5x)(7x - 14)$

64. $(6y^2 - 3y)(y + 7)$

ANSWERS

51. _____

52. _____

53. _____

54. _____

55. _____

56. _____

57. _____

58. _____

59. _____

60. _____

61. _____

62. _____

63. _____

64. _____

Skillscan (Section 4.4)

Multiply.

a. $(x - 1)(x + 1)$ **b.** $(a + 7)(a - 7)$

c. $(x - y)(x + y)$ **d.** $(2x - 5)(2x + 5)$

e. $(3a - b)(3a + b)$ **f.** $(5a - 4b)(5a + 4b)$

ANSWERS

1. 2   **3.** 8   **5.** x^2 **7.** a^3 **9.** $5x^4$ **11.** $2a^4$ **13.** $3xy$ **15.** $5b$ **17.** $3abc^2$
19. $(x + y)^2$ **21.** $4(2a + 1)$ **23.** $7(2m - 3n)$ **25.** $3m(2m - 3)$ **27.** $5s(2s + 1)$
29. $12x(x + 2)$ **31.** $5a^2(3a - 5)$ **33.** $6pq(1 + 3p)$ **35.** $7mn(m^2 - 3n^2)$
37. $6(x^2 - 3x + 5)$ **39.** $3a(a^2 + 2a - 4)$ **41.** $4m(1 + 2n - 4n^2)$ **43.** $7xy(x - 2 + 4y)$
45. $5r^2s^2(2r + 5 - 3s)$ **47.** $3a(3a^4 - 5a^3 + 7a^2 - 9)$ **49.** $5mn(3m^2n - 4m + 7n^2 - 2)$
51. $(x - 2)(x + 3)$ **53.** $(p - 2q)(p - q)$ **55.** $3(x + y)(x + y + 3)$ **57.** 10   **59.** $6x$
61. 6   **63.** $7x$ **a.** $x^2 - 1$ **b.** $a^2 - 49$ **c.** $x^2 - y^2$ **d.** $4x^2 - 25$ **e.** $9a^2 - b^2$
f. $25a^2 - 16b^2$

5.2   The Difference of Squares

OBJECTIVE

To factor a binomial that is a difference of two squares

In Section 4.4 we introduced with some special products. Recall the following formula for the product of a sum and difference of two terms:

$$(a + b)(a - b) = a^2 - b^2$$

This also means that a binomial of the form $a^2 - b^2$ (called a *difference of two squares*) has as its factors $a + b$ and $a - b$.

To use this idea for factoring, we can write

A "perfect-square" term has a coefficient that is a square (1, 4, 9, 16, 25, 36, etc.), and any variables will have exponents that are multiples of 2 (x^2, y^4, z^6, etc.).

$$a^2 - b^2 = (a + b)(a - b)$$

Example 1

Factor $x^2 - 16$.

Think $\quad x^2 - 4^2$

Since $x^2 - 16$ is a difference of squares, we have

You could also write $(x - 4)(x + 4)$. The order doesn't matter since multiplication is commutative.

$$x^2 - 16 = (x + 4)(x - 4)$$

CHECK YOURSELF 1

Factor $m^2 - 49$.

Any time an expression is a difference of two squares, it can be factored.

Example 2

Factor $4a^2 - 9$.

Think $\quad (2a)^2 - 3^2$

So

$$4a^2 - 9 = (2a)^2 - (3)^2$$
$$= (2a + 3)(2a - 3)$$

Factor $9b^2 - 25$.

The process for factoring a difference of squares doesn't change when more than one variable is involved.

Example 3

$(5a)^2 - (4b^2)^2$

Factor $25a^2 - 16b^4$.

$$25a^2 - 16b^4 = (5a + 4b^2)(5a - 4b^2)$$

CHECK YOURSELF 3

Factor $49c^4 - 9d^2$.

We will now consider an example that combines common-term factoring with difference-of-squares factoring. Note that the common factor is always removed as the *first step*.

Example 4

Factor $32x^2y - 18y^3$.

Note that $2y$ is a common factor, so

Step 1
Remove the GCF.

$$32x^2y - 18y^3 = 2y(16x^2 - 9y^2)$$
Difference of squares

Step 2
Factor the remaining binomial.

$$= 2y(4x + 3y)(4x - 3y)$$

CHECK YOURSELF 4

Factor $50a^3 - 8ab^2$.

CHECK YOURSELF ANSWERS

1. $(m + 7)(m - 7)$. **2.** $(3b + 5)(3b - 5)$.
3. $(7c^2 + 3d)(7c^2 - 3d)$. **4.** $2a(5a + 2b)(5a - 2b)$.

Name _____

Date _____

Build Your Skills

For each of the following binomials, state whether the binomial is a difference of squares.

1. $2x^2 + y^2$

2. $2x^2 - y^2$

3. $9a^2 - 16b^2$

4. $4m^2 - 64n^2$

5. $16r^2 + 4$

6. $p^2 - 45$

7. $16a^2 - 12b^3$

8. $9a^2b^2 - 16c^2d^2$

9. $a^2b^2 - 25$

10. $4a^3 - b^3$

Factor the following binomials.

11. $m^2 - n^2$

12. $r^2 - 9$

13. $x^2 - 49$

14. $c^2 - d^2$

15. $25 - x^2$

16. $64 - a^2$

17. $4a^2 - 9$

18. $p^2 - 25$

ANSWERS

1. _____

2. _____

3. _____

4. _____

5. _____

6. _____

7. _____

8. _____

9. _____

10. _____

11. _____

12. _____

13. _____

14. _____

15. _____

16. _____

17. _____

18. _____

19. _____

20. _____

21. _____

22. _____

23. _____

24. _____

25. _____

26. _____

27. _____

28. _____

29. _____

30. _____

31. _____

32. _____

33. _____

34. _____

35. _____

36. _____

37. _____

38. _____

19. $9w^2 - 25$

20. $9x^2 - 64$

21. $r^2 - 9s^2$

22. $49x^2 - y^2$

23. $9w^2 - 49z^2$

24. $25x^2 - 81y^2$

25. $16a^2 - 49b^2$

26. $64m^2 - 9n^2$

27. $x^4 - 36$

28. $y^6 - 49$

29. $x^2y^2 - 16$

30. $m^2n^2 - 64$

31. $25 - a^2b^2$

32. $49 - w^2z^2$

33. $r^4 - 4s^2$

34. $p^2 - 9q^4$

35. $81a^2 - 100b^6$

36. $64x^4 - 25y^4$

37. $18x^3 - 2xy^2$

38. $50a^2b - 2b^3$

39. $50m^3n - 18mn^3$

40. $32p^2q^2 - 8q^4$

41. $48a^2b^2 - 27b^4$

42. $20w^5 - 45w^3z^4$

Think About These

Factor each expression.

43. $x^2(x + y) - y^2(x + y)$

44. $a^2(b - c) - 16b^2(b - c)$

45. $2m^2(m - 2n) - 18n^2(m - 2n)$

46. $3a^3(2a + b) - 27ab^2(2a + b)$

47. Find a value for k so that $kx^2 - 25$ will have the factors $2x + 5$ and $2x - 5$.

48. Find a value for k so that $9m^2 - kn^2$ will have the factors $3m + 7n$ and $3m - 7n$.

49. Find a value for k so that $2x^3 - kxy^2$ will have the factors $2x$, $x - 3y$, and $x + 3y$.

50. Find a value for k so that $20a^3b - kab^3$ will have the factors $5ab$, $2a - 3b$, and $2a + 3b$.

39. _____

40. _____

41. _____

42. _____

43. _____

44. _____

45. _____

46. _____

47. _____

48. _____

49. _____

50. _____

a. _____

b. _____

c. _____

d. _____

e. _____

f. _____

g. _____

h. _____

Skillscan (Section 4.3)

Multiply.

a. $(x - 1)(x + 2)$

b. $(a - 3)(a + 2)$

c. $(x + 4)(x + 6)$

d. $(w + 1)(w + 7)$

e. $(b + 1)(b + 3)$

f. $(a + 1)(a - 4)$

g. $(x - 1)(x - 1)$

h. $(p - 2)(p - 5)$

ANSWERS

1. No **3.** Yes **5.** No **7.** No **9.** Yes **11.** $(m + n)(m - n)$ **13.** $(x + 7)(x - 7)$
15. $(5 + x)(5 - x)$ **17.** $(2a + 3)(2a - 3)$ **19.** $(3w + 5)(3w - 5)$ **21.** $(r + 3s)(r - 3s)$
23. $(3w + 7z)(3w - 7z)$ **25.** $(4a + 7b)(4a - 7b)$ **27.** $(x^2 + 6)(x^2 - 6)$
29. $(xy + 4)(xy - 4)$ **31.** $(5 + ab)(5 - ab)$ **33.** $(r^2 + 2s)(r^2 - 2s)$
35. $(9a + 10b^3)(9a - 10b^3)$ **37.** $2x(3x + y)(3x - y)$ **39.** $2mn(5m + 3n)(5m - 3n)$
41. $3b^2(4a + 3b)(4a - 3b)$ **43.** $(x + y)^2(x - y)$ **45.** $2(m - 2n)(m + 3n)(m - 3n)$ **47.** 4
49. 18 **a.** $x^2 + x - 2$ **b.** $a^2 - a - 6$ **c.** $x^2 + 10x + 24$ **d.** $w^2 + 8w + 7$
e. $b^2 + 4b + 3$ **f.** $a^2 - 3a - 4$ **g.** $x^2 - 2x + 1$ **h.** $p^2 - 7p + 10$

Factoring Trinomials—Part 1

OBJECTIVE

To factor trinomials of the form

$$x^2 + bx + c$$

The process used to factor here is frequently called the *trial-and-error method.* You'll see the reason for the name as you work through this section.

You learned how to find the product of any two binomials by using the FOIL method in Section 4.3. Since factoring is the reverse of multiplication, we now want to use that pattern to find the factors of certain trinomials.

Recall that to multiply two binomials, we have

$$(x + 2)(x + 3) = x^2 + 5x + 6$$

The product of the first terms ($x \cdot x$).

The sum of the products of the outer and inner terms ($3x$ and $2x$).

The product of the last terms ($2 \cdot 3$).

⟨**CAUTION**⟩

Not every trinomial can be written as the product of two binomials.

Suppose now that you are given $x^2 + 5x + 6$ and want to find its factors. First, you know that the factors of a trinomial may be two binomials. So write

$$x^2 + 5x + 6 = (\qquad)(\qquad)$$

Since the first term of the trinomial is x^2, the first terms of the binomial factors must be x and x. We now have

$$x^2 + 5x + 6 = (x \qquad)(x \qquad)$$

The product of the last terms must be 6. Since 6 is positive, the factors must have *like* signs. Here are the possibilities:

$$6 = 1 \cdot 6$$
$$= 2 \cdot 3$$
$$= (-1)(-6)$$
$$= (-2)(-3)$$

This means that the possible factors of the trinomial are

$$(x + 1)(x + 6)$$
$$(x + 2)(x + 3)$$
$$(x - 1)(x - 6)$$
$$(x - 2)(x - 3)$$

How do we tell which is the correct pair? From the FOIL pattern we know that the sum of the outer and inner products must equal the middle term of the trinomial, in this case $5x$. This is the crucial step!

POSSIBLE FACTORS	MIDDLE TERMS
$(x + 1)(x + 6)$	$7x$
$(x + 2)(x + 3)$	$5x$
$(x - 1)(x - 6)$	$-7x$
$(x - 2)(x - 3)$	$-5x$

The correct middle term!

So we know that the correct factorization is

$$x^2 + 5x + 6 = (x + 2)(x + 3)$$

Are there any clues so far that will make this process quicker? Yes, there is an important one that you may have spotted. We started with a trinomial that had a positive middle term and a positive last term. The negative pairs of factors for 6 led to negative middle terms. So you don't need to bother with the negative factors if the middle term and the last term of the trinomial are both positive.

Example 1

(a) Factor $x^2 + 9x + 8$.

Since the middle term and the last term of the trinomial are both positive, consider only the positive factors of 8, that is, $8 = 1 \cdot 8$ or $8 = 2 \cdot 4$.

POSSIBLE FACTORS	MIDDLE TERMS
$(x + 1)(x + 8)$	$9x$
$(x + 2)(x + 4)$	$6x$

If you are wondering why we didn't list $(x + 8)(x + 1)$ as a possibility, remember that multiplication is commutative. The order doesn't matter!

Since the first pair gives the correct middle term,

$$x^2 + 9x + 8 = (x + 1)(x + 8)$$

(b) Factor $x^2 + 12x + 20$.

POSSIBLE FACTORS	MIDDLE TERMS
$(x + 1)(x + 20)$	$21x$
$(x + 2)(x + 10)$	$12x$
$(x + 4)(x + 5)$	$9x$

The factors for 20 are

$20 = 1 \cdot 20$
$\quad = 2 \cdot 10$
$\quad = 4 \cdot 5$

so

$$x^2 + 12x + 20 = (x + 2)(x + 10)$$

CHECK YOURSELF 1

Factor.

1. $x^2 + 6x + 5$ **2.** $x^2 + 10x + 16$

Let's look at some examples in which the middle term of the trinomial is negative but the first and last terms are still positive. Consider

Positive   Positive

$$x^2 - 11x + 18$$

Negative

Since we want a negative middle term $(-11x)$, we use *two negative factors* for 18. Recall that the product of two negative numbers is positive.

Example 2

(*a*) Factor $x^2 - 11x + 18$.

The negative factors of 18 are

$18 = (-1)(-18)$
$\quad = (-2)(-9)$
$\quad = (-3)(-6)$

POSSIBLE FACTORS	MIDDLE TERMS
$(x - 1)(x - 18)$	$-19x$
$(x - 2)(x - 9)$	$-11x$
$(x - 3)(x - 6)$	$-9x$

So

$$x^2 - 11x + 18 = (x - 2)(x - 9)$$

(*b*) Factor $x^2 - 13x + 12$.

The negative factors of 12 are

$12 = (-1)(-12)$
$\quad = (-2)(-6)$
$\quad = (-3)(-4)$

POSSIBLE FACTORS	MIDDLE TERMS
$(x - 1)(x - 12)$	$-13x$
$(x - 2)(x - 6)$	$-8x$
$(x - 3)(x - 4)$	$-7x$

So

$$x^2 - 13x + 12 = (x - 1)(x - 12)$$

A few more clues: We have listed all the possible factors in the above examples. It really isn't necessary. Just work until you find the right pair. Also, with practice much of this work can be done mentally.

Factor.

1. $x^2 - 10x + 9$ **2.** $x^2 - 10x + 21$

Let's look now at the process of factoring a trinomial whose last term is negative. For instance, to factor $x^2 + 2x - 15$, we can start as before:

$$x^2 + 2x - 15 = (x \quad ?)(x \quad ?)$$

Note that the product of the last terms must be negative (-15 here). So we must choose factors that have different signs.

What are our choices for the factors of -15?

$$
\begin{aligned}
-15 &= (1)(-15) \\
&= (-1)(15) \\
&= (3)(-5) \\
&= (-3)(5)
\end{aligned}
$$

This means that the possible factors and the resulting middle terms are

Another clue: Some students prefer to look at the list of numerical factors rather than looking at the actual algebraic factors. Here you want the pair whose sum is 2, the coefficient of the middle term of the trinomial. That pair is -3 and 5, which leads us to the correct factors.

POSSIBLE FACTORS	MIDDLE TERMS
$(x + 1)(x - 15)$	$-14x$
$(x - 1)(x + 15)$	$14x$
$(x + 3)(x - 5)$	$-2x$
$(x - 3)(x + 5)$	$2x$

So $x^2 + 2x - 15 = (x - 3)(x + 5)$.

Let's work through some examples in which the constant term is negative.

Example 3

(*a*) Factor $x^2 - 5x - 6$.

First, list the factors of -6. Of course, one factor will be positive, and one will be negative.

You may be able to pick the factors directly from this list. You want the pair whose sum is -5 (the coefficient of the middle term).

$$
\begin{aligned}
-6 &= (1)(-6) \\
&= (-1)(6) \\
&= (2)(-3) \\
&= (-2)(3)
\end{aligned}
$$

For the trinomial, then, we have

POSSIBLE FACTORS	MIDDLE TERMS
$(x + 1)(x - 6)$	$-5x$
$(x - 1)(x + 6)$	$5x$
$(x + 2)(x - 3)$	$-x$
$(x - 2)(x + 3)$	x

So $x^2 - 5x - 6 = (x + 1)(x - 6)$.

(*b*) Factor $x^2 + 8xy - 9y^2$.

The process is similar if two variables are involved in the trinomial you are to factor. Start with

$$x^2 + 8xy - 9y^2 = (x \qquad ?)(x \qquad ?).$$

The product of the last terms must be $-9y^2$.

$$-9y^2 = (-y)(9y)$$
$$= (y)(-9y)$$
$$= (3y)(-3y)$$

POSSIBLE FACTORS	MIDDLE TERMS
$(x - y)(x + 9y)$	$8xy$
$(x + y)(x - 9y)$	$-8xy$
$(x + 3y)(x - 3y)$	0

So $x^2 + 8xy - 9y^2 = (x - y)(x + 9y)$.

CHECK YOURSELF 3

Factor.

1. $x^2 + 7x - 30$ **2.** $x^2 - 3xy - 10y^2$

As was pointed out in the last section, any time that we have a common factor, that factor should be removed *before* we try any other factoring technique. Consider the following example.

Example 4

(*a*) Factor $3x^2 - 21x + 18$.

$3x^2 - 21x + 18 = 3(x^2 - 7x + 6)$ Remove the common factor of 3.

We now factor the remaining trinomial. For $x^2 - 7x + 6$:

POSSIBLE FACTORS	MIDDLE TERMS
$(x - 2)(x - 3)$	$-5x$
$(x - 1)(x - 6)$	$-7x$

The correct middle term

So $3x^2 - 21x + 18 = 3(x - 1)(x - 6)$.

(b) Factor $2x^3 + 16x^2 - 40x$.

$2x^3 + 16x^2 - 40x = 2x(x^2 + 8x - 20)$ Remove the common factor of $2x$.

To factor the remaining trinomial, which is $x^2 + 8x - 20$, we have

POSSIBLE FACTORS	MIDDLE TERMS
$(x - 4)(x + 5)$	x
$(x - 5)(x + 4)$	$-x$
$(x - 10)(x + 2)$	$-8x$
$(x - 2)(x + 10)$	$8x$

The correct middle term

so $2x^3 + 16x^2 - 40x = 2x(x - 2)(x + 10)$.

CHECK YOURSELF 4

Factor.

1. $3x^2 - 3x - 36$ **2.** $4x^3 + 24x^2 + 32x$

One further comment: Have you wondered if all trinomials are factorable? Look at the trinomial

$x^2 + 2x + 6$

The only possible factors are $(x + 1)(x + 6)$ and $(x + 2)(x + 3)$. Neither pair is correct (you should check the middle terms), and so this trinomial does not have factors with integer coefficients. Of course, there are many others.

CHECK YOURSELF ANSWERS

1. (1) $(x + 1)(x + 5)$; (2) $(x + 2)(x + 8)$.
2. (1) $(x - 9)(x - 1)$; (2) $(x - 3)(x - 7)$.
3. (1) $(x + 10)(x - 3)$; (2) $(x + 2y)(x - 5y)$.
4. (1) $3(x - 4)(x + 3)$; (2) $4x(x + 2)(x + 4)$.

Name _____

Date _____

Build Your Skills

Complete each of the following statements.

1. $x^2 - 8x + 15 = (x - 3)(\quad)$

2. $y^2 - 3y - 18 = (y - 6)(\quad)$

3. $m^2 + 8m + 12 = (m + 2)(\quad)$

4. $x^2 - 10x + 24 = (x - 6)(\quad)$

5. $p^2 - 8p - 20 = (p + 2)(\quad)$

6. $a^2 + 9a - 36 = (a + 12)(\quad)$

7. $x^2 - 16x + 64 = (x - 8)(\quad)$

8. $w^2 - 12w - 45 = (w + 3)(\quad)$

9. $x^2 - 7xy + 10y^2 = (x - 2y)(\quad)$

10. $a^2 + 18ab + 81b^2 = (a + 9b)(\quad)$

Factor each of the following trinomials.

11. $x^2 + 8x + 15$

12. $x^2 - 11x + 24$

13. $x^2 - 11x + 28$

14. $y^2 - y - 20$

15. $s^2 + 13s + 30$

16. $b^2 + 14b + 33$

17. $a^2 - 2a - 48$

18. $x^2 - 17x + 60$

19. $x^2 - 8x + 7$

20. $x^2 + 7x - 18$

21. $m^2 + 3m - 28$

22. $a^2 + 10a + 25$

1. _____

2. _____

3. _____

4. _____

5. _____

6. _____

7. _____

8. _____

9. _____

10. _____

11. _____

12. _____

13. _____

14. _____

15. _____

16. _____

17. _____

18. _____

19. _____

20. _____

21. _____

22. _____

23. _____

24. _____

25. _____

26. _____

27. _____

28. _____

29. _____

30. _____

31. _____

32. _____

33. _____

34. _____

35. _____

36. _____

37. _____

38. _____

39. _____

40. _____

41. _____

42. _____

43. _____

44. _____

45. _____

46. _____

23. $x^2 - 6x - 40$

24. $x^2 - 11x + 10$

25. $x^2 - 14x + 49$

26. $s^2 - 4s - 32$

27. $p^2 - 10p - 24$

28. $x^2 - 11x - 60$

29. $x^2 + 5x - 66$

30. $a^2 + 2a - 80$

31. $c^2 + 19c + 60$

32. $t^2 - 4t - 60$

33. $n^2 + 5n - 50$

34. $x^2 - 16x + 63$

35. $x^2 + 7xy + 10y^2$

36. $x^2 - 8xy + 12y^2$

37. $a^2 - ab - 42b^2$

38. $m^2 - 8mn + 16n^2$

39. $x^2 - 13xy + 40y^2$

40. $r^2 - 9rs - 36s^2$

41. $b^2 + 6ab + 9a^2$

42. $x^2 + 3xy - 10y^2$

43. $x^2 - 2xy - 8y^2$

44. $u^2 + 6uv - 55v^2$

45. $25m^2 + 10mn + n^2$

46. $64m^2 - 16mn + n^2$

Factor each of the following trinomials completely. Factor out the greatest common factor first.

47. $3a^2 - 3a - 126$

48. $2c^2 + 2c - 60$

49. $r^3 + 7r^2 - 18r$

50. $m^3 + 5m^2 - 14m$

51. $2x^3 - 20x^2 - 48x$

52. $3p^3 + 48p^2 - 108p$

53. $x^2y - 9xy^2 - 36y^3$

54. $4s^4 - 20s^3t - 96s^2t^2$

55. $m^3 - 29m^2n + 120mn^2$

56. $2a^3 - 52a^2b + 96ab^2$

Think About These

Find a positive value for k for which each of the following can be factored.

57. $x^2 + kx + 8$

58. $x^2 + kx + 9$

59. $x^2 - kx + 16$

60. $x^2 - kx + 17$

61. $x^2 - kx - 5$

62. $x^2 - kx - 7$

63. $x^2 + 3x + k$

64. $x^2 + 5x + k$

65. $x^2 + 2x - k$

66. $x^2 + x - k$

ANSWERS

47. _____

48. _____

49. _____

50. _____

51. _____

52. _____

53. _____

54. _____

55. _____

56. _____

57. _____

58. _____

59. _____

60. _____

61. _____

62. _____

63. _____

64. _____

65. _____

66. _____

Skillscan (Section 4.3)

Multiply.

a. $(2x - 1)(2x + 3)$

b. $(3a - 1)(a + 4)$

c. $(x - 4)(2x - 3)$

d. $(2w - 11)(w + 2)$

e. $(y + 5)(2y + 9)$

f. $(2x + 1)(x - 12)$

g. $(p + 9)(2p + 5)$

h. $(3a - 5)(2a + 4)$

5.4 Factoring Trinomials—Part 2

OBJECTIVES

1. To factor a trinomial of the form

$$ax^2 + bx + c$$

2. To completely factor a trinomial

Factoring trinomials is more time-consuming when the coefficient of the first term is not 1. Look at the following multiplication.

$$(5x + 2)(2x + 3) = 10x^2 + 19x + 6$$

Factors   Factors
of $10x^2$ of 6

Do you see the additional problem? We must consider all possible factors of the first coefficient (10 in the example) as well as those of the third term (6 in our example).

There is no easy way out! You need to form all possible combinations of factors and then check the middle term until the proper pair is found. If this seems a bit like guesswork, you're almost right. In fact some call this process factoring by "trial and error."

We can simplify the work a bit by reviewing the sign patterns found in Section 5.3.

Any time the leading coefficient is negative, factor out a negative one from the trinomial. This will leave one of these cases.

SIGN PATTERNS FOR FACTORING TRINOMIALS

1. If all terms of a trinomial are positive, the signs between the terms in the binomial factors are both plus signs.
2. If the first and third terms of the trinomial are positive and the middle term is negative, the signs between the terms in the binomial factors are both minus signs.
3. If the third term of the trinomial is negative, the signs between the terms in the binomial factors are opposite (one is + and one is −).

Example 1

Factor $3x^2 + 14x + 15$.

First, list the possible factors of 3, the coefficient of the first term.

$$3 = 1 \cdot 3$$

Now list the factors of 15, the last term.

$$15 = 1 \cdot 15$$
$$= 3 \cdot 5$$

Because the signs of the trinomial are all positive, we know any factors will have the form

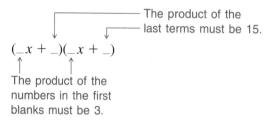

So the following are the possible factors and the corresponding middle terms:

POSSIBLE FACTORS	MIDDLE TERMS
$(x + 1)(3x + 15)$	$18x$
$(x + 15)(3x + 1)$	$46x$
$(3x + 3)(x + 5)$	$18x$
$(3x + 5)(x + 3)$	$14x$

The correct middle term

So

Take the time to multiply the binomial factors. This habit will ensure that you have an expression equivalent to the original problem.

$$3x^2 + 14x + 15 = (3x + 5)(x + 3)$$

CHECK YOURSELF 1

Factor.

1. $5x^2 + 14x + 8$ **2.** $3x^2 + 20x + 12$

Example 2

Factor $4x^2 - 11x + 6$.

Since only the middle term is negative, we know the factors have the form

$$(\_x - \_)(\_x - \_)$$

Both signs are negative.

Now look at the factors of the first coefficient and the last term.

$$4 = 1 \cdot 4 \qquad 6 = 1 \cdot 6$$
$$ = 2 \cdot 2 \qquad = 2 \cdot 3$$

This gives us the possible factors:

POSSIBLE FACTORS	MIDDLE TERMS
$(x - 1)(4x - 6)$	$-10x$
$(x - 6)(4x - 1)$	$-25x$
$(x - 2)(4x - 3)$	$-11x$

The correct middle term

Note that, in this example, we *stopped* as soon as the correct pair of factors was found. So

$4x^2 - 11x + 6 = (x - 2)(4x - 3)$

Again, at least mentally, check your work by multiplying the factors.

CHECK YOURSELF 2

Factor.

1. $2x^2 - 9x + 9$ **2.** $6x^2 - 17x + 10$

Let's factor a trinomial whose last term is negative.

Example 3

Factor $5x^2 + 6x - 8$.

Since the last term is negative, the factors have the form

$(\_x + \_)(\_x - \_)$

Consider the factors of the first coefficient and the last term.

$5 = 1 \cdot 5 \qquad 8 = 1 \cdot 8$
$\qquad\qquad\quad = 2 \cdot 4$

The possible factors are then

POSSIBLE FACTORS	MIDDLE TERMS
$(x + 1)(5x - 8)$	$-3x$
$(x + 8)(5x - 1)$	$39x$
$(5x + 1)(x - 8)$	$-39x$
$(5x + 8)(x - 1)$	$3x$
$(x + 2)(5x - 4)$	$6x$

Again we stop as soon as the correct pair of factors is found.

$5x^2 + 6x - 8 = (x + 2)(5x - 4)$

CHECK YOURSELF 3

Factor $4x^2 + 5x - 6$.

The same process is used to factor a trinomial with more than one variable.

Example 4

Factor $6x^2 + 7xy - 10y^2$.

The form of the factors must be

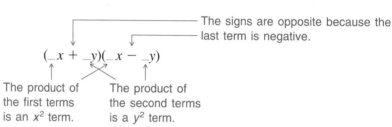

The signs are opposite because the last term is negative.

$(\_x + \_y)(\_x - \_y)$

The product of the first terms is an x^2 term.

The product of the second terms is a y^2 term.

Again look at the factors of the first and last coefficients.

$$6 = 1 \cdot 6 \qquad 10 = 1 \cdot 10$$
$$ = 2 \cdot 3 \qquad = 2 \cdot 5$$

Be certain that you have a pattern that matches up every possible pair of coefficients.

POSSIBLE FACTORS	MIDDLE TERMS
$(x + y)(6x - 10y)$	$-4xy$
$(x + 10y)(6x - y)$	$59xy$
$(6x + y)(x - 10y)$	$-59xy$
$(6x + 10y)(x - y)$	$4xy$
$(x + 2y)(6x - 5y)$	$7xy$

Once more, we stop as soon as the correct factors are found.

$$6x^2 + 7xy - 10y^2 = (x + 2y)(6x - 5y)$$

CHECK YOURSELF 4

Factor $15x^2 - 4xy - 4y^2$.

The next example illustrates a special kind of trinomial called a *perfect-square trinomial*.

Example 5

Factor $9x^2 + 12xy + 4y^2$.

Since all terms are positive, the form of the factors must be

$(\_x + \_y)(\_x + \_y)$

Consider the factors of the first and last coefficients.

$$9 = 9 \cdot 1 \qquad 4 = 4 \cdot 1$$
$$ = 3 \cdot 3 \qquad = 2 \cdot 2$$

POSSIBLE FACTORS	MIDDLE TERMS
$(x + y)(9x + 4y)$	$13xy$
$(x + 4y)(9x + y)$	$37xy$
$(3x + 2y)(3x + 2y)$	$12x$

So

Perfect-square trinomials can be factored by using previous methods. Recognizing the special pattern simply saves time.

$$9x^2 + 12xy + 4y^2 = (3x + 2y)(3x + 2y)$$
$$= (3x + 2y)^2$$

Square of $3x$ $2(3x)(2y)$ Square of $2y$

This trinomial is the result of squaring a binomial, thus the special name of perfect-square trinomial.

CHECK YOURSELF 5

1. $4x^2 + 28x + 49$ **2.** $16x^2 - 40xy + 25y^2$

Before we look at our next example, let's review one important point from Section 5.3. Recall that when you factor trinomials, you should not forget to look for a common factor as the first step. If there is a common factor, remove it and factor the remaining trinomial as before.

Example 6

Factor $18x^2 - 18x + 4$.

First look for a common factor in all three terms. Here that factor is 2, so write

$$18x^2 - 18x + 4 = 2(9x^2 - 9x + 2)$$

By our earlier methods, we can factor the remaining trinomial as

If you don't see why this is true, you need to use your pencil to work it out before you move on!

$$9x^2 - 9x + 2 = (3x - 1)(3x - 2)$$

So

$$18x^2 - 18x + 4 = 2(3x - 1)(3x - 2)$$

Don't forget the 2 that was factored out!

CHECK YOURSELF 6

Factor $16x^2 + 44x - 12$.

Let's look at an example in which the common factor includes a variable.

Example 7

Factor

$$6x^3 + 10x^2 - 4x$$

The common factor is $2x$.

So

$$6x^3 + 10x^2 - 4x = 2x(3x^2 + 5x - 2)$$
$$= 2x(3x - 1)(x + 2)$$

since

$$3x^2 + 5x - 2 = (3x - 1)(x + 2)$$

And so

Remember to include the monomial factor.

$$6x^3 + 10x^2 - 4x = 2x(3x - 1)(x + 2)$$

CHECK YOURSELF 7

Factor $6x^3 - 27x^2 + 30x$.

Strategies in Factoring

You have now had a chance to work with a variety of factoring techniques. Your success in factoring polynomials depends on your ability to recognize when to use which technique. Here are some guidelines to help you apply the factoring methods you have studied in this chapter.

FACTORING POLYNOMIALS

1. Look for a greatest common factor other than 1. If such a factor exists, factor out the GCF.
2. To continue the factoring process:
   a. If the polynomial that remains is a *binomial*, check to see if it is a difference of two squares. If it is, apply the difference-of-squares formula.
   b. If the polynomial that remains is a *trinomial*, try to factor the trinomial by the trial-and-error methods of Sections 5.3 and 5.4.

The following example illustrates the use of this strategy.

Example 8

(a) Factor $5m^2n + 20n$.

First, we see that the GCF is $5n$. Removing that factor gives

$$5m^2n + 20n = 5n(m^2 + 4)$$

Since the binomial that remains is a *sum* of squares and is *not* factorable, we have completed the factorization.

(b) Factor $3x^3 - 48x$.

First, we see that the GCF is $3x$. Factoring out $3x$ yields

$$3x^3 - 48x = 3x(x^2 - 16)$$

Now we see that the binomial that remains, $x^2 - 16$, is a difference of squares. We then complete the factorization as

$$3x^3 - 48x = 3x(x + 4)(x - 4)$$

(c) Factor $8r^2s + 20rs^2 - 12s^3$.

First, the GCF is $4s$, and we can write the original polynomial as

$$8r^2s + 20rs^2 - 12s^3 = 4s(2r^2 + 5rs - 3s^2)$$

Since the remaining polynomial is a trinomial, we can use the trial-and-error method to complete the factoring as

$$8r^2s + 20rs^2 - 12s^3 = 4s(2r - s)(r + 3s)$$

CHECK YOURSELF 8

Factor the following polynomials.

1. $8a^3 - 32ab^2$ **2.** $7x^3 + 7x^2y - 42xy^2$ **3.** $5m^4 + 15m^3 + 5m^2$

CHECK YOURSELF ANSWERS

1. (1) $(5x + 4)(x + 2)$; (2) $(3x + 2)(x + 6)$.
2. (1) $(2x - 3)(x - 3)$; (2) $(6x - 5)(x - 2)$.
3. $(4x - 3)(x + 2)$.
4. $(3x - 2y)(5x + 2y)$.
5. (1) $(2x + 7)^2$; (2) $(4x - 5y)^2$.
6. $4(4x - 1)(x + 3)$.
7. $3x(2x - 5)(x - 2)$.
8. (1) $8a(a + 2b)(a - 2b)$; (2) $7x(x + 3y)(x - 2y)$; (3) $5m^2(m^2 + 3m + 1)$.

Build Your Skills

Complete each of the following statements.

1. $4x^2 - 4x - 3 = (2x + 1)($ $)$

2. $3w^2 + 11w - 4 = (w + 4)($ $)$

3. $6a^2 + 13a + 6 = (2a + 3)($ $)$

4. $25y^2 - 10y + 1 = (5y - 1)($ $)$

5. $15x^2 - 16x + 4 = (3x - 2)($ $)$

6. $6m^2 + 5m - 4 = (3m + 4)($ $)$

7. $16a^2 + 8ab + b^2 = (4a + b)($ $)$

8. $6x^2 + 5xy - 4y^2 = (3x + 4y)($ $)$

9. $4m^2 + 5mn - 6n^2 = (m + 2n)($ $)$

10. $10p^2 - pq - 3q^2 = (5p - 3q)($ $)$

Factor each of the following polynomials.

11. $3x^2 + 7x + 2$

12. $5y^2 + 8y + 3$

13. $2w^2 + 13w + 15$

14. $3x^2 - 16x + 21$

15. $5x^2 - 16x + 3$

16. $2a^2 + 7a + 5$

17. $4x^2 - 12x + 5$

18. $2x^2 + 11x + 12$

19. $3x^2 - 5x - 2$

20. $4m^2 - 23m + 15$

21. $4p^2 + 19p - 5$

22. $5x^2 - 36x + 7$

ANSWERS

1. _____

2. _____

3. _____

4. _____

5. _____

6. _____

7. _____

8. _____

9. _____

10. _____

11. _____

12. _____

13. _____

14. _____

15. _____

16. _____

17. _____

18. _____

19. _____

20. _____

21. _____

22. _____

23. _____

24. _____

25. _____

26. _____

27. _____

28. _____

29. _____

30. _____

31. _____

32. _____

33. _____

34. _____

35. _____

36. _____

37. _____

38. _____

39. _____

40. _____

41. _____

42. _____

43. _____

44. _____

45. _____

46. _____

47. _____

48. _____

23. $6x^2 + 19x + 10$

24. $6x^2 - 7x - 3$

25. $15x^2 + x - 6$

26. $12w^2 + 19w + 4$

27. $6m^2 + 25m - 25$

28. $8x^2 - 6x - 9$

29. $9x^2 - 12x + 4$

30. $20x^2 - 23x + 6$

31. $12x^2 - 8x - 15$

32. $16a^2 + 40a + 25$

33. $3y^2 + 7y - 6$

34. $12x^2 + 11x - 15$

35. $8x^2 - 27x - 20$

36. $24v^2 + 5v - 36$

37. $2x^2 + 3xy + y^2$

38. $3x^2 - 5xy + 2y^2$

39. $5a^2 - 8ab - 4b^2$

40. $5x^2 + 7xy - 6y^2$

41. $9x^2 + 4xy - 5y^2$

42. $16x^2 + 32xy + 15y^2$

43. $6m^2 - 17mn + 12n^2$

44. $15x^2 - xy - 6y^2$

45. $36a^2 - 3ab - 5b^2$

46. $3q^2 - 17qr - 6r^2$

47. $x^2 + 4xy + 4y^2$

48. $25b^2 - 80bc + 64c^2$

Factor each of the following polynomials completely.

49. $20x^2 - 20x - 15$

50. $24x^2 - 18x - 6$

51. $8m^2 + 12m + 4$

52. $14x^2 - 20x + 6$

53. $15r^2 - 21rs + 6s^2$

54. $10x^2 + 5xy - 30y^2$

55. $2x^3 - 2x^2 - 4x$

56. $2y^3 + y^2 - 3y$

57. $2y^4 + 5y^3 + 3y^2$

58. $4z^3 - 18z^2 - 10z$

59. $36a^3 - 66a^2 + 18a$

60. $20n^4 - 22n^3 - 12n^2$

61. $9p^2 + 30pq + 21q^2$

62. $12x^2 + 2xy - 24y^2$

Think About These

Factor each of the following polynomials completely.

63. $10(x + y)^2 - 11(x + y) - 6$

64. $8(a - b)^2 + 14(a - b) - 15$

65. $5(x - 1)^2 - 15(x - 1) - 350$

66. $3(x + 1)^2 - 6(x + 1) - 45$

67. $15 + 29x - 48x^2$

68. $12 + 4a - 21a^2$

69. $-6x^2 + 19x - 15$

70. $-3s^2 - 10s + 8$

49. _____

50. _____

51. _____

52. _____

53. _____

54. _____

55. _____

56. _____

57. _____

58. _____

59. _____

60. _____

61. _____

62. _____

63. _____

64. _____

65. _____

66. _____

67. _____

68. _____

69. _____

70. _____

a. _____	
b. _____	
c. _____	
d. _____	
e. _____	
f. _____	
g. _____	
h. _____	

Skillscan (Section 3.4)

Solve each of the following equations.

a. $x + 2 = 0$ **b.** $x - 3 = 0$

c. $2x - 3 = 0$ **d.** $3x - 2 = 0$

e. $6x - 5 = 0$ **f.** $3x + 2 = 0$

g. $5x + 1 = 0$ **h.** $8x - 3 = 0$

ANSWERS

1. $2x - 3$ **3.** $3a + 2$ **5.** $5x - 2$ **7.** $4a + b$ **9.** $4m - 3n$ **11.** $(3x + 1)(x + 2)$

13. $(2w + 3)(w + 5)$ **15.** $(5x - 1)(x - 3)$ **17.** $(2x - 5)(2x - 1)$ **19.** $(3x + 1)(x - 2)$

21. $(4p - 1)(p + 5)$ **23.** $(3x + 2)(2x + 5)$ **25.** $(5x - 3)(3x + 2)$ **27.** $(6m - 5)(m + 5)$

29. $(3x - 2)(3x - 2)$ **31.** $(6x + 5)(2x - 3)$ **33.** $(3y - 2)(y + 3)$ **35.** $(8x + 5)(x - 4)$

37. $(2x + y)(x + y)$ **39.** $(5a + 2b)(a - 2b)$ **41.** $(9x - 5y)(x + y)$

43. $(3m - 4n)(2m - 3n)$ **45.** $(12a - 5b)(3a + b)$ **47.** $(x + 2y)^2$

49. $5(2x - 3)(2x + 1)$ **51.** $4(2m + 1)(m + 1)$ **53.** $3(5r - 2s)(r - s)$

55. $2x(x - 2)(x + 1)$ **57.** $y^2(2y + 3)(y + 1)$ **59.** $6a(3a - 1)(2a - 3)$

61. $3(p + q)(3p + 7q)$ **63.** $(5x + 5y + 2)(2x + 2y - 3)$ **65.** $5(x - 11)(x + 6)$

67. $(1 + 3x)(15 - 16x)$ **69.** $(3x - 5)(-2x + 3)$ **a.** -2 **b.** 3 **c.** $\dfrac{3}{2}$ **d.** $\dfrac{2}{3}$

e. $\dfrac{5}{6}$ **f.** $-\dfrac{2}{3}$ **g.** $-\dfrac{1}{5}$ **h.** $\dfrac{3}{8}$

© 1993 McGraw-Hill, Inc.

5.5

Solving Equations by Factoring

OBJECTIVE

To solve quadratic equations by factoring

There are many applications of our work with factoring. One important use of factoring is to solve certain types of equations. First we need to review an idea from arithmetic. If the product of two factors is 0, then one or both of the factors must be equal to 0. This is called the *zero-product principle*. In symbols,

Note: This rule applies only if a product is equal to 0. This will be very important later on.

If $a \cdot b = 0$, then $a = 0$ or $b = 0$ or both.

Let's use this principle to solve an equation.

Example 1

Note that we are now working with equations instead of expressions. Now we are looking for values for the variable that make the statement true.

Solve $(x - 3)(x + 2) = 0$.

Using the zero-product principle gives

If $(x - 3)(x + 2) = 0$, then

$x - 3 = 0$ or $x + 2 = 0$	Set each factor equal to 0.
$x = 3$ or $x = -2$	Solve each equation.

So 3 and -2 are the solutions for the equation.

CHECK YOURSELF 1

Solve $(x + 3)(x - 4) = 0$.

The left side of the equation of Example 1 was already in factored form. The next example illustrates a case in which factoring is necessary.

Example 2

Note: This equation is *not linear* because of the x^2 term.

Solve $x^2 - 6x + 5 = 0$.

$$x^2 - 6x + 5 = 0$$
$$(x - 5)(x - 1) = 0 \qquad \text{Factor on the left.}$$

Again set each of the factors equal to zero.

$$x - 5 = 0 \quad \text{or} \quad x - 1 = 0$$
$$x = 5 \qquad\qquad x = 1$$

The solutions for the equation are 5 and 1.

We can check the solutions as before by substituting the two values back into the original equation.

Letting x be 5: | Letting x be 1:
$5^2 - 6 \cdot 5 + 5 = 0$ | $1^2 - 6 \cdot 1 + 5 = 0$
$25 - 30 + 5 = 0$ | $1 - 6 + 5 = 0$
$0 = 0$ | $0 = 0$

Both solutions are verified.

CHECK YOURSELF 2

Solve $x^2 - 2x - 8 = 0$.

The equation of Example 2 has two solutions and has a special form.

Note: Standard form: The equation is *set equal to 0*, and the terms are written in *descending-exponent order*.

An equation with the form

$ax^2 + bx + c = 0$ where $a \neq 0$

is called a *quadratic equation in standard form*.

Quadratic equations must be in standard form before you factor, as the next example illustrates.

Example 3

Solve $2x^2 + 5x = 3$.

Be very careful! To use the zero-product principle, one side of the equation *must be zero*. So rewrite the equation in standard form by subtracting 3 from both sides. We now have

$$2x^2 + 5x - 3 = 0$$
$$(2x - 1)(x + 3) = 0 \qquad \text{Again factor on the left.}$$

Set each of the factors equal to zero.

$$2x - 1 = 0 \quad \text{or} \quad x + 3 = 0$$
$$2x = 1 \qquad\qquad x = -3$$
$$x = \frac{1}{2}$$

So $\dfrac{1}{2}$ and -3 are the two solutions for the equation. We will leave the check of these solutions to you. Be sure to return to the original equation to verify these results.

CHECK YOURSELF 3

Solve.

1. $3x^2 - 5x = 2$ **2.** $2x^2 + 3 = 7x$

The following examples show how other factoring techniques are used in solving quadratic equations.

Example 4

(*a*) Solve $x^2 + 5x = 0$.

Note that x is a common factor on the left. Factoring, we have

Note: When x is a common factor of the quadratic member, you must set x equal to 0 to find the zero solution of the equation.

$$x(x + 5) = 0$$

Set each factor equal to 0.

$$x = 0 \quad \text{or} \quad x + 5 = 0$$
$$x = -5$$

The solutions are 0 and -5.

(*b*) Solve $x^2 = 4x$.

First, write the equation in standard form (set equal to 0).

$$x^2 - 4x = 0$$
$$x(x - 4) = 0 \qquad \text{Factor on the left.}$$

$$x = 0 \quad \text{or} \quad x - 4 = 0$$
$$x = 4$$

The solutions are 0 and 4.

Solve.

1. $x^2 + 8x = 0$ **2.** $x^2 = 9x$

The following example illustrates how the difference-of-squares factoring technique is applied in solving quadratic equations.

Example 5

Solve $x^2 = 9$.

Again, write the equation in standard form.

$$x^2 - 9 = 0$$

The left side is a difference of squares, so we have

$$(x + 3)(x - 3) = 0$$

So

$$x + 3 = 0 \qquad \text{or} \qquad x - 3 = 0$$
$$x = -3 \qquad\qquad x = 3$$

The solutions are -3 and 3.

Solve $x^2 = 25$.

Example 6

Solve $3x^2 - 3x - 6 = 0$.

Note that there is a common factor of 3 on the left. Write

$$3(x^2 - x - 2) = 0$$
$$3(x - 2)(x + 1) = 0$$

Then, after dividing both sides by 3, we have

Note: On the right:

$$\frac{0}{3} = 0$$

$$(x - 2)(x + 1) = 0$$

So

$$x - 2 = 0 \qquad \text{or} \qquad x + 1 = 0$$
$$x = 2 \qquad\qquad\qquad x = -1$$

The solutions are 2 and -1.

CHECK YOURSELF 6

Solve $4x^2 + 14x = -6$.

The following rules summarize our work in solving quadratic equations by factoring.

TO SOLVE A QUADRATIC EQUATION BY FACTORING

STEP 1   Add or subtract the necessary terms on both sides of the equation so that the equation is in standard form (set equal to 0).

STEP 2   Factor the quadratic expression.

STEP 3   Set each factor which contains a variable equal to 0.

STEP 4   Solve the resulting equations to find the solutions.

STEP 5   Check each solution by substituting in the original equation.

Note: If the polynomial in step 2 is not factorable, you must use other methods for solving the equation. We will discuss other approaches in Chapter 10.

Strategies in Equation Solving

Keep in mind that the methods of this section deal with solving quadratic equations. Earlier, in Chapter 3, we developed techniques to solve linear equations.

An important skill in equation solving is pattern recognition, that is, knowing which method to apply by recognizing the form of an equation. Our final example will help you in that process.

Example 7

For each of the following equations, determine whether the given equation is linear or quadratic. Then find the solution for each equation.

The equation is *linear* because it can be written in the form

$$ax + b = 0$$

where $a \neq 0$. The variable x can appear only to the first power.

This equation is *quadratic* because x appears to the second power.

(*a*) $2x + 1 = 0$

The equation $2x + 1 = 0$ is *linear*. To solve the equation, we want to isolate the variable x on the left.

$$2x = -1$$

$$x = -\frac{1}{2}$$

(*b*) $x(2x + 1) = 0$

Multiplying on the left, we see that the equivalent equation, $2x^2 + x = 0$, is *quadratic*. The solutions can be found by the methods of this section.

$$x = 0 \quad \text{or} \quad x = -\frac{1}{2}$$

(*c*) $(3x + 1) + (x - 2) = 0$

Clearing parentheses and combining like terms, we see that the equivalent equation, $4x - 1 = 0$, is *linear*. The solution in this case is $x = \frac{1}{4}$.

(*d*) $(3x + 1)(x - 2) = 26$

Multiplying on the left and writing the equation in standard form, we see that the equivalent equation, $3x^2 - 5x - 28 = 0$, is *quadratic*.

The solutions, which can be found by factoring, are $-\frac{7}{3}$ and 4.

CHECK YOURSELF 7

Determine whether each equation is linear or quadratic. Then find all solutions.

1. $3x - 5 = 0$ **2.** $(2x - 1)(3x + 4) = 10$
3. $(2x - 1) + (3x - 5) = 0$ **4.** $(x^2 + 2) + (2x^2 - 5) = 0$

CHECK YOURSELF ANSWERS

1. -3, 4.   **2.** 4, -2.

3. (1) 2, $-\frac{1}{3}$; (2) 3, $\frac{1}{2}$.   **4.** (1) 0, -8; (2) 0, 9.

5. 5, -5.   **6.** -3, $-\frac{1}{2}$.

7. (1) Linear, $\frac{5}{3}$; (2) quadratic, -2, $\frac{7}{6}$; (3) linear, $\frac{6}{5}$; (4) quadratic, -1, 1.

Name _____

Date _____

Build Your Skills

Solve each of the following quadratic equations.

1. $(x - 1)(x - 2) = 0$

2. $(x - 5)(x + 3) = 0$

3. $(2x + 1)(x - 5) = 0$

4. $(3x - 2)(x - 5) = 0$

5. $x^2 - 2x - 3 = 0$

6. $x^2 + 5x + 4 = 0$

7. $x^2 - 7x + 6 = 0$

8. $x^2 + 3x - 10 = 0$

9. $x^2 + 8x + 15 = 0$

10. $x^2 - 3x - 18 = 0$

11. $x^2 + 4x - 21 = 0$

12. $x^2 - 12x + 32 = 0$

13. $x^2 - 6x = 7$

14. $x^2 + 6x = -8$

15. $x^2 + 2x = 15$

16. $x^2 = 7x - 10$

17. $2x^2 + 5x - 3 = 0$

18. $3x^2 + 7x + 2 = 0$

19. $4x^2 - 24x + 35 = 0$

20. $6x^2 + 11x - 10 = 0$

21. $4x^2 + 11x = -6$

22. $5x^2 + 2x = 3$

23. $5x^2 + 13x = 6$

24. $4x^2 = 13x + 12$

ANSWERS

1. _____
2. _____
3. _____
4. _____
5. _____
6. _____
7. _____
8. _____
9. _____
10. _____
11. _____
12. _____
13. _____
14. _____
15. _____
16. _____
17. _____
18. _____
19. _____
20. _____
21. _____
22. _____
23. _____
24. _____

25. $x^2 - 2x = 0$ **26.** $x^2 + 5x = 0$

27. $x^2 = -8x$ **28.** $x^2 = 7x$

29. $5x^2 - 15x = 0$ **30.** $4x^2 + 20x = 0$

31. $x^2 - 25 = 0$ **32.** $x^2 = 49$

33. $x^2 = 81$ **34.** $x^2 = 64$

35. $2x^2 - 18 = 0$ **36.** $3x^2 - 75 = 0$

37. $3x^2 + 24x + 45 = 0$ **38.** $4x^2 - 4x = 24$

39. $8x^2 + 20x = 12$ **40.** $9x^2 + 42x = 15$

41. $(x - 2)(x + 1) = 10$ **42.** $(x - 4)(x + 1) = 14$

Think About These

Translate each of the following statements to an equation. Let x represent the number in each case.

43. The square of a number minus 3 times that same number is 4.

44. The sum of the square of a number and 5 times that same number is 14.

340

45. If 4 times a number is added to the square of that same number, the sum is 12.

46. If a number is subtracted from 3 times the square of that number, the difference is 2.

Simplify each equation, label the equation as linear or quadratic, and then find all solutions.

47. $3x + 4 = 2x$

48. $3x + 4 = 2$

49. $(2x + 5)(x + 1) = 0$

50. $(2x + 5) + (x + 1) = 0$

51. $(2x - 1) - (x + 5) = 0$

52. $(2x - 1)(x + 5) = 13$

53. $x^2 + 4x - 5 = x^2$

54. $x^2 + 4x = 5$

55. $(x^2 + 2x - 3) + (x^2 + 10x - 29) = 0$

56. $x^2 + 2x - 3 = x^2 + 10x - 29$

57. $(x + 1)^2 = x^2 + 4$

58. $(2x + 1)^2 = 3x^2 + 13$

59. $x^2 - 3x - 5 = x + 7$

60. $x(x + 1) = (x + 2)^2$

Skillscan (Section 3.8)

Solve each of the following word problems. Show the equation used for each solution.

a. One number is 5 more than another. If the sum of the two numbers is 19, find the two numbers.

b. One number is 3 less than twice another. If the sum of the two numbers is 24, what are the two numbers?

ANSWERS

45. _____

46. _____

47. _____

48. _____

49. _____

50. _____

51. _____

52. _____

53. _____

54. _____

55. _____

56. _____

57. _____

58. _____

59. _____

60. _____

a. _____

b. _____

c. The sum of two consecutive even integers is 82. Find the two numbers.

d. If the sum of three consecutive odd integers is 57, what are the three integers?

e. The length of a rectangle is 5 centimeters (cm) more than its width. If the perimeter of the rectangle is 42 cm, find the length and width of the rectangle.

f. The length of a rectangle is 2 inches (in) more than 3 times its width. If the perimeter of the rectangle is 60 in, what are the dimensions of the rectangle?

ANSWERS

1. 1, 2   **3.** $-\dfrac{1}{2}$, 5   **5.** -1, 3   **7.** 1, 6   **9.** -3, -5 **11.** -7, 3   **13.** -1, 7

15. -5, 3   **17.** -3, $\dfrac{1}{2}$ **19.** $\dfrac{5}{2}$, $\dfrac{7}{2}$ **21.** $-\dfrac{3}{4}$, -2 **23.** -3, $\dfrac{2}{5}$ **25.** 0, 2

27. 0, -8 **29.** 0, 3   **31.** -5, 5   **33.** -9, 9   **35.** -3, 3   **37.** -5, -3

39. -3, $\dfrac{1}{2}$ **41.** -3, 4   **43.** $x^2 - 3x = 4$ **45.** $x^2 + 4x = 12$ **47.** Linear; -4

49. Quadratic; $-\dfrac{5}{2}$, -1 **51.** Linear; 6   **53.** Linear; $\dfrac{5}{4}$ **55.** Quadratic; -8, 2

57. Linear; $\dfrac{3}{2}$ **59.** Quadratic; -2, 6   **a.** 7, 12   **b.** 9, 15   **c.** 40, 42   **d.** 17, 19, 21

e. 13 cm, 8 cm   **f.** 7 in by 23 in

5.6 More Applications

OBJECTIVE

To solve word problems that can be represented by quadratic equations

Certain types of word problems will lead to quadratic equations. You will have to use your work from Section 5.5 to solve them. Look at the following example.

Example 1

Remember our five-step process for solving word problems:

1. Read the problem carefully. Then reread it to decide what you are asked to find.

2. Choose a letter to represent the unknown or unknowns.

3. Translate the problem to the language of algebra to form an equation.

4. Solve the equation.

One number is 3 more than another. If their product is 54, find the two numbers.

Step 1 You are asked to find the two unknown numbers.

Step 2 Let x be the smaller number.

Then $x + 3$ is the larger.

$\underbrace{}$
$\uparrow$
"3 more than x"

Step 3

$\underbrace{x(x + 3)}_{} = 54$
$\uparrow$
"Their product is 54."

Step 4

$x(x + 3) = 54$
$x^2 + 3x = 54$ Multiply on the left.

You should recognize that the equation is quadratic. It must be solved by the methods of Section 5.5.

$x^2 + 3x - 54 = 0$ Write the equation in standard form.
$(x + 9)(x - 6) = 0$ Factor.

$x + 9 = 0$ or $x - 6 = 0$
$x = -9$ $x = 6$

When word problems lead to quadratic equations, there may be two possible solutions to the equation. It is important to check, because one of the solutions may be eliminated by the conditions of the original problem. Here, both satisfy those conditions and so are valid solutions to the problem. The numbers are

Note that -9 and 6 are *not* the numbers asked for in the problem. The solutions are -9 and -6 *or* 6 and 9.

$$x = -9 \xleftarrow{\quad\quad} \begin{array}{c} \text{Smaller} \\ \text{number} \end{array} \xrightarrow{\quad\quad} x = 6$$

$$x + 3 = -6 \xleftarrow{\quad\quad} \begin{array}{c} \text{Larger} \\ \text{number} \end{array} \xrightarrow{\quad\quad} x + 3 = 9$$

5. Verify your solution by returning to the original problem.

Step 5   Since there are two separate solutions, both must be verified in the original problem. Since $(-9)(-6)$ and $6 \cdot 9$ are both 54, both pairs of numbers are solutions.

CHECK YOURSELF 1

The product of two consecutive integers is 90. What are the two integers? *Hint:*   Consecutive integers can be represented by x and $x + 1$.

Example 2

The sum of a number and its square is 30. What is the number?

Step 1   We want to find the unknown number.

Step 2   Let x be the number. Then the square of the number is x^2.

Step 3   Now write the equation.

$$\underbrace{x + x^2}_{\uparrow} = 30$$
The sum of the number and its square

Step 4   Writing the equation in standard form, we have

We factor and solve as before.

$$x^2 + x - 30 = 0$$
$$(x + 6)(x - 5) = 0$$
$$x + 6 = 0 \qquad \text{or} \qquad x - 5 = 0$$
$$x = -6 \qquad\qquad\qquad x = 5$$

Possible solutions to the problem are -6 and 5.

Step 5   Since we have two possible values, both must be checked.

$$
\begin{array}{c|c}
-6 + (-6)^2 = 30 & 5 + 5^2 = 30 \\
-6 + 36 = 30 & 5 + 25 = 30 \\
30 = 30 & 30 = 30
\end{array}
$$

Both solutions are verified.

CHECK YOURSELF 2

The square of an integer is 6 less than 7 times the integer. What is the integer?

Many problems involving geometric figures will also lead to quadratic equations that can be solved by factoring. The following example illustrates.

Example 3

A rectangle is 5 cm longer than it is wide. If the area of the rectangle is 84 cm^2, find the dimensions of the rectangle.

Step 1 You want to find the dimensions (the width and length) of the rectangle.

Step 2 Let x be the width of the rectangle; then $x + 5$ is the length. In solving geometric problems, always draw a sketch labeled with the variables assigned in this step. Here we would have

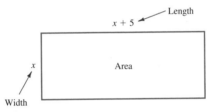

Step 3 Since the area of a rectangle is the product of its width and its length, we have

$$x(x + 5) = 84$$

Width Length Area

Step 4 Multiply and write the equation in standard form.

$$x^2 + 5x - 84 = 0$$
$$(x + 12)(x - 7) = 0$$
$$x + 12 = 0 \qquad \text{or} \qquad x - 7 = 0$$
$$x = -12 \qquad\qquad x = 7$$

This is not a possible solution—the width can't be negative!

The only possible solution is 7 cm. The width of the rectangle is 7 cm, and the length is 12 cm.

Step 5 If the width is 7 cm and the length is 12 cm, the area is 84 cm^2. We have verified the solution.

CHECK YOURSELF 3

The length of a rectangle is 3 in more than twice its width. If the area of the rectangle is 90 in^2, what are its dimensions?

CHECK YOURSELF ANSWERS

1. 9 and 10 *or* −10 and −9.   **2.** 1 or 6.
3. Width, 6 in; length, 15 in.

Name

Date

Solve each of the following word problems. Be sure to show the equations used.

1. One integer is 5 less than another. If the product of the numbers is 66, find the two numbers.

1. _____

2. One integer is 7 more than another. If the product of the numbers is 60, find the two numbers.

2. _____

3. One integer is 1 less than twice another. If the product of the integers is 120, what are the two integers?

3. _____

4. One integer is 2 more than 3 times another. If the product of the integers is 56, find the two integers.

4. _____

5. The product of two consecutive integers is 132. Find the two integers?

5. _____

6. If the product of two consecutive positive even integers is 120, find the two integers.

6. _____

7. The sum of an integer and its square is 72. What is the integer?

7. _____

8. The square of an integer is 56 more than that integer. What is the integer?

8. _____

9. The square of an integer is 20 more than 8 times that integer. What is the integer?

9. _____

10. An integer is added to 3 times its square. The sum is 52. What is the integer?

10. _____

11. If 2 times the square of an integer is increased by that integer, the sum is 55. Find the integer.

11. _____

12. If the square of an integer is 21 less than 10 times that integer, what is the integer?

12. _____

13. _____

13. One positive integer is 1 more than twice another. If the difference of their squares is 65, what are the two integers?

14. _____

14. One positive integer is 3 less than twice another. If the difference of their squares is 24, find the two integers.

15. _____

15. The sum of the squares of two consecutive positive even integers is 100. Find the two integers.

16. _____

16. If the product of two consecutive positive odd integers is 63, find the two integers.

17. _____

17. The sum of the squares of two consecutive integers is 61. Find the integers.

18. _____

18. The sum of the squares of two consecutive positive even integers is 164. Find the numbers.

19. _____

19. The sum of the squares of three consecutive positive odd integers is 83. What are the integers?

20. _____

20. The sum of the squares of three consecutive integers is 50. Find the three integers.

21. _____

21. Twice the square of a positive integer is 12 more than 5 times that integer. Find the integer.

22. _____

22. Find an integer such that 10 more than the square of the integer is 40 more than the integer.

23. _____

23. The length of a rectangle is 8 feet (ft) longer than its width. If the area of the rectangle is 65 square feet (ft^2), what are the dimensions of the rectangle?

24. The length of a rectangle is 4 cm longer than its width. If the area of the rectangle is 140 cm², find the length and width of the rectangle.

25. The width of a rectangle is 3 ft less than its length. If the area of the rectangle is 70 ft², what are the dimensions of the rectangle?

26. The length of a rectangle is 5 cm more than its width. If the area of the rectangle is 150 cm², find the dimensions of the rectangle.

27. The length of a rectangle is 1 in longer than twice its width. If the area of the rectangle is 105 in², what are the dimensions of the rectangle?

28. The length of a rectangle is 2 cm less than 3 times its width. If the area of the rectangle is 40 cm², find the dimensions of the rectangle.

29. The length of a rectangle is 2 cm more than 3 times its width. If the area of the rectangle is 85 cm², find the dimensions of the rectangle.

30. If the length of a rectangle is 3 ft less than twice its width, and the area of the rectangle is 54 ft², what are the dimensions of the rectangle?

31. If the sides of a square are increased by 3 in, the area is increased by 39 in². What were the dimensions of the original square?

32. If the sides of a square are decreased by 2 cm, the area is decreased by 36 cm². What were the dimensions of the original square?

33. The length of a rectangle is 3 in more than its width. If each dimension is increased by 2 in, the area of the rectangle is increased by 26 in². Find the dimensions of the original rectangle.

34. The length of a rectangle is 4 ft more than its width. If each dimension is increased by 3 ft, the area of the rectangle is increased by 51 ft². Find the dimensions of the original rectangle.

35. _____

36. _____

a. _____

b. _____

c. _____

d. _____

e. _____

f. _____

35. A boiler in a large power plant has a volume of 2250 m³ with dimensions 15 m × 15 m × 10 m. Because of air pollution problems the power company has been required to reduce the size of its boiler. If the company has decided to reduce the 15-m length and width by the same amount and the new boiler must have a volume of 1440 m³, by what amount must the length and width be reduced?

36. A boiler in a large power plant has a volume of 3468 m³ with dimensions 17 m × 17 m × 12 m. Because of air pollution problems the power company has been required to reduce the size of its boiler. If the company has decided to reduce the 17-m length and width by the same amount and the new boiler must have a volume of 2028 m³, by what amount must the length and width be reduced?

Skillscan (Section 3.5)

Solve each of the following equations for the indicated variable.

a. $2x - 3y = 12$ (for y)

b. $V - E + F = 2$ (for E)

c. $S = 2\pi rh$ (for h)

d. $s = \dfrac{1}{2}gt^2$ (for g)

e. $S = b^2 + \dfrac{1}{2}hb$ (for h)

f. $S = \pi r^2 + 2\pi rh$ (for h)

ANSWERS

1. 6, 11 or −6, −11   **3.** 8, 15   **5.** 11, 12 or −11, −12   **7.** −9 or 8   **9.** −2 or 10
11. 5   **13.** 4, 9   **15.** 6, 8   **17.** −6, −5 or 5, 6   **19.** 3, 5, 7   **21.** 4   **23.** 5 ft by
13 ft   **25.** 7 ft by 10 ft   **27.** 7 in by 15 in   **29.** 5 cm by 17 cm   **31.** 5 in by 5 in
33. 4 in by 7 in   **35.** 3 m   **a.** $\dfrac{2(x-6)}{3}$ **b.** $V + F - 2$ **c.** $\dfrac{S}{2\pi r}$ **d.** $\dfrac{2s}{t^2}$
e. $\dfrac{2S - 2b^2}{b}$ **f.** $\dfrac{S - \pi r^2}{2\pi r}$

More on Literal Equations

OBJECTIVES
1. To solve literal equations involving parentheses
2. To solve literal equations by factoring

Back in Section 3.5 you dealt with literal equations, that is, equations that contain more than one letter or variable. The idea was to solve the literal equations for some particular letter. Often you will have to use the distributive property to remove parentheses. Look at the following example.

Example 1

Solve $a(m + n) = c$ for m.

First, use the distributive property to remove the parentheses on the left.

$$a(m + n) = c$$
$$am + an = c$$

Remember that we want all terms involving m on one side and all other terms on the opposite side. So subtract an from both sides of the equation.

$$am + an - an = c - an$$
$$\text{or} \qquad am = c - an$$

Now divide both sides by a to isolate m.

$$\frac{am}{a} = \frac{c - an}{a}$$

or

$$m = \frac{c - an}{a}$$

CHECK YOURSELF 1

Solve $x(a - b) = c$ for a.

Example 2

Solve

You may recognize this as the formula for the area of a trapezoid.

$$A = \frac{h}{2}(B + b)$$

for B.

In this case the best first step is to multiply both sides of the equation by 2 to clear fractions.

$$2A = 2 \cdot \frac{h}{2}(B + b)$$

or

$$2A = h(B + b)$$

Now remove the parentheses.

$$2A = hB + hb$$

Subtract hb from both sides.

$$2A - hb = hB$$

Remember that we want to isolate B on one side. It makes no difference whether B is on the left or the right.

Divide by h to isolate B.

$$\frac{2A - hb}{h} = \frac{hB}{h}$$

$$\frac{2A - hb}{h} = B$$

or

$$B = \frac{2A - hb}{h}$$

CHECK YOURSELF 2

Solve $S = \dfrac{n}{2}(a + t)$ for a.

Solving literal equations may also require that you factor to isolate the specified letter. This will be the case whenever the letter that you are solving for appears in more than one term.

Example 3

Solve $ax = bx + ab$ for x.

First, subtract bx from both sides.

$$\underbrace{ax - bx}_{\uparrow} = ab$$

All terms with x are now on the left.

Since x appears in more than one term, factor.

$$x(a - b) = ab$$

Now divide both sides by $a - b$ to isolate x.

$$\frac{x(a - b)}{a - b} = \frac{ab}{a - b}$$

or

$$x = \frac{ab}{a - b}$$

CHECK YOURSELF 3

Solve $my = mn - ny$ for y.

The following series of steps summarizes our work in solving literal equations.

> **TO SOLVE A LITERAL EQUATION**
>
> STEP 1   If necessary, multiply both sides of the equation by the same factor to clear fractions.
>
> STEP 2   Remove any parentheses by multiplying.
>
> STEP 3   Add or subtract the same terms on both sides so that all terms involving the variable you are solving for are on one side of the equation and all other terms are on the opposite side.
>
> STEP 4   Factor if the variable you are solving for appears in more than one term.
>
> STEP 5   Divide both sides of the equation by the same expression to isolate the variable you are solving for.

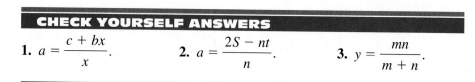

CHECK YOURSELF ANSWERS

1. $a = \dfrac{c + bx}{x}$.

2. $a = \dfrac{2S - nt}{n}$.

3. $y = \dfrac{mn}{m + n}$.

Name

Date

Build Your Skills

Solve each of the following equations for the indicated variable.

1. $a(x - y) = 3$ for x

2. $m(n + p) = 5$ for p

3. $P = 2(L + W)$ for L

4. $r = 3(s - t)$ for s

5. $C = \dfrac{5}{9}(F - 32)$ for F

6. $s = \dfrac{2}{3}(a + b)$ for b

7. $ax + bx = c$ for x

8. $my - ny = p$ for y

9. $am = bm + n$ for m

10. $pq = p - 2q$ for q

11. $m(a - b) = ab$ for b

12. $s(t + v) = tv$ for t

13. $m = \dfrac{a - b}{b}$ for b

14. $s = \dfrac{c + d}{d}$ for d

15. $t = a + (n - 1)d$ for n

16. $L = a(1 + ct)$ for c

17. $S = C - rC$ for C

18. $S = C + rC$ for C

19. $A = \dfrac{1}{2}h(B + b)$ for b

20. $A = P(1 + rt)$ for t

ANSWERS

1. _____
2. _____
3. _____
4. _____
5. _____
6. _____
7. _____
8. _____
9. _____
10. _____
11. _____
12. _____
13. _____
14. _____
15. _____
16. _____
17. _____
18. _____
19. _____
20. _____

21. _____

21. The efficient burning of fossil fuel is necessary to minimize the amount of air pollution caused by unburned fuel. Burning efficiency is related to the temperature at which the fuel is burned. If the most efficient temperature to burn coal in a large power plant is 1500° Celsius, what is this temperature in degrees Fahrenheit? $\left[C = \left(\dfrac{5}{9}\right)(F - 32).\right]$

22. _____

22. An automobile emits the least pollution when it is properly tuned because the fuel is being burned at the proper temperature and pressure. If your car burns most efficiently at 1700° Celsius, what is this temperature in degrees Fahrenheit? (See formula in problem 21 above.)

23. The volume of a cylinder is given by the formula $V = \pi r^2 h$, where r is the radius and h is the height. The scrubber on the stack of a power plant is to be in the shape of a cylinder. It is to have a capacity of 200 m^3 of gas per minute. If it must fit a stack with a radius of 5 m, how high must the scrubber be?

23. _____

24. The volume of a cylinder is given by the formula $V = \pi r^2 h$, where r is the radius and h is the height. The scrubber on the stack of a power plant is to be in the shape of a cylinder. It is to have a capacity of 175 m^3 of gas per minute. If it must fit a stack with a radius of 4 m, how high must the scrubber be?

24. _____

ANSWERS

1. $x = \dfrac{3 + ay}{a}$ **3.** $L = \dfrac{P - 2W}{2}$ **5.** $F = \dfrac{9C + 160}{5}$ **7.** $x = \dfrac{c}{a + b}$ **9.** $m = \dfrac{n}{a - b}$

11. $b = \dfrac{am}{a + m}$ **13.** $b = \dfrac{a}{m + 1}$ **15.** $n = \dfrac{t - a + d}{d}$ **17.** $C = \dfrac{S}{1 - r}$

19. $b = \dfrac{2A - hB}{h}$ **21.** 2732°F   **23.** ≈ 2.55 m

Direct and Inverse Variation

OBJECTIVES

1. To write an equation for a variation from a description
2. To solve problems of direct variation
3. To solve problems of indirect variation

Pedro makes $25 an hour as an electrician. If he works one hour, he makes $25; if he works 2 hours, he makes $50; and so on. We say his total pay varies directly with the number of hours worked.

DIRECT VARIATION

If y is a constant multiple of x, we write

$y = kx$ k is a constant

We say that y *varies directly* as x, or that y is *directly proportional* to x. The constant k is called the constant of variation.

Example 1

Marina earns $9 an hour as a tutor. Write the equation that describes the relationship between the number of hours she works and her pay.

Her pay (P) is equal to the rate of pay (r) times the number of hours worked (h), so

$P = r \cdot h$ or $P = 9h$

CHECK YOURSELF 1

Sorina is driving at a constant rate of 50 m/h. Write the equation that shows the distance she travels (d) in h hours.

Remember: k is the constant of variation.

If two things vary directly and values are given for x and y, we can find k.

Example 2

If y varies directly with x, and $y = 30$ when $x = 6$, find k.

Since y varies directly with x, we know from the definition that

$y = kx$

We need to find k. We do this by substituting 30 for y and 6 for x.

$30 = k(6)$ or $k = 5$

CHECK YOURSELF 2

If y varies directly with x and $y = 100$ when $x = 25$, find the constant of variation.

If two quantities are related so that an *increase* in the value of the first gives a proportional *decrease* in the value of the second, we say the variables *vary inversely* with each other.

INVERSE VARIATION

If y varies *inversely* as x, we write

$$y = \frac{k}{x} \qquad k \text{ is a constant}$$

We also say y is *inversely proportional* to x.

Example 3

If y varies inversely as x, and $y = 18$ when $x = \frac{1}{2}$, find the constant of variation.

From the definition, we have

$$y = \frac{k}{x}$$

Substituting 18 for y and $\frac{1}{2}$ for x, we get

$$18 = \frac{k}{\frac{1}{2}} \qquad \text{or} \qquad k = 9$$

CHECK YOURSELF 3

If y varies inversely as x, and $y = 35$ when $x = 4$, find the constant of variation.

CHECK YOURSELF ANSWERS

1. $d = 50\,h$ **2.** $k = 4$ or $y = 4x$ **3.** $k = 140$ or $y = \dfrac{140}{x}$

Name _____

Date _____

ANSWERS

Build Your Skills

1. Robin earns $14 per hour. How much does she earn in 7 h?

2. Kwang earns $8.50 per hour. How much does he earn in 30 h?

3. Lee is traveling at a constant rate of 45 m/hr. How far does she travel in 6 h?

4. An airplane is traveling at a constant rate of 550 m/h. How far does it travel in 5 h?

Translate each of the following statements of variation to an algebraic equation using k as the constant of variation.

5. s varies directly as x

6. r is inversely proportional to s

7. V is directly proportional to t

8. m varies indirectly as n

9. y varies indirectly as the square of x

10. V varies directly as the cube of r

Find k, the constant of variation.

11. y varies directly with x; $y = 54$ when $x = 6$

12. m varies directly with n; $m = 144$ when $n = 9$

1. _____

2. _____

3. _____

4. _____

5. _____

6. _____

7. _____

8. _____

9. _____

10. _____

11. _____

12. _____

13. _____

14. _____

15. _____

16. _____

17. _____

18. _____

13. V is directly proportional to h; $V = 96$ when $h = 6$

14. d is directly proportional to t; $d = 550$ when $t = 11$

15. m varies inversely to p; $m = 10$ when $p = 5$

16. y varies inversely as x; $y = 12$ when $x = \dfrac{1}{2}$

17. w is inversely proportional to t; $w = 3$ when $t = 12$

18. r is inversely proportional to t; $r = \dfrac{1}{2}$ when $t = 14$

ANSWERS

1. $98   **3.** 270 mi   **5.** $s = kx$ **7.** $V = kt$ **9.** $y = \dfrac{k}{x^2}$ **11.** $k = 9$ **13.** $k = 16$

15. $k = 50$ **17.** $k = 36$

Summary

Common-Term Factoring [5.1]

$4x^2$ is the greatest common monomial factor of $8x^4 - 12x^3 + 16x^2$.

Common Monomial Factor   A single term that is a factor of every term of the polynomial. The greatest common factor (GCF) is the common monomial factor that has the largest possible numerical coefficient and the largest possible exponents.

Factoring a Monomial from a Polynomial

$8x^4 - 12x^3 + 16x^2$
$= 4x^2(2x^2 - 3x + 4)$

1. Determine the greatest common factor.
2. Apply the distributive law in the form

$$ab + ac = a(b + c)$$

The greatest common factor

The Difference of Squares [5.2]

To factor: $16x^2 - 25y^2$:

Think: $(4x)^2 - (5y)^2$

so

$16x^2 - 25y^2$
$= (4x + 5y)(4x - 5y)$

To Factor a Difference of Squares   Use the following form:

$$a^2 - b^2 = (a + b)(a - b)$$

Factoring Trinomials [5.3 and 5.4]

For
$6x^2 - 19x + 10$
possible first terms are
$(6x\quad)(x\quad)$
or
$(3x\quad)(2x\quad)$

Forming Possible Binomial Factors   The product of the first terms of the binomial factors must equal the first term of the trinomial.

Possible last terms are
$(\quad 10)(\quad 1)$
or
$(\quad 5)(\quad 2)$

The product of the last terms of the binomial factors must equal the last term of the trinomial.

The signs in the factors are
$(\quad -)(\quad -)$

The signs in the binomial factors will follow the patterns developed in Section 5.3.

The correct binomial factors are
$(3x - 2)(2x - 5)$

To Factor a Trinomial   List all possible pairs of binomial factors. Find the correct pair of factors by calculating the sum of the inner and outer products. That sum must equal the middle term of the trinomial.

Solving Equations by Factoring [5.5]

To Solve a Quadratic Equation by Factoring

To solve:

$$x^2 + 7x = 30$$

$$x^2 + 7x - 30 = 0$$

$$(x + 10)(x - 3) = 0$$

$x + 10 = 0$ or $x - 3 = 0$

$x = -10$ and $x = 3$ are solutions.

1. Add or subtract the necessary terms on both sides of the equation so that the equation is in standard form (set equal to 0).
2. Factor the quadratic expression.
3. Set each factor equal to 0.
4. Solve the resulting equations to find the solutions.
5. Check each solution by substituting in the original equation.

Applications of Quadratic Equations [5.6]

Solving Word Problems   Certain types of word problems will lead to quadratic equations. Use the five-step process developed in Section 3.8.

In step 5 of the process (verify your solution), it is possible that one of the solutions resulting from the quadratic equation used will not satisfy the conditions of the original problem. Be sure to consider the possibility that one of the solutions must be discarded.

Solving Literal Equations [5.7]

To solve:

$$x = \frac{a + b}{a} \quad \text{for } a$$

$$ax = a + b$$

$$ax - a = b$$

1. If necessary, multiply both sides of the equation by the same term to clear fractions.
2. Remove any parentheses by multiplying.

3. Add or subtract the same terms on both sides so that all terms involving the variable you are solving for are on one side of the equation and all other terms are on the opposite side.

$$a(x - 1) = b$$

$$\frac{a(x - 1)}{x - 1} = \frac{b}{x - 1}$$

$$a = \frac{b}{x - 1}$$

4. Factor if the variable you are solving for appears in more than one term.

5. Divide both sides of the equation by the same expression to isolate the specified variable.

Direct and Inverse Variation [5.8]

y varies directly to x

$y = 45$ when $x = 9$

$y = kx$, so

$45 = k \cdot 9$

$k = 5$

Direct Variation   If y is a constant multiple of x, we write

$$y = kx \quad k \text{ is a constant}$$

We say that y *varies directly* as x, or that y is *directly proportional* to x. The constant k is called the constant of variation.

y varies inversely to x

$y = 5$ when $x = 6$

$y = \dfrac{k}{x}$, so

$5 = \dfrac{k}{6}$

$k = 30$

Inverse Variation   If y varies *inversely* as x, we write

$$y = \frac{k}{x} \quad k \text{ is a constant}$$

We also say y is *inversely proportional* to x.

Summary Exercises Chapter 5

This summary exercise set is provided to give you practice with each of the objectives of the chapter. Each exercise is keyed to the appropriate chapter section. The answers are provided in the instructor's manual. Your instructor will give you guidelines on how to best use these exercises in your instructional setting.

[5.1] Factor each of the following polynomials.

1. $18a + 24$

2. $9m^2 - 21m$

3. $24s^2t - 16s^2$

4. $18a^2b + 36ab^2$

5. $35s^3 - 28s^2$

6. $3x^3 - 6x^2 + 15x$

7. $18m^2n^2 - 27m^2n + 45m^2n^3$

8. $121x^8y^3 + 77x^6y^3$

9. $8a^2b + 24ab - 16ab^2$

10. $3x^2y - 6xy^3 + 9x^3y - 12xy^2$

11. $x(2x - y) + y(2x - y)$

12. $5(w - 3z) - w(w - 3z)$

[5.2] Factor each of the following binomials completely.

13. $p^2 - 49$

14. $25a^2 - 16$

15. $m^2 - 9n^2$

16. $16r^2 - 49s^2$

17. $25 - z^2$

18. $a^4 - 16b^2$

19. $25a^2 - 36b^2$

20. $x^6 - 4y^2$

21. $3w^3 - 12wz^2$

22. $16a^4 - 49b^2$

23. $2m^2 - 72n^4$

24. $3w^3z - 12wz^3$

[5.3] Factor each of the following trinomials completely.

25. $x^2 + 9x + 20$

26. $x^2 - 10x + 24$

27. $a^2 - a - 12$

28. $w^2 - 13w + 40$

29. $x^2 + 12x + 36$

30. $r^2 - 9r - 36$

31. $b^2 - 4bc - 21c^2$

32. $m^2n + 4mn - 32n$

33. $m^3 + 2m^2 - 35m$

34. $2x^2 - 2x - 40$

35. $3y^3 - 48y^2 + 189y$

36. $3b^3 - 15b^2 - 42b$

[5.4] Factor each of the following trinomials completely.

37. $3x^2 + 8x + 5$

38. $5w^2 + 13w - 6$

39. $2b^2 - 9b + 9$

40. $8x^2 + 2x - 3$ **41.** $10x^2 - 11x + 3$ **42.** $4a^2 + 7a - 15$

43. $9y^2 - 3yz - 20z^2$ **44.** $8x^2 + 14xy - 15y^2$ **45.** $8x^3 - 36x^2 - 20x$

46. $9x^2 - 15x - 6$ **47.** $6x^3 - 3x^2 - 9x$ **48.** $5w^2 - 25wz + 30z^2$

[5.5] Solve each of the following quadratic equations.

49. $(x - 1)(2x + 3) = 0$ **50.** $x^2 - 5x + 6 = 0$ **51.** $x^2 - 10x = 0$

52. $x^2 = 144$ **53.** $x^2 - 2x = 15$ **54.** $3x^2 - 5x - 2 = 0$

55. $4x^2 - 13x + 10 = 0$ **56.** $2x^2 - 3x = 5$ **57.** $3x^2 - 9x = 0$

58. $x^2 - 25 = 0$ **59.** $2x^2 - 32 = 0$ **60.** $2x^2 - x - 3 = 0$

[5.6] Solve each of the following applications. Be sure to show the equation used for the solution.

61. One integer is 8 less than another. If the product of the two integers is 84, what are the two integers?

62. The length of a rectangle is 3 cm less than twice its width, and the area of that rectangle is 35 cm^2. Find the length and width of the rectangle.

63. The sides of a square are increased by 3 ft, and this increases the area of the square by 33 ft^2. What was the length of a side of the original square?

[5.7] Solve each equation for the indicated variable.

64. $A = P(1 + rt)$ for t **65.** $A = \dfrac{1}{2}h(B + b)$ for B **66.** $p = \dfrac{x - y}{y}$ for y

[5.8] Solve for k, the constant of variation.

67. y varies directly as x; $y = 20$ when $x = 40$

68. y varies inversely as x; $y = 5$ when $x = 3$

Name

Section

Date

The purpose of this self-test is to help you check your progress and to review for a chapter test in class. Allow yourself about an hour to take the test. When you are done, check your answers in the back of the book. If you missed any problems, be sure to go back and review the appropriate sections in the chapter and the exercises that are provided.

Factor each of the following polynomials.

1. $12b + 18$

2. $9p^3 - 12p^2$

3. $5x^2 - 10x + 20$

4. $6a^2b - 18ab + 12ab^2$

Factor each of the following polynomials completely.

5. $a^2 - 25$

6. $64m^2 - n^2$

7. $49x^2 - 16y^2$

8. $32a^2b - 50b^3$

Factor each of the following polynomials completely.

9. $a^2 - 5a - 14$

10. $b^2 + 8b + 15$

11. $x^2 - 11x + 28$

12. $y^2 + 12yz + 20z^2$

Factor each of the following polynomials completely.

13. $2x^2 + 15x - 8$

14. $3w^2 + 10w + 7$

15. $8x^2 - 2xy - 3y^2$

16. $6x^3 + 3x^2 - 30x$

Solve each of the following equations for x.

17. $x^2 - 8x + 15 = 0$

18. $x^2 - 3x = 4$

19. $3x^2 + x - 2 = 0$

20. $4x^2 - 12x = 0$

ANSWERS

1. _____
2. _____
3. _____
4. _____
5. _____
6. _____
7. _____
8. _____
9. _____
10. _____
11. _____
12. _____
13. _____
14. _____
15. _____
16. _____
17. _____
18. _____
19. _____
20. _____

Solve the following word problems.

21. One integer is 3 less than twice another. If the product of the integers is 35, what are the two integers?

22. The length of a rectangle is 2 cm more than 3 times its width. If the area of the rectangle is 33 cm^2, what are the dimensions of the rectangle?

Solve for the variable listed after each of the following equations.

23. $P = 2(L + W)$ for W

24. $a = \dfrac{b + c}{c}$ for c

25. If m varies inversely with n and $m = 3$ when $n = 6$, find k, the constant of variation.

Retain Your Skills

This review covers selected topics from the first four chapters.

Simplify the expression.

1. $7x^2y + 3xy - 5x^2y + 2xy$

2. $\dfrac{27a^5b^7}{9ab}$

3. $(3x^2 - 2x + 5) + (x^2 + 3x - 2)$

4. $(5x^2 + 4x - 3) - (4x^2 - 5x - 1)$

Evaluate the expression.

5. $3(5 - 2)^2$

6. $|3| - |-4|$

Multiply.

7. $(a - 3b)(a + 3b)$

8. $(x - 2y)^2$

9. $(x - 2)(x + 5)$

10. $(a - 3)(a + 4)$

Divide.

11. $(9x^2 + 12x + 4) \div (3x + 2)$

12. $(3x^2 - 2) \div (x - 1)$

Evaluate each expression where $x = 3$, $y = -5$, and $z = 2$.

13. $5x - 2y$

14. $\dfrac{2x + y}{y + z}$

Solve each equation and check your results.

15. $7a - 3 = 6a + 8$

16. $\dfrac{2}{3}x = -22$

17. $x + 2 = 2x + 4$ −2

18. $x^2 - 3x = x^2 + 9$ −3

Solve the following word problems. Show the equation used for the solution.

19. 3 times a number decreased by 5 is 46. Find the number.   17

20. Juan's biology text cost \$5 more than his mathematics text. Together they cost \$81. Find the cost of the biology text.   \$43

CHAPTER 6

Algebraic Fractions

Pesticides and Alternatives

Worldwide approximately 2.3 million metric tons of pesticides are used each year. This translates to an average of nearly 0.5 kilogram (1 pound) for each person on earth. In 1990, global sales of pesticides reached $19 billion. The manufacture, sales, and distribution of pesticides are major components of the world economy.

In the United States, approximately 600 biologically active ingredients and 1200 inert ingredients are mixed to form about 50,000 different pesticide products. Of these products, two-thirds are herbicides, nearly one-quarter are insecticides, and one-tenth are fungicides. About 70 percent of the insecticides and 80 percent of the herbicides used in the United States are applied to four crops—wheat, soybeans, corn, and cotton.

Pesticides have gained worldwide acceptance because they provide many benefits. Insecticides such as DDT have been used for many years to control insect pests that transmit diseases, such as malaria, plague, typhus, and sleeping sickness. Insecticides and herbicides are used in agriculture to increase the available world food supply and to decrease world hunger. Fungicides and fumigants increase the usable lifetimes of food supplies by retarding spoilage. Pesticides also tend to work more quickly and are cheaper than alternative control measures.

Pesticides are also major pollutants in the world's ecosystems. Pesticides kill plants and animals other than the pests they are intended for. Pesticides pollute water systems. Some pests develop immunity to frequently used pesticides. Some pesticides such as DDT and its relatives can remain in the environment for many years beyond the time necessary to do their intended job. Some pesticides have been linked to cancer and other health problems in humans. A chemical leak at a pesticide manufacturing plant in Bhopal, India, killed 3700 people and injured 300,000 others in 1984. These health and environmental problems caused by pesticides have forced some people to question their widespread use. These people ask if we cannot find alternatives to pesticides that lack these negative side effects.

More and more people are advocating a switch from chemical pesticides to a more organic approach to raising and protecting our food supply. By using predator insects such as ladybugs, praying mantises, and wasps, farmers and gardeners can reduce the amount of chemicals they release into the environment. Crop rotation has been used to control pests by changing their food supply on a regular basis. Intercropping, or mixing different crops together in the same fields, also helps control pest populations by creating different environments that will not allow large numbers of problem plants and animals to become established.

A new pest management technique called *integrated pest management* (IPM) is being promoted as an alternative to chemical pest control. IPM is a management strategy rather than an attempt to eliminate problem-causing plants and animals. Instead of trying to kill all the plants and animals that cause problems, an IPM approach attempts to keep pest populations below the level where they cause bad economic or health effects. Although IPM incorporates the use of some pesticides, its primary control measures are nonchemical.

IPM involves the evaluation of each crop and its related pest species as an ecological system. Then a control program is developed that includes various cultivation, biological, and chemical control methods applied in the best sequence and timing to achieve the best results. Fields are carefully monitored for damage, and appropriate control measures are applied only when pests reach an economically damaging level.

This approach saves the farmers money by reducing their need for chemicals and for the machinery to apply them, and it protects the environment by reducing the chemical impact on the ecosystems. This also protects the health of farmworkers by reducing their exposure to chemicals in the fields where they work. By introducing natural control methods, farmers can actually help restore areas to more natural systems and thus help reclaim land that has become unproductive.

Although a sudden change to organic methods may reduce production, a slow conversion from chemicals to organic pest control methods usually will not cause much economic loss. Yields per hectare may drop, but costs also fall so that profits usually remain relatively stable. There is also evidence that as the soils recover over time, they become more productive and crop yields may even improve.

As the environmental impacts of heavy chemical usage become more widely known, more and more people prefer more blemishes on their food and fewer blemishes on their environment. Consumers are changing their attitudes, and this is causing the food producers to change as well. These changes would certainly appear to be changes for the good.

Algebraic Fractions—An Introduction

OBJECTIVES

1. To review the language of fractions
2. To determine the excluded values for the variables of an algebraic fraction

In arithmetic you learned about fractions, or rational numbers. Recall that a rational number is the quotient of two integers $\dfrac{p}{q}$, where q is not equal to 0. For example, $\dfrac{2}{3}$, $-\dfrac{4}{5}$, $\dfrac{12}{7}$, and $\dfrac{5}{1}$ are all rational numbers. We now want to extend the idea of fractions to algebra. All that you learned in arithmetic about fractions will be very helpful here.

Algebraic fractions are also called *rational expressions*. Note the similarity to rational numbers.

Let's define an *algebraic fraction*. It is an expression of the form

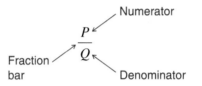

where P and Q are polynomials and Q cannot have the value 0.

Recall that a divisor (the denominator here) cannot have the value 0, or else the division is undefined. See Section 2.5 for details.

The condition that Q, the polynomial in the denominator, cannot be 0 means that certain values for the variable will have to be excluded.

Example 1

In the following algebraic fractions, what values for x must be excluded?

(a) $\dfrac{x}{5}$. Here x can have any value, so none need to be excluded.

(b) $\dfrac{3}{x}$. If $x = 0$, then $\dfrac{3}{x}$ is undefined; 0 is the excluded value.

(c) $\dfrac{5}{x - 2}$. If $x = 2$, then $\dfrac{5}{x - 2} = \dfrac{5}{2 - 2} = \dfrac{5}{0}$, which is undefined so 2 is the excluded value.

(d) $\dfrac{4}{(x + 3)(x - 4)}$. If $x = -3$, then

Again, if $x + 3 = 0$

$$x = -3$$

or if $x - 4 = 0$

$$x = 4$$

$$\frac{4}{(x + 3)(x - 4)} = \frac{4}{(-3 + 3)(-3 - 4)} = \frac{4}{0 \cdot -7} = \frac{4}{0}$$

371

which is undefined. Try $x = 4$; you will see that the same thing happens.

Both -3 and 4 will make the denominator 0 and must be excluded.

CHECK YOURSELF 1

What values for x, if any, must be excluded?

1. $\dfrac{x}{7}$ **2.** $\dfrac{5}{x}$ **3.** $\dfrac{7}{x - 5}$ **4.** $\dfrac{2x}{(x - 6)(x + 1)}$

In some cases you will have to factor the denominator to see the restrictions on the values for the variable.

Example 2

What values for x must be excluded in this fraction?

$$\frac{3}{x^2 - 6x - 16}$$

Factoring the denominator, we have

$$\frac{3}{x^2 - 6x - 16} = \frac{3}{(x - 8)(x + 2)}$$

Letting $x - 8 = 0$ or $x + 2 = 0$, we see that 8 and -2 make the denominator 0 so both 8 and -2 must be excluded.

CHECK YOURSELF 2

What values for x must be excluded on the following fraction?

$$\frac{5}{x^2 - 3x - 10}$$

CHECK YOURSELF ANSWERS

1. (1) None; (2) 0; (3) 5; (4) 6, -1.
2. 5 and -2.

Name

Date

Build Your Skills

What values for x, if any, must be excluded in each of the following algebraic fractions?

1. $\dfrac{x}{9}$

2. $\dfrac{7}{x}$

3. $\dfrac{12}{x}$

4. $\dfrac{x}{6}$

5. $\dfrac{3}{x-2}$

6. $\dfrac{x-1}{5}$

7. $\dfrac{-5}{x+4}$

8. $\dfrac{4}{x+3}$

9. $\dfrac{x-5}{2}$

10. $\dfrac{x-1}{x-5}$

11. $\dfrac{3x}{(x+1)(x-2)}$

12. $\dfrac{5x}{(x-3)(x+7)}$

13. $\dfrac{x-1}{(2x-1)(x+3)}$

14. $\dfrac{x+3}{(3x+1)(x-2)}$

15. $\dfrac{3}{x^2-4}$

16. $\dfrac{4x}{x^2-2x-3}$

17. $\dfrac{x+1}{x^2-6x+5}$

18. $\dfrac{2x-3}{x^2-25}$

ANSWERS

1. _____

2. _____

3. _____

4. _____

5. _____

6. _____

7. _____

8. _____

9. _____

10. _____

11. _____

12. _____

13. _____

14. _____

15. _____

16. _____

17. _____

18. _____

19. _____

20. _____

21. _____

22. _____

23. _____

24. _____

a. _____

b. _____

c. _____

d. _____

e. _____

f. _____

g. _____

h. _____

19. $\dfrac{2x-1}{3x^2+x-2}$

20. $\dfrac{3x+1}{4x^2-11x+6}$

Think About These

What values for x must be excluded in each denominator?

21. $\dfrac{15}{x-y}$

22. $\dfrac{x+3y}{3x+y}$

23. $\dfrac{14+y}{2x-y}$

24. $\dfrac{8}{x+y}$

Skillscan (Section 1.5)

Divide.

a. $\dfrac{12a}{3}$

b. $\dfrac{35w}{7}$

c. $\dfrac{15c^5}{3c^2}$

d. $\dfrac{-48x^5}{3x^3}$

e. $\dfrac{56m^3}{-8m}$

f. $\dfrac{-72p^5}{-9p^2}$

g. $\dfrac{100r^3s^4}{4r^2s^3}$

h. $\dfrac{-84x^4y^6}{7x^2y^2}$

ANSWERS

1. None   **3.** 0   **5.** 2   **7.** -4 **9.** None   **11.** $-1, 2$ **13.** $-3, \dfrac{1}{2}$ **15.** $-2, 2$

17. 1, 5   **19.** $-1, \dfrac{2}{3}$ **21.** $x \neq y$ **23.** $x \neq \dfrac{y}{2}$ **a.** $4a$ **b.** $5w$ **c.** $5c^3$

d. $-16x^2$ **e.** $-7m^2$ **f.** $8p^3$ **g.** $25rs$ **h.** $-12x^2y^4$

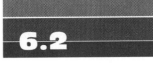

6.2 Writing Algebraic Fractions in Simplest Form

OBJECTIVE
To write algebraic fractions in simplest form

As we pointed out in the last section, much of our work with algebraic fractions will be similar to your work in arithmetic. For instance, in algebra, as in arithmetic, many fractions name the same number. You will remember from arithmetic that

$$\frac{1}{4} = \frac{1 \cdot 2}{4 \cdot 2} = \frac{2}{8}$$

or

$$\frac{1}{4} = \frac{1 \cdot 3}{4 \cdot 3} = \frac{3}{12}$$

So $\frac{1}{4}$, $\frac{2}{8}$, and $\frac{3}{12}$ all name the same number. They are called *equivalent fractions*. These examples illustrate what is called the *fundamental principle of fractions*. In algebra it becomes

> For polynomials P, Q, and R,
>
> $$\frac{P}{Q} = \frac{PR}{QR} \qquad \text{where } Q \neq 0 \text{ and } R \neq 0$$

This principle allows us to multiply or divide the numerator and denominator of a fraction by the same nonzero polynomial. The result will be an expression that is equivalent to the original one.

Our objective in this section is to simplify algebraic fractions by using the fundamental principle. In algebra, as in arithmetic, to write a fraction in simplest form, you divide the numerator and denominator of the fraction by all common factors. The numerator and denominator of the resulting fraction will have no common factors other than 1, and the fraction is then in *simplest form*. The following rule summarizes this procedure.

Note that step 2 uses the fact that $\frac{R}{R} = 1$, if $R \neq 0$.

> **TO WRITE ALGEBRAIC FRACTIONS IN SIMPLEST FORM**
>
> STEP 1   Factor the numerator and denominator.
>
> STEP 2   Divide the numerator and denominator by all common factors.
>
> STEP 3   The resulting fraction will be in lowest terms.

Example 1

(a) Write $\dfrac{18}{30}$ in simplest form.

This is the same as dividing both the numerator and denominator of $\dfrac{18}{30}$ by 6.

$$\frac{18}{30} = \frac{2 \cdot 3 \cdot 3}{2 \cdot 3 \cdot 5} = \frac{\cancel{2} \cdot \cancel{3} \cdot 3}{\cancel{2} \cdot \cancel{3} \cdot 5} = \frac{3}{5}$$

Divide by the common factors. The slash lines indicate that we have divided the numerator and denominator by 2 and by 3.

(b) Write $\dfrac{4x^3}{6x}$ in simplest form.

$$\frac{4x^3}{6x} = \frac{\cancel{2} \cdot 2 \cdot \cancel{x} \cdot x \cdot x}{\cancel{2} \cdot 3 \cdot \cancel{x}} = \frac{2x^2}{3}$$

(c) Write $\dfrac{15x^3y^2}{20xy^4}$ in simplest form.

$$\frac{15x^3y^2}{20xy^4} = \frac{3 \cdot \cancel{5} \cdot \cancel{x} \cdot x \cdot x \cdot \cancel{y} \cdot \cancel{y}}{4 \cdot \cancel{5} \cdot \cancel{x} \cdot \cancel{y} \cdot \cancel{y} \cdot y \cdot y} = \frac{3x^2}{4y^2}$$

(d) Write $\dfrac{3a^2b}{9a^3b^2}$ in simplest form.

When all the factors in the numerator have been divided by a common factor, the value of the numerator is 1.

$$\frac{3a^2b}{9a^3b^2} = \frac{\cancel{3} \cdot \cancel{a} \cdot \cancel{a} \cdot \cancel{b}}{\cancel{3} \cdot 3 \cdot \cancel{a} \cdot \cancel{a} \cdot a \cdot \cancel{b} \cdot b} = \frac{1}{3ab}$$

(e) Write $\dfrac{10a^5b^4}{2a^2b^3}$ in simplest form.

$$\frac{10a^5b^4}{2a^2b^3} = \frac{5 \cdot \cancel{2} \cdot \cancel{a} \cdot \cancel{a} \cdot a \cdot a \cdot a \cdot \cancel{b} \cdot \cancel{b} \cdot \cancel{b} \cdot b}{\cancel{2} \cdot \cancel{a} \cdot \cancel{a} \cdot \cancel{b} \cdot \cancel{b} \cdot \cancel{b}} = \frac{5a^3b}{1} = 5a^3b$$

CHECK YOURSELF 1

Write each fraction in simplest form.

1. $\dfrac{30}{66}$ 2. $\dfrac{5x^4}{15x}$ 3. $\dfrac{12xy^4}{18x^3y^2}$ 4. $\dfrac{5m^2n}{10m^3n^3}$ 5. $\dfrac{12a^4b^6}{2a^3b^4}$

In fact you will see most of the methods of this chapter center on our factoring work of the last chapter.

In simplifying arithmetic fractions, common factors are generally easy to recognize. With algebraic fractions, the factoring techniques you studied in Chapter 5 will have to be used as the *first step* in determining those factors.

Example 2

Write each fraction in simplest form.

(a) $\dfrac{2x - 4}{x^2 - 4} = \dfrac{2(x - 2)}{(x + 2)(x - 2)}$ Factor the numerator and denominator.

$ = \dfrac{2\cancel{(x-2)}}{(x + 2)\cancel{(x-2)}}$ Divide by the common factor $x - 2$. The lines drawn indicate that we have divided by that common factor.

$ = \dfrac{2}{x + 2}$

(b) $\dfrac{3x^2 - 3}{x^2 - 2x - 3} = \dfrac{3(x - 1)\cancel{(x + 1)}}{(x - 3)\cancel{(x + 1)}}$

$ = \dfrac{3(x - 1)}{x - 3}$

(c) $\dfrac{2x^2 + x - 6}{2x^2 - x - 3} = \dfrac{(x + 2)\cancel{(2x - 3)}}{(x + 1)\cancel{(2x - 3)}}$

$ = \dfrac{x + 2}{x + 1}$

Be Careful!   The expression $\dfrac{x + 2}{x + 1}$ is already in simplest form. Students are often tempted to divide as follows:

$\dfrac{\cancel{x} + 2}{\cancel{x} + 1}$ is *not equal* to $\dfrac{2}{1}$

The x's are *terms* in the numerator and denominator. They *cannot* be divided out. Only *factors* can be divided. The fraction

$$\dfrac{x + 2}{x + 1}$$

is in its simplest form.

<⟨CAUTION⟩>

Pick any value, other than 0, for x and substitute. You will quickly see that

$$\dfrac{x + 2}{x + 1} \neq \dfrac{2}{1}$$

CHECK YOURSELF 2

Write each fraction in simplest form.

1. $\dfrac{5x - 15}{x^2 - 9}$

2. $\dfrac{a^2 - 5a + 6}{3a^2 - 6a}$

3. $\dfrac{3x^2 + 14x - 5}{3x^2 + 2x - 1}$

4. $\dfrac{5p - 15}{p^2 - 4}$

Remember the rules for signs in division. The quotient of a positive number and a negative number is always negative. So there are three equivalent ways to write such a quotient. For instance,

$$\frac{-2}{3} = \frac{2}{-3} = -\frac{2}{3}$$

Note: $\dfrac{-2}{3}$, with the negative sign in the numerator, is the most common way to write the quotient.

The quotient of two positive numbers or two negative numbers is always positive. For example,

$$\frac{-2}{-3} = \frac{2}{3}$$

Example 3

Write each fraction in simplest form.

In part (a) the final quotient is written in the most common way with the minus sign in the numerator.

$(a)\ \dfrac{6x^2}{-3xy} = \dfrac{2 \cdot \cancel{3} \cdot \cancel{x} \cdot x}{(-1) \cdot \cancel{3} \cdot \cancel{x} \cdot y} = \dfrac{2x}{-y} = \dfrac{-2x}{y}$

$(b)\ \dfrac{-5a^2b}{-10b^2} = \dfrac{(-1) \cdot \cancel{5} \cdot a \cdot a \cdot \cancel{b}}{(-1) \cdot 2 \cdot \cancel{5} \cdot \cancel{b} \cdot b} = \dfrac{a^2}{2b}$

CHECK YOURSELF 3

Write each fraction in simplest form.

1. $\dfrac{8x^3y}{-4xy^2}$ **2.** $\dfrac{-16a^4b^2}{-12a^2b^5}$

It is sometimes necessary to factor a binomial before simplifying the fraction.

Example 4

Write each fraction in simplest form.

$(a)\ \dfrac{6x^2 + 2x}{2x^2 + 12x} = \dfrac{2x(3x + 1)}{2x(x + 6)} = \dfrac{3x + 1}{x + 6}$

$(b)\ \dfrac{x^2 - 4}{x^2 + 6x + 8} = \dfrac{(x + 2)(x - 2)}{(x + 2)(x + 4)} = \dfrac{x - 2}{x + 4}$

CHECK YOURSELF 4

Simplify each fraction.

1. $\dfrac{3x^3 - 6x^2}{9x^4 - 3x^2}$ **2.** $\dfrac{x^2 - 9}{x^2 - 12x + 27}$

Reducing certain algebraic fractions will be easier with the following result. First, verify for yourself that

$$5 - 8 = -(8 - 5)$$

In general, it is true that

$$a - b = -(b - a)$$

or, by dividing both sides of the equation by $b - a$,

$$\frac{a - b}{b - a} = \frac{-(b - a)}{b - a}$$

So dividing by $b - a$ on the right, we have

Remember a and b cannot be divided out since they are not factors.

$$\frac{a - b}{b - a} = -1$$

Let's look at some applications of that result.

Example 5

Write each fraction in simplest form.

$(a)\ \dfrac{2x - 4}{4 - x^2} = \dfrac{2(x - 2)}{(2 + x)(2 - x)}$ $\longleftarrow$ This is equal. to -1.

$\qquad\qquad = \dfrac{2(-1)}{2 + x} = \dfrac{-2}{2 + x}$

$(b)\ \dfrac{9 - x^2}{x^2 + 2x - 15} = \dfrac{(3 + x)(3 - x)}{(x + 5)(x - 3)}$ $\longleftarrow$ This is equal to -1.

$\qquad\qquad = \dfrac{(3 + x)(-1)}{x + 5}$

$\qquad\qquad = \dfrac{-x - 3}{x + 5}$

Write each fraction in simplest form.

1. $\dfrac{3x - 9}{9 - x^2}$

2. $\dfrac{x^2 - 6x - 27}{81 - x^2}$

1. (1) $\dfrac{5}{11}$; (2) $\dfrac{x^3}{3}$; (3) $\dfrac{2y^2}{3x^2}$; (4) $\dfrac{1}{2mn^2}$; (5) $6ab^2$.

2. (1) $\dfrac{5}{x + 3}$; (2) $\dfrac{a - 3}{3a}$; (3) $\dfrac{x + 5}{x + 1}$; (4) $\dfrac{5(p - 3)}{p^2 - 4}$.

3. (1) $\dfrac{-2x^2}{y}$; (2) $\dfrac{4a^2}{3b^3}$.

4. (1) $\dfrac{x - 2}{3x^2 - 1}$; (2) $\dfrac{x + 3}{x - 9}$.

5. (1) $\dfrac{-3}{x + 3}$; (2) $\dfrac{-x - 3}{x + 9}$.

Build Your Skills

Write each fraction in simplest form.

1. $\dfrac{14}{21}$

2. $\dfrac{35}{40}$

3. $\dfrac{40}{90}$

4. $\dfrac{45}{75}$

5. $\dfrac{4x^5}{6x^2}$

6. $\dfrac{10x^2}{15x^4}$

7. $\dfrac{9x^3}{27x^6}$

8. $\dfrac{25w^6}{20w^2}$

9. $\dfrac{10a^2b^5}{25ab^2}$

10. $\dfrac{18x^4y^3}{24x^2y^3}$

11. $\dfrac{42x^3y}{14xy^3}$

12. $\dfrac{18pq}{45p^2q^2}$

13. $\dfrac{2xyw^2}{6x^2y^3w^3}$

14. $\dfrac{3c^2d^2}{6bc^3d^3}$

15. $\dfrac{10x^5y^5}{2x^3y^4}$

16. $\dfrac{3bc^6d^3}{bc^3d}$

1. _____

2. _____

3. _____

4. _____

5. _____

6. _____

7. _____

8. _____

9. _____

10. _____

11. _____

12. _____

13. _____

14. _____

15. _____

16. _____

17. _____

18. _____

19. _____

20. _____

21. _____

22. _____

23. _____

24. _____

25. _____

26. _____

27. _____

28. _____

29. _____

30. _____

31. _____

32. _____

382

17. $\dfrac{-4m^3n}{6mn^2}$

18. $\dfrac{-15x^3y^3}{-20xy^4}$

19. $\dfrac{-8ab^3}{-16a^3b}$

20. $\dfrac{14x^2y}{-21xy^4}$

21. $\dfrac{8r^2s^3t}{-16rs^4t^3}$

22. $\dfrac{-10a^3b^2c^3}{15ab^4c}$

23. $\dfrac{3x + 18}{5x + 30}$

24. $\dfrac{4x - 28}{5x - 35}$

25. $\dfrac{3x - 6}{5x - 15}$

26. $\dfrac{x^2 - 25}{3x - 15}$

27. $\dfrac{6a - 24}{a^2 - 16}$

28. $\dfrac{5x - 5}{x^2 - 4}$

29. $\dfrac{x^2 + 2x + 1}{5x + 5}$

30. $\dfrac{4w^2 - 8w}{w^2 + w - 6}$

31. $\dfrac{x^2 - 5x - 14}{x^2 - 49}$

32. $\dfrac{y^2 - 9}{y^2 + 8y + 15}$

33. $\dfrac{2m^2 + 3m - 5}{2m^2 + 11m + 15}$

34. $\dfrac{6x^2 - x - 2}{3x^2 - 5x + 2}$

35. $\dfrac{p^2 + 2pq - 15q^2}{p^2 - 25q^2}$

36. $\dfrac{4r^2 - 25s^2}{2r^2 + 3rs - 20s^2}$

37. $\dfrac{2x - 10}{25 - x^2}$

38. $\dfrac{3a - 12}{16 - a^2}$

39. $\dfrac{25 - a^2}{a^2 + a - 30}$

40. $\dfrac{2x^2 - 7x + 3}{9 - x^2}$

41. $\dfrac{x^2 + xy - 6y^2}{4y^2 - x^2}$

42. $\dfrac{16z^2 - w^2}{2w^2 - 5wz - 12z^2}$

43. $\dfrac{x^2 + 4x + 4}{x + 2}$

44. $\dfrac{4x^2 + 12x + 9}{2x + 3}$

Think About These

Simplify.

45. $\dfrac{xy - 2y + 4x - 8}{2y + 6 - xy - 3x}$

46. $\dfrac{ab - 3a + 5b - 15}{15 + 3a^2 - 5b - a^2b}$

ANSWERS

33. _____

34. _____

35. _____

36. _____

37. _____

38. _____

39. _____

40. _____

41. _____

42. _____

43. _____

44. _____

45. _____

46. _____

Skillscan (Appendix 1)

Perform the indicated operations.

a. $\dfrac{2}{3} \cdot \dfrac{4}{5}$

b. $\dfrac{5}{6} \cdot \dfrac{4}{11}$

c. $\dfrac{4}{7} \div \dfrac{8}{5}$

d. $\dfrac{1}{6} \div \dfrac{7}{9}$

e. $\dfrac{5}{8} \cdot \dfrac{16}{15}$

f. $\dfrac{15}{21} \div \dfrac{10}{7}$

g. $\dfrac{15}{8} \cdot \dfrac{24}{25}$

h. $\dfrac{28}{16} \div \dfrac{21}{20}$

ANSWERS

1. $\dfrac{2}{3}$ **3.** $\dfrac{4}{9}$ **5.** $\dfrac{2x^3}{3}$ **7.** $\dfrac{1}{3x^3}$ **9.** $\dfrac{2ab^3}{5}$ **11.** $\dfrac{3x^2}{y^2}$ **13.** $\dfrac{1}{3xy^2w}$ **15.** $5x^2y$

17. $\dfrac{-2m^2}{3n}$ **19.** $\dfrac{b^2}{2a^2}$ **21.** $\dfrac{-r}{2st^2}$ **23.** $\dfrac{3}{5}$ **25.** $\dfrac{3(x-2)}{5(x-3)}$ **27.** $\dfrac{6}{a+4}$ **29.** $\dfrac{x+1}{5}$

31. $\dfrac{x+2}{x+7}$ **33.** $\dfrac{m-1}{m+3}$ **35.** $\dfrac{p-3q}{p-5q}$ **37.** $\dfrac{-2}{x+5}$ **39.** $\dfrac{-a-5}{a+6}$ **41.** $\dfrac{-x-3y}{2y+x}$

43. $x+2$ **45.** $\dfrac{-(y+4)}{y+3}$ **a.** $\dfrac{8}{15}$ **b.** $\dfrac{10}{33}$ **c.** $\dfrac{5}{14}$ **d.** $\dfrac{3}{14}$ **e.** $\dfrac{2}{3}$ **f.** $\dfrac{1}{2}$

g. $\dfrac{9}{5}$ **h.** $\dfrac{5}{3}$

Multiplying and Dividing Algebraic Fractions

OBJECTIVES

1. To write the product of algebraic fractions in simplest form
2. To write the quotient of algebraic fractions in simplest form

In arithmetic, you found the product of two fractions by multiplying the numerators and multiplying the denominators. For example,

$$\frac{2}{5} \cdot \frac{3}{7} = \frac{2 \cdot 3}{5 \cdot 7} = \frac{6}{35}$$

In symbols, we have

P, Q, R, and *S* again represent polynomials.

$$\frac{P}{Q} \cdot \frac{R}{S} = \frac{PR}{QS} \qquad \text{where } Q \neq 0 \text{ and } S \neq 0$$

It is easiest to divide numerator and denominator by any common factors *before* multiplying. Consider the following.

Divide by the common factors of 3 and 4. The alternative is to multiply *first:*

$$\frac{3}{8} \cdot \frac{4}{9} = \frac{12}{72}$$

and then reduce to lowest terms

$$\frac{12}{72} = \frac{1}{6}$$

$$\frac{3}{8} \cdot \frac{4}{9} = \frac{3 \cdot 4}{8 \cdot 9} = \frac{1}{6}$$

In algebra we multiply fractions in exactly the same way.

> **TO MULTIPLY ALGEBRAIC FRACTIONS**
>
> STEP 1 Factor the numerators and denominators.
>
> STEP 2 Divide the numerator and denominator by any common factors.
>
> STEP 3 Write the product of the remaining factors in the numerator over the product of the remaining factors in the denominator.

The following example illustrates.

Example 1

Multiply the following fractions.

Divide by the common factors of 5, x^2, and *y.*

$$(a) \quad \frac{2x^3}{5y^2} \cdot \frac{10y}{3x^2} = \frac{2x^3 \cdot 10y}{5y^2 \cdot 3x^2} = \frac{4x}{3y}$$

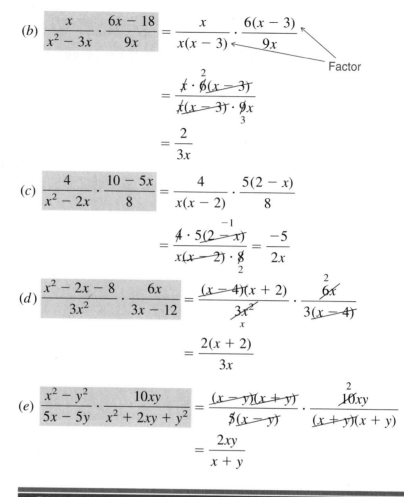

(b) $\dfrac{x}{x^2-3x} \cdot \dfrac{6x-18}{9x} = \dfrac{x}{x(x-3)} \cdot \dfrac{6(x-3)}{9x}$

Factor

Divide by the common factors of 3, x, and $x-3$.

$$= \dfrac{x \cdot \overset{2}{6}(x-3)}{x(x-3) \cdot \underset{3}{9x}}$$

$$= \dfrac{2}{3x}$$

Note:

$$\dfrac{2-x}{x-2} = \dfrac{-(x-2)}{x-2} = -1$$

(c) $\dfrac{4}{x^2-2x} \cdot \dfrac{10-5x}{8} = \dfrac{4}{x(x-2)} \cdot \dfrac{5(2-x)}{8}$

$$= \dfrac{4 \cdot 5\overset{-1}{(2-x)}}{x(x-2) \cdot \underset{2}{8}} = \dfrac{-5}{2x}$$

Divide by the common factors of $x-4$, x, and 3.

(d) $\dfrac{x^2-2x-8}{3x^2} \cdot \dfrac{6x}{3x-12} = \dfrac{(x-4)(x+2)}{\underset{x}{3x^2}} \cdot \dfrac{\overset{2}{6x}}{3(x-4)}$

$$= \dfrac{2(x+2)}{3x}$$

(e) $\dfrac{x^2-y^2}{5x-5y} \cdot \dfrac{10xy}{x^2+2xy+y^2} = \dfrac{(x-y)(x+y)}{5(x-y)} \cdot \dfrac{\overset{2}{10xy}}{(x+y)(x+y)}$

$$= \dfrac{2xy}{x+y}$$

CHECK YOURSELF 1

Multiply.

1. $\dfrac{3x^2}{5y^2} \cdot \dfrac{10y^5}{15x^3}$ 2. $\dfrac{5x+15}{x} \cdot \dfrac{2x^2}{x^2+3x}$ 3. $\dfrac{x}{2x-6} \cdot \dfrac{3x-x^2}{2}$

4. $\dfrac{3x-15}{6x^2} \cdot \dfrac{2x}{x^2-25}$ 5. $\dfrac{x^2-5x-14}{4x^2} \cdot \dfrac{8x}{x^2-49}$

You can also use your experience from arithmetic in dividing fractions. Recall that to divide fractions, we *invert the divisor* (the *second* fraction) and multiply. For example,

Recall, $\dfrac{6}{5}$ is the reciprocal of $\dfrac{5}{6}$.

$$\dfrac{2}{3} \div \dfrac{5}{6} = \dfrac{2}{3} \cdot \dfrac{6}{5} = \dfrac{2 \cdot 6}{3 \cdot 5} = \dfrac{4}{5}$$

In symbols, we have

Once more P, Q, R, and S are polynomials.

$$\dfrac{P}{Q} \div \dfrac{R}{S} = \dfrac{P}{Q} \cdot \dfrac{S}{R} = \dfrac{PS}{QR}$$

where $Q \neq 0$, $R \neq 0$, and $S \neq 0$.

Division of algebraic fractions is done in exactly the same way.

> **TO DIVIDE ALGEBRAIC FRACTIONS**
>
> STEP 1   Invert the divisor.
>
> STEP 2   Proceed, using the steps for multiplying algebraic fractions.

The following example illustrates.

Example 2

Divide the following.

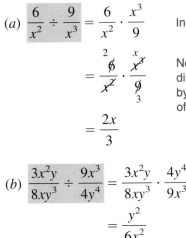

(a) $\dfrac{6}{x^2} \div \dfrac{9}{x^3} = \dfrac{6}{x^2} \cdot \dfrac{x^3}{9}$ Invert the divisor and multiply.

$$= \dfrac{\overset{2}{\cancel{6}}}{\cancel{x^2}} \cdot \dfrac{\overset{x}{\cancel{x^3}}}{\underset{3}{\cancel{9}}}$$ No simplification can be done until the divisor is inverted. Then divide by the common factors of 3 and x^2.

$$= \dfrac{2x}{3}$$

(b) $\dfrac{3x^2y}{8xy^3} \div \dfrac{9x^3}{4y^4} = \dfrac{3x^2y}{8xy^3} \cdot \dfrac{4y^4}{9x^3}$

$$= \dfrac{y^2}{6x^2}$$

(c) $\dfrac{2x + 4y}{9x - 18y} \div \dfrac{4x + 8y}{3x - 6y} = \dfrac{2x + 4y}{9x - 18y} \cdot \dfrac{3x - 6y}{4x + 8y}$

$$= \dfrac{\cancel{2(x+2y)}}{\underset{3}{\cancel{9(x-2y)}}} \cdot \dfrac{\cancel{3(x-2y)}}{\underset{2}{\cancel{4(x+2y)}}}$$

$$= \dfrac{1}{6}$$

Factor all numerators and denominators *before* dividing out any common factors.

(d) $\dfrac{x^2 - x - 6}{2x - 6} \div \dfrac{x^2 - 4}{4x^2} = \dfrac{x^2 - x - 6}{2x - 6} \cdot \dfrac{4x^2}{x^2 - 4}$

$$= \dfrac{\cancel{(x-3)}(x+2)}{\cancel{2(x-3)}} \cdot \dfrac{\overset{2}{\cancel{4x^2}}}{\cancel{(x+2)}(x - 2)}$$

$$= \dfrac{2x^2}{x - 2}$$

Divide.

1. $\dfrac{4}{x^5} \div \dfrac{12}{x^3}$

2. $\dfrac{5xy^2}{7x^3y} \div \dfrac{10y^2}{14x^3}$

3. $\dfrac{3x - 9y}{2x + 10y} \div \dfrac{x^2 - 3xy}{4x + 20y}$

4. $\dfrac{x^2 - 9}{4x} \div \dfrac{x^2 - 2x - 15}{2x - 10}$

Before we conclude this section, let's review why the invert-and-multiply rule works for dividing fractions. We will use an example from arithmetic for the explanation. Suppose that we want to divide as follows:

$$\frac{3}{5} \div \frac{2}{3} \tag{1}$$

We can write

$$\underbrace{\frac{3}{5} \div \frac{2}{3}}_{(1)} = \frac{\dfrac{3}{5}}{\dfrac{2}{3}} = \frac{\dfrac{3}{5} \cdot \dfrac{3}{2}}{\dfrac{2}{3} \cdot \dfrac{3}{2}} \qquad \text{We are multiplying by 1.}$$

Interpret the division as a fraction.

$$= \frac{\dfrac{3}{5} \cdot \dfrac{3}{2}}{1}$$

$$\frac{2}{3} \cdot \frac{3}{2} = 1$$

$$= \underbrace{\frac{3}{5} \cdot \frac{3}{2}}_{(2)}$$

We then have

$$(1) \qquad \frac{3}{5} \div \frac{2}{3} = \frac{3}{5} \cdot \frac{3}{2} \qquad (2)$$

Comparing expressions (1) and (2), you should see the rule for dividing fractions. Invert the fraction that follows the division symbol and multiply.

1. (1) $\dfrac{2y^3}{5x}$; (2) 10; (3) $-\dfrac{x^2}{4}$; (4) $\dfrac{1}{x(x+5)}$; (5) $\dfrac{2(x+2)}{x(x+7)}$.

2. (1) $\dfrac{1}{3x^2}$; (2) $\dfrac{x}{y}$; (3) $\dfrac{6}{x}$; (4) $\dfrac{x-3}{2x}$.

Name

Date

Build Your Skills

Multiply.

1. $\dfrac{2}{5} \cdot \dfrac{10}{18}$

2. $\dfrac{7}{12} \cdot \dfrac{3}{28}$

3. $\dfrac{x}{3} \cdot \dfrac{y}{4}$

4. $\dfrac{w}{7} \cdot \dfrac{5}{4}$

5. $\dfrac{3a}{2} \cdot \dfrac{4}{a^2}$

6. $\dfrac{5x^3}{3x} \cdot \dfrac{9}{20x}$

7. $\dfrac{3x^3y}{10xy^3} \cdot \dfrac{5xy^2}{9xy^2}$

8. $\dfrac{8xy^5}{5x^3y^2} \cdot \dfrac{15y^2}{16xy^3}$

9. $\dfrac{-4ab^2}{15a^3} \cdot \dfrac{25ab}{-16b^3}$

10. $\dfrac{-7xy^2}{12x^2y} \cdot \dfrac{24x^3y^5}{-21x^2y^7}$

11. $\dfrac{-3m^3n}{10mn^3} \cdot \dfrac{5mn^2}{-9mn^3}$

12. $\dfrac{3x}{2x-6} \cdot \dfrac{x^2-3x}{6}$

13. $\dfrac{x^2+5x}{3x^2} \cdot \dfrac{10x}{5x+25}$

14. $\dfrac{x^2-3x-10}{5x} \cdot \dfrac{15x^2}{3x-15}$

ANSWERS

1. _____

2. _____

3. _____

4. _____

5. _____

6. _____

7. _____

8. _____

9. _____

10. _____

11. _____

12. _____

13. _____

14. _____

15. $\dfrac{p^2 - 8p}{4p} \cdot \dfrac{12p^2}{p^2 - 64}$

16. $\dfrac{a^2 - 81}{a^2 + 9a} \cdot \dfrac{5a^2}{a^2 - 7a - 18}$

17. $\dfrac{m^2 - 4m - 21}{3m^2} \cdot \dfrac{m^2 + 7m}{m^2 - 49}$

18. $\dfrac{2x^2 - x - 3}{3x^2 + 7x + 4} \cdot \dfrac{3x^2 - 11x - 20}{4x^2 - 9}$

19. $\dfrac{4r^2 - 1}{2r^2 - 9r - 5} \cdot \dfrac{3r^2 - 13r - 10}{9r^2 - 4}$

20. $\dfrac{a^2 + ab}{2a^2 - ab - 3b^2} \cdot \dfrac{4a^2 - 9b^2}{5a^2 - 4ab}$

21. $\dfrac{x^2 - 4y^2}{x^2 - xy - 6y^2} \cdot \dfrac{7x^2 - 21xy}{5x - 10y}$

22. $\dfrac{a^2 - 9b^2}{a^2 + ab - 6b^2} \cdot \dfrac{6a^2 - 12ab}{7a - 21b}$

23. $\dfrac{2x - 6}{x^2 + 2x} \cdot \dfrac{3x}{3 - x}$

24. $\dfrac{3x - 15}{x^2 + 3x} \cdot \dfrac{4x}{5 - x}$

Divide.

25. $\dfrac{5}{8} \div \dfrac{15}{16}$

26. $\dfrac{4}{9} \div \dfrac{12}{18}$

27. $\dfrac{5}{x^2} \div \dfrac{10}{x}$

28. $\dfrac{w^2}{3} \div \dfrac{w}{9}$

29. $\dfrac{4x^2y^2}{9x^3} \div \dfrac{8y^2}{27xy}$

30. $\dfrac{8x^3y}{27xy^3} \div \dfrac{16x^3y}{45y}$

31. $\dfrac{3x + 6}{8} \div \dfrac{5x + 10}{6}$

32. $\dfrac{x^2 - 2x}{4x} \div \dfrac{6x - 12}{8}$

33. $\dfrac{4a - 12}{5a + 15} \div \dfrac{8a^2}{a^2 + 3a}$

34. $\dfrac{6p - 18}{9p} \div \dfrac{3p - 9}{p^2 + 2p}$

35. $\dfrac{x^2 + 2x - 8}{9x^2} \div \dfrac{x^2 - 16}{3x - 12}$

36. $\dfrac{16x}{4x^2 - 16} \div \dfrac{4x - 24}{x^2 - 4x - 12}$

37. $\dfrac{x^2 - 9}{2x^2 - 6x} \div \dfrac{2x^2 + 5x - 3}{4x^2 - 1}$

38. $\dfrac{2m^2 - 5m - 7}{4m^2 - 9} \div \dfrac{5m^2 + 5m}{2m^2 + 3m}$

39. $\dfrac{a^2 - 9b^2}{4a^2 + 12ab} \div \dfrac{a^2 - ab - 6b^2}{12ab}$

40. $\dfrac{r^2 + 2rs - 15s^2}{r^3 + 5r^2s} \div \dfrac{r^2 - 9s^2}{5r^3}$

41. $\dfrac{x^2 - 16y^2}{3x^2 - 12xy} \div (x^2 + 4xy)$

42. $\dfrac{p^2 - 4pq - 21q^2}{4p - 28q} \div (2p^2 + 6pq)$

43. $\dfrac{x - 7}{2x + 6} \div \dfrac{21 - 3x}{x^2 + 3x}$

44. $\dfrac{x - 4}{x^2 + 2x} \div \dfrac{16 - 4x}{3x + 6}$

Think About These

Perform the indicated operations.

45. $\dfrac{x^2 - 2x - 8}{2x - 8} \cdot \dfrac{x^2 + 5x}{x^2 + 5x + 6} \div \dfrac{x^2 + 2x - 15}{x^2 - 9}$

46. $\dfrac{14x - 7}{x^2 + 3x - 4} \cdot \dfrac{x^2 + 6x + 8}{2x^2 + 5x - 3} \div \dfrac{x^2 + 2x}{x^2 + 2x - 3}$

33. _____

34. _____

35. _____

36. _____

37. _____

38. _____

39. _____

40. _____

41. _____

42. _____

43. _____

44. _____

45. _____

46. _____

47. $\dfrac{x^2 + 5x}{3x - 6} \cdot \dfrac{x^2 - 4}{3x^2 + 15x} \cdot \dfrac{6x}{x^2 + 6x + 8}$

48. $\dfrac{m^2 - n^2}{m^2 - mn} \cdot \dfrac{6m}{2m^2 + mn - n^2} \cdot \dfrac{8m - 4n}{12m^2 + 12mn}$

Skillscan (Appendix A.1)

Perform the indicated operations.

a. $\dfrac{3}{10} + \dfrac{4}{10}$

b. $\dfrac{5}{8} - \dfrac{4}{8}$

c. $\dfrac{5}{12} - \dfrac{1}{12}$

d. $\dfrac{7}{16} + \dfrac{3}{16}$

e. $\dfrac{7}{20} + \dfrac{9}{20}$

f. $\dfrac{13}{8} - \dfrac{5}{8}$

g. $\dfrac{11}{6} - \dfrac{2}{6}$

h. $\dfrac{5}{9} + \dfrac{7}{9}$

ANSWERS

1. $\dfrac{2}{9}$ **3.** $\dfrac{xy}{12}$ **5.** $\dfrac{6}{a}$ **7.** $\dfrac{x^2}{6y^2}$ **9.** $\dfrac{5}{12a}$ **11.** $\dfrac{m^2}{6n^3}$ **13.** $\dfrac{2}{3}$ **15.** $\dfrac{3p^2}{p + 8}$

17. $\dfrac{m + 3}{3m}$ **19.** $\dfrac{2r - 1}{3r - 2}$ **21.** $\dfrac{7x}{5}$ **23.** $\dfrac{-6}{x + 2}$ **25.** $\dfrac{2}{3}$ **27.** $\dfrac{1}{2x}$ **29.** $\dfrac{3y}{2}$

31. $\dfrac{9}{20}$ **33.** $\dfrac{a - 3}{10a}$ **35.** $\dfrac{x - 2}{3x^2}$ **37.** $\dfrac{2x + 1}{2x}$ **39.** $\dfrac{3b}{a + 2b}$ **41.** $\dfrac{1}{3x^2}$ **43.** $\dfrac{-x}{6}$

45. $\dfrac{x}{2}$ **47.** $3\dfrac{2x}{(x + 4)}$ **a.** $\dfrac{7}{10}$ **b.** $\dfrac{1}{8}$ **c.** $\dfrac{1}{3}$ **d.** $\dfrac{5}{8}$ **e.** $\dfrac{4}{5}$ **f.** 1 **g.** $\dfrac{3}{2}$

h. $\dfrac{4}{3}$

6.4 Adding and Subtracting Like Fractions

OBJECTIVE

To write the sum or difference of like fractions as a single fraction in lowest terms

You probably remember from arithmetic that *like fractions* are fractions that have the same denominator. The same is true in algebra.

$\dfrac{2}{5}, \dfrac{12}{5}$, and $\dfrac{4}{5}$ are like fractions.

$\dfrac{x}{3}, \dfrac{y}{3}$, and $\dfrac{z-5}{3}$ are like fractions.

$\dfrac{3x}{2}, \dfrac{x}{4}$, and $\dfrac{3x}{8}$ are unlike fractions.

The fractions have different denominators.

$\dfrac{3}{x}, \dfrac{2}{x^2}$, and $\dfrac{x+1}{x^3}$ are unlike fractions.

In arithmetic, the sum or difference of like fractions was found by adding or subtracting the numerators and writing the result over the common denominator. For example,

$$\frac{3}{11} + \frac{5}{11} = \frac{3+5}{11} = \frac{8}{11}$$

In symbols we have

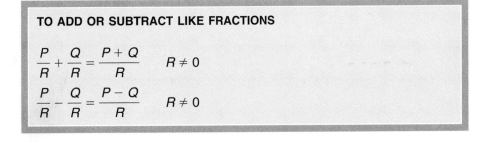

TO ADD OR SUBTRACT LIKE FRACTIONS

$$\frac{P}{R} + \frac{Q}{R} = \frac{P+Q}{R} \qquad R \neq 0$$

$$\frac{P}{R} - \frac{Q}{R} = \frac{P-Q}{R} \qquad R \neq 0$$

Adding or subtracting like fractions in algebra is just as straightforward. You can use the following steps.

TO ADD OR SUBTRACT LIKE ALGEBRAIC FRACTIONS

STEP 1 Add or subtract the numerators.

STEP 2 Write the sum or difference over the common denominator.

STEP 3 Write the resulting fraction in simplest form.

393

Example 1

Add or subtract as indicated. Express your results in simplest form.

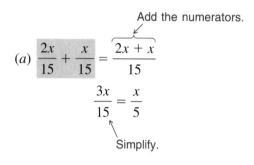

Add the numerators.

$$(a) \quad \frac{2x}{15} + \frac{x}{15} = \frac{2x + x}{15}$$

$$\frac{3x}{15} = \frac{x}{5}$$

Simplify.

Subtract the numerators.

$$(b) \quad \frac{5y}{6} - \frac{y}{6} = \frac{5y - y}{6}$$

$$= \frac{4y}{6} = \frac{2y}{3}$$

Simplify.

$$(c) \quad \frac{3}{x} + \frac{5}{x} = \frac{3 + 5}{x} = \frac{8}{x}$$

$$(d) \quad \frac{9}{a^2} - \frac{7}{a^2} = \frac{9 - 7}{a^2} = \frac{2}{a^2}$$

$$(e) \quad \frac{7}{2ab} - \frac{5}{2ab} = \frac{7 - 5}{2ab}$$

$$= \frac{2}{2ab}$$

$$= \frac{1}{ab}$$

CHECK YOURSELF 1

Add or subtract as indicated.

1. $\dfrac{3a}{10} + \dfrac{2a}{10}$ 2. $\dfrac{7b}{8} - \dfrac{3b}{8}$ 3. $\dfrac{4}{x} + \dfrac{3}{x}$ 4. $\dfrac{5}{3xy} - \dfrac{2}{3xy}$

If polynomials are involved in the numerators or denominators, the process is exactly the same.

Example 2

Add or subtract as indicated. Express your results in simplest form.

(a) $\dfrac{5}{x+3} + \dfrac{2}{x+3} = \dfrac{5+2}{x+3} = \dfrac{7}{x+3}$

(b) $\dfrac{4x}{x-4} - \dfrac{16}{x-4} = \dfrac{4x-16}{x-4}$

Factor and simplify.

$= \dfrac{4(x-4)}{x-4} = 4$

The final answer is always written in simplest form.

(c) $\dfrac{a-b}{3} + \dfrac{2a+b}{3} = \dfrac{(a-b)+(2a+b)}{3}$

$= \dfrac{a-b+2a+b}{3}$

$= \dfrac{3a}{3} = a$

Be sure to enclose the second numerator in parentheses!

(d) $\dfrac{3x+y}{2x} - \dfrac{x-3y}{2x} = \dfrac{(3x+y)-(x-3y)}{2x}$

Change both signs.

$= \dfrac{3x+y-x+3y}{2x}$

$= \dfrac{2x+4y}{2x}$

$= \dfrac{2(x+2y)}{2x}$ Factor and divide by the common factor of 2.

$= \dfrac{x+2y}{x}$

(e) $\dfrac{3x-5}{x^2+x-2} - \dfrac{2x-4}{x^2+x-2} = \dfrac{3x-5-(2x-4)}{x^2+x-2}$ Put second numerator in parentheses.

Change both signs.

$= \dfrac{3x-5-2x+4}{x^2+x-2}$

$= \dfrac{x-1}{x^2+x-2}$

$= \dfrac{(x-1)}{(x+2)(x-1)}$ Factor and divide by the common factor of $x-1$.

$= \dfrac{1}{x+2}$

$(f) \ \dfrac{2x + 7y}{x + 3y} - \dfrac{x + 4y}{x + 3y} = \dfrac{(2x + 7y) - (x + 4y)}{x + 3y}$

Change both signs.

$= \dfrac{2x + 7y - x - 4y}{x + 3y}$

$= \dfrac{x + 3y}{x + 3y} = 1$

CHECK YOURSELF 2

Add or subtract as indicated.

1. $\dfrac{4}{x - 5} - \dfrac{2}{x - 5}$

2. $\dfrac{3x}{x + 3} + \dfrac{9}{x + 3}$

3. $\dfrac{5x - y}{3y} - \dfrac{2x - 4y}{3y}$

4. $\dfrac{5x + 8}{x^2 - 2x - 15} - \dfrac{4x + 5}{x^2 - 2x - 15}$

CHECK YOURSELF ANSWERS

1. (1) $\dfrac{a}{2}$; (2) $\dfrac{b}{2}$; (3) $\dfrac{7}{x}$; (4) $\dfrac{1}{xy}$.

2. (1) $\dfrac{2}{x - 5}$; (2) 3; (3) $\dfrac{x + y}{y}$; (4) $\dfrac{1}{x - 5}$.

Name _____

Date _____

Build Your Skills

Add or subtract as indicated. Express your results in simplest form.

1. $\dfrac{8}{15} + \dfrac{2}{15}$

2. $\dfrac{7}{12} - \dfrac{5}{12}$

3. $\dfrac{11}{20} - \dfrac{6}{20}$

4. $\dfrac{5}{9} + \dfrac{7}{9}$

5. $\dfrac{x}{8} + \dfrac{3x}{8}$

6. $\dfrac{5y}{16} + \dfrac{7y}{16}$

7. $\dfrac{7a}{10} - \dfrac{3a}{10}$

8. $\dfrac{5x}{12} - \dfrac{x}{12}$

9. $\dfrac{5}{x} + \dfrac{3}{x}$

10. $\dfrac{9}{y} - \dfrac{3}{y}$

11. $\dfrac{8}{w} - \dfrac{2}{w}$

12. $\dfrac{7}{z} + \dfrac{9}{z}$

13. $\dfrac{2}{xy} + \dfrac{3}{xy}$

14. $\dfrac{8}{ab} + \dfrac{4}{ab}$

ANSWERS

1. _____

2. _____

3. _____

4. _____

5. _____

6. _____

7. _____

8. _____

9. _____

10. _____

11. _____

12. _____

13. _____

14. _____

ANSWERS

15. _____

16. _____

17. _____

18. _____

19. _____

20. _____

21. _____

22. _____

23. _____

24. _____

25. _____

26. _____

27. _____

28. _____

29. _____

30. _____

15. $\dfrac{6}{5cd} + \dfrac{4}{5cd}$

16. $\dfrac{3}{2cd} + \dfrac{5}{2cd}$

17. $\dfrac{6}{x-3} + \dfrac{2}{x-3}$

18. $\dfrac{5}{x+2} - \dfrac{3}{x+2}$

19. $\dfrac{2x}{x-2} - \dfrac{4}{x-2}$

20. $\dfrac{7w}{w+3} + \dfrac{21}{w+3}$

21. $\dfrac{8p}{p+4} + \dfrac{32}{p+4}$

22. $\dfrac{5a}{a-3} - \dfrac{15}{a-3}$

23. $\dfrac{x^2}{x+4} + \dfrac{3x-4}{x+4}$

24. $\dfrac{x^2}{x-3} - \dfrac{9}{x-3}$

25. $\dfrac{m^2}{m-5} - \dfrac{25}{m-5}$

26. $\dfrac{s^2}{s+3} + \dfrac{2s-3}{s+3}$

27. $\dfrac{a-1}{3} + \dfrac{2a-5}{3}$

28. $\dfrac{y+2}{5} + \dfrac{4y+8}{5}$

29. $\dfrac{3x-1}{4} - \dfrac{x+7}{4}$

30. $\dfrac{4x+2}{3} - \dfrac{x-1}{3}$

31. $\dfrac{3m-2}{5m} + \dfrac{2m+12}{5m}$

32. $\dfrac{5x-2y}{3y} - \dfrac{2x+y}{3y}$

33. $\dfrac{4w-7}{w-5} - \dfrac{2w+3}{w-5}$

34. $\dfrac{3b-8}{b-6} + \dfrac{b-16}{b-6}$

35. $\dfrac{x-7}{x^2-x-6} + \dfrac{2x-2}{x^2-x-6}$

36. $\dfrac{5a-12}{a^2-8a+15} - \dfrac{3a-2}{a^2-8a+15}$

Think About These

Add or subtract as indicated. Express your results in simplest form.

37. $\dfrac{5}{x-3} - \dfrac{3}{3-x}$

38. $\dfrac{-10}{5-y} + \dfrac{3}{y-5}$

39. $\dfrac{x}{9-x} + \dfrac{9}{x-9}$

40. $\dfrac{-5}{y-5} - \dfrac{y}{5-y}$

41. $\dfrac{x^2}{x^2-x-6} - \dfrac{6}{(x-3)(x+2)} + \dfrac{x}{6+x-x^2}$

42. $\dfrac{-12}{x^2+2-12} + \dfrac{x^2}{(x+4)(x-3)} - \dfrac{x}{12-x-x^2}$

ANSWERS

31. _____

32. _____

33. _____

34. _____

35. _____

36. _____

37. _____

38. _____

39. _____

40. _____

41. _____

42. _____

Skillscan (Appendix 1)

a. $\dfrac{3}{4} + \dfrac{1}{2}$

b. $\dfrac{5}{6} - \dfrac{2}{3}$

c. $\dfrac{7}{10} - \dfrac{3}{5}$

d. $\dfrac{5}{8} + \dfrac{3}{4}$

e. $\dfrac{5}{6} + \dfrac{3}{8}$

f. $\dfrac{7}{8} - \dfrac{3}{5}$

g. $\dfrac{9}{10} - \dfrac{2}{15}$

h. $\dfrac{5}{12} + \dfrac{7}{18}$

ANSWERS

1. $\dfrac{2}{3}$ 3. $\dfrac{1}{4}$ 5. $\dfrac{x}{2}$ 7. $\dfrac{2a}{5}$ 9. $\dfrac{8}{x}$ 11. $\dfrac{6}{w}$ 13. $\dfrac{5}{xy}$ 15. $\dfrac{2}{cd}$ 17. $\dfrac{8}{x-3}$

19. 2   21. 8   23. $x-1$ 25. $m+5$ 27. $a-2$ 29. $\dfrac{x-4}{2}$ 31. $\dfrac{m+2}{m}$

33. 2   35. $\dfrac{3}{x+2}$ 37. $\dfrac{8}{x-3}$ 39. -1 41. 1   a. $\dfrac{5}{4}$ b. $\dfrac{1}{6}$ c. $\dfrac{1}{10}$

d. $\dfrac{11}{8}$ e. $\dfrac{29}{24}$ f. $\dfrac{11}{40}$ g. $\dfrac{23}{30}$ h. $\dfrac{29}{36}$

6.5 Adding and Subtracting Unlike Fractions

OBJECTIVE

To write the sum or difference of unlike fractions as a single fraction in simplest form

Adding or subtracting *unlike fractions* (fractions that do not have the same denominator) requires a bit more work than adding or subtracting the like fractions of the previous section. When the denominators are not the same, we must use the idea of the *lowest common denominator* (LCD). Each fraction is "built up" to an equivalent fraction having the LCD as a denominator. You can then add or subtract as before.

Let's review with an example from arithmetic.

Example 1

Add $\dfrac{5}{9} + \dfrac{1}{6}$.

Step 1   Find the LCD. Factor each denominator.

$9 = 3 \cdot 3 \longleftarrow$ 3 appears twice.
$6 = 2 \cdot 3$

To form the LCD, include each factor the greatest number of times it appears in any single denominator. Use one 2, since 2 appears only once in the factorization of 6. Use two 3s since 3 appears twice in the factorization of 9. Thus the LCD for the fractions is $2 \cdot 3 \cdot 3 = 18$.

Step 2   "Build up" each fraction to an equivalent fraction with the LCD as the denominator. Do this by multiplying the numerator and denominator of the given fractions by the same number.

Do you see that this uses the fundamental principle in the following form?

$$\frac{P}{Q} = \frac{PR}{QR}$$

$$\frac{5}{9} = \frac{5 \cdot 2}{9 \cdot 2} = \frac{10}{18}$$

$$\frac{1}{6} = \frac{1 \cdot 3}{6 \cdot 3} = \frac{3}{18}$$

Step 3   Add the fractions.

$$\frac{5}{9} + \frac{1}{6} = \frac{10}{18} + \frac{3}{18} = \frac{13}{18}$$

$\dfrac{13}{18}$ is in simplest form, and so we are done!

CHECK YOURSELF 1

Add.

1. $\dfrac{1}{6} + \dfrac{3}{8}$

2. $\dfrac{3}{10} + \dfrac{4}{15}$

The process is exactly the same in algebra. We can summarize the steps with the following rule:

TO ADD OR SUBTRACT UNLIKE FRACTIONS

STEP 1   Find the lowest common denominator of all the fractions.

STEP 2   Convert each fraction to an equivalent fraction with the LCD as a denominator.

STEP 3   Add or subtract the like fractions formed in step 2.

STEP 4   Write the sum or difference in simplest form.

Example 2

(*a*) Add $\dfrac{3}{2x} + \dfrac{4}{x^2}$.

Step 1   Factor the denominators.

$$2x = 2 \cdot x$$
$$x^2 = x \cdot x$$

The original denominators are not multiplied (together) to form the LCD.

The LCD must contain the factors 2 and x. The factor x must appear *twice* because it appears twice as a factor in the second denominator.

   The LCD is $2 \cdot x \cdot x$, or $2x^2$.

Step 2

$$\frac{3}{2x} = \frac{3 \cdot x}{2x \cdot x} = \frac{3x}{2x^2}$$

$$\frac{4}{x^2} = \frac{4 \cdot 2}{x^2 \cdot 2} = \frac{8}{2x^2}$$

Step 3

$$\frac{3}{2x} + \frac{4}{x^2} = \frac{3x}{2x^2} + \frac{8}{2x^2} = \frac{3x + 8}{2x^2}$$

The sum is in simplest form.

(b) Subtract $\dfrac{4}{3x^2} - \dfrac{3}{2x^3}$.

Step 1   Factor the denominators.

$3x^2 = 3 \cdot x \cdot x$ The factor x must appear
$2x^3 = 2 \cdot x \cdot x \cdot x$ 3 times. Do you see why?

The LCD must contain the factors 2, 3, and x. The LCD is

$2 \cdot 3 \cdot x \cdot x \cdot x$ or $6x^3$

Step 2

Both the numerator and the denominator must be multiplied by the same quantity.

$$\frac{4}{3x^2} = \frac{4 \cdot 2x}{3x^2 \cdot 2x} = \frac{8x}{6x^3}$$

$$\frac{3}{2x^3} = \frac{3 \cdot 3}{2x^3 \cdot 3} = \frac{9}{6x^3}$$

Step 3

$$\frac{4}{3x^2} - \frac{3}{2x^3} = \frac{8x}{6x^3} - \frac{9}{6x^3} = \frac{8x - 9}{6x^3}$$

The difference is in simplest form.

CHECK YOURSELF 2

Add or subtract as indicated.

1. $\dfrac{5}{x^2} + \dfrac{3}{x^3}$ **2.** $\dfrac{3}{5x} - \dfrac{1}{4x^2}$

We can also add fractions with more than one variable in the denominator. The next example shows this.

Example 3

Add $\dfrac{2}{3x^2y} + \dfrac{3}{4x^3}$.

Step 1   Factor the denominators.

$3x^2y = 3 \cdot x \cdot x \cdot y$
$4x^3 = 2 \cdot 2 \cdot x \cdot x \cdot x$

The LCD is $12x^3y$. Do you see why?

Step 2

$$\frac{2}{3x^2y} = \frac{2 \cdot 4x}{3x^2y \cdot 4x} = \frac{8x}{12x^3y}$$

$$\frac{3}{4x^3} = \frac{3 \cdot 3y}{4x^3 \cdot 3y} = \frac{9y}{12x^3y}$$

Step 3

The y in the numerator and that in the denominator cannot be divided out since they are not factors.

$$\frac{2}{3x^2y} + \frac{3}{4x^3} = \frac{8x}{12x^3y} + \frac{9y}{12x^3y}$$

$$= \frac{8x + 9y}{12x^3y}$$

CHECK YOURSELF 3

Add.

$$\frac{2}{3x^2y} + \frac{1}{6xy^2}$$

Fractions with binomials in the denominator can also be added by taking the same approach. The next example illustrates.

Example 4

(*a*) Add $\dfrac{5}{x} + \dfrac{2}{x - 1}$.

Step 1   The LCD must have factors of x and $x - 1$. The LCD is $x(x - 1)$.

Step 2

$$\frac{5}{x} = \frac{5(x - 1)}{x(x - 1)}$$

$$\frac{2}{x - 1} = \frac{2x}{x(x - 1)}$$

Step 3

$$\frac{5}{x} + \frac{2}{x - 1} = \frac{5(x - 1)}{x(x - 1)} + \frac{2x}{x(x - 1)}$$

$$= \frac{5x - 5 + 2x}{x(x - 1)}$$

$$= \frac{7x - 5}{x(x - 1)}$$

(b) Subtract $\dfrac{3}{x-2} - \dfrac{4}{x+2}$.

Step 1 The LCD must have factors of $x-2$ and $x+2$. The LCD is $(x-2)(x+2)$.

Step 2

Multiply numerator and denominator by $x+2$.

Multiply numerator and denominator by $x-2$.

$$\frac{3}{x-2} = \frac{3(x+2)}{(x-2)(x+2)}$$

$$\frac{4}{x+2} = \frac{4(x-2)}{(x+2)(x-2)}$$

Step 3

$$\frac{3}{x-2} - \frac{4}{x+2} = \frac{3(x+2) - 4(x-2)}{(x+2)(x-2)}$$

Note the sign changes.

$$= \frac{3x+6-4x+8}{(x+2)(x-2)}$$

$$= \frac{-x+14}{(x+2)(x-2)}$$

CHECK YOURSELF 4

Add or subtract as indicated.

1. $\dfrac{3}{x+2} + \dfrac{5}{x}$ **2.** $\dfrac{4}{x+3} - \dfrac{2}{x-3}$

The following examples will show how factoring must sometimes be used in forming the LCD.

Example 5

(a) Add $\dfrac{3}{2x-2} + \dfrac{5}{3x-3}$.

Step 1 Factor the denominators.

$2x-2 = 2(x-1)$
$3x-3 = 3(x-1)$

CAUTION

$x-1$ is not used twice in forming the LCD.

The LCD must have factors of 2, 3, and $x-1$. The LCD is $2 \cdot 3(x-1)$, or $6(x-1)$.

Step 2

$$\frac{3}{2x - 2} = \frac{3}{2(x - 1)} = \frac{3 \cdot 3}{2(x - 1) \cdot 3} = \frac{9}{6(x - 1)}$$

$$\frac{5}{3x - 3} = \frac{5}{3(x - 1)} = \frac{5 \cdot 2}{3(x - 1) \cdot 2} = \frac{10}{6(x - 1)}$$

Step 3

$$\frac{3}{2x - 2} + \frac{5}{3x - 3} = \frac{9}{6(x - 1)} + \frac{10}{6(x - 1)}$$

$$= \frac{9 + 10}{6(x - 1)}$$

$$= \frac{19}{6(x - 1)}$$

(b) Subtract $\dfrac{3}{2x - 4} - \dfrac{6}{x^2 - 4}$.

Step 1 Factor the denominators.

$$2x - 4 = 2(x - 2)$$
$$x^2 - 4 = (x + 2)(x - 2)$$

The LCD must have factors of 2, $x - 2$, and $x + 2$. The LCD is $2(x - 2)(x + 2)$.

Step 2

Multiply numerator and denominator by $x + 2$.

$$\frac{3}{2x - 4} = \frac{3}{2(x - 2)} = \frac{3(x + 2)}{2(x - 2)(x + 2)}$$

Multiply numerator and denominator by 2.

$$\frac{6}{x^2 - 4} = \frac{6}{(x + 2)(x - 2)} = \frac{6 \cdot 2}{2(x + 2)(x - 2)} = \frac{12}{2(x + 2)(x - 2)}$$

Step 3

$$\frac{3}{2x - 4} - \frac{6}{x^2 - 4} = \frac{3(x + 2) - 12}{2(x - 2)(x + 2)}$$

Remove the parentheses and combine like terms in the numerator.

$$= \frac{3x + 6 - 12}{2(x - 2)(x + 2)}$$

$$= \frac{3x - 6}{2(x - 2)(x + 2)}$$

Step 4 Simplify the difference.

Factor the numerator and divide by the common factor $x - 2$.

$$\frac{3x - 6}{2(x - 2)(x + 2)} = \frac{3\cancel{(x - 2)}}{2\cancel{(x - 2)}(x + 2)} = \frac{3}{2(x + 2)}$$

(c) Subtract $\dfrac{5}{x^2 - 1} - \dfrac{2}{x^2 + 2x + 1}$.

Step 1 Factor the denominators.

$$x^2 - 1 = (x - 1)(x + 1)$$
$$x^2 + 2x + 1 = (x + 1)(x + 1)$$

The LCD is $(x - 1)(x + 1)(x + 1)$.

Two factors are needed.

Step 2

$$\frac{5}{(x - 1)(x + 1)} = \frac{5(x + 1)}{(x - 1)(x + 1)(x + 1)}$$

$$\frac{2}{(x + 1)(x + 1)} = \frac{2(x - 1)}{(x + 1)(x + 1)(x - 1)}$$

Step 3

Remove the parentheses and simplify in the numerator.

$$\frac{5}{x^2 - 1} - \frac{2}{x^2 + 2x + 1} = \frac{5(x + 1) - 2(x - 1)}{(x - 1)(x + 1)(x + 1)}$$

$$= \frac{5x + 5 - 2x + 2}{(x - 1)(x + 1)(x + 1)}$$

$$= \frac{3x + 7}{(x - 1)(x + 1)(x + 1)}$$

CHECK YOURSELF 5

Add or subtract as indicated.

1. $\dfrac{5}{2x + 2} + \dfrac{1}{5x + 5}$ **2.** $\dfrac{3}{x^2 - 9} - \dfrac{1}{2x - 6}$

3. $\dfrac{4}{x^2 - x - 2} - \dfrac{3}{x^2 + 4x + 3}$

Recall from Section 6.2 that

$$a - b = -(b - a)$$

Let's see how this can be used in adding or subtracting algebraic fractions.

Example 6

Add $\dfrac{4}{x-5} + \dfrac{2}{5-x}$.

Rather than try a denominator of $(x-5)(5-x)$, let's simplify first.

Replace $5-x$ with $-(x-5)$.

We now use the fact that

$$\dfrac{a}{-b} = -\dfrac{a}{b}$$

$$\dfrac{4}{x-5} + \dfrac{2}{5-x} = \dfrac{4}{x-5} + \dfrac{2}{-(x-5)}$$

$$= \dfrac{4}{x-5} - \dfrac{2}{x-5}$$

The LCD is now $x-5$, and we can combine the fractions as

$$= \dfrac{4-2}{x-5}$$

$$= \dfrac{2}{x-5}$$

CHECK YOURSELF 6

Subtract.

$$\dfrac{3}{x-3} - \dfrac{1}{3-x}$$

CHECK YOURSELF ANSWERS

1. (1) $\dfrac{13}{24}$; (2) $\dfrac{17}{30}$. **2.** (1) $\dfrac{5x+3}{x^3}$; (2) $\dfrac{12x-5}{20x^2}$.

3. $\dfrac{4y+x}{6x^2y^2}$.

4. (1) $\dfrac{8x+10}{x(x+2)}$; (2) $\dfrac{2x-18}{(x+3)(x-3)}$.

5. (1) $\dfrac{27}{10(x+1)}$; (2) $\dfrac{-1}{2(x+3)}$; (3) $\dfrac{x+18}{(x+1)(x-2)(x+3)}$.

6. $\dfrac{4}{x-3}$.

Name

Date

Build Your Skills

Add or subtract as indicated. Express your result in simplest form.

1. $\dfrac{5}{8} + \dfrac{3}{5}$

2. $\dfrac{7}{9} - \dfrac{1}{6}$

3. $\dfrac{11}{15} - \dfrac{3}{10}$

4. $\dfrac{7}{8} + \dfrac{5}{6}$

5. $\dfrac{y}{4} + \dfrac{3y}{5}$

6. $\dfrac{5x}{6} - \dfrac{2x}{3}$

7. $\dfrac{7a}{3} - \dfrac{a}{7}$

8. $\dfrac{3m}{4} + \dfrac{m}{9}$

9. $\dfrac{3}{x} - \dfrac{4}{5}$

10. $\dfrac{5}{x} + \dfrac{2}{3}$

11. $\dfrac{5}{a} + \dfrac{a}{5}$

12. $\dfrac{y}{3} - \dfrac{3}{y}$

13. $\dfrac{5}{m} + \dfrac{3}{m^2}$

14. $\dfrac{4}{x^2} - \dfrac{3}{x}$

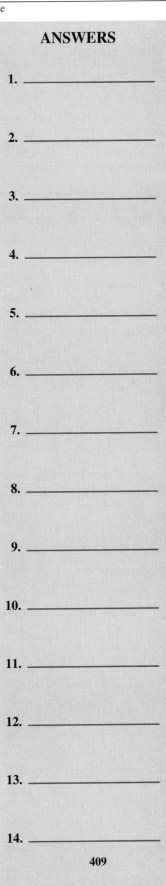

ANSWERS

1. _____

2. _____

3. _____

4. _____

5. _____

6. _____

7. _____

8. _____

9. _____

10. _____

11. _____

12. _____

13. _____

14. _____

15. _____

16. _____

17. _____

18. _____

19. _____

20. _____

21. _____

22. _____

23. _____

24. _____

25. _____

26. _____

27. _____

28. _____

29. _____

30. _____

15. $\dfrac{1}{x^2} - \dfrac{2}{3x}$

16. $\dfrac{5}{2w} + \dfrac{7}{w^3}$

17. $\dfrac{3}{5s} + \dfrac{2}{s^2}$

18. $\dfrac{7}{a^2} - \dfrac{3}{7a}$

19. $\dfrac{3}{4b^2} + \dfrac{5}{3b^3}$

20. $\dfrac{4}{5x^3} - \dfrac{3}{2x^2}$

21. $\dfrac{x}{x + 2} + \dfrac{2}{5}$

22. $\dfrac{3}{4} - \dfrac{a}{a - 1}$

23. $\dfrac{y}{y - 4} - \dfrac{3}{4}$

24. $\dfrac{m}{m + 3} + \dfrac{2}{3}$

25. $\dfrac{4}{x} + \dfrac{3}{x + 1}$

26. $\dfrac{2}{x} - \dfrac{1}{x - 2}$

27. $\dfrac{5}{a - 1} - \dfrac{2}{a}$

28. $\dfrac{4}{x + 2} + \dfrac{3}{x}$

29. $\dfrac{4}{2x - 3} + \dfrac{2}{3x}$

30. $\dfrac{7}{2y - 1} - \dfrac{3}{2y}$

31. $\dfrac{2}{x + 1} + \dfrac{3}{x + 3}$

32. $\dfrac{5}{x - 1} + \dfrac{2}{x + 2}$

33. $\dfrac{4}{y - 2} - \dfrac{1}{y + 1}$

34. $\dfrac{5}{x + 4} - \dfrac{3}{x - 1}$

35. $\dfrac{2}{b - 3} + \dfrac{3}{2b - 6}$

36. $\dfrac{4}{a + 5} - \dfrac{3}{4a + 20}$

37. $\dfrac{x}{x + 4} - \dfrac{2}{3x + 12}$

38. $\dfrac{x}{x - 3} + \dfrac{5}{2x - 6}$

39. $\dfrac{4}{3m + 3} + \dfrac{1}{2m + 2}$

40. $\dfrac{3}{5y - 5} - \dfrac{2}{3y - 3}$

41. $\dfrac{4}{5x - 10} - \dfrac{1}{3x - 6}$

42. $\dfrac{2}{3w + 3} + \dfrac{5}{2w + 2}$

43. $\dfrac{5}{4b + 8} - \dfrac{b}{5b + 10}$

44. $\dfrac{4}{3b - 9} + \dfrac{3b}{4b - 12}$

ANSWERS

31. _____

32. _____

33. _____

34. _____

35. _____

36. _____

37. _____

38. _____

39. _____

40. _____

41. _____

42. _____

43. _____

44. _____

45. _____

46. _____

47. _____

48. _____

49. _____

50. _____

51. _____

52. _____

53. _____

54. _____

55. _____

56. _____

57. _____

58. _____

45. $\dfrac{y-1}{y+1} - \dfrac{y}{3y+3}$

46. $\dfrac{x+2}{x-2} - \dfrac{x}{3x-6}$

47. $\dfrac{3}{x^2-4} + \dfrac{2}{x+2}$

48. $\dfrac{4}{x-2} + \dfrac{3}{x^2-x-2}$

49. $\dfrac{3x}{x^2-3x+2} - \dfrac{1}{x-2}$

50. $\dfrac{a}{a^2-1} - \dfrac{4}{a+1}$

51. $\dfrac{2x}{x^2-5x+6} + \dfrac{4}{x-2}$

52. $\dfrac{7a}{a^2+a-12} - \dfrac{4}{a+4}$

53. $\dfrac{2}{3x-3} + \dfrac{1}{4x+4}$

54. $\dfrac{2}{5w+10} - \dfrac{3}{2w-4}$

55. $\dfrac{4}{3a-9} - \dfrac{3}{2a+4}$

56. $\dfrac{2}{3b-6} + \dfrac{3}{4b+8}$

57. $\dfrac{5}{x^2-16} - \dfrac{3}{x^2-x-12}$

58. $\dfrac{3}{x^2+4x+3} - \dfrac{1}{x^2-9}$

59. $\dfrac{2}{y^2 + y - 6} + \dfrac{3y}{y^2 - 2y - 15}$

60. $\dfrac{2a}{a^2 - a - 12} - \dfrac{3}{a^2 - 2a - 8}$

61. $\dfrac{6x}{x^2 - 9} - \dfrac{5x}{x^2 + x - 6}$

62. $\dfrac{4y}{y^2 + 6y + 5} + \dfrac{2y}{y^2 - 1}$

63. $\dfrac{3}{a - 7} + \dfrac{2}{7 - a}$

64. $\dfrac{5}{x - 5} - \dfrac{3}{5 - x}$

65. $\dfrac{2x}{2x - 3} - \dfrac{1}{3 - 2x}$

66. $\dfrac{9m}{3m - 1} + \dfrac{3}{1 - 3m}$

Think About These

67. $\dfrac{1}{a - 3} - \dfrac{1}{a + 3} + \dfrac{2a}{a^2 - 9}$

68. $\dfrac{1}{p + 1} + \dfrac{1}{p - 3} - \dfrac{4}{p^2 - 2p - 3}$

69. $\dfrac{2x^2 + 3x}{x^2 - 2x - 63} + \dfrac{7 - x}{x^2 - 2x - 63} - \dfrac{x^2 - 3x + 21}{x^2 - 2x - 63}$

70. $-\dfrac{3 - 2x^2}{x^2 - 9x + 20} - \dfrac{4x^2 + 2x + 1}{x^2 - 9x + 20} + \dfrac{2x^2 + 3x}{x^2 - 9x + 20}$

ANSWERS

59. _____

60. _____

61. _____

62. _____

63. _____

64. _____

65. _____

66. _____

67. _____

68. _____

69. _____

70. _____

Skillscan (Section 6.3)

Multiply.

a. $\dfrac{3}{4} \cdot 8$

b. $\dfrac{7}{10} \cdot 20$

c. $\dfrac{4}{x^2} \cdot x^2$

d. $\dfrac{9}{w^2} \cdot w^3$

e. $\dfrac{1}{xy} \cdot xy^2$

f. $\dfrac{2}{a^2} \cdot a^2b^2$

g. $\dfrac{3}{pq} \cdot p^2q^2$

h. $\dfrac{2}{a^2b} \cdot a^2b^2$

ANSWERS

1. $\dfrac{49}{40}$ **3.** $\dfrac{13}{30}$ **5.** $\dfrac{17y}{20}$ **7.** $\dfrac{46a}{21}$ **9.** $\dfrac{15 - 4x}{5x}$ **11.** $\dfrac{25 + a^2}{5a}$ **13.** $\dfrac{5m + 3}{m^2}$

15. $\dfrac{3 - 2x}{3x^2}$ **17.** $\dfrac{3s + 10}{5s^2}$ **19.** $\dfrac{9b + 20}{12b^3}$ **21.** $\dfrac{7x + 4}{5(x + 2)}$ **23.** $\dfrac{y + 12}{4(y - 4)}$

25. $\dfrac{7x + 4}{x(x + 1)}$ **27.** $\dfrac{3a + 2}{a(a - 1)}$ **29.** $\dfrac{2(8x - 3)}{3x(2x - 3)}$ **31.** $\dfrac{5x + 9}{(x + 1)(x + 3)}$

33. $\dfrac{3(y + 2)}{(y - 2)(y + 1)}$ **35.** $\dfrac{7}{2(b - 3)}$ **37.** $\dfrac{3x - 2}{3(x + 4)}$ **39.** $\dfrac{11}{6(m + 1)}$ **41.** $\dfrac{7}{15(x - 2)}$

43. $\dfrac{25 - 4b}{20(b + 2)}$ **45.** $\dfrac{2y - 3}{3(y + 1)}$ **47.** $\dfrac{2x - 1}{(x - 2)(x + 2)}$ **49.** $\dfrac{2x + 1}{(x - 1)(x - 2)}$ **51.** $\dfrac{6}{x - 3}$

53. $\dfrac{11x + 5}{12(x - 1)(x + 1)}$ **55.** $\dfrac{-a + 43}{6(a - 3)(a + 2)}$ **57.** $\dfrac{2x + 3}{(x + 4)(x - 4)(x + 3)}$

59. $\dfrac{3y^2 - 4y - 10}{(y + 3)(y - 2)(y - 5)}$ **61.** $\dfrac{x}{(x - 3)(x - 2)}$ **63.** $\dfrac{1}{a - 7}$ **65.** $\dfrac{2x + 1}{2x - 3}$

67. $\dfrac{2}{a - 3}$ **69.** $\dfrac{x - 2}{x - 9}$ **a.** 6 **b.** 14 **c.** 4 **d.** $9w$ **e.** y **f.** $2b^2$ **g.** $3pq$

h. $2b$

6.6 Complex Fractions

OBJECTIVE
To simplify complex algebraic fractions

A fraction that has a fraction in its numerator, in its denominator, or in both is called a *complex fraction*.

$$\frac{\dfrac{5}{6}}{\dfrac{3}{4}} \qquad \frac{\dfrac{4}{x}}{\dfrac{3}{x^2}} \quad \text{and} \quad \frac{\dfrac{a+2}{3}}{\dfrac{a-2}{5}}$$

are all complex fractions.

There are two methods for simplifying a complex fraction. To develop the first, remember that we can always multiply the numerator and the denominator of a fraction by the same nonzero term.

This is the fundamental principle of fractions.

$$\frac{P}{Q} = \frac{P \cdot R}{Q \cdot R} \qquad \text{where } Q \neq 0 \text{ and } R \neq 0$$

For our first approach to simplifying a complex fraction, multiply the numerator and denominator by the LCD of all fractions that appear within the complex fraction.

Example 1

Simplify $\dfrac{\dfrac{3}{4}}{\dfrac{5}{8}}$.

The LCD of $\dfrac{3}{4}$ and $\dfrac{5}{8}$ is 8. So multiply the numerator and denominator by 8.

$$\frac{\dfrac{3}{4}}{\dfrac{5}{8}} = \frac{\dfrac{3}{4} \cdot 8}{\dfrac{5}{8} \cdot 8} = \frac{3 \cdot 2}{5 \cdot 1} = \frac{6}{5}$$

Simplify.

1. $\dfrac{\dfrac{4}{7}}{\dfrac{3}{7}}$ **2.** $\dfrac{\dfrac{3}{8}}{\dfrac{5}{6}}$

The same method can be used to simplify a complex fraction when variables are involved in the expression. Consider the following example.

Example 2

Simplify $\dfrac{\dfrac{5}{x}}{\dfrac{10}{x^2}}$.

The LCD of $\dfrac{5}{x}$ and $\dfrac{10}{x^2}$ is x^2, so multiply the numerator and denominator by x^2.

Be sure to write the result in simplest form.

$$\frac{\dfrac{5}{x}}{\dfrac{10}{x^2}} = \frac{\left(\dfrac{5}{x}\right)x^2}{\left(\dfrac{10}{x^2}\right)x^2} = \frac{5x}{10} = \frac{x}{2}$$

Simplify.

1. $\dfrac{\dfrac{6}{x^3}}{\dfrac{9}{x^2}}$ **2.** $\dfrac{\dfrac{m^4}{15}}{\dfrac{m^3}{20}}$

We may also have a sum or a difference in the numerator or denominator of a complex fraction. The simplification steps are exactly the same. Consider the following example.

Example 3

Simplify $\dfrac{1 + \dfrac{x}{y}}{1 - \dfrac{x}{y}}$.

The LCD of 1, $\dfrac{x}{y}$, 1, and $\dfrac{x}{y}$ is y, so multiply the numerator and denominator by y.

Note the use of the distributive property to multiply *each term* in the numerator and in the denominator by y.

$$\frac{1 + \dfrac{x}{y}}{1 - \dfrac{x}{y}} = \frac{\left(1 + \dfrac{x}{y}\right)y}{\left(1 - \dfrac{x}{y}\right)y} = \frac{1 \cdot y + \dfrac{x}{y} \cdot y}{1 \cdot y - \dfrac{x}{y} \cdot y}$$

$$= \frac{y + x}{y - x}$$

CHECK YOURSELF 3

Simplify.

$$\frac{\dfrac{x}{y} - 2}{\dfrac{x}{y} + 2}$$

Our second method for simplifying complex fractions uses the fact that

To divide by a fraction, we invert the divisor (it *follows* the division sign) and multiply.

$$\frac{\dfrac{P}{Q}}{\dfrac{R}{S}} = \frac{P}{Q} \div \frac{R}{S} = \frac{P}{Q} \cdot \frac{S}{R}$$

To use this method, we must write the numerator and denominator of the complex fraction as single fractions. We can then divide the numerator by the denominator as before.

Example 4

Simplify $\dfrac{1 + \dfrac{x}{y}}{1 - \dfrac{x}{y}}$.

We have written the numerator and denominator as single fractions. Be sure you see how this is done.

$$\frac{1 + \dfrac{x}{y}}{1 - \dfrac{x}{y}} = \frac{\dfrac{y + x}{y}}{\dfrac{y - x}{y}}$$

$$= \frac{y + x}{y} \div \frac{y - x}{y}$$

$$= \frac{y + x}{y} \cdot \frac{y}{y - x}$$

$$= \frac{y + x}{y - x}$$

Of course, the answer is the same as the one we obtained in Example 3. You can use whichever method you find easier in a particular problem.

CHECK YOURSELF 4

Using the second method, simplify $\dfrac{\dfrac{x}{y} - 2}{\dfrac{x}{y} + 2}$.

Let's look at one more example of simplifying a complex fraction by the second method.

Example 5

Simplify $\dfrac{4 - \dfrac{y^2}{x^2}}{2 + \dfrac{y}{x}}$.

In this approach we must first work *separately* in the numerator and denominator to form single fractions.

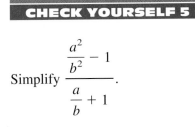

$$\frac{4 - \dfrac{y^2}{x^2}}{2 + \dfrac{y}{x}} = \frac{\dfrac{4x^2 - y^2}{x^2}}{\dfrac{2x + y}{x}}$$

$$= \frac{4x^2 - y^2}{x^2} \cdot \frac{x}{2x + y}$$ Invert the divisor (the denominator) and multiply.

$$= \frac{(2x - y)(2x + y)}{x^2} \cdot \frac{x}{2x + y}$$ Factor and divide by the common factors of $2x + y$ and x.

$$= \frac{2x - y}{x}$$

CHECK YOURSELF 5

Simplify $\dfrac{\dfrac{a^2}{b^2} - 1}{\dfrac{a}{b} + 1}$.

The following algorithm summarizes our work with the two methods of simplifying complex fractions.

TO SIMPLIFY COMPLEX FRACTIONS

METHOD 1

STEP 1. Multiply the numerator and denominator of the complex fraction by the LCD of all the fractions that appear within the complex fraction.
STEP 2. Write the resulting fraction in simplest form.

METHOD 2

STEP 1. Write the numerator and denominator of the complex fraction as single fractions, if necessary.
STEP 2. Invert the denominator and multiply as before, writing the resulting fraction in simplest form.

CHECK YOURSELF ANSWERS

1. (1) $\dfrac{4}{3}$; (2) $\dfrac{9}{20}$. **2.** (1) $\dfrac{2}{3x}$; (2) $\dfrac{4m}{3}$. **3.** $\dfrac{x-2y}{x+2y}$.

4. $\dfrac{x-2y}{x+2y}$. **5.** $\dfrac{a-b}{b}$.

Name _____

Date _____

Build Your Skills

Simplify each complex fraction.

1. $\dfrac{\frac{2}{3}}{\frac{6}{8}}$

2. $\dfrac{\frac{5}{6}}{\frac{10}{15}}$

3. $\dfrac{1 + \frac{1}{2}}{2 + \frac{1}{4}}$

4. $\dfrac{1 + \frac{3}{4}}{2 - \frac{1}{8}}$

5. $\dfrac{2 + \frac{1}{3}}{3 - \frac{1}{5}}$

6. $\dfrac{2 + \frac{3}{5}}{1 + \frac{3}{10}}$

7. $\dfrac{\frac{2}{3} + \frac{1}{2}}{\frac{3}{4} - \frac{1}{3}}$

8. $\dfrac{\frac{3}{4} + \frac{1}{2}}{\frac{7}{8} - \frac{1}{4}}$

9. $\dfrac{\frac{x}{8}}{\frac{x^2}{4}}$

10. $\dfrac{\frac{m^2}{10}}{\frac{m^3}{15}}$

11. $\dfrac{\frac{3}{a}}{\frac{2}{a^2}}$

12. $\dfrac{\frac{6}{x^2}}{\frac{9}{x^3}}$

ANSWERS

1. _____

2. _____

3. _____

4. _____

5. _____

6. _____

7. _____

8. _____

9. _____

10. _____

11. _____

12. _____

ANSWERS

13. _____

14. _____

15. _____

16. _____

17. _____

18. _____

19. _____

20. _____

21. _____

22. _____

23. _____

24. _____

25. _____

26. _____

13. $\dfrac{\dfrac{y+1}{y}}{\dfrac{y-1}{2y}}$

14. $\dfrac{\dfrac{w+3}{4w}}{\dfrac{w-3}{2w}}$

15. $\dfrac{2-\dfrac{1}{x}}{2+\dfrac{1}{x}}$

16. $\dfrac{3+\dfrac{1}{a}}{3-\dfrac{1}{a}}$

17. $\dfrac{3-\dfrac{x}{y}}{\dfrac{6}{y}}$

18. $\dfrac{2+\dfrac{x}{y}}{\dfrac{4}{y}}$

19. $\dfrac{2+\dfrac{p}{q}}{1+\dfrac{p}{q}}$

20. $\dfrac{\dfrac{m}{n}-3}{\dfrac{m}{n}+3}$

21. $\dfrac{a^2-1}{1-\dfrac{1}{a}}$

22. $\dfrac{1+\dfrac{1}{2x}}{4x^2-1}$

23. $\dfrac{\dfrac{x^2}{y^2}-1}{\dfrac{x}{y}+1}$

24. $\dfrac{\dfrac{a}{b}+2}{\dfrac{a^2}{b^2}-4}$

25. $\dfrac{1+\dfrac{3}{x}-\dfrac{4}{x^2}}{1+\dfrac{2}{x}-\dfrac{3}{x^2}}$

26. $\dfrac{1-\dfrac{2}{r}-\dfrac{8}{r^2}}{1-\dfrac{1}{r}-\dfrac{6}{r^2}}$

27. $\dfrac{\dfrac{1}{x} + \dfrac{1}{y}}{\dfrac{2}{x} - \dfrac{2}{y}}$

28. $\dfrac{\dfrac{3}{a} - \dfrac{3}{b}}{\dfrac{1}{a} + \dfrac{1}{b}}$

29. $\dfrac{\dfrac{2}{x} - \dfrac{1}{xy}}{\dfrac{1}{xy} + \dfrac{2}{y}}$

30. $\dfrac{\dfrac{1}{xy} + \dfrac{2}{x}}{\dfrac{3}{y} - \dfrac{1}{xy}}$

31. $\dfrac{\dfrac{x^2}{y} + 2x + y}{\dfrac{1}{y^2} - \dfrac{1}{x^2}}$

32. $\dfrac{\dfrac{x}{y} + 1 - \dfrac{2y}{x}}{\dfrac{1}{y^2} - \dfrac{4}{x^2}}$

Think About These

33. $\dfrac{\dfrac{2}{x - 1} + 1}{1 - \dfrac{3}{x - 1}}$

34. $\dfrac{\dfrac{3}{a + 2} - 1}{1 + \dfrac{2}{a + 2}}$

35. $\dfrac{1 - \dfrac{1}{y - 1}}{y - \dfrac{8}{y + 2}}$

36. $\dfrac{1 + \dfrac{1}{x + 2}}{x - \dfrac{18}{x - 3}}$

37. $1 + \dfrac{1}{1 + \dfrac{1}{x}}$

38. $1 + \dfrac{1}{1 - \dfrac{1}{y}}$

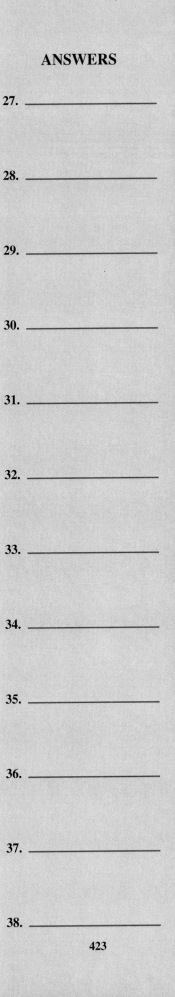

ANSWERS

27. _____

28. _____

29. _____

30. _____

31. _____

32. _____

33. _____

34. _____

35. _____

36. _____

37. _____

38. _____

39. _____

40. _____

41. _____

a. _____

b. _____

c. _____

d. _____

e. _____

f. _____

39. Herbicides constitute two-thirds of all pesticides used in the United States. Insecticides are one-fourth of all pesticides used in the United States. The ratio of herbicides to insecticides used in the United States can be written $\dfrac{2}{3} \div \dfrac{1}{4}$. Write this ratio in simplest form.

40. Fungicides account for one-tenth of the pesticides used in the United States. Insecticides account for one-fourth of all pesticides used in the United States. The ratio of fungicides to insecticides used in the United States can be written $\dfrac{1}{10} \div \dfrac{1}{4}$. Write this ratio in simplest form.

41. The ratio of insecticides to herbicides applied to wheat, soybeans, corn, and cotton can be expressed as $\dfrac{7X}{10} \div \dfrac{4X}{5}$. Simplify this ratio.

Skillscan (Section 3.4)

Solve each of the following equations.

a. $x + 8 = 10$

b. $5x - 4 = 2$

c. $3x + 8 = 4$

d. $3(x - 2) - 4 = 5$

e. $4(2x + 1) - 3 = -23$

f. $4(2x - 5) - 3(3x + 1) = -8$

ANSWERS

1. $\dfrac{8}{9}$ **3.** $\dfrac{2}{3}$ **5.** $\dfrac{5}{6}$ **7.** $\dfrac{14}{5}$ **9.** $\dfrac{1}{2x}$ **11.** $\dfrac{3a}{2}$ **13.** $\dfrac{2(y + 1)}{y - 1}$ **15.** $\dfrac{2x - 1}{2x + 1}$

17. $\dfrac{3y - x}{6}$ **19.** $\dfrac{2q + p}{q + p}$ **21.** $a(a + 1)$ **23.** $\dfrac{x - y}{y}$ **25.** $\dfrac{x + 4}{x + 3}$ **27.** $\dfrac{y + x}{2(y - x)}$

29. $\dfrac{2y - 1}{1 + 2x}$ **31.** $\dfrac{x^2 y(x + y)}{x - y}$ **33.** $\dfrac{x + 1}{x - 4}$ **35.** $\dfrac{y + 2}{(y - 1)(y + 4)}$ **37.** $\dfrac{2x + 1}{x + 1}$

39. $\dfrac{8}{3}$ **41.** $\dfrac{7}{8}$ **a.** 2 **b.** $\dfrac{6}{5}$ **c.** $-\dfrac{4}{3}$ **d.** 5 **e.** -3 **f.** -15

6.7 Equations Involving Fractions

OBJECTIVE
To solve fractional equations

In Chapter 3 you learned how to solve a variety of equations. We now want to extend that work to the solution of *fractional equations*. These are equations that involve algebraic fractions as one or more of their terms.

To solve a fractional equation, we multiply each term of the equation by the LCD of any fractions. The resulting equation should be equivalent to the original equation and be cleared of all fractions.

The resulting equation *will* be equivalent unless a solution results that makes a denominator in the original equation 0. More about this later!

Example 1

Solve

$$\frac{x}{2} - \frac{1}{3} = \frac{2x + 3}{6} \tag{1}$$

This equation has three terms: $\frac{x}{2}$, $-\frac{1}{3}$, and $\frac{2x + 3}{6}$. The sign of the term is not used to find the LCD.

The LCD for $\frac{x}{2}$, $\frac{1}{3}$, and $\frac{2x + 3}{6}$ is 6. Multiply *each* term by 6.

$$6 \cdot \frac{x}{2} - 6 \cdot \frac{1}{3} = 6\left(\frac{2x + 3}{6}\right)$$

or

By the multiplication property of equality, this equation is equivalent to the original equation, labeled (1).

$$3x - 2 = 2x + 3 \tag{2}$$

Solving as before, we have

$$3x - 2x = 3 + 2$$

or

$$x = 5$$

To check, substitute 5 for x in the *original* equation.

$$\frac{5}{2} - \frac{1}{3} \stackrel{?}{=} \frac{2 \cdot 5 + 3}{6}$$

$$\frac{13}{6} = \frac{13}{6} \quad \text{(True)}$$

⟨CAUTION⟩

Be Careful! Many students have difficulty because they don't distinguish between adding or subtracting *expressions* (as we did in Sections 6.4 and 6.5) and solving equations (illustrated in the above example). In the *expression*

$$\frac{x+1}{2} + \frac{x}{3}$$

we want to add the two fractions to form a single fraction. In the *equation*

$$\frac{x+1}{2} = \frac{x}{3} + 1$$

we want to solve for x.

CHECK YOURSELF 1

Solve and check.

$$\frac{x}{4} - \frac{1}{6} = \frac{4x-5}{12}$$

The steps of the solution illustrated in Example 1 are summarized in the following rule.

The equation that is formed in step 2 can be solved by the methods of Sections 3.4 and 5.5.

> **TO SOLVE A FRACTIONAL EQUATION**
>
> **STEP 1**   Remove the fractions in the equation by multiplying each term by the LCD of all the fractions.
>
> **STEP 2**   Solve the equation resulting from step 1 as before.
>
> **STEP 3**   Check your solution in the *original equation*.

We can also solve fractional equations with variables in the denominator by using the above algorithm. The following example illustrates.

Example 2

Solve

$$\frac{7}{4x} - \frac{3}{x^2} = \frac{1}{2x^2}$$

The factor x appears twice in the LCD.

The LCD of the three terms in the equation is $4x^2$, and so we multiply each term by $4x^2$.

$$4x^2 \cdot \frac{7}{4x} - 4x^2 \cdot \frac{3}{x^2} = 4x^2 \cdot \frac{1}{2x^2}$$

Simplifying, we have

$$7x - 12 = 2$$
$$7x = 14$$
$$x = 2$$

We'll leave the check to you. Be sure to return to the original equation.

CHECK YOURSELF 2

Solve and check.

$$\frac{5}{2x} - \frac{4}{x^2} = \frac{7}{2x^2}$$

The process of solving fractional equations is exactly the same when binomials are involved in the denominators.

Example 3

(*a*) Solve

There are three terms.

$$\frac{x}{x - 3} - 2 = \frac{1}{x - 3}$$

The LCD is $x - 3$, and so we multiply each term by $x - 3$.

Each of the terms is multiplied by $x - 3$.

$$(x - 3) \cdot \left(\frac{x}{x - 3}\right) - 2(x - 3) = (x - 3) \cdot \left(\frac{1}{x - 3}\right)$$

Simplifying, we have

$$x - 2(x - 3) = 1$$

<CAUTION>

Be careful of the signs!

$$x - 2x + 6 = 1$$
$$-x = -5$$
$$x = 5$$

To check, substitute 5 for x in the original equation.

(*b*) Solve

Recall that

$$x^2 - 9 = (x - 3)(x + 3)$$

$$\frac{3}{x - 3} - \frac{7}{x + 3} = \frac{2}{x^2 - 9}$$

In factored form, the three denominators are $x - 3$, $x + 3$, and $(x + 3)(x - 3)$. This means that the LCD is $(x + 3)(x - 3)$, and so we multiply:

$$(x + 3)(x - 3)\left(\frac{3}{x - 3}\right) - (x + 3)(x - 3)\left(\frac{7}{x + 3}\right)$$

$$= (x + 3)(x - 3)\left(\frac{2}{x^2 - 9}\right)$$

Simplifying, we have

$$3(x + 3) - 7(x - 3) = 2$$
$$3x + 9 - 7x + 21 = 2$$
$$-4x + 30 = 2$$
$$-4x = -28$$
$$x = 7$$

CHECK YOURSELF 3

Solve and check.

1. $\dfrac{x}{x - 5} - 2 = \dfrac{2}{x - 5}$

2. $\dfrac{4}{x - 4} - \dfrac{3}{x + 1} = \dfrac{5}{x^2 - 3x - 4}$

You should be aware of one further problem in dealing with fractional equations. The following example shows that possibility.

Example 4

Solve

$$\frac{x}{x - 2} - 7 = \frac{2}{x - 2}$$

The LCD is $x - 2$, and so we multiply each term by $x - 2$.

$$(x - 2)\left(\frac{x}{x - 2}\right) - 7(x - 2) = (x - 2)\left(\frac{2}{x - 2}\right)$$

Simplifying, we have

$$x - 7x + 14 = 2$$
$$-6x = -12$$
$$x = 2$$

Now, when we try to check our result, we have

2 is substituted for x in the original equation.

$$\frac{2}{2-2} - 7 = \frac{2}{2-2} \qquad \text{or} \qquad \frac{2}{0} - 7 = \frac{2}{0}$$

These terms are undefined.

What went wrong? Remember that two of the terms in our original equation were $\dfrac{x}{x-2}$ and $\dfrac{2}{x-2}$. The variable x cannot have the value 2 because 2 is an excluded value (it makes the denominator 0). So our original equation has *no solution*.

CHECK YOURSELF 4

Solve if possible.

$$\frac{x}{x+3} - 6 = \frac{-3}{x+3}$$

Equations involving fractions may also lead to quadratic equations, as the following example illustrates.

Example 5

Solve

$$\frac{x}{x-4} = \frac{15}{x-3} - \frac{2x}{x^2 - 7x + 12}$$

The LCD is $(x-4)(x-3)$. Multiply each term by $(x-4)(x-3)$.

$$\frac{x}{(x-4)}(x-4)(x-3)$$
$$= \frac{15}{(x-3)}(x-4)(x-3) - \frac{2x}{(x-4)(x-3)}(x-4)(x-3)$$

Simplifying, we have

$$x(x-3) = 15(x-4) - 2x$$

Multiply to clear of parentheses:

$$x^2 - 3x = 15x - 60 - 2x$$

In standard form, the equation is

$$x^2 - 16x + 60 = 0 \qquad \text{or} \qquad (x-6)(x-10) = 0$$

Note that this equation is *quadratic*. It can be solved by the methods of Section 5.5.

Setting the factors to 0, we have

$x - 6 = 0$ or $x - 10 = 0$

$x = 6$ $x = 10$

So $x = 6$ and $x = 10$ are possible solutions. We will leave the check of *each* solution to you.

CHECK YOURSELF 5

Solve and check.

$$\frac{3x}{x + 2} - \frac{2}{x + 3} = \frac{36}{x^2 + 5x + 6}$$

Strategies in Equation Solving

As the examples of this section have illustrated, *whenever* an equation involves algebraic fractions, the *first step* of the solution is to clear the equation of fractions by multiplication.

The following algorithm summarizes our work in solving equations that involve algebraic fractions.

TO SOLVE AN EQUATION INVOLVING FRACTIONS

STEP 1   Remove the fractions appearing in the equation by multiplying each term by the LCD of all the fractions.

STEP 2   Solve the equation resulting from step 1. If the equation is linear, use the methods of Section 3.4 for the solution. If the equation is quadratic, use the methods of Section 5.5.

STEP 3   Check all solutions by substitution in the *original equation*. Be sure to discard any *extraneous* solutions, that is, solutions that would result in a zero denominator in the original equation.

CHECK YOURSELF ANSWERS

1. 3.   **2.** 3.   **3.** (1) 8; (2) -11.

4. No solution   **5.** $x = -5$ or $x = \dfrac{8}{3}$.

Name

Date

Build Your Skills

Solve each of the following equations for x.

1. $\dfrac{x}{2} + 3 = 6$

2. $\dfrac{x}{3} - 2 = 1$

3. $\dfrac{x}{2} - \dfrac{x}{3} = 2$

4. $\dfrac{x}{6} - \dfrac{x}{8} = 1$

5. $\dfrac{x}{5} - \dfrac{1}{3} = \dfrac{x - 7}{3}$

6. $\dfrac{x}{6} + \dfrac{3}{4} = \dfrac{x - 1}{4}$

7. $\dfrac{x}{4} - \dfrac{1}{5} = \dfrac{4x + 3}{20}$

8. $\dfrac{x}{12} - \dfrac{1}{6} = \dfrac{2x - 7}{12}$

9. $\dfrac{3}{x} + 2 = \dfrac{7}{x}$

10. $\dfrac{4}{x} - 3 = \dfrac{16}{x}$

11. $\dfrac{4}{x} + \dfrac{3}{4} = \dfrac{10}{x}$

12. $\dfrac{3}{x} = \dfrac{5}{3} - \dfrac{7}{x}$

13. $\dfrac{5}{2x} - \dfrac{1}{x} = \dfrac{9}{2x^2}$

14. $\dfrac{4}{3x} + \dfrac{1}{x} = \dfrac{14}{3x^2}$

15. $\dfrac{2}{x - 3} + 1 = \dfrac{7}{x - 3}$

16. $\dfrac{x}{x + 1} + 2 = \dfrac{14}{x + 1}$

17. $\dfrac{12}{x + 3} = \dfrac{x}{x + 3} + 2$

18. $\dfrac{5}{x - 3} + 3 = \dfrac{x}{x - 3}$

1. _____

2. _____

3. _____

4. _____

5. _____

6. _____

7. _____

8. _____

9. _____

10. _____

11. _____

12. _____

13. _____

14. _____

15. _____

16. _____

17. _____

18. _____

ANSWERS

19. _____

20. _____

21. _____

22. _____

23. _____

24. _____

25. _____

26. _____

27. _____

28. _____

29. _____

30. _____

31. _____

32. _____

33. _____

34. _____

35. _____

36. _____

37. _____

38. _____

19. $\dfrac{3}{x-5} + 4 = \dfrac{2x+5}{x-5}$

20. $\dfrac{24}{x+5} - 2 = \dfrac{x+2}{x+5}$

21. $\dfrac{2}{x+3} + \dfrac{1}{2} = \dfrac{x+6}{x+3}$

22. $\dfrac{6}{x-5} - \dfrac{2}{3} = \dfrac{x-9}{x-5}$

23. $\dfrac{x}{3x+12} + \dfrac{x-1}{x+4} = \dfrac{5}{3}$

24. $\dfrac{x}{4x-12} - \dfrac{x-4}{x-3} = \dfrac{1}{8}$

25. $\dfrac{x}{x-3} - 2 = \dfrac{3}{x-3}$

26. $\dfrac{x}{x-5} + 2 = \dfrac{5}{x-5}$

27. $\dfrac{x-1}{x+3} - \dfrac{x-3}{x} = \dfrac{3}{x^2+3x}$

28. $\dfrac{x}{x-2} - \dfrac{x+1}{x} = \dfrac{8}{x^2-2x}$

29. $\dfrac{1}{x-2} - \dfrac{2}{x+2} = \dfrac{2}{x^2-4}$

30. $\dfrac{1}{x+4} + \dfrac{1}{x-4} = \dfrac{12}{x^2-16}$

31. $\dfrac{5}{x-4} = \dfrac{1}{x+2} - \dfrac{2}{x^2-2x-8}$

32. $\dfrac{11}{x+2} = \dfrac{5}{x^2-x-6} + \dfrac{1}{x-3}$

33. $\dfrac{3}{x-1} - \dfrac{1}{x+9} = \dfrac{18}{x^2+8x-9}$

34. $\dfrac{2}{x+2} = \dfrac{3}{x+6} + \dfrac{9}{x^2+8x+12}$

35. $\dfrac{3}{x+3} + \dfrac{25}{x^2+x-6} = \dfrac{5}{x-2}$

36. $\dfrac{5}{x+6} + \dfrac{2}{x^2+7x+6} = \dfrac{3}{x+1}$

37. $\dfrac{7}{x-5} - \dfrac{3}{x+5} = \dfrac{40}{x^2-25}$

38. $\dfrac{3}{x-3} - \dfrac{18}{x^2-9} = \dfrac{5}{x+3}$

39. $\dfrac{2x}{x-3} + \dfrac{2}{x-5} = \dfrac{3x}{x^2 - 8x + 15}$

40. $\dfrac{x}{x-4} = \dfrac{5x}{x^2 - x - 12} - \dfrac{3}{x+3}$

41. $\dfrac{2x}{x+2} = \dfrac{5}{x^2 - x - 6} - \dfrac{1}{x-3}$

42. $\dfrac{3x}{x-1} = \dfrac{2}{x-2} - \dfrac{2}{x^2 - 3x + 2}$

43. $\dfrac{7}{x-2} + \dfrac{16}{x+3} = 3$

44. $\dfrac{5}{x-2} + \dfrac{6}{x+2} = 2$

45. $\dfrac{11}{x-3} - 1 = \dfrac{10}{x+3}$

46. $\dfrac{17}{x-4} - 2 = \dfrac{10}{x+2}$

Think About These

Some of the following are expressions, and others are equations. In each case identify which it is and then combine or solve.

47. $\dfrac{5}{x-2} + \dfrac{3}{x+3} = \dfrac{8}{x^2 + x - 6}$

48. $\dfrac{7}{x+5} + \dfrac{4}{x-1}$

49. $\dfrac{x-1}{x} - \dfrac{x+2}{3x} = 0$

50. $\dfrac{x-3}{2x} - \dfrac{2x+5}{x^2}$

51. A farmer uses $1\dfrac{1}{3}$ times as much herbicide as she does insecticide and one-fourth as much fungicide as insecticide. She uses a total of $10\dfrac{1}{3}$ liters (L) of pesticides. Use the following equation to determine how much of each type of pesticide the farmer uses.

$$x + \dfrac{4x}{3} + \dfrac{x}{4} = 10\dfrac{1}{3}$$

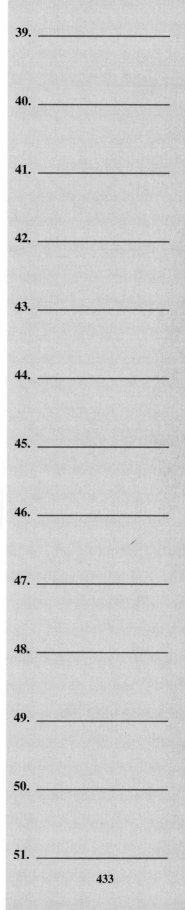

ANSWERS

39. _____

40. _____

41. _____

42. _____

43. _____

44. _____

45. _____

46. _____

47. _____

48. _____

49. _____

50. _____

51. _____

433

52. An organic fertilizer recipe calls for a mixture of bone meal, limestone, and wood ash. The recipe says for each unit of bone meal you should use three-fourths unit of limestone and one-fifth unit of wood ash. Use the following equation to determine how much of each ingredient you will need if you require $9\frac{3}{4}$ kilograms (kg) of the fertilizer.

$$x + \frac{3x}{4} + \frac{x}{5} = 9\frac{3}{4}$$

Skillscan (Section 4.6)

Write the equation necessary to solve each of the following problems. Then solve the equation.

a. One number is 5 less than another. If 3 times the first number is 9 more than the second number, find the two numbers.

b. One number is 3 more than another. If twice the smaller number is 5 less than the larger number, what are the two numbers?

c. The sum of an integer and 3 times the next consecutive integer is 47. What are the two integers?

d. 5 times an integer is one more than 4 times the next consecutive odd integer. Find the two integers.

e. Claudia rowed upstream for 3 hours (h). While she was rowing downstream, her speed was 2 miles per hour (mi/h) faster and the trip took 2 h. What was her speed each way?

f. Joe left the city at 11 A.M., heading west at a rate of 55 mi/h. An hour later Jeanine headed east at a rate of 45 mi/h. At what time will they be 305 mi apart?

ANSWERS

1. 6   **3.** 12   **5.** 15   **7.** 7   **9.** 2   **11.** 8   **13.** 3   **15.** 8   **17.** 2   **19.** 11
21. -5 **23.** -23 **25.** No solution   **27.** 6   **29.** 4   **31.** -4 **33.** -5
35. No solution   **37.** $-\frac{5}{2}$ **39.** $-\frac{1}{2}$, 6   **41.** $-\frac{1}{2}$ **43.** $-\frac{1}{3}$, 7   **45.** -8, 9
47. Equation, $-\frac{1}{8}$ **49.** Equation, $\frac{5}{2}$ **51.** 4 L insecticide, 1 L fungicide, $5\frac{1}{3}$ L herbicide
a. 7, 12   **b.** -2, 1   **c.** 11, 12   **d.** 9, 11   **e.** 4 mi/h, 6 mi/h   **f.** 2:30 P.M.

More Applications

OBJECTIVE
To solve word problems that lead to fractional equations

Many word problems will lead to fractional equations that must be solved by using the methods of the last section. The five steps in solving word problems are, of course, the same as you saw earlier.

Example 1

If one-third of a number is added to three-fourths of that same number, the sum is 26. Find the number.

Step 1 Read the problem carefully. You want to find the unknown number.

Step 2 Choose a letter to represent the unknown. Let x be the unknown number.

Step 3 Form an equation.

The equation expresses the relationship between the two numbers.

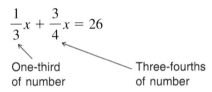

$$\frac{1}{3}x + \frac{3}{4}x = 26$$

One-third of number — Three-fourths of number

Step 4 Solve the equation. Multiply each term of the equation by 12, the LCD.

$$12 \cdot \frac{1}{3}x + 12 \cdot \frac{3}{4}x = 12 \cdot 26$$

Simplifying yields

$$4x + 9x = 312$$
$$13x = 312$$
$$x = 24$$

Be sure to answer the question raised in the problem.

The number is 24.

Step 5 Check your solution by returning to the *original problem*. If the number is 24, we have

$$\frac{1}{3} \cdot 24 + \frac{3}{4} \cdot 24 = 8 + 18 = 26$$

and the solution is verified.

CHECK YOURSELF 1

The sum of two-fifths of a number and one-half of that number is 18. Find the number.

Number problems that involve reciprocals can be solved by using fractional equations. The next example illustrates.

Example 2

One number is twice another number. If the sum of their reciprocals is $\dfrac{3}{10}$, what are the two numbers?

Step 1 You want to find the two numbers.

Step 2 Let x be one number. Then $2x$ is the other number.

Twice the first

Step 3

The reciprocal of a fraction is the fraction obtained by switching the numerator and denominator.

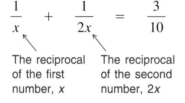

$$\frac{1}{x} + \frac{1}{2x} = \frac{3}{10}$$

The reciprocal of the first number, x

The reciprocal of the second number, $2x$

Step 4 The LCD of the fractions is $10x$, and so we multiply by $10x$.

$$10x\left(\frac{1}{x}\right) + 10x\left(\frac{1}{2x}\right) = 10x\left(\frac{3}{10}\right)$$

Simplifying, we have

$$10 + 5 = 3x$$
$$15 = 3x$$
$$5 = x$$

x was one number, and $2x$ was the other.

The numbers are 5 and 10.

Step 5

Again check the result by returning to the original problem. If the numbers are 5 and 10, we have

The sum of the reciprocals is $\frac{3}{10}$.

$$\frac{1}{5} + \frac{1}{10} = \frac{2+1}{10} = \frac{3}{10}$$

One number is 3 times another. If the sum of their reciprocals is $\frac{2}{9}$, find the two numbers.

The solution of many motion problems will also involve fractional equations. Remember that the key equation for solving all motion problems relates the distance traveled, the speed or rate, and the time:

$$d = r \cdot t$$

Often we will use this equation in different forms by solving for r or for t. So

$$r = \frac{d}{t} \quad \text{or} \quad t = \frac{d}{r}$$

Example 3

Vince took 1 h longer to drive 180 mi than he did on a trip of 135 mi. If his speed was the same both times, how long did each trip take?

Step 1 You want to find the times taken for the 180-mi trip and for the 135-mi trip.

Note: It is often helpful to choose your variable to "suggest" the unknown quantity—here t for time.

Step 2 Let t be the time for the 135-mi trip (in hours).

1 h longer

Then $t + 1$ is the time for the 180-mi trip.

It is often helpful to arrange the information in tabular form such as that shown.

Remember that rate is distance divided by time. The rightmost column is formed by using that relationship.

	DISTANCE	TIME	RATE
135-mi trip	135	t	$\frac{135}{t}$
180-mi trip	180	$t+1$	$\frac{180}{t+1}$

Step 3 To form an equation, remember that the speed (or rate) for each trip was the same. That is the *key* idea. You can equate the rates for the two trips that were found in step 2. From the rightmost column of the above table, if the rates are equal, we can write

$$\frac{135}{t} = \frac{180}{t+1}$$

Step 4 To solve the above equation, multiply each term by $t(t+1)$, the LCD of the fractions.

$$\cancel{t}(t+1)\left(\frac{135}{\cancel{t}}\right) = t\cancel{(t+1)}\left(\frac{180}{\cancel{t+1}}\right)$$

Simplifying, we have

$$135(t+1) = 180t$$
$$135t + 135 = 180t$$
$$135 = 45t$$
$$t = 3 \text{ h}$$

The time for the 135-mi trip was 3 h, and the time for the 180-mi trip was 4 h. We'll leave the check to you.

CHECK YOURSELF 3

Cynthia took 1 h longer to bicycle 60 mi than she did on a trip of 45 mi. If her speed was the same each time, find the time for each trip.

The next example uses the $d = r \cdot t$ relationship to find the speed.

Example 4

A train makes a trip of 300 mi in the same time that a bus can travel 250 mi. If the speed of the train is 10 mi/h faster than the speed of the bus, find the speed of each.

Step 1 You want to find the speeds of the train and of the bus.

Step 2 Let r be the speed (or rate) of the bus (in miles per hour).

Then $\underbrace{r + 10}$ is the rate of the train.

10 mi/h faster

Again let's form a table of the information.

	DISTANCE	RATE	TIME
Train	300	$r + 10$	$\dfrac{300}{r + 10}$
Bus	250	r	$\dfrac{250}{r}$

Remember that time is distance divided by rate. Here the rightmost column is found by using that relationship.

Step 3 To form an equation, remember that the times for the train and bus are the same. We can equate the expressions for time found in step 2. Again, working from the rightmost column, we have

$$\frac{250}{r} = \frac{300}{r + 10}$$

Step 4 We multiply each term by $r(r + 10)$, the LCD of the fractions.

$$\cancel{r}(r + 10)\left(\frac{250}{\cancel{r}}\right) = r\cancel{(r + 10)}\left(\frac{300}{\cancel{r + 10}}\right)$$

Simplifying, we have

$$250(r + 10) = 300r$$
$$250r + 2500 = 300r$$
$$2500 = 50r$$
$$r = 50 \text{ mi/h}$$

Make sure to find the rates of both vehicles.

The rate of the bus is 50 mi/h, and the rate of the train is 60 mi/h. You can check this result.

CHECK YOURSELF 4

A car makes a trip of 280 mi in the same time that a truck travels 245 mi. If the speed of the truck is 5 mi/h slower than that of the car, find the speed of each.

A final group of applications involves fractions in decimal form. Mixture problems often use percentages, and those percentages can be written as decimals. The following example illustrates.

Example 5

A solution of antifreeze is 20% alcohol. How much pure alcohol must be added to 12 quarts (qt) of the solution to make a 40% solution?

Step 1   You want to find the number of quarts of pure alcohol that must be added.

Step 2   Let x be the number of quarts of pure alcohol to be added.

Step 3   To form our equation, note that the amount of alcohol present before mixing *must be the same* as the amount in the combined solution.

A picture will help.

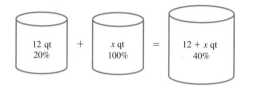

So

Express the percentages as decimals in the equation.

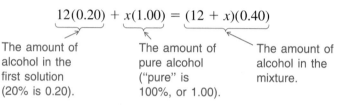

$$12(0.20) + x(1.00) = (12 + x)(0.40)$$

The amount of alcohol in the first solution (20% is 0.20).

The amount of pure alcohol ("pure" is 100%, or 1.00).

The amount of alcohol in the mixture.

Step 4   Most students prefer to clear the decimals at this stage. It's easy here—multiplying by 100 will move the decimal point *two places to the right*. We then have

$$12(20) + x(100) = (12 + x)(40)$$
$$240 + 100x = 480 + 40x$$
$$60x = 240$$
$$x = 4 \text{ qt}$$

CHECK YOURSELF 5

How much pure alcohol must be added to 500 cubic centimeters (cm^3) of a 40% alcohol mixture to make a solution that is 80% alcohol?

CHECK YOURSELF ANSWERS

1. The number is 20.
2. The numbers are 6 and 18.
3. 60-mi trip: 4 h; 45-mi trip: 3 h.
4. Car: 40 mi/h; truck: 35 mi/h.
5. 1000 cm^3.

ANSWERS

Build Your Skills

Solve the following word problems. Be sure to write the equation used for the solution.

1. If two-thirds of a number is added to one-half of that number, the sum is 35. Find the number.

2. If one-third of a number is subtracted from three-fourths of that number, the difference is 15. What is the number?

3. If one-fourth of a number is subtracted from two-fifths of a number, the difference is 3. Find that number.

4. If five-sixths of a number is added to one-fifth of the number, the sum is 31. What is the number?

5. If one-third of an integer is added to one-half of the next consecutive integer, the sum is 13. What are the two integers?

6. If one-half of one integer is subtracted from three-fifths of the next consecutive integer, the difference is 3. What are the two integers?

7. One number is twice another number. If the sum of their reciprocals is $\dfrac{1}{4}$, find the two numbers.

8. One number is 3 times another. If the sum of their reciprocals is $\dfrac{1}{6}$, find the two numbers.

9. One number is 4 times another. If the sum of their reciprocals is $\dfrac{5}{12}$, find the two numbers.

1. _____

2. _____

3. _____

4. _____

5. _____

6. _____

7. _____

8. _____

9. _____

10. _____

10. One number is 3 times another. If the sum of their reciprocals is $\dfrac{4}{15}$, what are the two numbers?

11. _____

11. One number is 5 times another number. If the sum of their reciprocals is $\dfrac{6}{35}$, what are the two numbers?

12. _____

12. One number is 4 times another. The sum of their reciprocals is $\dfrac{5}{24}$. What are the two numbers?

13. _____

13. If the reciprocal of 5 times a number is subtracted from the reciprocal of that number, the result is $\dfrac{4}{25}$. What is the number?

14. _____

14. If the reciprocal of a number is added to 4 times the reciprocal of that number, the result is $\dfrac{5}{9}$. Find that number.

15. _____

15. Lee can ride his bicycle 50 mi in the same time it takes him to drive 125 mi. If his driving rate is 30 mi/h faster than his rate bicycling, find each rate.

16. _____

16. Tina can run 12 mi in the same time that it takes her to bicycle 72 mi. If her bicycling rate is 20 mi/h faster than her running rate, find each rate.

17. _____

17. An express bus can travel 275 mi in the same time that it takes a local bus to travel 225 mi. If the rate of the express bus is 10 mi/h faster than that of the local bus, find the rate for each bus.

18. _____

18. A light plane took 1 h longer to travel 450 mi on the first portion of a trip than it took to fly 300 mi on the second. If the speed was the same for each portion, what was the flying time for each part of the trip?

19. A passenger train can travel 325 mi in the same time a freight train takes to travel 200 mi. If the speed of the passenger train is 25 mi/h faster than the speed of the freight, find the speed of each.

20. A small business jet took 1 h longer to fly 810 mi on the first part of a flight than to fly 540 mi on the second portion. If the jet's rate was the same for each leg of the flight, what was the flying time for each leg?

21. Charles took 2 h longer to drive 240 mi on the first day of a vacation trip than to drive 144 mi on the second day. If his average driving rate was the same on both days, what was his driving time for each of the days?

22. Julie took 2 h longer to drive 360 mi on the first day of a trip than she took to drive 270 mi on the second day. If her speed was the same both days, what was the driving time each day?

23. An airplane took 3 h longer to fly 1200 mi than it took for a flight of 480 mi. If the plane's rate was the same on each trip, what was the time of each flight?

24. A train travels 80 mi in the same time that a light plane can fly 280 mi. If the speed of the plane is 100 mi/h faster than that of the train, find each of the rates.

25. Jan and Ed took a canoeing trip, traveling 6 mi upstream against a 2 mi/h current. They then returned to the same point downstream. If their entire trip took 4 h, how fast can they paddle in still water? *Hint:* If r is their rate (in miles per hour) in still water, their rate upstream is $r - 2$.

26. A plane flies 720 mi against a steady 30 mi/h headwind and then returns to the same point with the wind. If the entire trip takes 10 h, what is the plane's speed in still air?

27. How much pure alcohol must be added to 40 ounces (oz) of a 25% solution to produce a mixture that is 40% alcohol?

ANSWERS

19. _____

20. _____

21. _____

22. _____

23. _____

24. _____

25. _____

26. _____

27. _____

28. _____

29. _____

30. _____

31. _____

32. _____

33. _____

a. _____

b. _____

c. _____

d. _____

e. _____

f. _____

g. _____

h. _____

28. How many centiliters (cL) of pure acid must be added to 200 cL of a 40% acid solution to produce a 50% solution?

29. How many centiliters of a 25% acid solution must be added to 100 cL of a 4% acid solution to make a 10% acid solution?

30. A storeroom has an 8% acid solution and a 30% solution. How many milliliters (mL) of the 30% solution should be added to 500 mL of the 8% solution to produce a solution that is 20% acid?

31. If a farmer already has 75 L of 50% herbicide but he needs a 60% herbicide for his soybean crop, how much pure herbicide concentrate must he add to get the required mixture?

32. A pilot has a 100-L pesticide tank on her airplane. How much 25% pesticide and 10% pesticide must she mix to make a full load of 15% pesticide?

33. An organic farmer has 200 kilograms (kg) of 20 percent manure compost that he wishes to use in a new mixture of 30 percent manure compost. How much pure manure must he add to the old compost to get his desired mixture?

Skillscan (Section 3.3)

Solve the following equations.

a. $\dfrac{x}{4} = 7$　　　　**b.** $\dfrac{x}{8} = 9$　　　　**c.** $\dfrac{x}{5} = \dfrac{8}{10}$　　　　**d.** $\dfrac{x}{2} = \dfrac{3}{4}$

e. $\dfrac{x}{9} = \dfrac{7}{3}$　　　　**f.** $\dfrac{x}{6} = \dfrac{28}{8}$　　　　**g.** $\dfrac{2x}{6} = \dfrac{4}{3}$　　　　**h.** $\dfrac{3x}{5} = \dfrac{9}{10}$

ANSWERS

1. 30　　**3.** 20　　**5.** 15, 16　　**7.** 6, 12　　**9.** 3, 12　　**11.** 7, 35　　**13.** 5

15. 20 mi/h bicycling, 50 mi/h driving　　**17.** 55 mi/h, 45 mi/h　　**19.** 40 mi/h, 65 mi/h

21. 5 h, 3 h　　**23.** 5 h, 2 h　　**25.** 4 mi/h　　**27.** 10 oz　　**29.** 40 cL　　**31.** 18.75 L

33. 28.6 kg　　**a.** 28　　**b.** 72　　**c.** 4　　**d.** $\dfrac{3}{2}$　　**e.** 21　　**f.** 21　　**g.** 4　　**h.** $\dfrac{3}{2}$

Ratio and Proportion

OBJECTIVES
1. To solve a proportion for an unknown
2. To apply proportions to the solution of word problems

To begin this section, let's return to an equation that was developed in the previous section. In Example 3, we had to solve the equation

$$\frac{135}{t} = \frac{180}{t+1}$$

Such an equation is said to be in *proportion form,* or more simply it is called a *proportion.* This type of equation occurs often enough in algebra that it is worth developing some special methods for its solution. First, we will need some definitions.

A *ratio* is a means of comparing two quantities. A ratio can be written as a fraction. For instance, the ratio of 2 to 3 can be written as $\frac{2}{3}$.

A statement that two ratios are equal is called a *proportion.* A proportion has the form

$$\frac{a}{b} = \frac{c}{d}$$

In the proportion above, *a* and *d* are called the *extremes* of the proportion, and *b* and *c* are called the *means.*

A useful property of proportions is easily developed. If

$$\frac{a}{b} = \frac{c}{d}$$

bd is the LCD of the denominators.

and we multiply both sides by $b \cdot d$, then

$$\left(\frac{a}{b}\right)bd = \left(\frac{c}{d}\right)bd \qquad \text{or} \qquad ad = bc$$

If $\dfrac{a}{b} = \dfrac{c}{d}$ then $ad = bc$

In words,

In any proportion, the product of the extremes (*ad*) is equal to the product of the means (*bc*).

Since a proportion is a special kind of fractional equation, this rule gives us an alternative approach to solving equations that are in the proportion form.

Example 1

Solve the equations for x.

The extremes are x and 15.
The means are 5 and 12.

(a) $\dfrac{x}{5} = \dfrac{12}{15}$

Set the product of the extremes equal to the product of the means.

$15x = 5 \cdot 12$

$15x = 60$

$x = 4$

Our solution is 4. You can check as before, by substituting in the original proportion.

(b) $\dfrac{x + 3}{10} = \dfrac{x}{7}$

Set the product of the extremes equal to the product of the means.

$7(x + 3) = 10x$

$7x + 21 = 10x$

$21 = 3x$

$7 = x$

We will leave the checking of this result to the reader.

CHECK YOURSELF 1

Solve for x.

1. $\dfrac{x}{8} = \dfrac{3}{4}$

2. $\dfrac{x - 1}{9} = \dfrac{x + 2}{12}$

There are many types of applications that lead to proportions in their solution. Typically these applications will involve a common ratio, such as miles to gallons or miles to hours, and they can be solved with three basic steps.

TO SOLVE AN APPLICATION BY USING PROPORTIONS

STEP 1   Assign a variable to represent the unknown quantity.

STEP 2   Write a proportion, using the known and unknown quantities. Be sure each ratio involves the same units.

STEP 3   Solve the proportion written in step 2 for the unknown quantity.

The following examples illustrate.

Example 2

A car uses 3 gallons (gal) of gas to travel 105 mi. At that mileage rate, how many gallons will be used on a trip of 385 mi?

Step 1   Assign a variable to represent the unknown quantity. Let x be the number of gallons of gas that will be used on the 385-mi trip.

Step 2   Write a proportion. Note that the ratio of miles to gallons must stay the same.

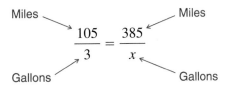

Step 3   Solve the proportion. The product of the extremes is equal to the product of the means.

$$105x = 3 \cdot 385$$
$$105x = 1155$$
$$\frac{105x}{105} = \frac{1155}{105}$$
$$x = 11 \text{ gal}$$

So 11 gal of gas will be used for the 385-mi trip.

To verify your solution, you can return to the original problem and check that the two ratios involved are the equivalent.

CHECK YOURSELF 2

A car uses 8 liters (L) of gasoline in traveling 100 kilometers (km). At that rate, how many liters of gas will be used on a trip of 250 km?

Proportions can also be used to solve problems in which a quantity is divided by using a specific ratio. The next example shows how.

Example 3

A piece of wire 60 inches (in) long is to be cut into two pieces whose lengths have the ratio 5 to 7. Find the length of each piece.

Step 1 Let x represent the length of the shorter piece. Then $60 - x$ is the length of the longer piece.

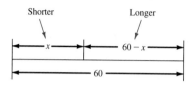

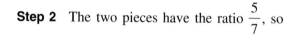

Step 2 The two pieces have the ratio $\dfrac{5}{7}$, so

On the left and right, we have the ratio of the length of the shorter piece to that of the longer piece.

$$\frac{x}{60 - x} = \frac{5}{7}$$

Step 3 Solving as before, we get

$$7x = (60 - x)5$$
$$7x = 300 - 5x$$
$$12x = 300$$
$$x = 25 \qquad \text{(shorter piece)}$$
$$60 - x = 35 \qquad \text{(longer piece)}$$

CHECK YOURSELF 3

A board 21 feet (ft) long is to be cut into two pieces so that the ratio of their lengths is 3 to 4. Find the lengths of the two pieces.

CHECK YOURSELF ANSWERS

1. (1) 6; (2) 10. **2.** 20 L. **3.** 9 ft, 12 ft.

Name _____

Date _____

Build Your Skills

Solve each of the following equations for x.

1. $\dfrac{x}{7} = \dfrac{8}{14}$

2. $\dfrac{3}{x} = \dfrac{9}{15}$

3. $\dfrac{5}{8} = \dfrac{20}{x}$

4. $\dfrac{x}{10} = \dfrac{9}{30}$

5. $\dfrac{x+1}{5} = \dfrac{20}{25}$

6. $\dfrac{2}{5} = \dfrac{x-2}{20}$

7. $\dfrac{3}{5} = \dfrac{x-1}{20}$

8. $\dfrac{5}{x-3} = \dfrac{15}{21}$

9. $\dfrac{x}{6} = \dfrac{x+5}{16}$

10. $\dfrac{x-2}{x+2} = \dfrac{12}{20}$

11. $\dfrac{x}{x-3} = \dfrac{10}{7}$

12. $\dfrac{x}{8} = \dfrac{x+3}{16}$

ANSWERS

1. _____

2. _____

3. _____

4. _____

5. _____

6. _____

7. _____

8. _____

9. _____

10. _____

11. _____

12. _____

13. _____

14. _____

15. _____

16. _____

17. _____

18. _____

19. _____

20. _____

21. _____

13. $\dfrac{2}{x-1} = \dfrac{6}{x+9}$

14. $\dfrac{3}{x-3} = \dfrac{4}{x-5}$

15. $\dfrac{1}{x-2} = \dfrac{7}{x^2-4}$

16. $\dfrac{1}{x-2} = \dfrac{3}{x^2-4x+4}$

Think About These

17. A speed of 60 mi/h corresponds to 88 feet per second (ft/s). If a light plane's speed is 150 mi/h, what is its speed in feet per second?

18. John completed 24 pages of his reading assignment in 60 minutes (min). If the assignment is 100 pages long, how long will the entire assignment take if John reads at the same rate?

19. A car uses 5 gal of gasoline on a trip of 160 mi. At the same mileage rate, how much gasoline will a 384-mi trip require?

20. A car uses 12 L of gasoline in traveling 150 kilometers (km). At that rate, how many liters of gasoline will be used on a trip of 400 km?

21. Susan earns $6500 commission in 20 weeks in her new sales position. At that rate, how much will she earn in 1 year (52 weeks)?

22. Kevin earned $165 interest for 1 year on an investment of $1500. At that same rate, what amount of interest would be earned by an investment of $2500?

23. A board 24 ft long is to be divided into two pieces whose lengths have the ratio 5 to 7. Find the length of each piece.

24. A 90-ft piece of cable is to be cut into two pieces whose lengths have the ratio 4 to 5. Find the length of each piece.

25. A brother and sister are to divide an inheritance of $12,000 in the ratio of 2 to 3. What amount will each receive?

26. Carlos wants to invest a total of $5600 in two savings accounts in the ratio of 3 to 4. What amount should he invest in each account?

27. In 1984, a chemical leak at a pesticide manufacturing plant in Bhopal, India, killed 3700 people and injured 300,000 others. Use a ratio to determine the approximate number of people injured for each person that was killed.

28. An experimental field of 170 hectares is to be managed with both chemical and organic methods. If the ratio of organically managed hectares to chemically managed hectares is 9 to 8, how many hectares will be in each category?

29. _____

30. _____

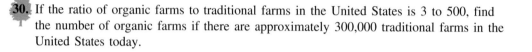

29. A company is selling a natural insect control which mixes ladybug beetles and praying mantises in the ratio of 7 to 4. If there are a total of 110 insects per package, how many of each type of insect is in a package?

30. If the ratio of organic farms to traditional farms in the United States is 3 to 500, find the number of organic farms if there are approximately 300,000 traditional farms in the United States today.

ANSWERS

1. 4   **3.** 32   **5.** 3   **7.** 13   **9.** 3   **11.** 10   **13.** 6   **15.** 5   **17.** 220 ft/s
19. 12 gal   **21.** $16,900   **23.** 10 ft, 14 ft   **25.** $4800, $7200   **27.** $x = 81$
29. 70 ladybugs, 40 praying mantises

Summary

Algebraic Fractions [6.1]

Algebraic Fractions   These have the form

$\dfrac{x^2 - 3x}{x - 2}$ is an algebraic fraction. The variable x cannot have the value 2.

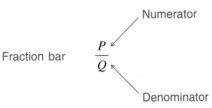

Fraction bar

Numerator

$\dfrac{P}{Q}$

Denominator

where P and Q are polynomials and Q cannot have the value 0.

Simplifying Algebraic Fractions [6.2]

Simplest Form   A fraction is in simplest form if its numerator and denominator have no common factors other than 1.

$\dfrac{x + 2}{x - 1}$ is in simplest form.

$$\dfrac{x^2 - 4}{x^2 - 2x - 8}$$

$$= \dfrac{(x - 2)(x + 2)}{(x - 4)(x + 2)}$$

$$= \dfrac{(x - 2)\cancel{(x + 2)}}{(x - 4)\cancel{(x + 2)}}$$

$$= \dfrac{x - 2}{x - 4}$$

To Write in Simplest Form

1. Factor the numerator and denominator.
2. Divide the numerator and denominator by all common factors.
3. The resulting fraction will be in simplest form.

Multiplying and Dividing Fractions [6.3]

Multiplying Fractions

$$\dfrac{2}{3} \cdot \dfrac{4}{5} = \dfrac{2 \cdot 4}{3 \cdot 5} = \dfrac{8}{15}$$

$$\dfrac{P}{Q} \cdot \dfrac{R}{S} = \dfrac{PR}{QS}$$

where $Q \neq 0$ and $S \neq 0$.

Multiplying Algebraic Fractions

$$\dfrac{2x - 4}{x^2 - 4} \cdot \dfrac{x^2 + 2x}{6x + 18}$$

$$= \dfrac{2(x - 2)}{(x - 2)(x + 2)} \cdot \dfrac{x(x + 2)}{6(x + 3)}$$

$$= \dfrac{2\cancel{(x - 2)}}{\cancel{(x - 2)}\cancel{(x + 2)}} \cdot \dfrac{x\cancel{(x + 2)}}{6(x + 3)}$$

$$= \dfrac{x}{3(x + 3)}$$

1. Factor the numerators and denominators.
2. Divide the numerator and denominator by any common factors.
3. Write the product of the remaining factors in the numerator over the product of the remaining factors in the denominator.

Dividing Fractions

$$\dfrac{4}{9} \div \dfrac{8}{12}$$

$$= \dfrac{4}{9} \cdot \dfrac{12}{8} = \dfrac{2}{3}$$

$$\dfrac{P}{Q} \div \dfrac{R}{S} = \dfrac{P}{Q} \cdot \dfrac{S}{R}$$

where $Q \neq 0$, $R \neq 0$, and $S \neq 0$. In words, invert the divisor (the second fraction) and multiply.

Dividing Algebraic Fractions

1. Invert the fraction following the division symbol.
2. Multiply the fractions.

$$\frac{3x}{2x-6} \div \frac{9x^2}{x^2-9}$$

$$= \frac{3x}{2x-6} \cdot \frac{x^2-9}{9x^2}$$

$$= \frac{3x}{2(x-3)} \cdot \frac{(x+3)(x-3)}{9x^2}$$

$$= \frac{x+3}{6x}$$

Adding and Subtracting Algebraic Fractions [6.4 and 6.5]

$$\frac{2x}{x^2+3x} + \frac{6}{x^2+3x}$$

$$= \frac{2x+6}{x^2+3x}$$

$$= \frac{2(x+3)}{x(x+3)} = \frac{2}{x}$$

For $\dfrac{2}{x^2+2x+1}$

and $\dfrac{3}{x^2+x}$

Factor:

$$x^2+2x+1 = (x+1)(x+1)$$

$$x^2+x = x(x+1)$$

The LCD is $x(x+1)(x+1)$

$$\frac{2}{x^2+2x+1} - \frac{3}{x^2+x}$$

$$= \frac{2x}{x(x+1)(x+1)}$$

$$- \frac{3(x+1)}{x(x+1)(x+1)}$$

$$= \frac{2x-3x-3}{x(x+1)(x+1)}$$

$$= \frac{-x-3}{x(x+1)(x+1)}$$

Like Fractions

1. Add or subtract the numerators.
2. Write the sum or difference over the common denominator.
3. Write the resulting fraction in simplest form.

The Lowest Common Denominator   Finding the LCD:

1. Factor each denominator.
2. Write each factor the greatest number of times it appears in any single denominator.
3. The LCD is the product of the factors found in step 2.

Unlike Fractions   To add or subtract unlike fractions:

1. Find the LCD.
2. Convert each fraction to an equivalent fraction with the LCD as a common denominator.
3. Add or subtract the like fractions formed.
4. Write the sum or difference in simplest form.

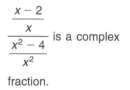 is a complex fraction.

Complex Fractions [6.6]

Complex Fractions   Fractions that have a fraction in their numerator, their denominator, or both.

Simplifying   To simplify a complex fraction, you can apply either of the following methods.

METHOD 1

$$\frac{\left(\dfrac{x-2}{x}\right)x^2}{\left(\dfrac{x^2-4}{x^2}\right)x^2}$$

$$=\frac{(x-2)x}{x^2-4}$$

$$=\frac{x(x-2)}{(x+2)(x-2)}$$

$$=\frac{x}{(x+2)}$$

1. Multiply the numerator and denominator by the LCD of all fractions in the complex fraction.
2. Simplify the resulting fraction, writing the result in simplest form, if necessary.

METHOD 2

$$\frac{1-\dfrac{x}{y}}{1-\dfrac{x^2}{y^2}}$$

$$=\frac{\dfrac{y-x}{y}}{\dfrac{y^2-x^2}{y^2}}$$

$$=\frac{y-x}{y}\cdot\frac{y^2}{y^2-x^2}$$

$$=\frac{y-x}{y}\cdot\frac{y^2}{(y-x)(y+x)}$$

$$=\frac{y}{y+x}$$

1. Write the numerator and denominator as single fractions if necessary.
2. Invert the denominator and multiply as before, writing the result in simplest form.

Fractional Equations [6.7]

To Solve

$$\frac{2}{x-2}-\frac{3}{x+2}=\frac{2}{x^2-4}$$

Multiply by $(x-2)(x+2)$.
We then have

$$2(x+2)-3(x-2)=2$$

Solving, we get

$$x=8$$

1. Remove the fractions in the equation by multiplying *each term* of the equation by the LCD of all the fractions.
2. Solve the resulting equation as before.
3. Check your solution in the *original equation*.

Ratio and Proportion [6.9]

Ratio   A means of comparing two quantities. A ratio can be written as a fraction.

$\dfrac{2}{3}$ is the ratio of 2 to 3.

$\dfrac{a}{b}$ is the ratio of a to b.

Proportion   A statement that two ratios are equal. The form is

$$\frac{2}{3}=\frac{8}{12}$$

is a proportion; 2 and 12 are the extremes, 3 and 8 are the means.

$$\frac{a}{b}=\frac{c}{d}$$

where a and d are the extremes and b and c are the means.

Solve:

$$\frac{x-2}{3} = \frac{x+2}{7}$$

Set the product of the extremes equal to the product of the means.

$$7(x - 2) = 3(x + 2)$$

$$7x - 14 = 3x + 6$$

$$4x = 20$$

$$x = 5$$

The Proportion Rule

If $\dfrac{a}{b} = \dfrac{c}{d}$ then $ad = bc$

In words, the product of the extremes is equal to the product of the means. This rule can be applied in solving fractional equations that are in the proportion form.

Summary Exercises Chapter 6

This summary exercise set is provided to give you practice with each of the objectives of the chapter. Each exercise is keyed to the appropriate chapter section. The answers are provided in the instructor's manual.

[6.1] What values for x, if any, must be excluded in the following algebraic fractions?

1. $\dfrac{x}{5}$

2. $\dfrac{3}{x-4}$

3. $\dfrac{2}{(x+1)(x-2)}$

4. $\dfrac{7}{x^2-16}$

5. $\dfrac{x-1}{x^2+3x+2}$

6. $\dfrac{2x+3}{3x^2+x-2}$

[6.2] Write each fraction in simplest form.

7. $\dfrac{6a^2}{9a^3}$

8. $\dfrac{-12x^4y^3}{18x^2y^2}$

9. $\dfrac{w^2-25}{2w-8}$

10. $\dfrac{3x^2+11x-4}{2x^2+11x+12}$

11. $\dfrac{m^2-2m-3}{9-m^2}$

12. $\dfrac{3c^2-2cd-d^2}{6c^2+2cd}$

[6.3] Multiply or divide as indicated.

13. $\dfrac{6x}{5}\cdot\dfrac{10}{18x^2}$

14. $\dfrac{-2a^2}{ab^3}\cdot\dfrac{3ab^2}{-4ab}$

15. $\dfrac{2x+6}{x^2-9}\cdot\dfrac{x^2-3x}{4}$

16. $\dfrac{a^2+5a+4}{2a^2+2a}\cdot\dfrac{a^2-a-12}{a^2-16}$

17. $\dfrac{3p}{5}\div\dfrac{9p^2}{10}$

18. $\dfrac{8m^3}{5mn}\div\dfrac{12m^2n^2}{15mn^3}$

19. $\dfrac{x^2+7x+10}{x^2+5x}\div\dfrac{x^2-4}{2x^2-7x+6}$

20. $\dfrac{2w^2+11w-21}{w^2-49}\div(4w-6)$

21. $\dfrac{a^2b+2ab^2}{a^2-4b^2}\div\dfrac{4a^2b}{a^2-ab-2b^2}$

22. $\dfrac{2x^2+6x}{4x}\cdot\dfrac{6x+12}{x^2+2x-3}\div\dfrac{x^2-4}{x^2-3x+2}$

[6.4] Add or subtract as indicated.

23. $\dfrac{x}{9} + \dfrac{2x}{9}$

24. $\dfrac{7a}{15} - \dfrac{2a}{15}$

25. $\dfrac{8}{x + 2} + \dfrac{3}{x + 2}$

26. $\dfrac{y - 2}{5} - \dfrac{2y + 3}{5}$

27. $\dfrac{7r - 3s}{4r} + \dfrac{r - s}{4r}$

28. $\dfrac{x^2}{x - 4} - \dfrac{16}{x - 4}$

29. $\dfrac{5w - 6}{w - 4} - \dfrac{3w + 2}{w - 4}$

30. $\dfrac{x + 3}{x^2 - 2x - 8} + \dfrac{2x + 3}{x^2 - 2x - 8}$

[6.5] Add or subtract as indicated.

31. $\dfrac{5x}{6} + \dfrac{x}{3}$

32. $\dfrac{3y}{10} - \dfrac{2y}{5}$

33. $\dfrac{5}{2m} - \dfrac{3}{m^2}$

34. $\dfrac{x}{x - 3} - \dfrac{2}{3}$

35. $\dfrac{4}{x - 3} - \dfrac{1}{x}$

36. $\dfrac{2}{s + 5} + \dfrac{3}{s + 1}$

37. $\dfrac{5}{w - 5} - \dfrac{2}{w - 3}$

38. $\dfrac{4x}{2x - 1} + \dfrac{2}{1 - 2x}$

39. $\dfrac{2}{3x - 3} - \dfrac{5}{2x - 2}$

40. $\dfrac{4y}{y^2 - 8y + 15} + \dfrac{6}{y - 3}$

41. $\dfrac{3a}{a^2 + 5a + 4} + \dfrac{2a}{a^2 - 1}$

42. $\dfrac{3x}{x^2 + 2x - 8} - \dfrac{1}{x - 2} + \dfrac{1}{x + 4}$

[6.6] Simplify the complex fractions.

43. $\dfrac{\dfrac{x^2}{12}}{\dfrac{x^3}{8}}$

44. $\dfrac{3 + \dfrac{1}{a}}{3 - \dfrac{1}{a}}$

45. $\dfrac{1 + \dfrac{x}{y}}{1 - \dfrac{x}{y}}$

46. $\dfrac{1 + \dfrac{1}{p}}{p^2 - 1}$

47. $\dfrac{\dfrac{1}{m} - \dfrac{1}{n}}{\dfrac{1}{m} + \dfrac{1}{n}}$

48. $\dfrac{2 - \dfrac{x}{y}}{4 - \dfrac{x^2}{y^2}}$

49. $\dfrac{\dfrac{2}{a + 1} + 1}{1 - \dfrac{4}{a + 1}}$

50. $\dfrac{\dfrac{a}{b} - 1 - \dfrac{2b}{a}}{\dfrac{1}{b^2} - \dfrac{1}{a^2}}$

[6.7] Solve the following equations for x.

51. $\dfrac{x}{4} - \dfrac{x}{5} = 2$

52. $\dfrac{13}{4x} + \dfrac{3}{x^2} = \dfrac{5}{2x}$

53. $\dfrac{x}{x - 2} + 1 = \dfrac{x + 4}{x - 2}$

54. $\dfrac{x}{x - 4} - 3 = \dfrac{4}{x - 4}$

55. $\dfrac{x}{2x - 6} - \dfrac{x - 4}{x - 3} = \dfrac{1}{8}$

56. $\dfrac{7}{x} - \dfrac{1}{x - 3} = \dfrac{9}{x^2 - 3x}$

57. $\dfrac{x}{x - 5} = \dfrac{3x}{x^2 - 7x + 10} + \dfrac{8}{x - 2}$

58. $\dfrac{6}{x + 5} + 1 = \dfrac{3}{x - 5}$

59. $\dfrac{24}{x + 2} - 2 = \dfrac{2}{x - 3}$

[6.8] Solve the following applications.

60. If two-fifths of a number is added to one-half of that number, the sum is 27. Find the number.

61. One number is 3 times another. If the sum of their reciprocals is $\dfrac{1}{3}$, what are the two numbers?

62. If the reciprocal of 4 times a number is subtracted from the reciprocal of that number, the result is $\dfrac{1}{8}$. What is the number?

63. Robert made a trip of 240 mi. Returning by a different route, he found that the distance was only 200 mi, but traffic slowed his speed by 8 mi/h. If the trip took the same time in both directions, what was Robert's rate each way?

64. On the first day of a vacation trip, Jovita drove 225 mi. On the second day it took her 1 h longer to drive 270 mi. If her average speed was the same both days, how long did she drive each day?

65. A light plane flies 700 mi against a steady 20 mi/h headwind and then returns to the same point with the wind. If the entire trip took 12 h, what was the speed of the plane in still air?

66. How much pure alcohol should be added to 300 milliliters (mL) of a 30% solution to obtain a 40% solution?

67. A chemist has a 10% acid solution and a 40% solution. How much of the 40% solution should be added to 300 mL of the 10% solution to produce a mixture with a concentration of 20%?

[6.9] Solve the following proportion problems.

68. $\dfrac{x-3}{8} = \dfrac{x-2}{10}$

69. $\dfrac{1}{x-3} = \dfrac{7}{x^2-x-6}$

70. Melina wants to invest a total of $10,800 in two types of savings accounts. If she wants the ratio of the amounts deposited in the two accounts to be 4 to 5, what amount should she invest in each account?

The purpose of this self-test is to help you check your progress and to review for a chapter test in class. Allow yourself about an hour to take the test. When you are done, check your answers in the back of the book. If you missed any problems, be sure to go back and review the appropriate sections in the chapter and the exercises that are provided.

ANSWERS

What values for x, if any, must be excluded in the following algebraic fractions?

1. $\dfrac{8}{x - 4}$

2. $\dfrac{3}{x^2 - 9}$

Write each fraction in simplest form.

3. $\dfrac{-21x^5y^3}{28xy^5}$

4. $\dfrac{4a - 24}{a^2 - 6a}$

5. $\dfrac{3x^2 + x - 2}{3x^2 - 8x + 4}$

Multiply or divide as indicated.

6. $\dfrac{3pq^2}{5pq^3} \cdot \dfrac{20p^2q}{21q}$

7. $\dfrac{x^2 - 3x}{5x^2} \cdot \dfrac{10x}{x^2 - 4x + 3}$

8. $\dfrac{2x^2}{3xy} \div \dfrac{8x^2y}{9xy}$

9. $\dfrac{3m - 9}{m^2 - 2m} \div \dfrac{m^2 - m - 6}{m^2 - 4}$

Add or subtract as indicated.

10. $\dfrac{3a}{8} + \dfrac{5a}{8}$

11. $\dfrac{2x}{x + 3} + \dfrac{6}{x + 3}$

12. $\dfrac{7x - 3}{x - 2} - \dfrac{2x + 7}{x - 2}$

13. $\dfrac{x}{3} + \dfrac{4x}{5}$

14. $\dfrac{3}{s} - \dfrac{2}{s^2}$

15. $\dfrac{5}{x - 2} - \dfrac{1}{x + 3}$

16. $\dfrac{6}{w - 2} + \dfrac{9w}{w^2 - 7w + 10}$

1. _____

2. _____

3. _____

4. _____

5. _____

6. _____

7. _____

8. _____

9. _____

10. _____

11. _____

12. _____

13. _____

14. _____

15. _____

16. _____

17. _____

18. _____

19. _____

20. _____

21. _____

22. _____

23. _____

24. _____

25. _____

Simplify the complex fractions.

17. $\dfrac{\dfrac{x^2}{18}}{\dfrac{x^3}{12}}$

18. $\dfrac{2 - \dfrac{m}{n}}{4 - \dfrac{m^2}{n^2}}$

Solve the following equations for x.

19. $\dfrac{x}{3} - \dfrac{x}{4} = 3$

20. $\dfrac{x}{x + 3} + 1 = \dfrac{3x - 6}{x + 3}$

21. $\dfrac{5}{x} - \dfrac{x - 3}{x + 2} = \dfrac{22}{x^2 + 2x}$

Solve the following applications.

22. One number is 3 times another. If the sum of their reciprocals is $\dfrac{1}{3}$, find the two numbers.

23. Mark drove 250 mi to visit Sandra. Returning by a shorter route, he found that the trip was only 225 mi, but traffic slowed his speed by 5 mi/h. If the two trips took exactly the same time, what was his rate each way?

Solve the following proportion problems.

24. $\dfrac{x - 1}{5} = \dfrac{x + 2}{8}$

25. A 55-ft cable is to be cut into two pieces whose lengths have the ratio 4 to 7. Find the lengths of the two pieces.

This test is provided to help you in the process of review of the previous chapters. Answers are provided in the back of the book. If you missed any problems, be sure to go back and review the appropriate chapter sections.

Perform each of the indicated operations.

1. $(5x^2 - 9x + 3) + (3x^2 + 2x - 7)$ **2.** $(8a^2 - 3a) - (9a^2 + 7)$

3. Subtract $9w^2 + 5w$ from the sum of $8w^2 - 3w$ and $2w^2 - 4$.

4. $7xy(4x^2y - 2xy + 3xy^2)$ **5.** $(3s - 7)(5s + 4)$ **6.** $(3a - b)(2a^2 + ab - b^2)$

7. $\dfrac{5x^3y - 10x^2y^2 + 15xy^2}{-5xy}$ **8.** $\dfrac{4x^2 + 6x - 4}{2x - 1}$ **9.** $\dfrac{3x^3 - 6x + 17}{3x + 6}$

Solve the following equation for x.

10. $5 - 3(2x - 7) = 8 - 4x$

Solve the following applications.

11. A bank teller has 83 $5 and $10 bills with a value of $695. How many bills of each denomination does she have?

12. A light plane makes a trip between two cities, against a steady headwind, in 7 h. Returning with the wind, the plane can travel 20 mi/h faster and makes the trip in 6 h. What is the plane's speed in each direction?

Factor each of the following polynomials completely.

13. $24a^3 - 16a^2$ **14.** $7m^2n - 21mn - 49mn^2$

1. _____

2. _____

3. _____

4. _____

5. _____

6. _____

7. _____

8. _____

9. _____

10. _____

11. _____

12. _____

13. _____

14. _____

Factor each of the following binomials completely.

15. $a^2 - 64b^2$ **16.** $5p^3 - 80pq^2$

Factor each of the following trinomials completely.

17. $a^2 - 14a + 48$ **18.** $2w^3 - 8w^2 - 42w$ **19.** $3r^2 + 5rs - 28s^2$

Solve each of the following equations.

20. $x^2 - 9x + 20 = 0$ **21.** $2x^2 - 32 = 0$ **22.** $21x^2 - 28x = 28$

Solve the following applications.

23. Twice the square of a positive integer is 35 more than 9 times that integer. What is the integer?

24. The length of a rectangle is 2 inches (in) more than 3 times its width. If the area of the rectangle is 85 inches squared (in^2), find the dimensions of the rectangle.

Solve the following equation for the indicated variable.

25. $S_n = \dfrac{n}{2}(a_1 + a_n)$ for a_1

What value(s) of x, if any, must be excluded in the following algebraic fraction?

26. $\dfrac{3x - 1}{2x^2 - x - 3}$

Write each fraction in simplest form.

27. $\dfrac{m^2 - 4m}{3m - 12}$ **28.** $\dfrac{a^2 - 49}{3a^2 + 22a + 7}$

Perform the indicated operations.

29. $\dfrac{3x^2 + 9x}{x^2 - 9} \cdot \dfrac{2x^2 - 9x + 9}{2x^3 - 3x^2}$ **30.** $\dfrac{4w^2 - 25}{2w^2 - 5w} \div (6w + 15)$

31. $\dfrac{4}{3r} + \dfrac{1}{2r^2}$

32. $\dfrac{2}{x-3} - \dfrac{5}{3x+9}$

33. $\dfrac{3y}{y^2 - 7y + 12} + \dfrac{9}{y-3}$

Simplify the complex fractions.

34. $\dfrac{1 - \dfrac{1}{x}}{2 + \dfrac{1}{x}}$

35. $\dfrac{3 - \dfrac{m}{n}}{9 - \dfrac{m^2}{n^2}}$

Solve the following equations for x.

36. $\dfrac{5}{3x} + \dfrac{1}{x^2} = \dfrac{5}{2x}$

37. $\dfrac{10}{x-3} - 2 = \dfrac{5}{x+3}$

Solve the following applications.

38. If the reciprocal of 5 times a number is subtracted from the reciprocal of that number, the result is $\dfrac{2}{5}$. What is the number?

39. Jennifer drove 260 mi to attend a business conference. In returning from the conference along a different route, the trip was only 240 mi, but traffic slowed her speed by 4 mi/h. If her driving time was the same both ways, what was her speed each way?

40. A letter-quality printer can print 80 form letters in 50 min. At that rate, how long will it take the printer to complete a job requiring 200 letters?

ANSWERS

31. _____

32. _____

33. _____

34. _____

35. _____

36. _____

37. _____

38. _____

39. _____

40. _____

CHAPTER 7

Graphing Linear Equations and Inequalities

The ENVIRONMENT

The Three R's: Reduce, Reuse, Recycle

We are hearing a lot today about the need to recycle. Communities across the country are setting up recycling programs, some mandated by law and others voluntary. These programs vary considerably in what they recycle. Some may recycle only bottle glass and metal cans while other "full-line" facilities may recycle plastics, cardboard, office papers, grocery bags, motor oil, used appliances, or other marketable materials as well as the traditional glass and metal containers.

The newly increased emphasis on recycling used materials is a move in the right direction, but it is really addressing the wrong end of a good solid waste reduction strategy. Recycling should be the third or fourth option in a truly complete solid waste reduction program.

Let's assume that the goal of recycling is to lessen the demand for landfills and, at the same time, reduce the consumption of natural resources. Then we need to think about creating less waste that needs to be recycled. We can do this by applying the principle known as the three R's: *Reduce, reuse, and then recycle.*

The principle of the three R's begins with *reducing* the amount of material we use that needs to be thrown away. Since packaging makes up about one-third of our household waste, this might be a good place to start. Purchasing goods with little or no packaging will greatly reduce the amount of waste generated by a home. Buying in bulk is a good way to get rid of excess packaging. Buying locally produced goods lessens the need for shipping containers. Buying in season reduces the need for packaging needed for preservation. Reducing the amount of goods we buy by purchasing only what we need when we need it reduces the amount of resources used.

After we've done all we can to reduce our level of consumption, we must *reuse* the things we buy as much as possible. Buying used goods and borrowing seldom used items such as tools and machines both reduce consumption of resources. Buying durable instead of disposable products creates less waste. Reclaiming and reusing industrial processing materials will reduce wastes generated by industry. Purchasing refillable bottles and jars, using grocery bags more than once, and washing aluminum foil to be used again are all ways to reuse resources.

To reuse also means finding new uses for objects besides what they were designed for. This will take some creativity, but we need to get beyond using cardboard boxes as toys and bottles and cans as pencil holders. Some work is being done with bottles and jars as building materials, and used tires are being used to create artificial reefs in some coastal areas. But much more needs to be done in this area.

Finally, we come to the *recycling* step. Because recycling is usually meant as remanufacturing, we want to employ this strategy as a last resort. Even so, it is preferable to putting waste in a landfill, so it is an important part of our conservation strategy. Using recycled materials to remanufacture products not only conserves the raw materials necessary to make certain goods but also is almost always an energy conservation measure. Aluminum made from recycled metal requires only a fraction of the energy needed to make aluminum from bauxite. Paper products made from recycled paper take less energy than those made from raw wood fiber. Recycling also creates less pollution of the air, water, and land in many manufacturing processes.

The three R's are a greatly simplified view of the recycling process, but it is a good beginning. If you are new to the idea of recycling, it is an easy rule to remember and gives you an outline for your personal strategy to recycle: Reduce, reuse, and recycle are steps in the right direction.

Solutions of Equations in Two Variables

OBJECTIVES
1. To find solutions for an equation in two variables
2. To use the ordered-pair notation to write solutions for equations in two variables

Recall that an equation is two expressions connected by an equal sign.

We discussed finding solutions for equations in Chapter 3. Recall that a solution is a value for the variable that "satisfies" the equation, or makes the equation a true statement. For instance, if

$$2x + 5 = 13$$

then 4 is a solution for the equation. This is true because, when we replace x with 4, we have

$$2 \cdot 4 + 5 = 13$$
$$8 + 5 = 13$$
$$13 = 13 \qquad \text{(A true statement)}$$

We now want to consider *equations in two variables*. An example is

$$x + y = 5$$

What will the solution look like? It is not going to be a single number, because there are two variables. Here the solution will be a pair of numbers—one value for each of the variables, x and y. Suppose that x has the value 3. In the equation $x + y = 5$, you can substitute 3 for x.

$$3 + y = 5$$

Solving for y gives

$$y = 2$$

An equation in two variables "pairs" two numbers, one for x and one for y.

So the pair of values $x = 3$ and $y = 2$ satisfies the equation because

$$3 + 2 = 5.$$

That pair of numbers is then a *solution* for the equation in two variables.

How many such pairs are there? Choose any value for x (or for y). You can always find the other *paired* or *corresponding* value in an equation of this form. We say that there are an *infinite* number of pairs that will satisfy the equation. Each of these pairs is a solution. We will find some other solutions for the equation $x + y = 5$ in the following example.

Example 1

For the equation $x + y = 5$, find (*a*) y if $x = 5$ and (*b*) x if $y = 4$.

(*a*) If $x = 5$,

$$5 + y = 5 \qquad \text{or} \qquad y = 0$$

(*b*) If $y = 4$,

$$x + 4 = 5 \qquad \text{or} \qquad x = 1$$

So the pairs $x = 5$, $y = 0$ and $x = 1$, $y = 4$ are also solutions.

CHECK YOURSELF 1

For the equation $2x + 3y = 26$,

1. If $x = 4$, $y = ?$ **2.** If $y = 0$, $x = ?$

To simplify writing the pairs that satisfy an equation, we use the *ordered-pair* notation. The numbers are written in parentheses and are separated by a comma. For example, we know that the values $x = 3$ and $y = 2$ satisfy the equation $x + y = 5$. So we write the pair as

$$(3, 2)$$

The *x* coordinate   The *y* coordinate

The first number of the pair is *always* the value for x and is called the x *coordinate*. The second number of the pair is *always* the value for y and is the y *coordinate*.

Using this ordered-pair notation, we can say that $(3, 2)$, $(5, 0)$, and $(1, 4)$ are all *solutions* for the equation $x + y = 5$. Each pair gives values for x and y that will satisfy the equation.

<CAUTION>

$(3, 2)$ means $x = 3$ and $y = 2$.
$(2, 3)$ means $x = 2$ and $y = 3$.
$(3, 2)$ and $(2, 3)$ are entirely different. That's why we call them *ordered pairs*.

Example 2

Which of the ordered pairs (*a*) $(2, 5)$, (*b*) $(5, -1)$, and (*c*) $(3, 4)$ are solutions for the equation $2x + y = 9$?

(*a*) To check whether $(2, 5)$ is a solution, let $x = 2$ and $y = 5$ and see if the equation is satisfied.

$$2x + y = 9 \qquad \text{The original equation}$$

$$\begin{matrix} x & y \\ \downarrow & \downarrow \end{matrix}$$

$$2 \cdot 2 + 5 = 9 \qquad \text{Substitute 2 for } x \text{ and 5 for } y.$$

$$4 + 5 = 9$$

$$9 = 9 \qquad \text{(True)}$$

(2, 5) is a solution because a *true statement* results.

(2, 5) is a solution for the equation.

(*b*) For $(5, -1)$, let $x = 5$ and $y = -1$.

$$2 \cdot 5 - 1 = 9$$

$$10 - 1 = 9$$

$$9 = 9 \qquad \text{(True)}$$

So $(5, -1)$ is a solution.

(*c*) For $(3, 4)$, let $x = 3$ and $y = 4$. Then

$$2 \cdot 3 + 4 = 9$$

$$6 + 4 = 9$$

$$10 = 9 \qquad (\textit{Not} \text{ a true statement})$$

So $(3, 4)$ is *not* a solution for the equation.

CHECK YOURSELF 2

Which of the ordered pairs $(3, 4)$, $(4, 3)$, $(1, -2)$, and $(0, -5)$ are solutions for the following equation?

$$3x - y = 5$$

If the equation contains only one variable, then the missing variable can take on any value.

Example 3

Which of the ordered pairs, $(2, 0)$, $(0, 2)$, $(5, 2)$, $(2, 5)$, and $(2, -1)$ are solutions for the equation $x = 2$?

A solution is any ordered pair in which the x coordinate is 2. That makes $(2, 0)$, $(2, 5)$, and $(2, -1)$ solutions for the given equation.

CHECK YOURSELF 3

Which of the ordered pairs $(3, 0)$, $(0, 3)$, $(3, 3)$, $(-1, 3)$, and $(3, -1)$ are solutions for the equation $y = 3$?

When an ordered pair is presented, the first number is always the x coordinate and the second number the y coordinate.

Example 4

Complete the ordered pairs (a) $(9, \ )$, (b) $(\ , -1)$, (c) $(0, \ )$, and (d) $(\ , 0)$ for the equation $x - 3y = 6$.

The x coordinate is sometimes called the *abscissa* and the y coordinate the *ordinate*.

(a) The first number, 9, appearing in $(9, \ )$ represents the x value, so to complete the pair $(9, \ )$, substitute 9 for x and then solve for y.

$$9 - 3y = 6$$
$$-3y = -3$$
$$y = 1$$

$(9, 1)$ is a solution.

(b) To complete the pair $(\ , -1)$, let y be -1 and solve for x.

$$x - 3(-1) = 6$$
$$x + 3 = 6$$
$$x = 3$$

$(3, -1)$ is a solution.

(c) To complete the pair $(0, \ )$, let x be 0.

$$0 - 3y = 6$$
$$-3y = 6$$
$$y = -2$$

$(0, -2)$ is a solution.

(d) To complete the pair $(\ , 0)$, let y be 0.

$$x - 3 \cdot 0 = 6$$
$$x - 0 = 6$$
$$x = 6$$

$(6, 0)$ is a solution.

CHECK YOURSELF 4

Complete the ordered pairs below so that each is a solution for the equation $2x + 5y = 10$.
(10,), (, 4), (0,), and (, 0)

Example 5

Find four solutions for the equation

$$2x + y = 8$$

Generally you'll want to pick values for x (or for y) so that the resulting equation in one variable is easy to solve.

In this case the values used to form the solutions are *up to you*. You can assign any value for x (or for y). We'll demonstrate with some possible choices.

Let $x = 2$:

$$2 \cdot 2 + y = 8$$
$$4 + y = 8$$
$$y = 4$$

$(2, 4)$ is a solution.

Let $y = 6$:

$$2x + 6 = 8$$
$$2x = 2$$
$$x = 1$$

$(1, 6)$ is a solution.

Let $x = 0$:

$$2 \cdot 0 + y = 8$$
$$y = 8$$

The solutions (0, 8) and (4, 0) will have special significance later in graphing. They are also easy to find!

$(0, 8)$ is a solution.

Let $y = 0$:

$$2x + 0 = 8$$
$$2x = 8$$
$$x = 4$$

$(4, 0)$ is a solution.

CHECK YOURSELF 5

Find four solutions for $x - 3y = 12$.

CHECK YOURSELF ANSWERS

1. (1) $y = 6$; (2) $x = 13$.

2. $(3, 4)$, $(1, -2)$, and $(0, -5)$ are solutions.

3. $(0, 3)$, $(3, 3)$, and $(-1, 3)$ are solutions.

4. $(10, -2)$, $(-5, 4)$, $(0, 2)$, and $(5, 0)$.

5. $(6, -2)$, $(3, -3)$, $(0, -4)$, and $(12, 0)$ are four possibilities.

Name

Date

Build Your Skills

Determine which of the ordered pairs are solutions for the given equation.

1. $x + y = 6$ $(4, 2),\ (-2, 4),\ (0, 6),\ (-3, 9)$

2. $x - y = 12$ $(13, 1),\ (13, -1),\ (12, 0),\ (6, 6)$

3. $2x - y = 8$ $(5, 2),\ (4, 0),\ (0, 8),\ (6, 4)$

4. $x + 5y = 20$ $(10, -2),\ (10, 2),\ (20, 0),\ (25, -1)$

5. $3x + y = 6$ $(2, 0),\ (2, 3),\ (0, 2),\ (1, 3)$

6. $x - 2y = 8$ $(8, 0),\ (0, 4),\ (5, -1),\ (10, -1)$

7. $2x - 3y = 6$ $(0, 2),\ (3, 0),\ (6, 2),\ (0, -2)$

8. $8x + 4y = 16$ $(2, 0),\ (6, -8),\ (0, 4),\ (6, -6)$

9. $3x - 2y = 12$ $(4, 0),\ \left(\dfrac{2}{3}, -5\right),\ (0, 6),\ \left(5, \dfrac{3}{2}\right)$

10. $3x + 4y = 12$ $(-4, 0),\ \left(\dfrac{2}{3}, \dfrac{5}{2}\right),\ (0, 3),\ \left(\dfrac{2}{3}, 2\right)$

11. $y = 4x$ $(0, 0),\ (1, 3),\ (2, 8),\ (8, 2)$

12. $y = 2x - 1$ $(0, -2),\ (0, -1),\ \left(\dfrac{1}{2}, 0\right),\ (3, -5)$

ANSWERS

1. _____

2. _____

3. _____

4. _____

5. _____

6. _____

7. _____

8. _____

9. _____

10. _____

11. _____

12. _____

13. _____

14. _____

15. _____

16. _____

17. _____

18. _____

19. _____

20. _____

21. _____

22. _____

23. _____

24. _____

13. $x = 3$ $(3, 5), (0, 3), (3, 0), (3, 7)$

14. $y = 5$ $(0, 5), (3, 5), (-2, -5), (5, 5)$

Complete the ordered pairs so that each is a solution for the given equation.

15. $x + y = 10$ $(5, \quad), (\quad, 3), (0, \quad), (\quad, 0)$

16. $x - y = 5$ $(\quad, 3), (9, \quad), (0, \quad), (\quad, 0)$

17. $2x + y = 6$ $(3, \quad), (\quad, 6), (\quad, -2), (0, \quad)$

18. $x + 4y = 8$ $(0, \quad), (\quad, 2), (4, \quad), (\quad, 0)$

19. $5x - y = 15$ $(\quad, 0), (2, \quad), (4, \quad), (\quad, -5)$

20. $x - 3y = 9$ $(0, \quad), (12, \quad), (\quad, 0), (\quad, -2)$

21. $3x - 2y = 12$ $(\quad, 0), (\quad, -6), (2, \quad), (\quad, 3)$

22. $2x + 5y = 20$ $(0, \quad), (5, \quad), (\quad, 0), (\quad, 6)$

23. $y = 3x + 9$ $(\quad, 0), \left(\dfrac{2}{3}, \quad\right), (0, \quad), \left(-\dfrac{2}{3}, \quad\right)$

24. $3x + 4y = 12$ $(0, \quad), \left(\quad, \dfrac{3}{4}\right), (\quad, 0), \left(\dfrac{8}{3}, \quad\right)$

25. $y = 3x - 4$ $(0, \ \ )$, $(\ , 5)$, $(\ , 0)$, $\left(\dfrac{5}{3}, \ \ \right)$

26. $y = -2x + 5$ $(0, \ \ )$, $(\ , 5)$, $\left(\dfrac{3}{2}, \ \ \right)$, $(\ , 1)$

Find four solutions for each of the following equations. **Note:** Your answers may vary from those shown in the text.

27. $x - y = 7$

28. $x + y = 18$

29. $2x - y = 6$

30. $3x - y = 12$

31. $x + 4y = 8$

32. $x + 3y = 12$

33. $2x - 5y = 10$

34. $2x + 7y = 14$

35. $y = 4x + 1$

36. $y = 6x - 2$

37. $x = -3$

38. $y = 5$

Think About These

An equation in three variables has an ordered triple as a solution. For example, $(1, 2, 2)$ is a solution to the equation $x + 2y - z = 3$. Complete the ordered-triple solutions for each equation.

39. $x + y + z = 0$ $(2, -3, \ \ )$

40. $x + y + z = 0$ $(1, \ \ , 5)$

41. $2x + y + z = 2$ $(-2, \ \ , 1)$

25. _____

26. _____

27. _____

28. _____

29. _____

30. _____

31. _____

32. _____

33. _____

34. _____

35. _____

36. _____

37. _____

38. _____

39. _____

40. _____

41. _____

42. _____

43. _____

44. _____

a. _____

b. _____

c. _____

d. _____

e. _____

f. _____

g. _____

h. _____

i. _____

j. _____

42. $2x + y + z = 2$ $\quad$ (, $-1, 3$)

43. $x + y - z = 1$ $\quad$ (4, , 3)

44. $x + y - z = 1$ $\quad$ ($-2, 1,$)

Skillscan (Section 2.1)

Plot points with the following coordinates on the number line shown below.

a. -3 $\qquad$ **b.** 7 $\qquad$ **c.** 0 $\qquad$ **d.** -8 $\qquad$ **e.** $\dfrac{3}{2}$

Give the coordinates of each of the following points.

f. A $\qquad$ **g.** B $\qquad$ **h.** C $\qquad$ **i.** D $\qquad$ **j.** E

ANSWERS

1. $(4, 2), (0, 6), (-3, 9)$ $\qquad$ **3.** $(5, 2), (4, 0), (6, 4)$ $\qquad$ **5.** $(2, 0), (1, 3)$

7. $(3, 0), (6, 2), (0, -2)$ $\qquad$ **9.** $(4, 0), \left(\dfrac{2}{3}, -5\right), \left(5, -\dfrac{3}{2}\right)$ $\qquad$ **11.** $(0, 0), (2, 8)$

13. $(3, 5), (3, 0), (3, 7)$ $\qquad$ **15.** $5, 7, 10, 10$ $\qquad$ **17.** $0, 0, 4, 6$ $\qquad$ **19.** $3, -5, 5, 2$

21. $4, 0, -3, 6$ $\qquad$ **23.** $-3, 11, 9, 7$ $\qquad$ **25.** $-4, 3, \dfrac{4}{3}, 1$ $\qquad$ **27.** $(0, -7), (2, -5), (4, -3), (6, -1)$

29. $(0, -6), (3, 0), (6, 6), (9, 12)$ $\qquad$ **31.** $(8, 0), (-4, 3), (0, 2), (4, 1)$

33. $(-5, -4), (0, -2), (5, 0), (10, 2)$ $\qquad$ **35.** $(0, 1), (1, 5), (2, 9), (3, 13)$

37. $(-3, 0), (-3, 1), (-3, 2), (-3, 3)$ $\qquad$ **39.** $(2, -3, 1)$ $\qquad$ **41.** $(-2, 5, 1)$ $\qquad$ **43.** $(4, 0, 3)$

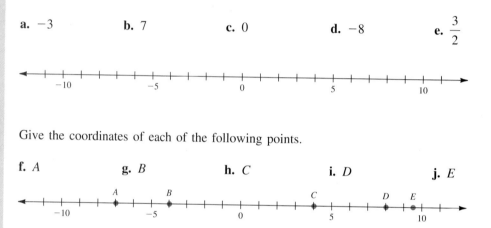

a–e.

f. -7 $\qquad$ **g.** -4 $\qquad$ **h.** 4 $\qquad$ **i.** 8 $\qquad$ **j.** $\dfrac{19}{2}$

7.2

The Rectangular Coordinate System

OBJECTIVES

1. To graph or plot points corresponding to ordered pairs
2. To give the coordinates of a point in a plane
3. To find the midpoint of two points

In Section 7.1 we saw that ordered pairs could be used to write the solutions of equations in two variables. The next step is to graph those ordered pairs as points in a plane.

Since there are two numbers (one for *x* and one for *y*), we will need two number lines. One line is drawn horizontally, and the other is drawn vertically; their point of intersection (at their respective zero points) is called the *origin*. The horizontal line is called the *x axis*, while the vertical line is called the *y axis*. Together the lines form the *rectangular coordinate system*.

The axes divide the plane into four regions called *quadrants*, which are numbered (usually by Roman numerals) counterclockwise from the upper right.

This system is also called the *cartesian* coordinate system, named in honor of its inventor, René Descartes (1596–1650) a French mathematician and philosopher.

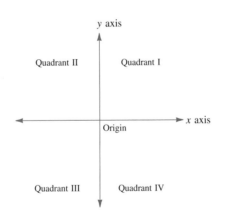

The origin is the point with coordinates (0, 0).

We now want to establish a correspondence between ordered pairs of numbers (*x*, *y*) and points in the plane.

First, for any ordered pair

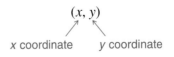

1. If the *x* coordinate is

Positive, the point corresponding to that pair is located *x* units to the *right* of the *y* axis.

Negative, the point is *x* units to the *left* of the *y* axis.

Zero, the point is on the *y* axis.

479

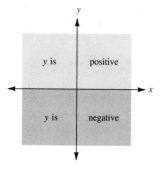

2. If the *y* coordinate is

Positive, the point is *y* units *above* the *x* axis.

Negative, the point is *y* units *below* the *x* axis.

Zero, the point is on the *x* axis.

The next example illustrates how to use these guidelines to give coordinates to points in the plane.

Example 1

Give the coordinates for the given points.

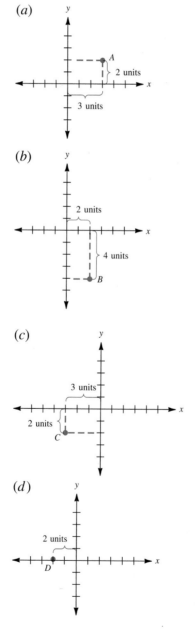

Remember: The *x* coordinate gives the horizontal distance from the origin. The *y* coordinate gives the vertical distance from the *x* axis.

(*a*)

Point *A* is 3 units to the *right* of the origin and 2 units *above* the *x* axis. Point *A* has coordinates (3, 2).

(*b*)

Point *B* is 2 units to the *right* of the origin and 4 units *below* the *x* axis. Point *B* has coordinates (2, −4).

(*c*)

Point *C* is 3 units to the *left* of the origin and 2 units *below* the *x* axis. *C* has coordinates (−3, −2).

(*d*)

Point *D* is 2 units to the *left* of the origin and *on* the *x* axis. Point *D* has coordinates (−2, 0).

CHECK YOURSELF 1

Give the coordinates of points P, Q, R, and S.

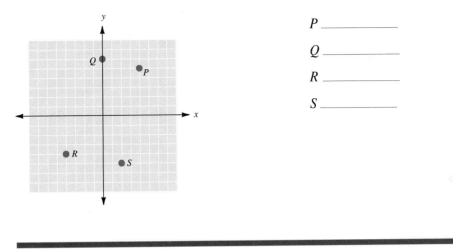

P _____

Q _____

R _____

S _____

Reversing the process above will allow us to graph (or plot) a point in the plane given the coordinates of the point. You can use the following steps.

The graphing of individual points is sometimes called *point plotting*.

TO GRAPH A POINT IN THE PLANE

STEP 1   Start at the origin.

STEP 2   Move right or left according to the value of the x coordinate.

STEP 3   Move up or down according to the value of the y coordinate.

Example 2

(a) Graph the point corresponding to the ordered pair (4, 3).

Move 4 units to the right on the x axis. Then move 3 units up from the point you stopped at on the x axis. This locates the point corresponding to (4, 3).

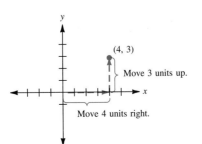

(*b*) Graph the point corresponding to the ordered pair $(-5, 2)$.

In this case move 5 units *left* (because the x coordinate is negative) and then 2 units *up*.

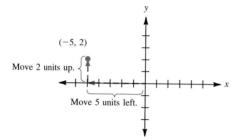

(*c*) Graph the point corresponding to $(-4, -2)$.

Here move 4 units *left* and then 2 units *down* (the y coordinate is negative).

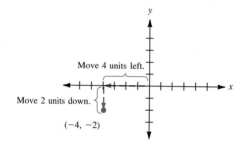

Any point on an axis will have 0 for one of its coordinates.

(*d*) Graph the point corresponding to $(0, -3)$.

There is *no* horizontal movement because the x coordinate is 0. Move 3 units *down*.

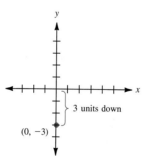

(*e*) Graph the point corresponding to $(5, 0)$.

Move 5 units *right*. The desired point is on the x axis because the y coordinate is 0.

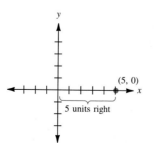

CHECK YOURSELF 2

Graph the points corresponding to $M(4, 3)$, $N(-2, 4)$, $P(-5, -3)$, and $Q(0, -3)$.

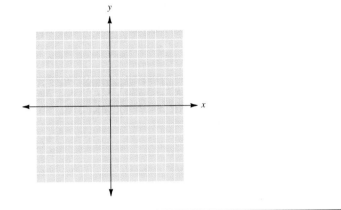

Every two points have a *midpoint*. The midpoint of A and B is the point on line $\overleftrightarrow{AB}$ that is equal distance from A and B. The following formula can be used to find the midpoint.

The x coordinate for the midpoint is the average of the x's, the y coordinate is the average of the y's.

> The midpoint M of points $A(x_1, y_1)$ and $B(x_2, y_2)$ is found with the formula
>
> $$M = \left(\frac{x_1 + x_2}{2}, \frac{y_1 + y_2}{2} \right)$$

Example 3

(*a*) Find the midpoint of $(2, 0)$ and $(10, 0)$.

$$M = \left(\frac{2 + 10}{2}, \frac{0 + 0}{2} \right) = (6, 0)$$

(*b*) Find the midpoint of $(5, 7)$ and $(-1, -3)$.

$$M = \left(\frac{5 + (-1)}{2}, \frac{7 + (-3)}{2} \right) = (2, 2)$$

(*c*) Find the midpoint of $(3, -5)$ and $(-2, -2)$.

$$M = \left(\frac{3 + (-2)}{2}, \frac{(-5) + (-2)}{2} \right) = \left(\frac{1}{2}, \frac{-7}{2} \right)$$

CHECK YOURSELF 3

Find the midpoint for each pair of points.

1. $(0, 6)$ and $(0, -4)$.   **2.** $(3, -6)$ and $(-5, 4)$.
3. $(-1, -5)$ and $(-2, 8)$.

CHECK YOURSELF ANSWERS

1. $P(4, 5)$, $Q(0, 6)$, $R(-4, -4)$, and $S(2, -5)$

2.

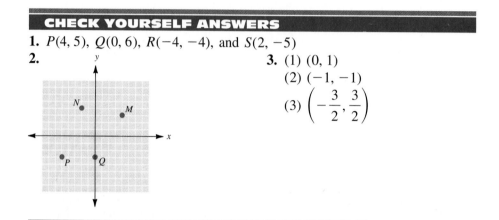

3. (1) $(0, 1)$
(2) $(-1, -1)$
(3) $\left(-\dfrac{3}{2}, \dfrac{3}{2} \right)$

Name _____

Date _____

ANSWERS

Build Your Skills

1–5. Give the coordinates of the points graphed below.

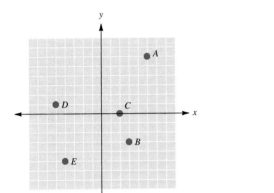

1. A   **2.** B

3. C   **4.** D

5. E

6–10. Give the coordinates of the points graphed below.

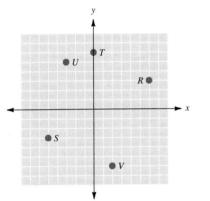

6. R   **7.** S

8. T   **9.** U

10. V

Find the midpoint for each pair of points.

11. $(0, 7)$ and $(0, -3)$

12. $(0, 4)$ and $(4, 0)$

13. $(2, -3)$ and $(4, -7)$

14. $(-3, 7)$ and $(-1, -3)$

15. $(2, -8)$ and $(-1, -5)$

16. $(-3, 5)$ and $(2, -6)$

1. _____
2. _____
3. _____
4. _____
5. _____
6. _____
7. _____
8. _____
9. _____
10. _____
11. _____
12. _____
13. _____
14. _____
15. _____
16. _____

17. _____

18. _____

19. _____

20. _____

21. _____

22. _____

23. _____

24. _____

25. _____

26. _____

27. _____

28. _____

17–22. Plot points with the following coordinates on the graph below.

17. $M(5, 3)$

18. $N(0, -3)$

19. $P(-2, 6)$

20. $Q(5, 0)$

21. $R(-4, -6)$

22. $S(-3, -4)$

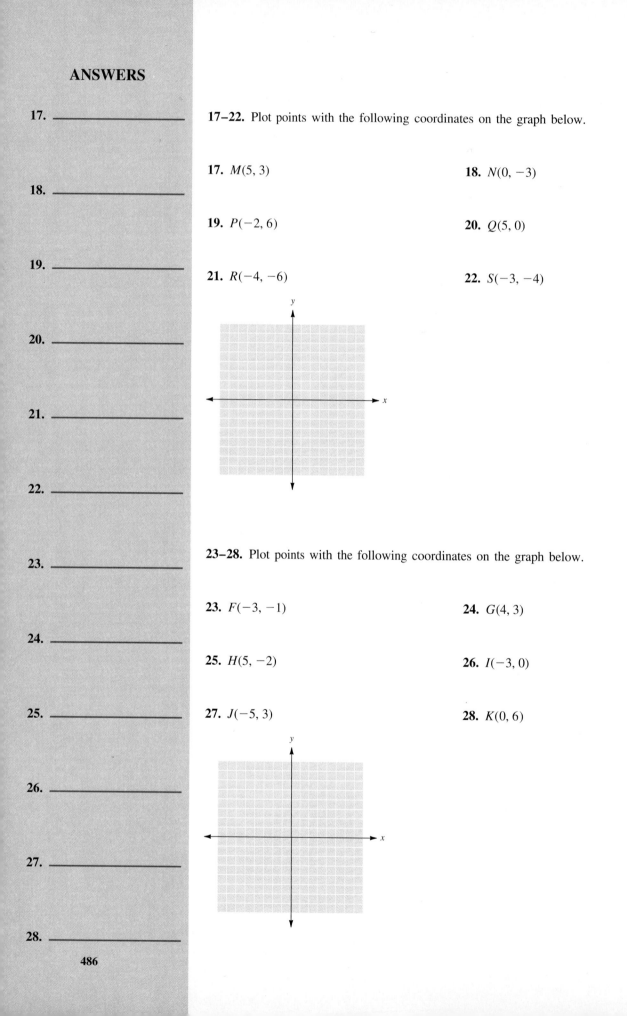

23–28. Plot points with the following coordinates on the graph below.

23. $F(-3, -1)$

24. $G(4, 3)$

25. $H(5, -2)$

26. $I(-3, 0)$

27. $J(-5, 3)$

28. $K(0, 6)$

29. Graph points with coordinates (2, 3), (3, 4), and (4, 5) below. What do you observe? Can you give the coordinates of another point with the same property?

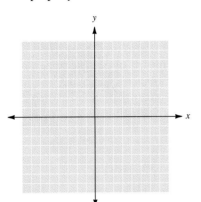

30. Graph points with coordinates (−1, 4), (0, 3), and (1, 2) below. What do you observe? Can you give the coordinates of another point with the same property?

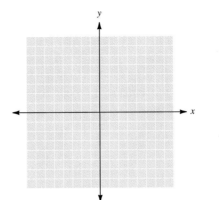

31. Graph points with coordinates (−1, 3), (0, 0), and (1, −3) below. What do you observe? Can you give the coordinates of another point with the same property?

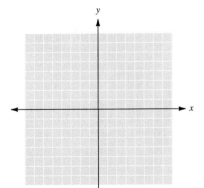

32. Graph points with coordinates (1, 5), (−1, 3), and (−3, 1) below. What do you observe? Can you give the coordinates of another point with the same property?

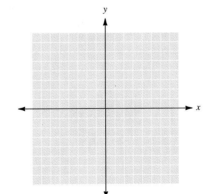

33. _____

34. _____

35. _____

36. _____

A local plastics company is sponsoring a plastics recycling contest for the local community. The focus of the contest is plastic milk, juice, and water jugs. The company will award $200 plus the current market price of the jugs collected to the group that collects the most jugs in a single month. The number of jugs collected and the amount of money won can be represented as an ordered pair.

33. In April, group A collected 1500 pounds (lb) of jugs to win first place. The prize for that month was $350. On the graph below, x represents the pounds of jugs and y represents the amount of money the group won. Graph the point that represents the winner for April.

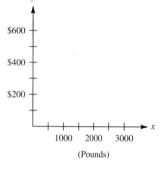

(Pounds)

34. In May, group B collected 2300 lb of jugs to win first place. The prize for the month was $430. Graph the point that represents the May winner on the same axis as you used for Exercise 33.

35. In June, group C collected 1200 lb of jugs to win the contest. The prize for the month was $320. Graph the point that represents the June winner on the same axis as you used before.

36. What can you determine about the points you graphed in questions 33 to 35. Can you discover the amount of prize money won in July if the winning group collected 1800 lb of jugs?

Skillscan (Section 3.4)

Solve each of the following equations.

a. $2x - 2 = 6$

b. $2 - 5x = 12$

c. $7y + 10 = -11$

d. $-3 + 5x = 1$

e. $6 - 3x = 8$

f. $-4y + 6 = 3$

ANSWERS

1. $(5, 6)$ **3.** $(2, 0)$ **5.** $(-4, -5)$ **7.** $(-5, -3)$ **9.** $(-3, 5)$ **11.** $(0, 2)$

13. $(3, -5)$ **15.** $\left(\dfrac{1}{2}, \dfrac{-13}{2}\right)$

17–27.

29. The points lie on a line; $(1, 2)$

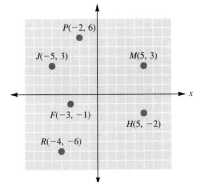

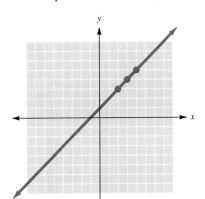

31. The points lie on a line; $(2, -6)$

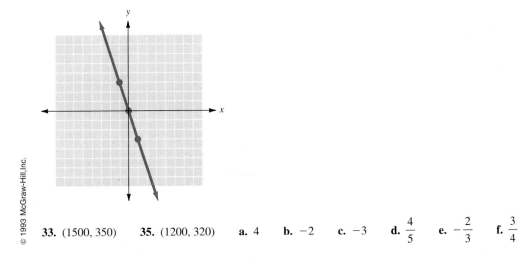

33. $(1500, 350)$ **35.** $(1200, 320)$ **a.** 4 **b.** -2 **c.** -3 **d.** $\dfrac{4}{5}$ **e.** $-\dfrac{2}{3}$ **f.** $\dfrac{3}{4}$

7.3 Graphing Linear Equations

OBJECTIVES
1. To graph linear equations by plotting points
2. To graph linear equations by the intercept method
3. To graph linear equations by solving the equation for *y*

We are now ready to combine our work of the last two sections. In Section 7.1 you learned to write the solutions of equations in two variables as ordered pairs. Then, in Section 7.2, these ordered pairs were graphed in the plane. Putting these ideas together will let us graph certain equations. The following example illustrates.

Example 1

Graph $x + 2y = 4$.

We are going to find three solutions for the equation. We'll point out why shortly.

Step 1   Find some solutions for $x + 2y = 4$.

To find solutions, we choose any convenient values for *x*, say $x = 0$, $x = 2$, and $x = 4$. Given these values for *x*, we can substitute and then solve for the corresponding value for *y*. So

If $x = 0$, then $y = 2$, so (0, 2) is a solution.

If $x = 2$, then $y = 1$, so (2, 1) is a solution.

If $x = 4$, then $y = 0$, so (4, 0) is a solution.

A handy way to show this information is in a table such as this:

The table is just a convenient way to display the information. It is the same as writing (0, 2), (2, 1), and (4, 0).

x	y
0	2
2	1
4	0

Step 2   We now graph the solutions found in step 1.

$x + 2y = 4$

x	y
0	2
2	1
4	0

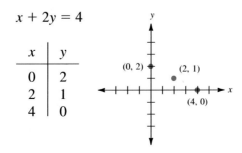

What pattern do you see? It appears that the three points lie on a straight line, and that is in fact the case.

491

Step 3 Draw a straight line through the three points graphed in step 2.

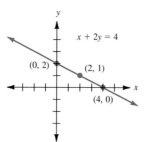

The arrows on the end of the line mean that the line extends indefinitely in either direction.

The graph is a "picture" of the solutions for the given equation.

The line shown is the *graph* of the equation $x + 2y = 4$. It represents *all* the ordered pairs that are solutions (an infinite number) for that equation.

Every ordered pair which is a solution will have its graph on this line. Any point on the line will have coordinates that are a solution for the equation.

Note: Why did we suggest finding *three* solutions in step 1? Two points determine a line, so technically you need only two. The third point that we find is a check to catch any possible errors.

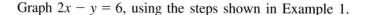

CHECK YOURSELF 1

Graph $2x - y = 6$, using the steps shown in Example 1.

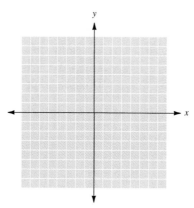

Let's summarize. An equation that can be written in the form

$$Ax + By = C$$

where A, B, and C are real numbers and A and B cannot both be 0 is called a *linear equation in two variables*. The graph of this equation is a *straight line*.

The steps of graphing:

TO GRAPH A LINEAR EQUATION

STEP 1   Find at least three solutions for the equation, and put your results in tabular form.

STEP 2   Graph the solutions found in step 1.

STEP 3   Draw a straight line through the points determined in step 2 to form the graph of the equation.

Example 2

Graph $y = 3x$.

Step 1   Some solutions are

Let $x = 0$, 1, and 2, and substitute to determine the corresponding y values. Again the choices for x are simply convenient. Other values for x would serve the same purpose.

x	y
0	0
1	3
2	6

Step 2   Graph the points.

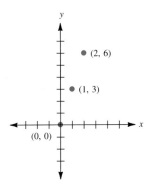

Step 3   Draw a line through the points.

Notice that connecting any two of these points produces the same line.

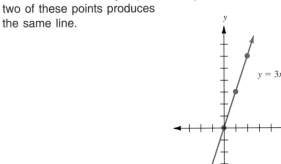

CHECK YOURSELF 2

Graph the equation $y = -2x$ after completing the table of values.

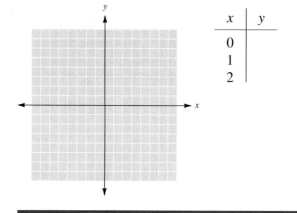

x	y
0	
1	
2	

Let's work through another example of graphing a line from its equation.

Example 3

Graph $y = 2x + 3$.

Step 1 Some solutions are

x	y
0	3
1	5
2	7

Step 2 Graph the points corresponding to these values.

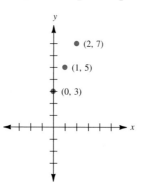

Step 3 Draw a line through the points.

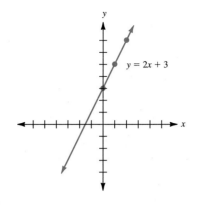

Graph the equation $y = 3x - 2$ after completing the table of values.

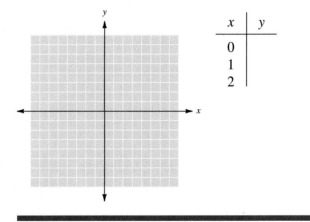

x	y
0	
1	
2	

In graphing equations, particularly when fractions are involved, a careful choice of values for x can simplify the process. Consider the following example.

Example 4

Graph

$$y = \frac{3}{2}x - 2$$

As before, we want to find solutions for the given equation by picking convenient values for x. Note that in this case, choosing *multiples of 2* will avoid fractional values for y and make the plotting of those solutions much easier. For instance, here we might choose values of -2, 0, and 2 for x.

Step 1

If $x = -2$,

$$y = \frac{3}{2}(-2) - 2$$

$$= -3 - 2 = -5$$

If $x = 0$,

Suppose we do *not* choose a multiple of 2, say, $x = 3$. Then

$$y = \frac{3}{2}(0) - 2$$

$$y = \frac{3}{2}(3) - 2$$

$$= 0 - 2 = -2$$

$$= \frac{9}{2} - 2$$

If $x = 2$,

$$= \frac{5}{2}$$

$$y = \frac{3}{2}(2) - 2$$

$\left(3, \dfrac{5}{2}\right)$ is still a valid

$$= 3 - 2 = 1$$

solution, but we must graph a point with fractional coordinates.

In tabular form, the solutions are

x	y
-2	-5
0	-2
2	1

Step 2   Graph the points determined above.

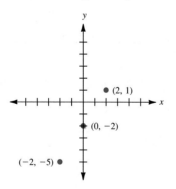

Step 3   Draw a line through the points.

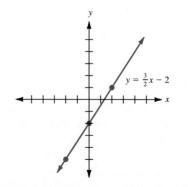

CHECK YOURSELF 4

Graph the equation $y = -\dfrac{1}{3}x + 3$ after completing the table of values.

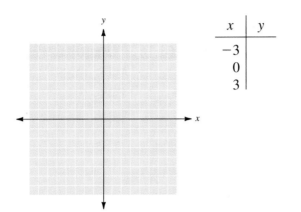

x	y
-3	
0	
3	

Some special cases of linear equations are illustrated in Examples 5 and 6.

Example 5

Graph $x = 3$.

The equation $x = 3$ is equivalent to $x + 0 \cdot y = 3$. Let's look at some solutions.

If $y = 1$,   If $y = 4$,   If $y = -2$,

$x + 0 \cdot 1 = 3$ $x + 0 \cdot 4 = 3$ $x + 0(-2) = 3$

$\quad\quad x = 3$ $\quad\quad x = 3$ $\quad\quad x = 3$

In tabular form,

x	y
3	1
3	4
3	-2

What do you observe? The variable x has the value 3, regardless of the value of y. Look at the graph.

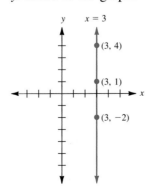

The graph of $x = 3$ is a vertical line crossing the x axis at $(3, 0)$.

Note that graphing (or plotting) points in this case is not really necessary. Simply recognize that the graph of $x = 3$ *must* be a vertical line (parallel to the y axis) which intercepts the x axis at 3.

CHECK YOURSELF 5

Graph the equation $x = -2$.

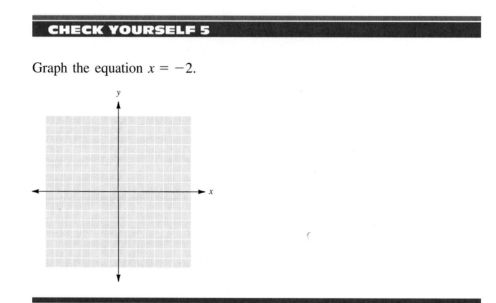

Here is a related example.

Example 6

Graph $y = 4$.

Since $y = 4$ is equivalent to $0 \cdot x + y = 4$, any value for x paired with 4 for y will form a solution. A table of values might be

x	y
-2	4
0	4
2	4

Here is the graph.

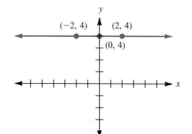

This time the graph is a horizontal line that crosses the *y* axis at (0, 4). Again the graphing of points is not required. The graph of $y = 4$ *must* be horizontal (parallel to the *x* axis) and intercepts the *y* axis at 4.

CHECK YOURSELF 6

Graph the equation $y = -3$.

To summarize our work in the previous two examples:

> **VERTICAL AND HORIZONTAL LINES**
>
> **1.** The graph of $x = a$ is a *vertical line* crossing the *x* axis at (*a*, 0).
> **2.** The graph of $y = b$ is a *horizontal line* crossing the *y* axis at (0, *b*).

To simplify the graphing of certain linear equations, some students prefer the *intercept* method of graphing. This method makes use of the fact that the solutions that are easiest to find are those with an *x* coordinate or a *y* coordinate of 0. For instance, let's graph the equation

$$4x + 3y = 12$$

First, let $x = 0$ and solve for *y*.

With practice all this can be done mentally, which is the big advantage of this method.

$$4 \cdot 0 + 3y = 12$$
$$3y = 12$$
$$y = 4$$

So (0, 4) is one solution. Now we let $y = 0$ and solve for *x*.

$$4x + 3 \cdot 0 = 12$$
$$4x = 12$$
$$x = 3$$

A second solution is (3, 0).

The two points corresponding to these solutions can now be used to graph the equation.

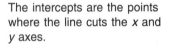

The intercepts are the points where the line cuts the *x* and *y* axes.

The number 3 is called the *x intercept,* and the number 4 is the *y intercept* of the graph. Using these points to draw the graph gives the name to this method. Let's look at a second example of graphing by the intercept method.

Example 7

Graph $3x - 5y = 15$, using the intercept method.

To find the *x* intercept, let $y = 0$.

$3x - 5 \cdot 0 = 15$
$\qquad x = 5$ ⟵ The *x* intercept

To find the *y* intercept, let $x = 0$.

$3 \cdot 0 - 5y = 15$
$\qquad y = -3$ ⟵ The *y* intercept

So $(5, 0)$ and $(0, -3)$ are solutions for the equation, and we can use the corresponding points to graph the equation.

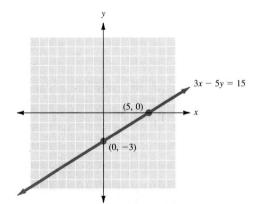

CHECK YOURSELF 7

Graph $4x + 5y = 20$, using the intercept method.

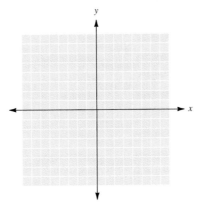

This all looks quite easy, and for many equations it is. What are the drawbacks?

For one, you don't have a third checkpoint, and it is possible for errors to occur. You can, of course, still find a third point (other than the two intercepts) to be sure your graph is correct. A second difficulty arises when the x and y intercepts are very close to each other (or are actually the same point—the origin). For instance, if we have the equation

$$3x + 2y = 1$$

the intercepts are $\left(\dfrac{1}{3}, 0\right)$ and $\left(0, \dfrac{1}{2}\right)$. It is hard to draw a line accurately through these intercepts, so choose other solutions farther away from the origin for your points.

Let's summarize the steps of graphing by the intercept method for appropriate equations.

> **GRAPHING A LINE BY THE INTERCEPT METHOD**
>
> STEP 1 To find the x intercept: Let $y = 0$, then solve for x.
>
> STEP 2 To find the y intercept: Let $x = 0$, then solve for y.
>
> STEP 3 Graph the x and y intercepts.
>
> STEP 4 Draw a straight line through the intercepts.

A third method of graphing linear equations involves ''solving the equation for y.'' The reason we use this extra step is that it often will make finding solutions for the equation much easier. Let's look at an example.

Finding the third "checkpoint" is always a good idea.

Example 8

Graph $2x + 3y = 6$.

Remember that solving for y means that we want to leave y isolated on the left.

Rather than finding solutions for the equation in this form, we solve for y.

$2x + 3y = 6$

$\quad\quad 3y = 6 - 2x$ } Subtract 2x.

$\quad\quad\quad y = \dfrac{6 - 2x}{3}$ } Divide by 3.

or $\quad y = 2 - \dfrac{2}{3}x$

Again, to pick convenient values for x, we suggest you look at the equation carefully. Here, for instance, picking multiples of 3 for x will make the work much easier.

Now find your solutions by picking convenient values for x.

If $x = -3$,

$y = 2 - \dfrac{2}{3}(-3)$

$\quad = 2 + 2 = 4$

So $(-3, 4)$ is a solution.

If $x = 0$,

$y = 2 - \dfrac{2}{3} \cdot 0$

$\quad = 2$

So $(0, 2)$ is a solution.

If $x = 3$,

$y = 2 - \dfrac{2}{3} \cdot 3$

$\quad = 2 - 2 = 0$

So $(3, 0)$ is a solution.

We can now plot the points that correspond to these solutions and form the graph of the equation as before.

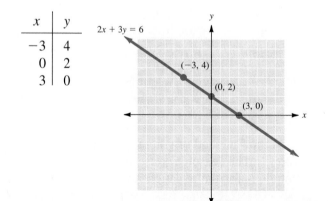

x	y
-3	4
0	2
3	0

CHECK YOURSELF 8

Graph the equation $5x + 2y = 10$. Solve for y to determine solutions.

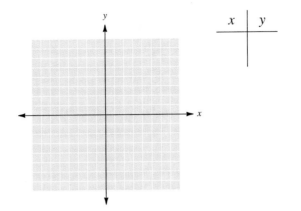

CHECK YOURSELF ANSWERS

1.

x	y
1	-4
2	-2
3	0

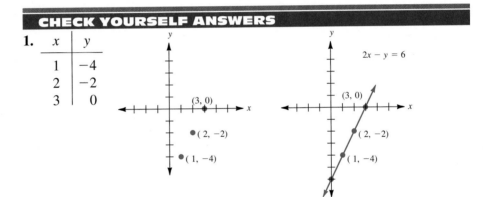

2.

x	y
0	0
1	-2
2	-4

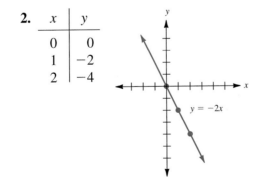

3.

x	y
0	-2
1	1
2	4

4.

x	y
-3	4
0	3
3	2

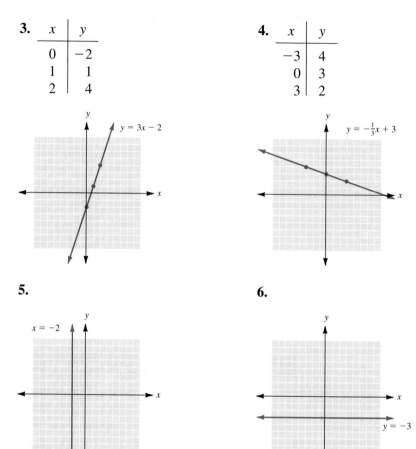

$y = 3x - 2$

$y = -\frac{1}{3}x + 3$

5.

$x = -2$

6.

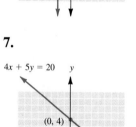

$y = -3$

7.

$4x + 5y = 20$

$(0, 4)$

$(5, 0)$

8.

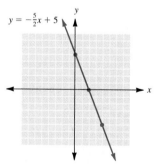

$y = -\frac{5}{2}x + 5$

Build Your Skills

Graph each of the following equations.

1. $x + y = 6$

2. $x - y = 5$

3. $x - y = -3$

4. $x + y = -3$

5. $2x + y = 2$

6. $x - 2y = 6$

7. $3x + y = 0$

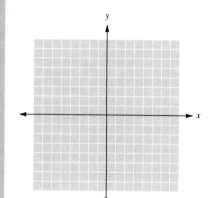

8. $3x - y = 6$

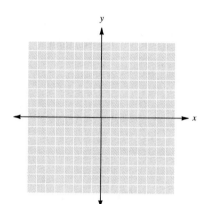

9. $x + 4y = 8$

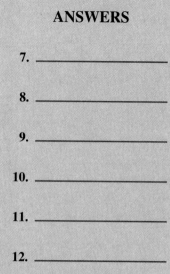

10. $2x - 3y = 6$

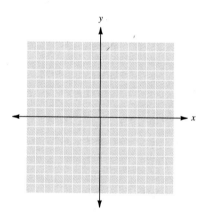

11. $y = 5x$

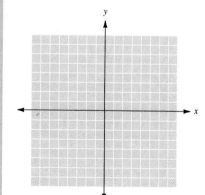

12. $y = -4x$

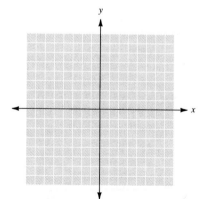

13. $y = 2x - 1$

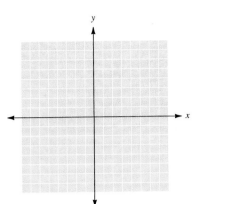

14. $y = 4x + 3$

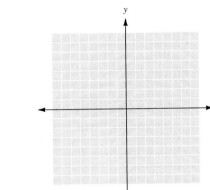

15. $y = -3x + 1$

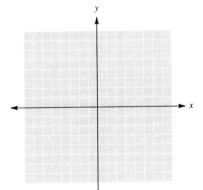

16. $y = -3x - 3$

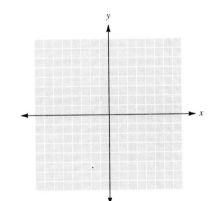

17. $y = \dfrac{1}{3}x$

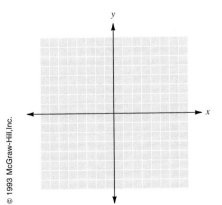

18. $y = -\dfrac{1}{4}x$

ANSWERS

13. _____

14. _____

15. _____

16. _____

17. _____

18. _____

19. $y = \dfrac{2}{3}x - 3$

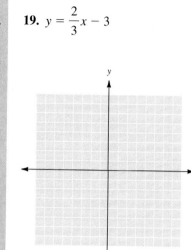

20. $y = \dfrac{3}{4}x + 2$

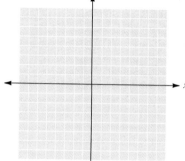

21. $x = 5$

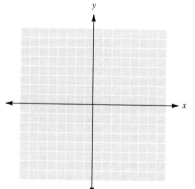

22. $y = -3$

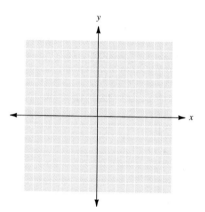

23. $y = 1$

24. $x = -2$

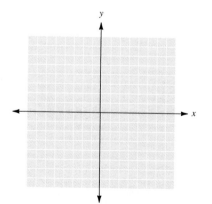

Graph each of the following equations, using the intercept method.

25. $x - 2y = 4$

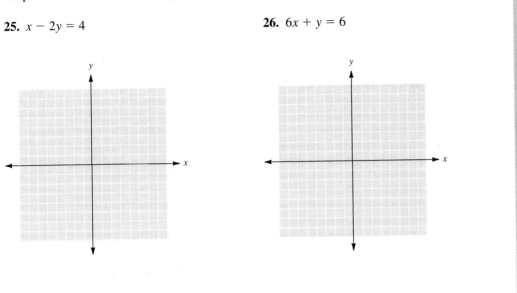

26. $6x + y = 6$

27. $5x + 2y = 10$

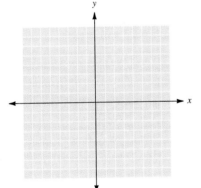

28. $2x + 3y = 6$

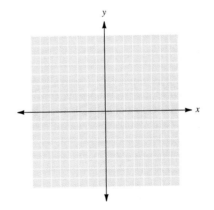

29. $3x + 5y = 15$

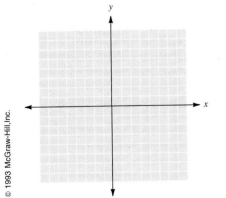

30. $4x + 3y = 12$

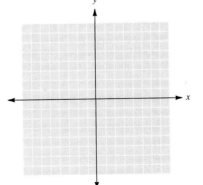

ANSWERS

25. _____

26. _____

27. _____

28. _____

29. _____

30. _____

31. _____

32. _____

33. _____

34. _____

35. _____

36. _____

Graph each of the following equations by first solving for y.

31. $x + 3y = 6$

32. $x - 2y = 6$

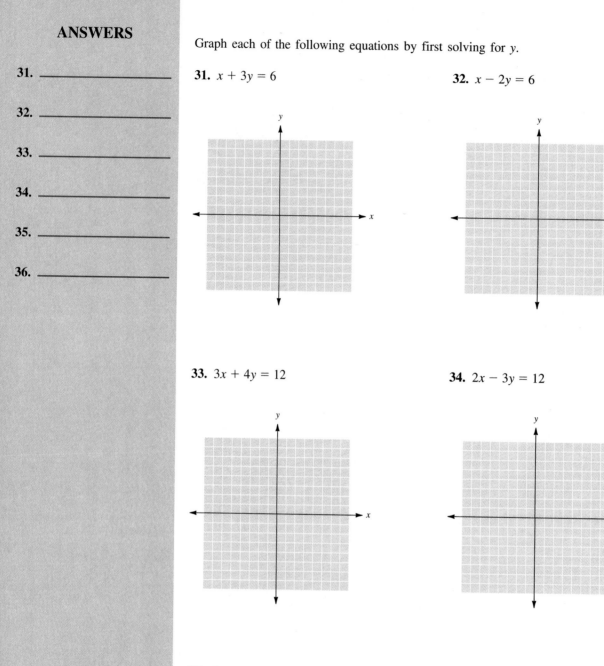

33. $3x + 4y = 12$

34. $2x - 3y = 12$

35. $5x - 4y = 20$

36. $7x + 3y = 21$

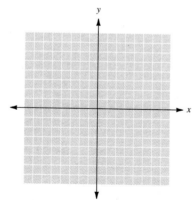
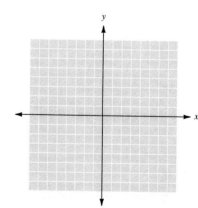

Think About These

Write an equation that describes the following relationships between x and y. Then graph each relationship.

37. y is twice x.

38. y is 3 times x.

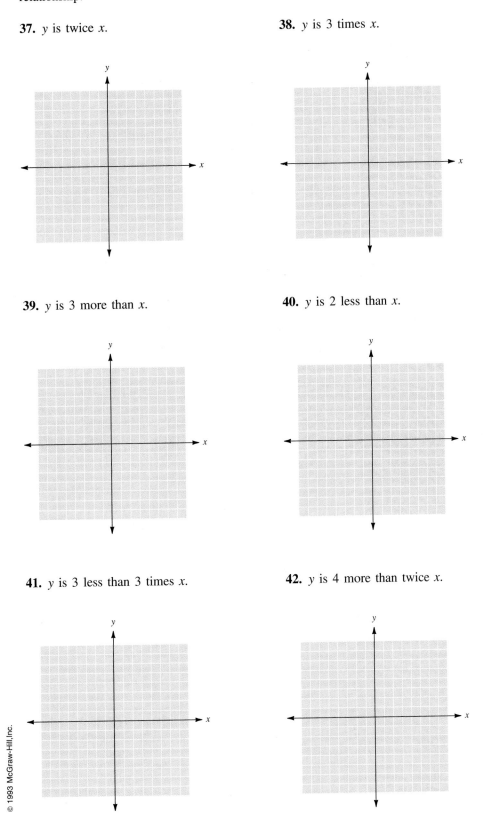

39. y is 3 more than x.

40. y is 2 less than x.

41. y is 3 less than 3 times x.

42. y is 4 more than twice x.

37. _____

38. _____

39. _____

40. _____

41. _____

42. _____

43. _____

44. _____

45. _____

46. _____

47. _____

43. The difference of x and the product of 4 and y is 12.

44. The difference of twice x and y is 6.

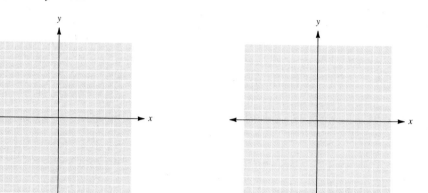

Graph each pair of equations on the same axes. Give the coordinates of the point where the lines intersect.

45. $x + y = 4$
$x - y = 2$

46. $x - y = 3$
$x + y = 5$

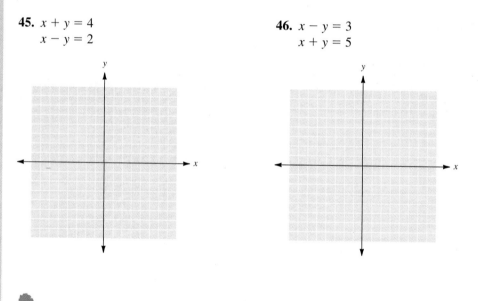

47. The equation $y = 0.10x + 200$ describes the amount of winnings a group earns for collecting plastic jugs in the recycling contest discussed in Section 7.2. Sketch the graph of the line on the axes below.

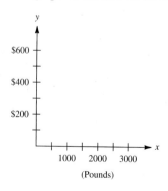

(Pounds)

512

48. The contest sponsor will award a prize only if the winning group in the contest collects 100 lb of jugs or more. Use your graph in Exercise 47 to determine the minimum prize possible.

A Scout troop wants to raise some money by recycling newspapers. They decide to rent a truck for a weekend and to collect the newspapers from homes in the neighborhood. The market price for recycled newsprint is currently $15 dollars per ton.

The equation $y = 15x - 100$ describes the amount of money the troop will make, where y is the amount of money made in dollars, x is the number of tons of newsprint collected, and 100 is the cost in dollars to rent the truck.

49. Using the axes below, draw a graph that represents the relationship between newsprint collected and money earned.

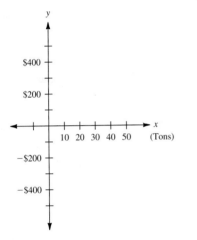

50. Because the truck is costing the troop $100, how many tons of newspapers must the troop collect to break even on this project?

51. If the troop members collect 16 tons of newsprint, how much money will they earn?

52. Six months later the price of newsprint is $17 dollars a ton, and the cost to rent the truck has risen to $125. Write the equation that describes the amount of money the troop might make at that time.

Graph each set of equations on the same coordinate system. Do the lines intersect? What are the *y* intercepts?

53. $y = 3x$
$y = 3x + 4$
$y = 3x - 5$

54. $y = -2x$
$y = -2x + 3$
$y = -2x - 5$

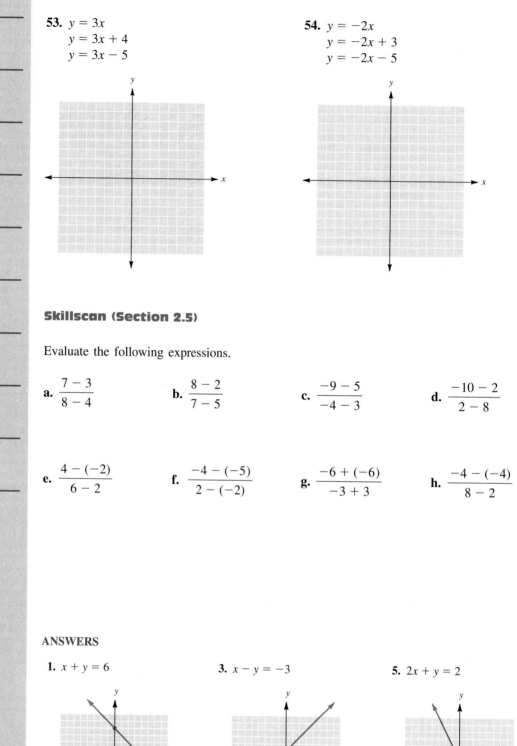

Skillscan (Section 2.5)

Evaluate the following expressions.

a. $\dfrac{7 - 3}{8 - 4}$

b. $\dfrac{8 - 2}{7 - 5}$

c. $\dfrac{-9 - 5}{-4 - 3}$

d. $\dfrac{-10 - 2}{2 - 8}$

e. $\dfrac{4 - (-2)}{6 - 2}$

f. $\dfrac{-4 - (-5)}{2 - (-2)}$

g. $\dfrac{-6 + (-6)}{-3 + 3}$

h. $\dfrac{-4 - (-4)}{8 - 2}$

ANSWERS

1. $x + y = 6$

3. $x - y = -3$

5. $2x + y = 2$

7. $3x + y = 0$

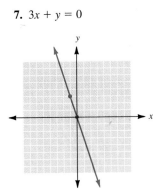

9. $x + 4y = 8$

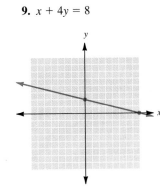

11. $y = 5x$

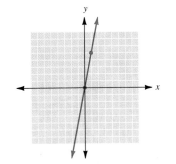

13. $y = 2x - 1$

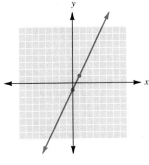

15. $y = -3x + 1$

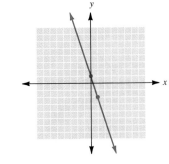

17. $y = \dfrac{1}{3}x$

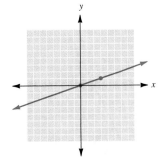

19. $y = \dfrac{2}{3}x - 3$

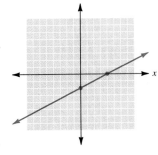

21. $x = 5$

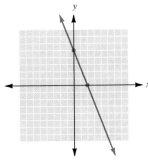

23. $y = 1$

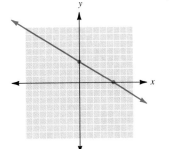

25. $x - 2y = 4$

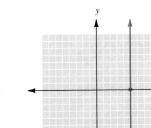

27. $5x + 2y = 10$

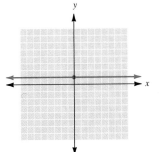

29. $3x + 5y = 15$

31. $y = 2 - \dfrac{x}{6}$

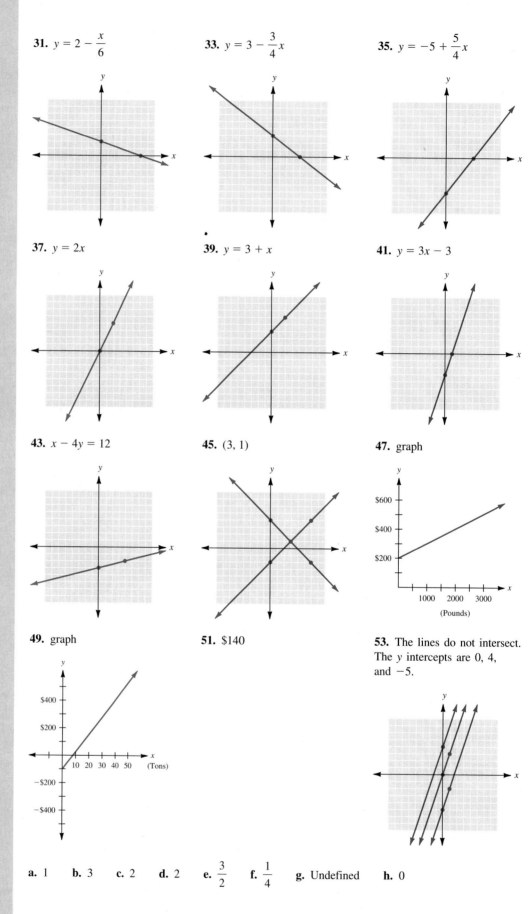

33. $y = 3 - \dfrac{3}{4}x$

35. $y = -5 + \dfrac{5}{4}x$

37. $y = 2x$

39. $y = 3 + x$

41. $y = 3x - 3$

43. $x - 4y = 12$

45. $(3, 1)$

47. graph

(Pounds)

49. graph

51. $140

53. The lines do not intersect. The y intercepts are 0, 4, and -5.

(Tons)

a. 1   **b.** 3   **c.** 2   **d.** 2   **e.** $\dfrac{3}{2}$ **f.** $\dfrac{1}{4}$ **g.** Undefined   **h.** 0

The Slope of a Line

OBJECTIVES
1. To find the slope of a line through two given points
2. To use the slope-intercept form for a line

We saw in Section 7.3 that the graph of an equation such as

$$y = 2x + 3$$

is a straight line. In this section we want to develop an important idea related to the equation of a line and its graph, called the *slope* of a line. Finding the slope of a line gives us a numerical measure of the "steepness" or inclination of that line.

Recall that an equation such as $y = 2x + 3$ is a *linear equation in two variables*. Its graph is always a straight line.

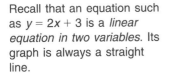

To find the slope of a line, we first let $P(x_1, y_1)$ and $Q(x_2, y_2)$ be any two distinct points on that line. The *horizontal change* (or the change in x) between the points is $x_2 - x_1$. The *vertical change* (or the change in y) between the points is $y_2 - y_1$.

We call the ratio of the vertical change, $y_2 - y_1$, to the horizontal change, $x_2 - x_1$, the *slope* of the line as we move along the line from P to Q. That ratio is usually denoted by the letter m, and so we have the following formula:

Note: x_1 is read "x sub 1," x_2 is read "x sub 2," and so on. The 1 in x_1 and the 2 in x_2 are called *subscripts*.

The difference $x_2 - x_1$ is sometimes called the *run* between points P and Q. The difference $y_2 - y_1$ is called the *rise*. So the slope may be thought of as "rise over run."

THE SLOPE OF A LINE

If $P(x_1, y_1)$ and $Q(x_2, y_2)$ are any two points on a line, then m, the slope of the line, is given by

$$m = \frac{\text{vertical change}}{\text{horizontal change}} = \frac{y_2 - y_1}{x_2 - x_1} \qquad \text{where } x_2 \neq x_1$$

This definition provides exactly the numerical measure of "steepness" that we want. If a line "rises" as we move from left to right, the slope will be positive—the steeper the line, the larger the numerical value of the slope. If the line "falls" from left to right, the slope will be negative.

Let's proceed to some examples.

Example 1

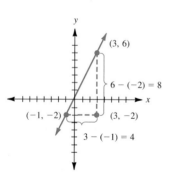

Find the slope of the line containing points with coordinates (1, 2) and (5, 4).

Let $P(x_1, y_1) = (1, 2)$ and $Q(x_2, y_2) = (5, 4)$. By the definition above, we have

$$m = \frac{y_2 - y_1}{x_2 - x_1} = \frac{4 - 2}{5 - 1} = \frac{2}{4} = \frac{1}{2}$$

Note: We would have found the same slope if we had reversed P and Q and subtracted in the other order. In that case, $P(x_1, y_1) = (5, 4)$ and $Q(x_2, y_2) = (1, 2)$, so

$$m = \frac{2 - 4}{1 - 5} = \frac{-2}{-4} = \frac{1}{2}$$

It makes no difference which point is labeled (x_1, y_1) and which is (x_2, y_2); the resulting slope will be the same. You must simply stay with your choice once it is made and *not* reverse the order of the subtraction in your calculations.

CHECK YOURSELF 1

Find the slope of the line containing points with coordinates (2, 3) and (5, 5).

By now you should be comfortable subtracting negative numbers. Let's apply that skill to finding a slope.

Example 2

Find the slope of the line containing points with the coordinates $(-1, -2)$ and $(3, 6)$.

Again, applying the definition, we have

$$m = \frac{6 - (-2)}{3 - (-1)} = \frac{6 + 2}{3 + 1} = \frac{8}{4} = 2$$

The figure on the next page compares the slopes found in the two previous examples. Line l_1, from Example 1, had slope $\frac{1}{2}$. Line l_2, from Example 2, had slope 2. Do you see the idea of slope measuring steepness? The greater the slope, the more steeply the line is inclined upward.

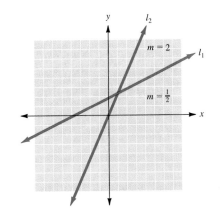

Find the slope of the line containing points with coordinates $(-1, 2)$ and $(2, 7)$. Draw a sketch of this line and the line of Check Yourself 1. Compare the lines and the two slopes.

Let's look at another fact about slope in our next example.

Example 3

Find the slope of the line containing points with coordinates $(-2, 3)$ and $(1, -3)$.

By the definition,

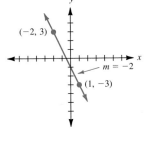

$$m = \frac{-3 - 3}{1 - (-2)} = \frac{-6}{3} = -2$$

This line has a *negative* slope. The line *falls* as we move from left to right.

Find the slope of the line containing points with coordinates $(-1, 3)$ and $(1, -3)$.

A line with a slope of 0 is especially important in mathematics.

Example 4

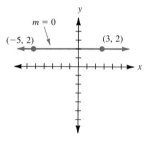

Find the slope of the line containing points with coordinates $(-5, 2)$ and $(3, 2)$.

By the definition,

$$m = \frac{2 - 2}{3 - (-5)} = \frac{0}{8} = 0$$

The slope of the line is 0. In fact, that will be the case for any horizontal line. Since any two points on the line have the same y coordinate, the vertical change $y_2 - y_1$ must always be 0, and so the resulting slope is 0.

CHECK YOURSELF 4

Find the slope of the line containing points with coordinates $(-2, -4)$ and $(3, -4)$.

Since division by 0 is undefined, it is possible to have a line with an undefined slope.

Example 5

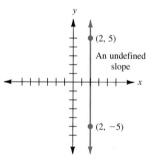

Find the slope of the line containing points with coordinates $(2, -5)$ and $(2, 5)$.

By the definition,

$$m = \frac{5 - (-5)}{2 - 2} = \frac{10}{0} \qquad \text{which is undefined}$$

We say that the vertical line has an undefined slope. On a vertical line, any two points have the same x coordinate. This means that the horizontal change $x_2 - x_1$ must always be 0 and since division by 0 is undefined, the slope of a vertical line will always be undefined.

CHECK YOURSELF 5

Find the slope of the line containing points with the coordinates $(-3, -5)$ and $(-3, 2)$.

The following sketch will summarize the results of the previous examples.

As the slope gets closer to 0, the line gets "flatter."

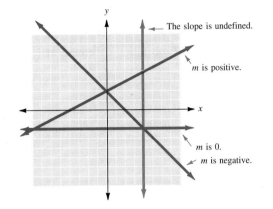

The slope is undefined.

m is positive.

m is 0.

m is negative.

Four lines are illustrated in the figure. Note that

1. The slope of a line that rises from left to right is positive.
2. The slope of a line that falls from left to right is negative.
3. The slope of a horizontal line is 0.
4. A vertical line has an undefined slope.

We now want to consider finding the equation of a line when its slope and *y* intercept are known. Suppose that the *y* intercept of a line is *b*. Then the point at which the line crosses the *y* axis must have coordinates $(0, b)$. Look at the sketch below.

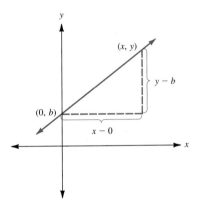

Now, using any other point $(x, \ y)$ on the line and using our definition of slope, we can write

Change in *y*.

$$m = \frac{y - b}{x - 0}$$ (1)

Change in *x*.

or

$$m = \frac{y - b}{x}$$ (2)

Multiplying both sides of equation (2) by x, we have

$$mx = y - b \tag{3}$$

Finally, adding b to both sides of equation (3) gives

$$mx + b = y$$

or

$$y = mx + b \tag{4}$$

We can summarize the above discussion as follows:

In this form, the equation is *solved for y*. The coefficient of x will give you the slope of the line, and the constant term gives the y intercept.

> **THE SLOPE-INTERCEPT FORM FOR A LINE**
>
> An equation of the line with slope m and y intercept b is
>
> $$y = mx + b$$

Example 6

(*a*) Find the slope and y intercept for the graph of the equation

$$y = 3x + 4$$
$$\quad\;\; m \quad\; b$$

The graph has slope 3 and y intercept 4.

(*b*) Find the slope and y intercept for the graph of the equation

$$y = -\frac{2}{3}x - 5$$
$$\qquad\; m \qquad b$$

The slope of the line is $-\dfrac{2}{3}$; the y intercept is -5.

CHECK YOURSELF 6

Find the slope and y intercept for the graph of each of the following equations.

1. $y = -3x - 7$

2. $y = \dfrac{3}{4}x + 5$

As the following example illustrates, we may have to solve for y as the first step in determining the slope and the y intercept for the graph of an equation.

Example 7

Find the slope and y intercept for the graph of the equation

$$3x + 2y = 6$$

First, we must solve the equation for y.

If we write the equation as
$$y = \frac{-3x + 6}{2}$$
it is more difficult to identify the slope and the intercept.

$$3x + 2y = 6$$

$$2y = -3x + 6 \qquad \text{Subtract } 3x \text{ from both sides.}$$

$$y = -\frac{3}{2}x + 3 \qquad \text{Divide each term by 2.}$$

The equation is now in slope-intercept form. The slope is $-\dfrac{3}{2}$, and the y intercept is 3.

CHECK YOURSELF 7

Find the slope and y intercept for the graph of the equation

$$2x - 5y = 10$$

As we mentioned earlier, knowing certain properties of a line (namely, its slope and y intercept) will also allow us to write the equation of the line by using the slope-intercept form. The following example illustrates.

Example 8

(a) Write the equation of a line with slope 3 and y intercept 5.

We know that $m = 3$ and $b = 5$. Using the slope-intercept form, we have

$$y = 3x + 5$$
$$\quad\; \uparrow \quad\;\; \uparrow$$
$$\quad\; m \quad\;\; b$$

which is the desired equation.

(b) Write the equation of a line with slope $-\dfrac{3}{4}$ and y intercept -3.

We know that $m = -\dfrac{3}{4}$ and $b = -3$. In this case,

$$y = -\overset{\overset{\displaystyle m}{\downarrow}}{\frac{3}{4}} x + \overset{\overset{\displaystyle b}{\downarrow}}{(-3)}$$

or

$$y = -\frac{3}{4}x - 3$$

which is the desired equation.

CHECK YOURSELF 8

Write the equation of a line

1. With slope -2 and y intercept 7

2. With slope $\dfrac{2}{3}$ and y intercept -3

    We can also use the slope and y intercept of a line in drawing its graph. Consider the following example.

Example 9

Graph the line with slope $\dfrac{2}{3}$ and y intercept 2.

Since the y intercept is 2, we begin by plotting the point $(0, 2)$. Now since the horizontal change (or run) is 3, we move 3 units to the right *from that y intercept.* Then since the vertical change (or rise) is 2, we move 2 units up to locate another point on the desired graph. Note that we will have located that second point at $(3, 4)$. The final step is to simply draw a line through that point and the y intercept.

Note:

$$m = \frac{2}{3} = \frac{\text{rise}}{\text{run}}$$

The line rises from left to right because the slope is positive.

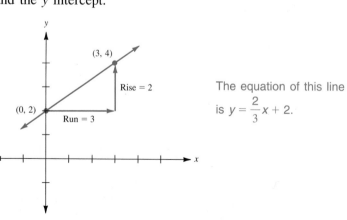

The equation of this line is $y = \dfrac{2}{3}x + 2$.

CHECK YOURSELF 9

Graph the equation of a line with slope $\dfrac{3}{5}$ and y intercept -2.

The following algorithm summarizes the use of graphing by the slope-intercept form.

GRAPHING BY USING THE SLOPE-INTERCEPT FORM

1. Write the original equation of the line in slope-intercept form.
2. Determine the slope m and the y intercept b.
3. Plot the y intercept at $(0, b)$.
4. Use m (the change in y over the change in x) to determine a second point on the desired line.
5. Draw a line through the two points determined above to complete the graph.

You have now seen two methods for graphing lines: the slope-intercept method (Section 7.4) and the intercept method (Section 7.3). When you are to graph a linear equation, you should first decide which is the appropriate method.

Example 10

Between the intercept method and the slope-intercept method, decide which would be more appropriate for graphing each equation.

(a) $2x - 5y = 10$

Because both intercepts are easy to find, you should choose the intercept method to graph this equation.

(b) $2x + y = 6$

This equation can be quickly graphed by either method. As it is written, you might choose the intercept method. It can, however, be rewritten as $y = -2x + 6$. In that case the slope-intercept method is more appropriate.

(c) $y = \dfrac{1}{4}x - 4$

Since the equation is in slope-intercept form, that is the more appropriate method to choose.

CHECK YOURSELF 10

Between the intercept method and the slope-intercept method, which would be more appropriate for graphing each equation?

1. $x + y = -2$ **2.** $3x - 2y = 12$ **3.** $y = -\dfrac{1}{2}x - 6$

CHECK YOURSELF ANSWERS

1. $m = \dfrac{2}{3}$.   **2.** $m = \dfrac{5}{3}$.

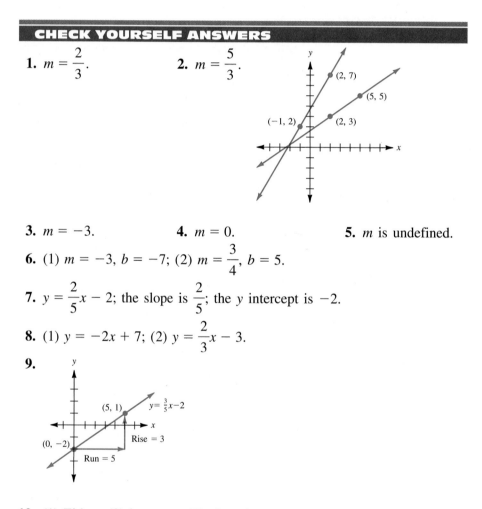

3. $m = -3$.   **4.** $m = 0$.   **5.** m is undefined.

6. (1) $m = -3$, $b = -7$; (2) $m = \dfrac{3}{4}$, $b = 5$.

7. $y = \dfrac{2}{5}x - 2$; the slope is $\dfrac{2}{5}$; the y intercept is -2.

8. (1) $y = -2x + 7$; (2) $y = \dfrac{2}{3}x - 3$.

9.

10. (1) Either; (2) intercept; (3) slope-intercept.

Name

Date

Build Your Skills

Find the slope of the line through the following pairs of points.

1. $(2, 3)$ and $(5, 6)$

2. $(3, 1)$ and $(7, 9)$

3. $(-1, -2)$ and $(1, 8)$

4. $(-2, 1)$ and $(0, 11)$

5. $(-2, 3)$ and $(3, 7)$

6. $(-3, -4)$ and $(3, -2)$

7. $(-3, 2)$ and $(2, -8)$

8. $(-6, 1)$ and $(2, -7)$

9. $(3, 3)$ and $(5, 0)$

10. $(-2, 4)$ and $(3, 1)$

11. $(5, -4)$ and $(5, 2)$

12. $(-5, 4)$ and $(2, 4)$

13. $(-4, -2)$ and $(3, 3)$

14. $(-5, -3)$ and $(-5, 2)$

15. $(-3, -4)$ and $(2, -4)$

16. $(-5, 7)$ and $(2, -2)$

17. $(-1, 7)$ and $(2, 3)$

18. $(-4, -2)$ and $(6, 4)$

Find the slope and y intercept of the line represented by each of the following equations.

19. $y = 2x + 3$

20. $y = -4x + 2$

21. $y = -5x - 4$

22. $y = 6x - 3$

ANSWERS

1. _____
2. _____
3. _____
4. _____
5. _____
6. _____
7. _____
8. _____
9. _____
10. _____
11. _____
12. _____
13. _____
14. _____
15. _____
16. _____
17. _____
18. _____
19. _____
20. _____
21. _____
22. _____

23. $y = \dfrac{3}{4}x + 1$

24. $y = -4x$

25. $y = \dfrac{2}{3}x$

26. $y = -\dfrac{3}{5}x - 2$

27. $4x + 3y = 12$

28. $2x + 5y = 10$

29. $y = 9$

30. $2x - 3y = 6$

31. $3x - 2y = 8$

32. $x = 5$

Write the equation of the line with given slope and y intercept. Then graph each line, using the slope and y intercept.

33. $m = 3$, $b = 5$

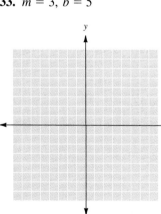

34. $m = -2$, $b = 4$

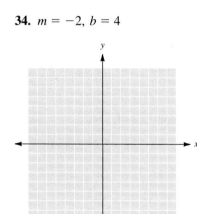

35. $m = -3$, $b = 4$

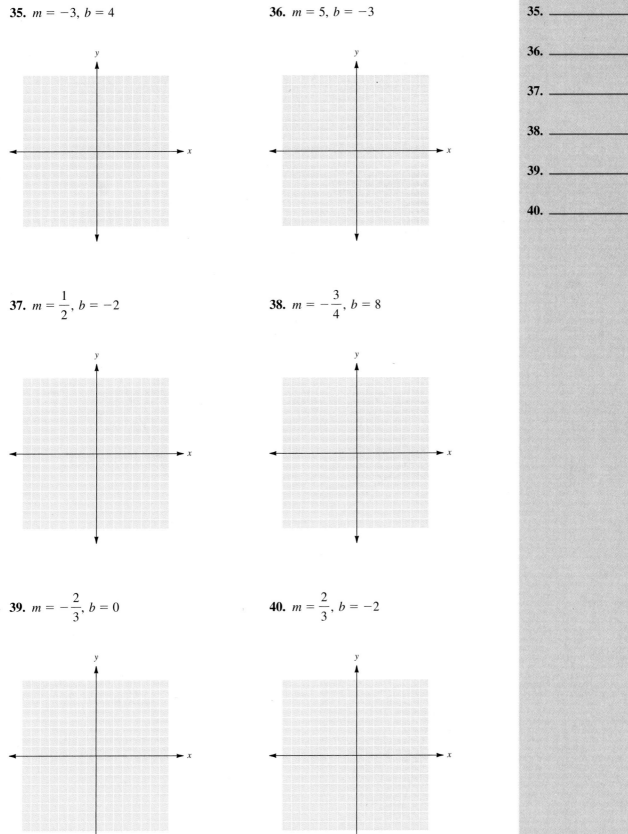

36. $m = 5$, $b = -3$

37. $m = \dfrac{1}{2}$, $b = -2$

38. $m = -\dfrac{3}{4}$, $b = 8$

39. $m = -\dfrac{2}{3}$, $b = 0$

40. $m = \dfrac{2}{3}$, $b = -2$

ANSWERS

35. _____

36. _____

37. _____

38. _____

39. _____

40. _____

41. $m = \dfrac{3}{4},\ b = 3$

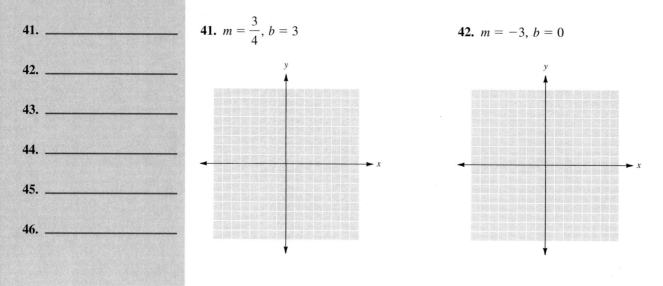

42. $m = -3,\ b = 0$

In Exercises 43 to 50, match the graph with one of the equations on the left.

(a) $y = 2x$

(b) $y = x + 1$

(c) $y = -x + 3$

(d) $y = 2x + 1$

(e) $y = -3x - 2$

(f) $y = \dfrac{2}{3}x + 1$

(g) $y = -\dfrac{3}{4}x + 1$

(h) $y = -4x$

43.

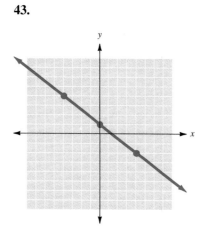

44.

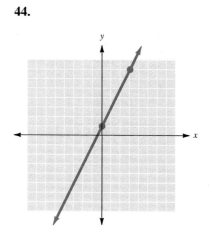

45.

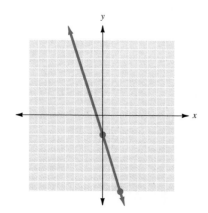

46.

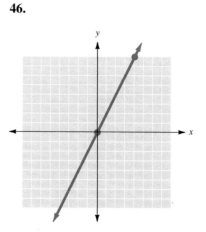

47.

48.

49.

50.

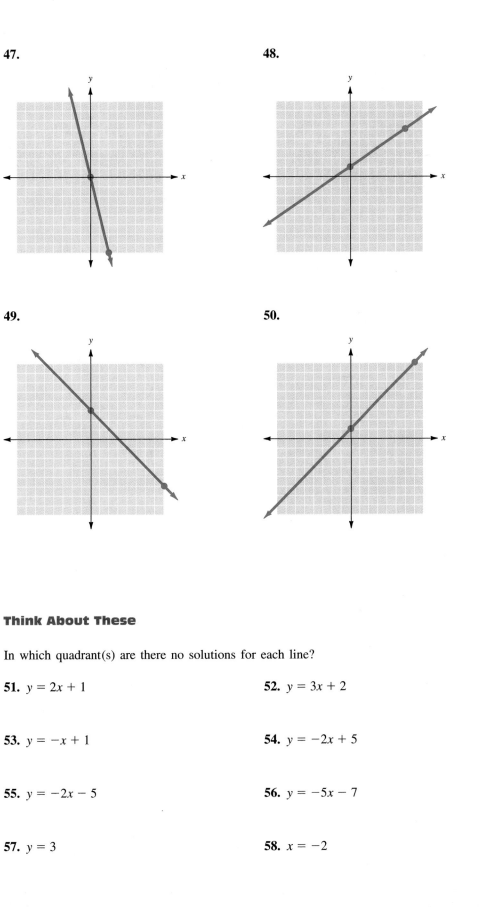

Think About These

In which quadrant(s) are there no solutions for each line?

51. $y = 2x + 1$

52. $y = 3x + 2$

53. $y = -x + 1$

54. $y = -2x + 5$

55. $y = -2x - 5$

56. $y = -5x - 7$

57. $y = 3$

58. $x = -2$

ANSWERS

47. _____

48. _____

49. _____

50. _____

51. _____

52. _____

53. _____

54. _____

55. _____

56. _____

57. _____

58. _____

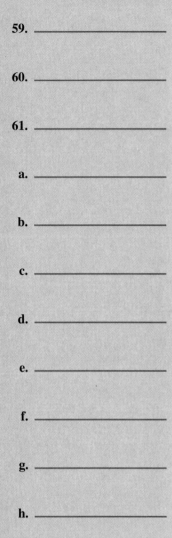

59. The equation $y = 0.10x + 200$ was used in Section 7.3 to describe the award money in a recycling contest. What are the slope and the y intercept for this equation? What does the slope of the line represent in the equation? What does the y intercept represent?

60. The equation $y = 15x - 100$ was used in Section 7.3 to describe the amount of money a Scout troop might earn from a paper drive. What are the slope and the y intercept for this equation?

61. In the equation in Exercise 60, what does the slope of the line represent to the Scout troop? What does the y intercept represent?

Skillscan (Section 3.7)

Graph each of the following inequalities.

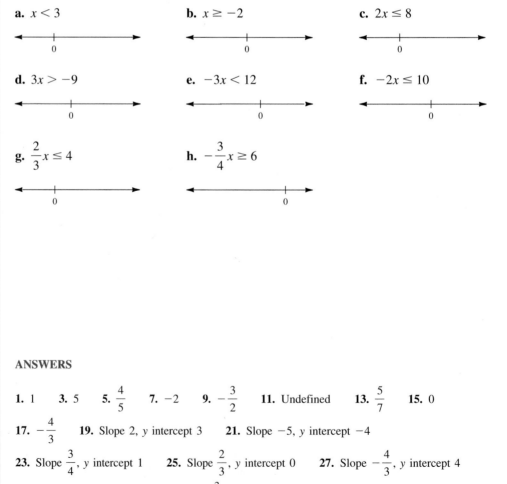

a. $x < 3$

b. $x \geq -2$

c. $2x \leq 8$

d. $3x > -9$

e. $-3x < 12$

f. $-2x \leq 10$

g. $\dfrac{2}{3}x \leq 4$

h. $-\dfrac{3}{4}x \geq 6$

ANSWERS

1. 1   **3.** 5   **5.** $\dfrac{4}{5}$ **7.** -2 **9.** $-\dfrac{3}{2}$ **11.** Undefined   **13.** $\dfrac{5}{7}$ **15.** 0

17. $-\dfrac{4}{3}$ **19.** Slope 2, y intercept 3   **21.** Slope -5, y intercept -4

23. Slope $\dfrac{3}{4}$, y intercept 1   **25.** Slope $\dfrac{2}{3}$, y intercept 0   **27.** Slope $-\dfrac{4}{3}$, y intercept 4

29. Slope 0, y intercept 9   **31.** Slope $\dfrac{3}{2}$, y intercept -4

33. $y = 3x + 5$

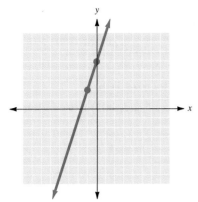

35. $y = -3x + 4$

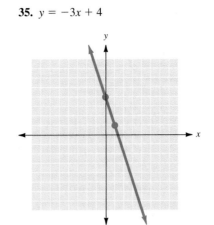

37. $y = \dfrac{1}{2}x - 2$

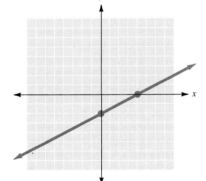

39. $y = -\dfrac{2}{3}x$

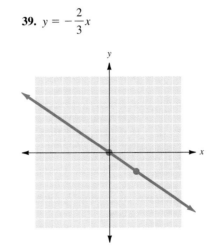

41. $y = \dfrac{3}{4}x + 3$

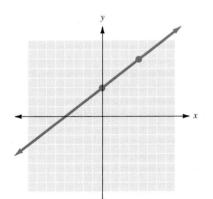

43. *g*   **45.** *e*   **47.** *h*   **49.** *c*   **51.** IV   **53.** III   **55.** I   **57.** III and IV

59. *m* = 0.10, the market price of the jugs; the *y* intercept is 200, the reward.

61. Slope represents the price of newsprint; the *y* intercept the cost of the truck.

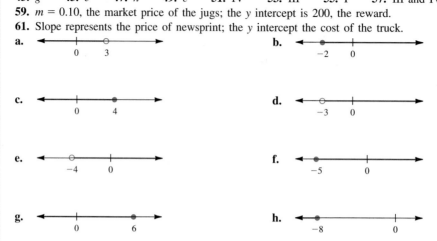

7.5 Graphing Linear Inequalities

OBJECTIVE

To graph a linear inequality in two variables

In Section 3.6 you learned to graph inequalities in one variable on a number line. We now want to extend our work with graphing to include linear inequalities in two variables.

First, consider a definition.

The inequality symbols ≤, >, and ≥ can also be used.

> An inequality that can be written in the form
>
> $Ax + By < C$
>
> where A and B are not both 0, is called a *linear inequality in two variables.*

Some examples of linear inequalities in two variables are

$$x + 3y > 6 \qquad y \le 3x + 1 \qquad 2x - y \ge 3$$

The *graph* of a linear inequality is always a region (actually a half-plane) of the plane whose boundary is a straight line. Let's look at an example of graphing such an inequality.

Example 1

Graph $2x + y < 4$.

First, replace the inequality symbol ($<$) with an equals sign. We then have $2x + y = 4$. This equation forms the *boundary line* of the graph of the original inequality. You can graph the line by any of the methods discussed earlier in this chapter.

Note: When equality is *not included* ($<$ or $>$), use a *dotted line* for the graph of the boundary line. This means that the line is not included in the graph of the linear inequality.

The boundary line for our inequality is shown below.

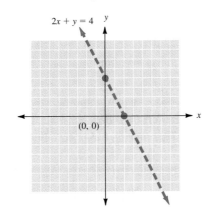

You can always use the origin for a test point unless the boundary line passes through the origin.

We see that the boundary line separates the plane into two regions, each of which is called a *half plane.*

We now need to choose the correct half plane. Choose any convenient test point not on the boundary line. The origin $(0, 0)$ is a good choice because it makes for easy calculation.

Substitute $x = 0$ and $y = 0$ into the inequality.

$$2 \cdot 0 + 0 < 4$$
$$0 + 0 < 4$$
$$0 < 4 \qquad \text{A true}$$
$$\text{statement}$$

Since the inequality is *true* for the test point, we shade the half plane containing that test point (here the origin). The origin and all other points *below* the boundary line then represent solutions for our original inequality.

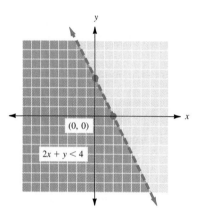

CHECK YOURSELF 1

Graph the inequality $x + 3y < 3$.

The process is similar when the boundary line is included in the solution.

Example 2

Graph $4x - 3y \geq 12$.

Again we replace the inequality symbol ($\geq$) with an equals sign to write the equation for our boundary line.

First, graph the boundary line, $4x - 3y = 12$.

Note: When equality *is included* ($\leq$ or $\geq$), use a *solid line* for the graph of the boundary line. This means the line is included in the graph of the linear inequality.

The graph of our boundary line (a solid line here) is shown below.

Although any of our graphing methods can be used here, the intercept method is probably the most efficient.

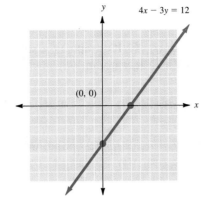

Again, we use $(0, 0)$ as a convenient test point. Substituting 0 for x and for y in the original inequality, we have

$$4 \cdot 0 - 3 \cdot 0 \geq 12$$
$$0 \geq 12 \quad \Big\} \quad \text{A false statement}$$

Since the inequality is *false* for the test point, we shade the half plane that does *not* contain that test point, here $(0, 0)$.

All points *on and below* the boundary line represent solutions for our original inequality.

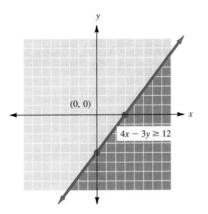

CHECK YOURSELF 2

Graph the inequality $3x + 2y \geq 6$.

Some applications will involve horizontal or vertical boundary lines.

Example 3

Graph $x \leq 5$.

The boundary line is $x = 5$. Its graph is a solid line because equality is included. Using $(0, 0)$ as a test point, we substitute 0 for x with the result

$0 \leq 5$ } A true statement

Since the inequality is *true* for the test point, we shade the half plane containing the origin.

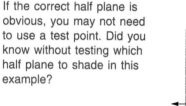

If the correct half plane is obvious, you may not need to use a test point. Did you know without testing which half plane to shade in this example?

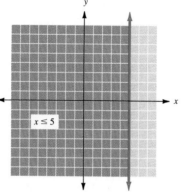

$x \leq 5$

CHECK YOURSELF 3

Graph the inequality $y < 2$.

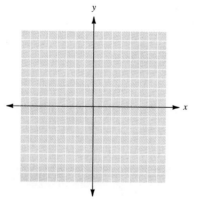

As we mentioned earlier, we may have to use a point other than the origin as our test point. The following example illustrates.

Example 4

Graph $2x + 3y < 0$.

The boundary line is $2x + 3y = 0$. Its graph is shown below.

We use a dotted line for our boundary line since equality is not included.

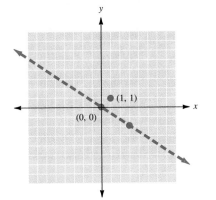

We cannot use $(0, 0)$ as our test point in this case. Do you see why?

Choose any other point *not* on the line. For instance, we have picked $(1, 1)$ as a test point. Substituting 1 for x and 1 for y gives

$$2 \cdot 1 + 3 \cdot 1 < 0$$
$$2 + 3 < 0$$
$$5 < 0 \quad \}\quad \text{A false statement}$$

Since the inequality is *false* at our test point, we shade the half plane *not* containing $(1, 1)$. This is shown in the following.

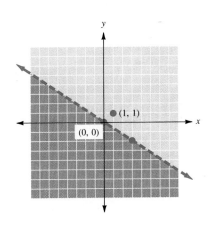

CHECK YOURSELF 4

Graph the inequality $x - 2y < 0$.

The following steps summarize our work in graphing linear inequalities in two variables.

> **TO GRAPH A LINEAR INEQUALITY**
>
> **STEP 1** Replace the inequality symbol with an equals sign to form the equation of the boundary line of the graph.
>
> **STEP 2** Graph the boundary line. Use a dotted line if equality is not included ($<$ or $>$). Use a solid line if equality is included ($\leq$ or $\geq$).
>
> **STEP 3** Choose any convenient test point *not* on the line.
>
> **STEP 4** If the inequality is *true* at the checkpoint, shade the half plane including the test point. If the inequality is *false* at the checkpoint, shade the half plane not including the test point.

CHECK YOURSELF ANSWERS

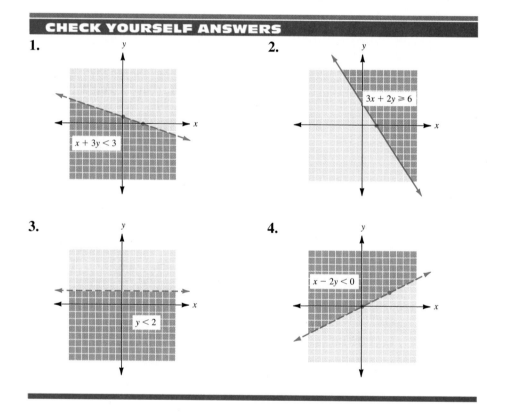

1.

$x + 3y < 3$

2.

$3x + 2y \geq 6$

3.

$y < 2$

4.

$x - 2y < 0$

Name

Date

Build Your Skills

In Exercises 1 to 8, we have graphed the boundary line for the linear inequality. Determine the correct half plane in each case, and complete the graph.

1. $x + y < 5$

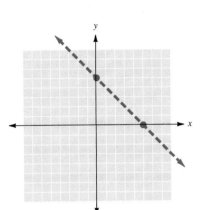

2. $x - y \geq 4$

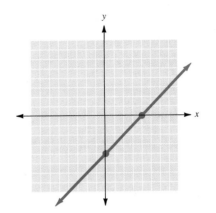

3. $x - 2y \geq 4$

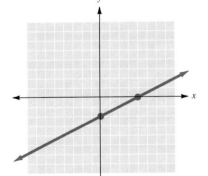

4. $2x + y < 6$

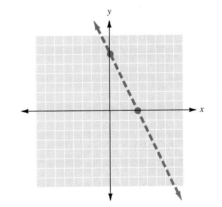

ANSWERS

1. _____

2. _____

3. _____

4. _____

5. $x \le -3$

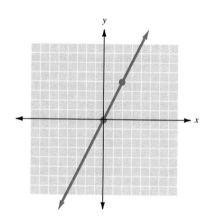

6. $y \ge 2x$

7. $y < 2x - 6$

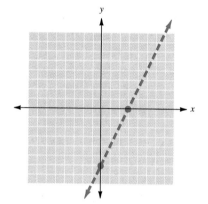

8. $y > 3$

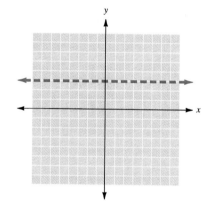

Graph each of the following inequalities.

9. $x + y < 3$

10. $x - y \ge 4$

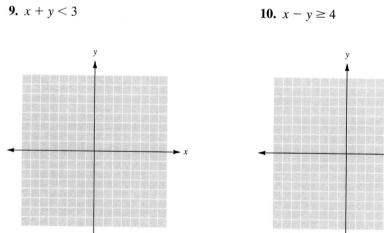

11. $x - y \leq 5$

12. $x + y > 5$

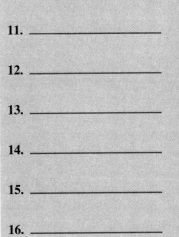
13. $2x + y < 6$

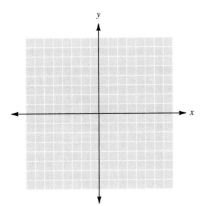

14. $3x + y \geq 6$

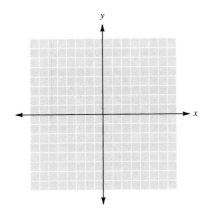

15. $x \leq 3$

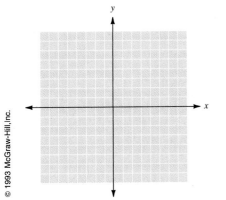

16. $4x + y \geq 4$

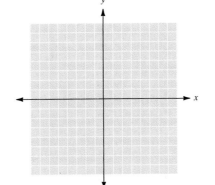

17. _____

18. _____

19. _____

20. _____

21. _____

22. _____

17. $x - 5y < 5$

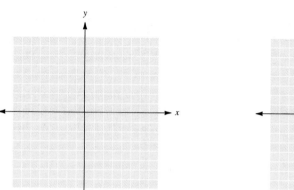

18. $y > 3$

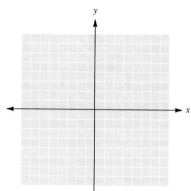

19. $y < -4$

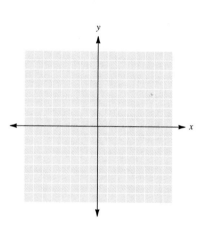

20. $4x + 3y > 12$

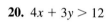

21. $2x - 3y \geq 6$

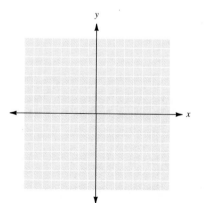

22. $x \geq -2$

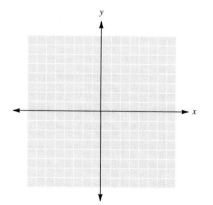

23. $3x + 2y \geq 0$

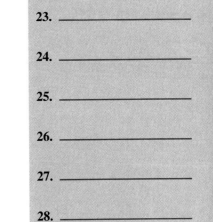

24. $3x + 5y < 15$

25. $5x + 2y > 10$

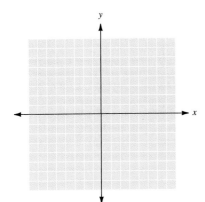

26. $x - 3y \geq 0$

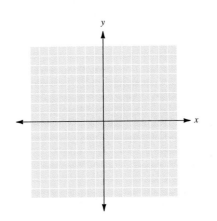

27. $y \leq 2x$

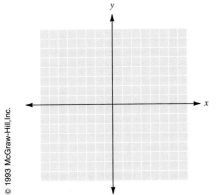

28. $3x - 4y < 12$

ANSWERS

23. _____

24. _____

25. _____

26. _____

27. _____

28. _____

ANSWERS

29. _____

30. _____

31. _____

32. _____

33. _____

34. _____

29. $y > 2x - 3$

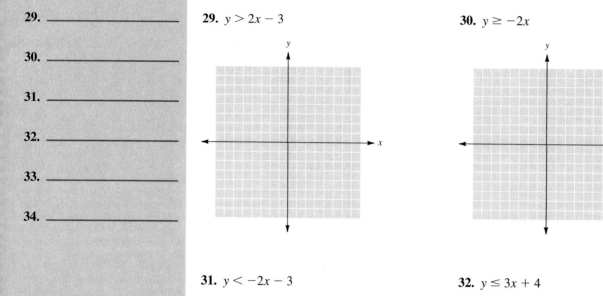

30. $y \geq -2x$

31. $y < -2x - 3$

32. $y \leq 3x + 4$

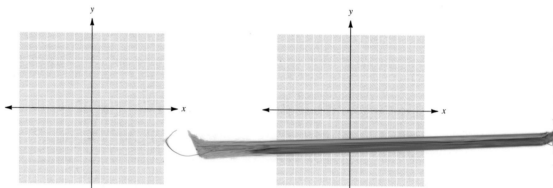

Think About These

Graph each of the following inequalities.

33. $2(x + y) - x > 6$

34. $3(x + y) - 2y < 3$

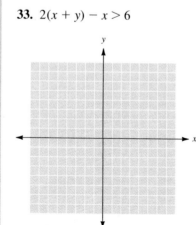

35. $4(x + y) - 3(x + y) \leq 5$

36. $5(2x + y) - 4(2x + y) \geq 4$

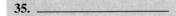

ANSWERS

1. $x + y < 5$

3. $x - 2y \geq 4$

5. $x \leq -3$

7. $y < 2x - 6$

9. $x + y < 3$

11. $x - y \leq 5$

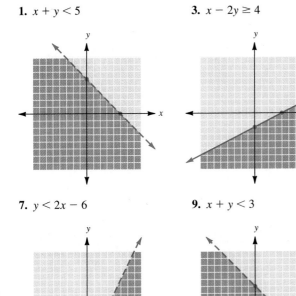

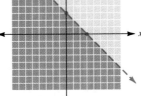

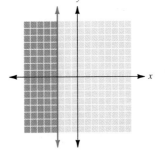

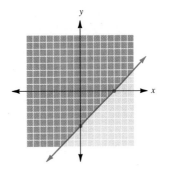

13. $2x + y < 6$

15. $x \le 3$

17. $x - 5y < 5$

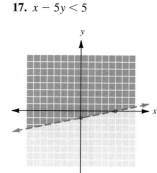

19. $y < -4$

21. $2x - 3y \ge 6$

23. $3x + 2y \ge 0$

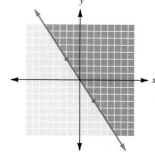

25. $5x + 2y > 10$

27. $y \le 2x$

29. $y > 2x - 3$

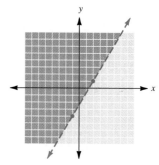

31. $y < -2x - 3$

33. $x + 2y > 6$

35. $x + y \le 5$

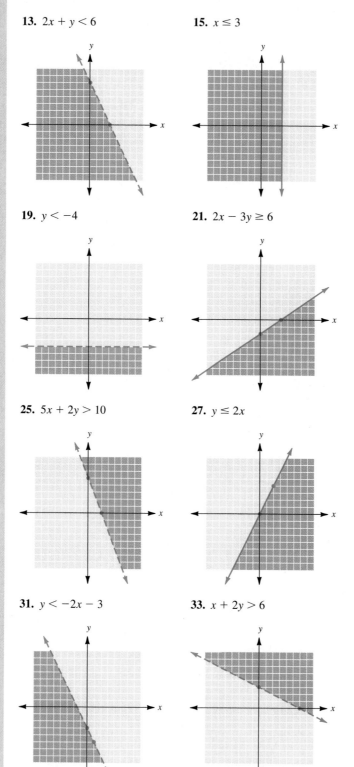
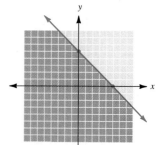

Summary

The Rectangular Coordinate System [7.1 and 7.2]

If $2x - y = 10$, (6, 2) is a solution for the equation, because substituting 6 for x and 2 for y gives a true statement.

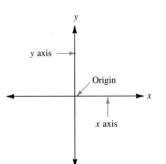

Solutions of Linear Equations   A pair of values that satisfy the equation. Solutions for linear equations in two variables are written as *ordered pairs*. An ordered pair has the form

$$(x, y)$$

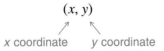

x coordinate   *y* coordinate

The Rectangular Coordinate System   A system formed by two perpendicular axes which intersect at a point called the *origin*. The horizontal line is called the *x axis*. The vertical line is called the *y axis*.

To graph the point corresponding to (2, 3):

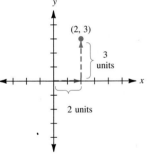

Graphing Points from Ordered Pairs   The coordinates of an ordered pair allow you to associate a point in the plane with every ordered pair.
   To graph a point in the plane,

1. Start at the origin.
2. Move right or left according to the value of the x coordinate, to the right if x is positive or to the left if x is negative.
3. Then move up or down according to the value of the y coordinate, up if y is positive and down if y is negative.

Midpoint   The midpoint of A and B is the point on line $\overleftrightarrow{AB}$ that is equal distance from A and B.

Graphing Linear Equations [7.3]

$2x + 3y = 4$ is a linear equation.

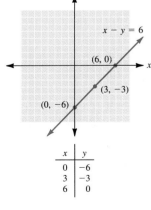

x	y
0	−6
3	−3
6	0

Linear Equation   An equation that can be written in the form

$$Ax + By = C$$

where A and B are not both 0.

Graphing Linear Equations

1. Find at least three solutions for the equation, and put your results in tabular form.
2. Graph the solutions found in step 1.
3. Draw a straight line through the points determined in step 2 to form the graph of the equation.

x and *y* **Intercepts of a Line**   The *x intercept* is the *x* coordinate of the point where the line intersects the *x* axis. The *y intercept* is the *y* coordinate of the point where the line intersects the *y* axis.

If $2x + 4y = 8$, $(4, 0)$ gives the x intercept and $(0, 2)$ gives the y intercept.
To graph $2x + 4y = 8$:

Graphing by the Intercept Method

1. Find the x intercept by letting $y = 0$. Then solve for x.
2. Find the y intercept by letting $x = 0$. Then solve for y.
3. Graph the x and y intercepts.
4. Draw a straight line through the intercepts.

Graphing by Solving for y

To graph $x + 2y = 6$, we can solve for y.

$$y = -\frac{1}{2}x + 3$$

1. Solve the given equation for y.
2. Use the equivalent equation (solved for y) to determine solutions.
3. Graph as before.

Some solutions are

x	y
-2	4
0	3
2	2

The Slope of a Line [7.4]

To find the slope of the line through $(-2, -3)$ and $(4, 6)$,

$$m = \frac{6 - (-3)}{4 - (-2)}$$

$$= \frac{6 + 3}{4 + 2}$$

$$= \frac{9}{6} = \frac{3}{2}$$

Slope   The slope of a line gives a numerical measure of the steepness of the line. The slope m of a line containing the distinct points in the plane $P(x_1, y_1)$ and $Q(x_2, y_2)$ is given by

$$m = \frac{y_2 - y_1}{x_2 - x_1} \qquad \text{where } x_2 \neq x_1$$

For the equation

$$y = \frac{2}{3}x - 3$$

the slope m is $\frac{2}{3}$ and b, the y intercept, is -3.

Slope-Intercept Form   The slope-intercept form for the equation of a line is

$$y = mx + b$$

where the line has slope m and y intercept b.

Graphing Linear Inequalities [7.5]

The Graphing Steps

To graph $x - 2y < 4$:
$x - 2y = 4$ is the boundary line. Using $(0, 0)$ as the checkpoint, we have

$0 - 2 \cdot 0 < 4$

$\qquad 0 < 4$ \qquad (true)

Shade the half plane that includes $(0, 0)$.

1. Replace the inequality symbol with an equals sign to form the equation of the boundary line of the graph.
2. Graph the boundary line. Use a dotted line if equality is not included ($<$ or $>$). Use a solid line if equality is included ($\leq$ or $\geq$).
3. Choose any convenient test point not on the line.
4. If the inequality is *true* at the checkpoint, shade the half plane including the test point. If the inequality is *false* at the checkpoint, shade the half plane that does not include the checkpoint.

Summary Exercises Chapter 7

This summary exercise set is provided to give you practice with each of the objectives of the chapter. Each exercise is keyed to the appropriate chapter section. The answers are provided in the instructor's manual.

[7.1] Determine which of the ordered pairs are solutions for the given equations.

1. $x - y = 6$ $(6, 0), (3, 3), (3, -3), (0, -6)$

2. $2x + y = 8$ $(4, 0), (2, 2), (2, 4), (4, 2)$

3. $2x + 3y = 6$ $(3, 0), (6, 2), (-3, 4), (0, 2)$

4. $2x - 5y = 10$ $(5, 0), \left(\dfrac{5}{2}, -1\right), \left(2, \dfrac{2}{5}\right), (0, -2)$

[7.1] Complete the ordered pairs so that each is a solution for the given equation.

5. $x + y = 8$ $(4, \ ), (\ , 8), (8, \ ), (6, \ )$

6. $x - 2y = 10$ $(0, \ ), (12, \ ), (\ , -2), (8, \ )$

7. $2x + 3y = 6$ $(3, \ ), (6, \ ), (\ , -4), (-3, \ )$

8. $y = 3x + 4$ $(2, \ ), (\ , 7), \left(\dfrac{1}{3}, \ \right), \left(\dfrac{4}{3}, \ \right)$

[7.1] Find four solutions for each of the following equations.

9. $x + y = 10$ **10.** $2x + y = 8$

11. $2x - 3y = 6$ **12.** $y = -\dfrac{3}{2}x + 2$

[7.2] Give the coordinates of the points graphed below.

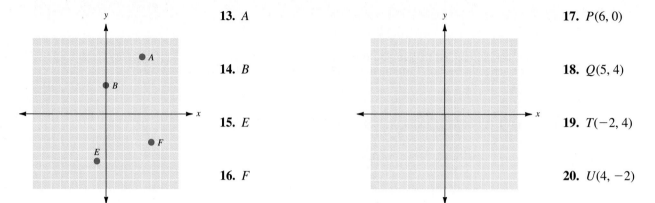

13. A

14. B

15. E

16. F

[7.2] Plot points with the coordinates shown.

17. $P(6, 0)$

18. $Q(5, 4)$

19. $T(-2, 4)$

20. $U(4, -2)$

[7.2] Find the midpoint for each pair of points.

21. $(0, 3)$ and $(0, 9)$

22. $(2, 4)$ and $(-6, 8)$

23. $(1, 5)$ and $(-3, -5)$

24. $(2, -7)$ and $(-3, -1)$

[7.3] Graph each of the following equations.

25. $x + y = 5$

26. $x - y = 6$

27. $y = 2x$

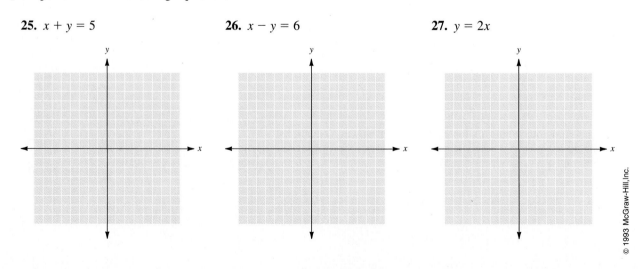

28. $y = -3x$

29. $y = \dfrac{3}{2}x$

30. $y = 3x + 2$

31. $y = 2x - 3$

32. $y = -3x + 4$

33. $y = \dfrac{2}{3}x + 2$

34. $3x - y = 3$

35. $2x + y = 6$

36. $3x + 2y = 12$

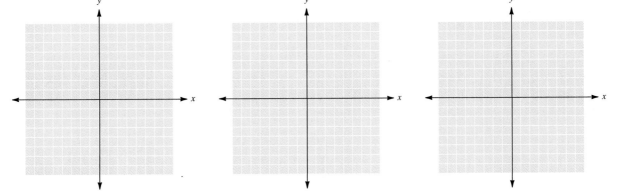

37. $3x - 4y = 12$ **38.** $x = 3$ **39.** $y = -2$

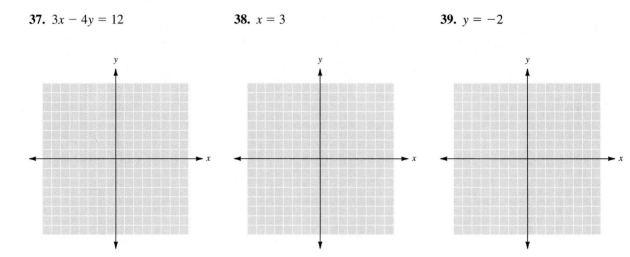

[7.3] Graph each of the following equations.

40. $5x - 3y = 15$ **41.** $4x + 3y = 12$

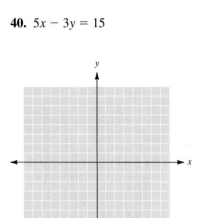
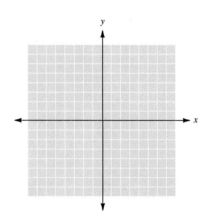

[7.3] Graph each equation by first solving for y.

42. $2x + y = 6$ **43.** $3x + 2y = 6$

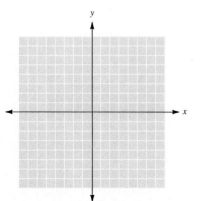
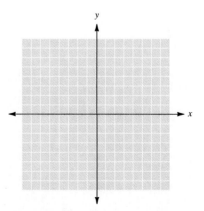

[7.4] Find the slope of the line through the following pairs of points.

44. $(3, 4)$ and $(5, 8)$ **45.** $(-2, 3)$ and $(1, -6)$ **46.** $(-2, 5)$ and $(2, 3)$

47. $(-5, -2)$ and $(1, 2)$ **48.** $(-2, 6)$ and $(5, 6)$ **49.** $(-3, 2)$ and $(-1, -3)$

50. $(-3, -6)$ and $(5, -2)$ **51.** $(-6, -2)$ and $(-6, 3)$

[7.4] Find the slope and y intercept of the line represented by each of the following equations.

52. $y = 2x + 5$ **53.** $y = -4x - 3$ **54.** $y = -\dfrac{3}{4}x$ **55.** $y = \dfrac{2}{3}x + 3$

56. $2x + 3y = 6$ **57.** $5x - 2y = 10$ **58.** $y = -3$ **59.** $x = 2$

[7.4] Write the equation of the line with the given slope and y intercept. Then graph each line, *using* the slope and y intercept.

60. $m = 2, b = 3$ **61.** $m = \dfrac{3}{4}, b = -2$ **62.** $m = -\dfrac{2}{3}, b = 2$

[7.5] Graph each of the following inequalities.

63. $x + y \leq 4$ **64.** $x - y > 5$ **65.** $2x + y < 6$

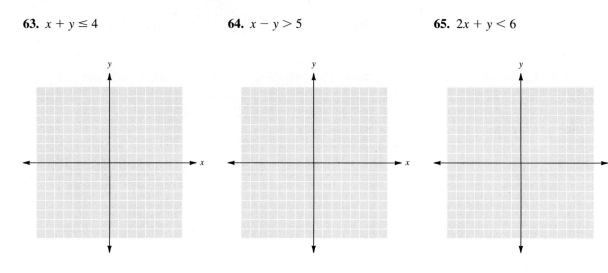

66. $2x - y \geq 6$ **67.** $x - 3y > 6$ **68.** $3x - y \leq 0$

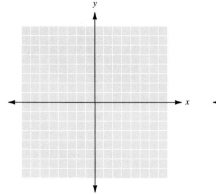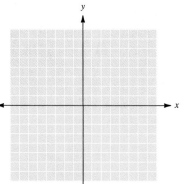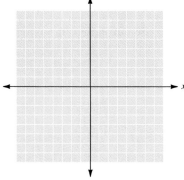

69. $x > 3$ **70.** $y \leq 2$

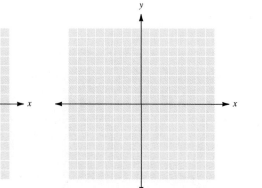

Self-Test for CHAPTER 7

Name

Section

Date

The purpose of this self-test is to help you check your progress and to review for a chapter test in class. Allow yourself about an hour to take the test. When you are done, check your answers in the back of the book. If you missed any problems, be sure to go back and review the appropriate sections in the chapter and the exercises that are provided.

Determine which of the ordered pairs are solutions for the given equations.

1. $x + y = 9$ $(3, 6), (9, 0), (3, 2)$ **2.** $4x - y = 16$ $(4, 0), (3, -1), (5, 4), (0, -8)$

Complete the ordered pairs so that each is a solution for the given equation.

3. $x + 3y = 12$ $(3, \ ), (\ , 2), (9, \ )$ **4.** $4x + 3y = 12$ $(3, \ ), (\ , 4), (\ , 3), \left(\dfrac{3}{4}, \ \right)$

Find four solutions for each of the following equations.

5. $x - y = 7$ **6.** $5x - 6y = 30$

Give the coordinates of the points graphed below.

7. A

8. B

9. C

Plot points with the coordinates shown.

10. $S(1, -2)$

11. $T(0, 3)$

12. $U(-2, -3)$

Graph each of the following equations.

13. $x + y = 4$ **14.** $y = 3x$ **15.** $y = \dfrac{3}{4}x - 4$

ANSWERS

1. _____
2. _____
3. _____
4. _____
5. _____
6. _____
7. _____
8. _____
9. _____
10. _____
11. _____
12. _____
13. _____
14. _____
15. _____

557

© 1993 McGraw-Hill, Inc.

16. _____

17. _____

18. _____

19. _____

20. _____

21. _____

22. _____

23. _____

24. _____

25. _____

16. $x + 3y = 6$ **17.** $2x + 5y = 10$ **18.** $y = -4$

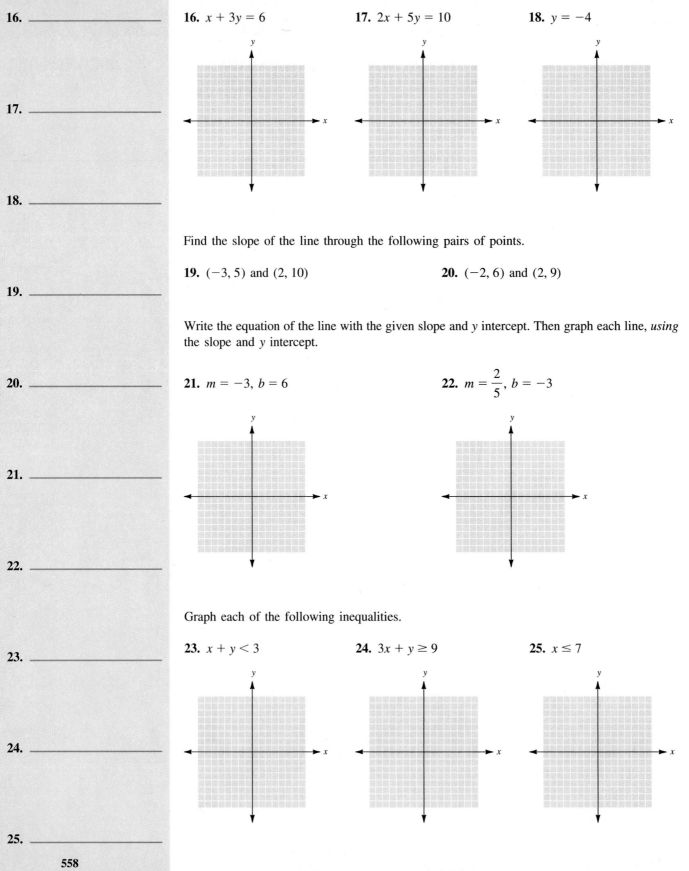

Find the slope of the line through the following pairs of points.

19. $(-3, 5)$ and $(2, 10)$ **20.** $(-2, 6)$ and $(2, 9)$

Write the equation of the line with the given slope and *y* intercept. Then graph each line, *using the slope and y intercept.*

21. $m = -3, b = 6$ **22.** $m = \dfrac{2}{5}, b = -3$

Graph each of the following inequalities.

23. $x + y < 3$ **24.** $3x + y \geq 9$ **25.** $x \leq 7$

Retain Your Skills

This review covers selected topics from the first six chapters.

Perform the indicated operation.

1. $x^2y - 4xy - x^2y + 2xy$

2. $\dfrac{12a^3b}{9ab}$

3. $(5x^2 - 2x + 1) - (3x^2 + 3x - 5)$

4. $(5a^2 + 6a) - (2a^2 - 1)$

Evaluate the expression.

5. $4 + 3(7 - 4)^2$

6. $|3 - 5| - |-4 + 3|$

Multiply.

7. $(x - 2y)(2x + 3y)$

8. $(x + 7)(x + 4)$

Divide.

9. $(2x^2 + 3x - 1) \div (x + 2)$

10. $(x^2 - 5) \div (x - 1)$

Solve each equation and check your results.

11. $4x - 3 = 2x + 5$

12. $2 - 3(2x + 1) = 11$

Factor each polynomial completely.

13. $x^2 - 5x - 14$

14. $3m^2n - 6mn^2 + 9mn$

15. $a^2 - 9b^2$

16. $2x^3 - 28x^2 + 96x$

Solve the following word problems. Show the equation used for the solution.

17. 2 more than 4 times a number is 30. Find the number.

18. If the reciprocal of 4 times a number is subtracted from the reciprocal of that number, the result is $\dfrac{3}{16}$. What is the number?

19. A speed of 60 mi/h corresponds to 88 ft/s. If a race car is traveling at 180 mi/h, what is its speed in feet per second?

20. The length of a rectangle is 3 in less than twice its width. If the area of the rectangle is 35 in^2, find the dimensions of the rectangle.

CHAPTER 8

Systems of Linear Equations

The ENVIRONMENT

Global Warming and the Greenhouse Effect

In 1827, Jean Baptiste Fourier, a French mathematician, first proposed the idea of the greenhouse effect. This theory says heat from the sun is trapped by certain gases in the earth's atmosphere and held close to the surface of the planet. This heat-trapping property of the atmosphere is one reason why life has evolved on earth. This "greenhouse effect" gets its name because it is the same effect experienced when heat is trapped by the glass in a greenhouse.

The greenhouse effect is a normal result of the structure of the atmosphere. Without the greenhouse effect, life as we know it would not exist. However, due to human activities, there is concern that the atmosphere may be holding in too much heat. If this is true, we may be altering the earth's climate, which could lead to unknown consequences.

Although there is much debate among the scientific community about what its effects might be, the idea of the greenhouse effect is nearly unanimously accepted. Also accepted is the fact that humans are causing a rapid increase in the amount of greenhouse gases in the atmosphere. The primary greenhouse gases are carbon dioxide (CO_2), chlorofluorocarbons (CFCs), methane (CH_4), and nitrogen oxides (NO_x). (There are a variety of nitrogen oxides. They are generally referred to as the NO_x group.) Carbon dioxide comes primarily from burning fossil fuels and deforestation. CFCs come primarily from leaky air conditioners and refrigerators, evaporation of industrial solvents, plastic foam production, and aerosol spray propellants. Sources of methane include landfills, livestock, bogs and marshes, rice paddies, some mining and drilling operations, storage tanks, and homes. And NO_x is released from the breakdown of fertilizers, livestock waste, and by burning organic debris.

Scientists agree that the greenhouse theory is correct and that the level of greenhouse gases in the atmosphere is increasing. Beyond this point, there is little agreement about what these ideas mean for the earth's future. The outcomes predicted range from large increases in the average global temperature and devastating climatic change to little or no temperature change and improved climate in the future. Even though there is still much disagreement, the most respected models indicate the following climatic changes are probable.

Surface temperatures around the world will increase an average of 1.5 to 4.5°C over the next 100 years. This may not seem like a major temperature change, but the last ice age was only 5°C cooler than today's average temperature.

Sea levels around the world will rise. This will be mainly caused by the expansion of seawater as it warms. Some unknown contribution may be made by the melting of glacial ice in the polar regions. Estimates of the amount of sea-level rise vary from a few centimeters to a few meters depending on the amount of glacial melting that occurs.

Global rainfall patterns will change significantly. Some regions will experience drying while others will experience increased moisture. This change in rainfall patterns will likely cause changes in the world's major food-producing areas. Areas currently producing the most food will probably become drier and less productive while countries farther from the equator are likely to become more productive.

If we assume some level of climate change is likely to occur if greenhouse gases continue to increase in the atmosphere, we need to be prepared to adjust to the new climate or we need to slow down the rate of change. Some experts believe we need to begin doing both immediately.

Suggested steps for slowing global warming include the following:

- Reduce and eventually ban the production and use of CFCs.
- Reduce our level of fossil-fuel use by increasing efficiency and shifting to energy sources which do not emit CO_2.
- Transfer energy and pollution control technology to developing countries rather than having them copy the polluting technologies of the developed world.
- Reduce methane in the atmosphere by capturing methane from landfills and using it as a cleaner-burning fuel. Reducing beef production will decrease methane produced by the animals and decrease the fossil-fuel inputs.
- Halt deforestation as soon as possible.
- Slow population growth in order to decrease the rising demand in all the other areas.

These are but a few of the changes we face if the projected global warming does occur. Most would be beneficial even if there were no major climate change. Although change is never easy, these suggested changes can be accomplished. Starting them now rather than later would probably be a good idea.

Systems of Linear Equations: Solving by Graphing

OBJECTIVE
To find the solution of a system of equations by graphing

From our work in Section 7.3, we know that an equation of the form $x + y = 3$ is a linear equation. Remember that its graph is a straight line. Often we will want to consider two equations together. They then form a *system of linear equations.* An example of such a system is

$$x + y = 3$$
$$3x - y = 5$$

You know that a solution for a linear equation in two variables is any ordered pair (x, y) that satisfies the equation. Often you will want to find a single ordered pair that satisfies both equations of the system. It is called the *solution for the system.* For instance, the solution for the system above is $(2, 1)$ because, replacing x with 2 and y with 1, we have

$x + y = 3$	$3x - y = 5$
$2 + 1 \stackrel{?}{=} 3$	$3 \cdot 2 - 1 \stackrel{?}{=} 5$
$3 = 3$	$6 - 1 \stackrel{?}{=} 5$
	$5 = 5$

There is no other ordered pair that satisfies both equations.

Since both statements are true, the ordered pair $(2, 1)$ satisfies both equations.

One approach to finding the solution for a system of linear equations is called the *graphical method.* Using this approach, we graph the two equations on the same coordinate system. The coordinates of the point where the lines intersect will be the solution for the given system.

Example 1

Solve the system by graphing.

$$x + y = 6$$
$$x - y = 4$$

Use the intercept method to graph each equation.

First, we determine solutions for the equations of our system. For $x + y = 6$, two solutions are $(6, 0)$ and $(0, 6)$. For $x - y = 4$, two solutions are $(4, 0)$ and $(0, -4)$. Using these intercepts, we graph the two equations. The lines intersect at the point $(5, 1)$.

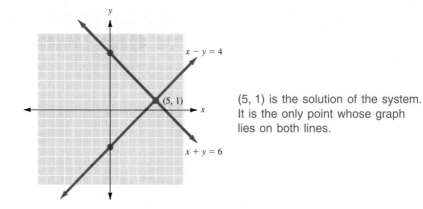

(5, 1) is the solution of the system. It is the only point whose graph lies on both lines.

Note: By substituting 5 for x and 1 for y into the two original equations, we can check that (5, 1) is indeed the solution for our system.

Both statements must be true for (5, 1) to be a solution for the system.

$$\begin{array}{ll} x + y = 6 & x - y = 4 \\ \overline{5 + 1 \stackrel{?}{=} 6} & \overline{5 - 1 \stackrel{?}{=} 4} \\ \quad\quad 6 = 6 & \quad\quad 4 = 4 \end{array}$$

CHECK YOURSELF 1

Solve the system by graphing.

$$2x - y = 4$$
$$x + y = 5$$

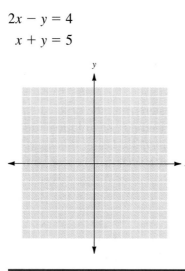

The next example shows how to graph a system when one of the equations represents a horizontal line.

Example 2

Solve the system by graphing.

$$3x + 2y = 6$$
$$y = 6$$

For $3x + 2y = 6$, two solutions are $(2, 0)$ and $(0, 3)$. These represent the x and y intercepts of the graph of the equation. The equation $y = 6$ represents a horizontal line that crosses the y axis at the point $(0, 6)$. Using these intercepts, we graph the two equations. The lines will intersect at the point $(-2, 6)$. So this is a solution to our system.

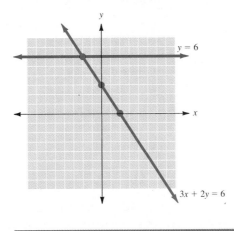

CHECK YOURSELF 2

Solve the system by graphing.

$4x + 5y = 20$
$\qquad y = 8$

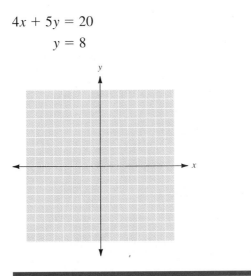

In both Examples 1 and 2 each system had exactly one solution. A system with one solution is called a *consistent* system. It is possible that a system of equations will have no solution. Look at the following.

Example 3

Solve by graphing.

$2x + y = 2$
$2x + y = 4$

We can graph the two lines as before. For $2x + y = 2$, two solutions are $(0, 2)$ and $(1, 0)$. For $2x + y = 4$, two solutions are $(0, 4)$ and $(2, 0)$. Using these intercepts, we graph the two equations.

Note: In slope-intercept form, our equations are

$y = -2x + 2$

and

$y = -2x + 4$

Both lines have slope -2.

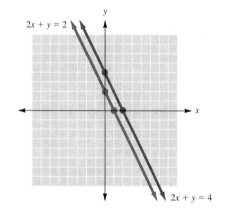

Notice that the slope for each of these lines is -2, but they have different y intercepts. This means that the lines are parallel (they will never intersect). Since the lines have no points in common, there is no ordered pair that will satisfy both equations. The system is called *inconsistent* and has no solution.

CHECK YOURSELF 3

Solve by graphing (if possible).

$x - 3y = 3$
$x - 3y = 6$

There is one more possibility for linear systems, as the following example illustrates.

Example 4

Solve by graphing.

$$x - 2y = 4$$
$$2x - 4y = 8$$

Note that multiplying the first equation by 2 results in the second equation.

Graphing as before and using the intercept method, we find

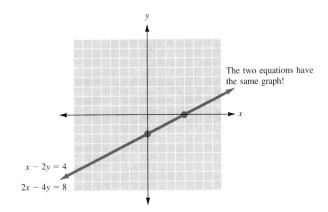

The two equations have the same graph!

$x - 2y = 4$

$2x - 4y = 8$

Since the graphs coincide, there are *infinitely many* solutions for this system. Every point on the graph of $x - 2y = 4$ is also on the graph of $2x - 4y = 8$, so any ordered pair satisfying $x - 2y = 4$ also satisfies $2x - 4y = 8$. This is called a *dependent* system, and any point on the line is a solution.

CHECK YOURSELF 4

Solve by graphing.

$$x + y = 4$$
$$2x + 2y = 8$$

The following summarizes our work in this section.

TO SOLVE A SYSTEM OF EQUATIONS BY GRAPHING

STEP 1   Graph both equations on the same coordinate system.

STEP 2   Determine the solution to the system as follows.
    (a) If the lines intersect at one point, the solution is the ordered pair corresponding to that point.
    (b) If the lines are parallel, there is no solution.
    (c) If the two equations have the same graph, then the system has infinitely many solutions and any ordered pair that corresponds to a point on the line is a solution.

STEP 3   Check the solution in both equations, if necessary.

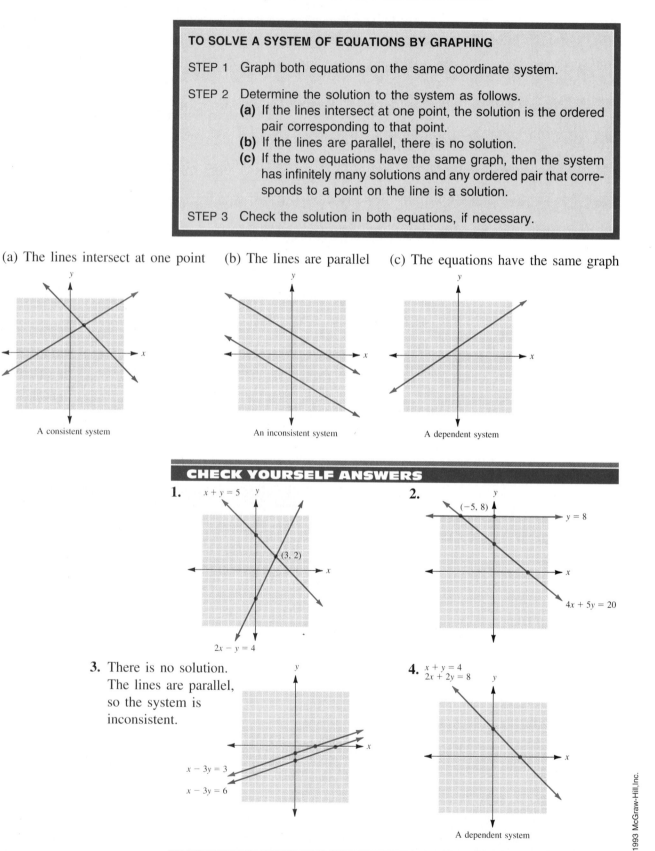

(a) The lines intersect at one point

A consistent system

(b) The lines are parallel

An inconsistent system

(c) The equations have the same graph

A dependent system

CHECK YOURSELF ANSWERS

1. $x + y = 5$ $(3, 2)$ $2x - y = 4$

2. $(-5, 8)$ $y = 8$ $4x + 5y = 20$

3. There is no solution. The lines are parallel, so the system is inconsistent.
$x - 3y = 3$
$x - 3y = 6$

4. $x + y = 4$
$2x + 2y = 8$
A dependent system

Name

Date

Build Your Skills

Solve each of the following systems by graphing.

1. $x + y = 6$
$x - y = 4$

2. $x - y = 8$
$x + y = 2$

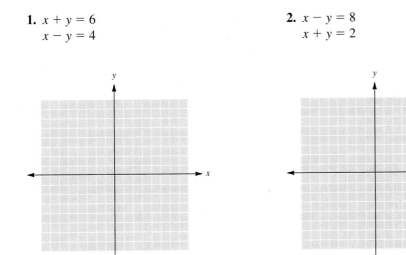

3. $-x + y = 3$
$x + y = 5$

4. $x + y = 7$
$-x + y = 3$

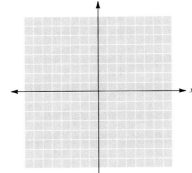

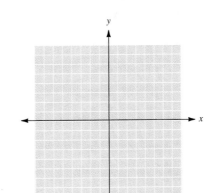

5. $x + 2y = 4$
　　$x - y = 1$

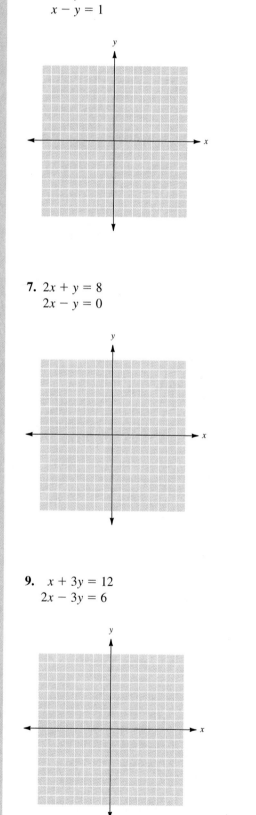

6. $3x + y = 6$
　　$x + y = 4$

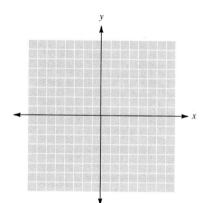

7. $2x + y = 8$
　　$2x - y = 0$

8. $x - 2y = -2$
　　$x + 2y = 6$

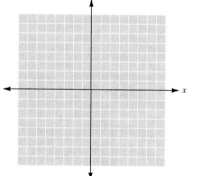

9. $x + 3y = 12$
　　$2x - 3y = 6$

10. $2x - y = 4$
　　$2x - y = 6$

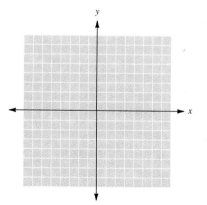

570

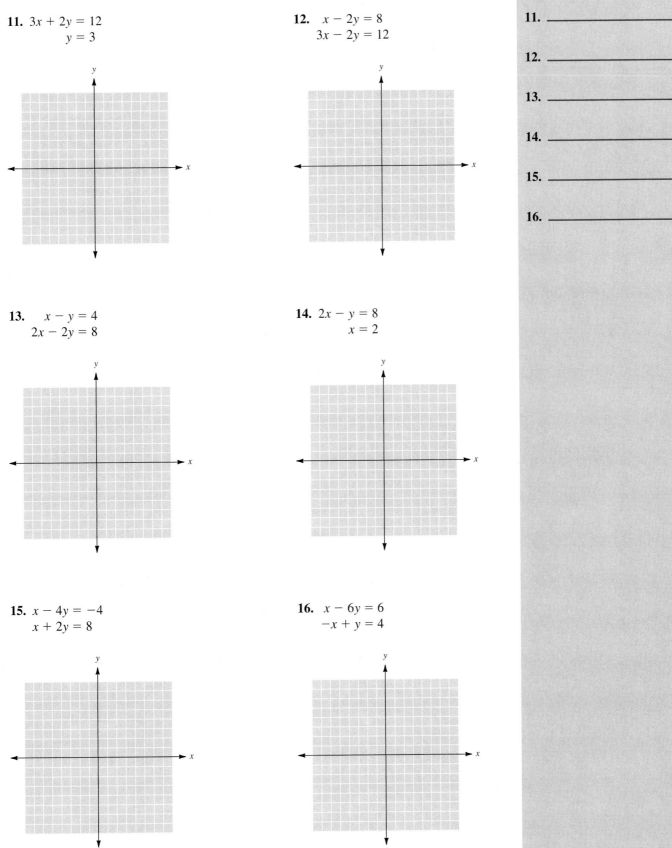

11. $3x + 2y = 12$
        $y = 3$

12. $x - 2y = 8$
      $3x - 2y = 12$

13. $x - y = 4$
     $2x - 2y = 8$

14. $2x - y = 8$
       $x = 2$

15. $x - 4y = -4$
     $x + 2y = 8$

16. $x - 6y = 6$
     $-x + y = 4$

ANSWERS

11. _____

12. _____

13. _____

14. _____

15. _____

16. _____

17. _____

18. _____

19. _____

20. _____

21. _____

22. _____

17. $3x - 2y = 6$
$2x - y = 5$

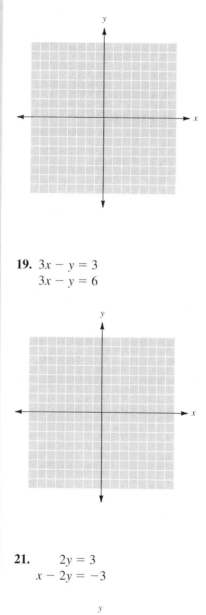

18. $4x + 3y = 12$
$x + y = 2$

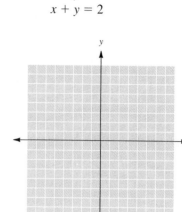

19. $3x - y = 3$
$3x - y = 6$

20. $3x - 6y = 9$
$x - 2y = 3$

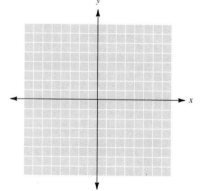

21. $2y = 3$
$x - 2y = -3$

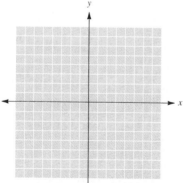

22. $x + y = -6$
$-x + 2y = 6$

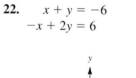

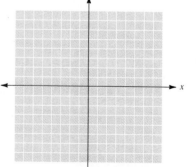

572

23. $x = 4$
  $y = -6$

24. $x = -3$
  $y = 5$

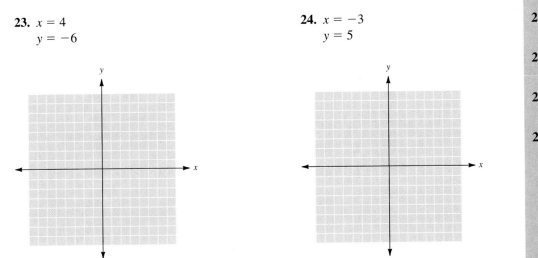

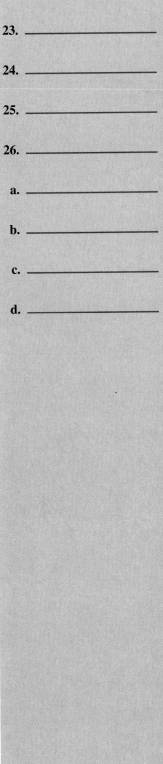
Think About These

25. Find values for m and b in the following system so that the solution to the system is $(1, 2)$.

$$mx + 3y = 8$$
$$-3x + 4y = b$$

26. Find values for m and b in the following system so that the solution to the system is $(-3, 4)$.

$$5x + 7y = b$$
$$mx + y = 22$$

Skillscan (Section 1.4)

Simplify each of the following expressions.

a. $(2x + y) + (x - y)$

b. $(x + y) + (-x + y)$

c. $(3x + 2y) + (-3x - 3y)$

d. $(x - 5y) + (2x + 5y)$

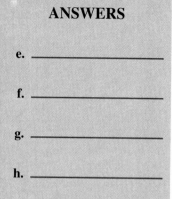
e. $2(x + y) + (3x - 2y)$

f. $2(2x - y) + (-4x - 3y)$

g. $3(2x + y) + 2(-3x + y)$

h. $3(2x - 4y) + 4(x + 3y)$

ANSWERS

1. $\left.\begin{array}{r} x + y = 6 \\ x - y = 4 \end{array}\right\}(5, 1)$

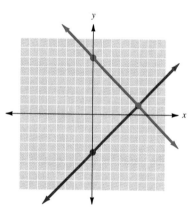

3. $\left.\begin{array}{r} -x + y = 3 \\ x + y = 5 \end{array}\right\}(1, 4)$

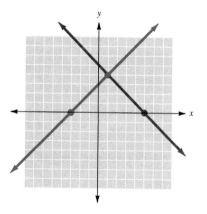

5. $\left.\begin{array}{r} x + 2y = 4 \\ x - y = 1 \end{array}\right\}(2, 1)$

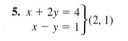 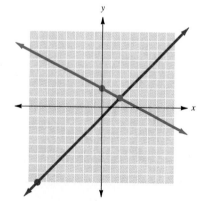

7. $\left.\begin{array}{r} 2x + y = 8 \\ 2x - y = 0 \end{array}\right\}(2, 4)$

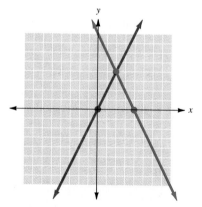

574

9. $\left.\begin{array}{r} x + 3y = 12 \\ 2x - 3y = 6 \end{array}\right\}$ (6, 2)

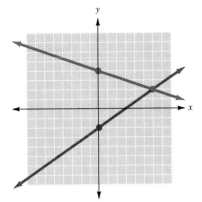

11. $\left.\begin{array}{r} 3x + 2y = 12 \\ y = 3 \end{array}\right\}$ (2, 3)

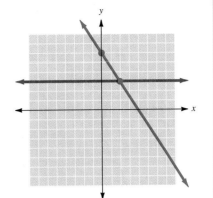

13. $\left.\begin{array}{r} x - y = 4 \\ 2x - 2y = 8 \end{array}\right\}$ Dependent

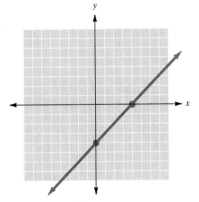

15. $\left.\begin{array}{r} x - 4y = -4 \\ x + 2y = 8 \end{array}\right\}$ (4, 2)

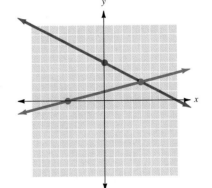

17. $\left.\begin{array}{r} 3x - 2y = 6 \\ 2x - y = 5 \end{array}\right\}$ (4, 3)

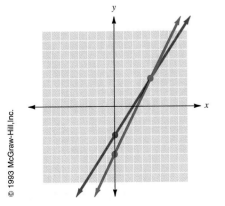

19. $\left.\begin{array}{r} 3x - y = 3 \\ 3x - y = 6 \end{array}\right\}$ Inconsistent

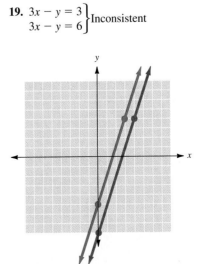

21. $\left.\begin{array}{r} 2y = 3 \\ x - 2y = -3 \end{array}\right\} \left(0, \dfrac{3}{2}\right)$

23. $\left.\begin{array}{r} x = 4 \\ y = -6 \end{array}\right\} (4, -6)$

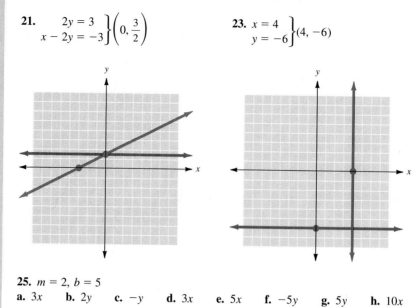

25. $m = 2,\ b = 5$

a. $3x$ **b.** $2y$ **c.** $-y$ **d.** $3x$ **e.** $5x$ **f.** $-5y$ **g.** $5y$ **h.** $10x$

Systems of Linear Equations: Solving by Adding

OBJECTIVE
To find the solution of a system of linear equations by adding

The graphical method of solving equations, shown in Section 8.1, has two definite disadvantages. First, it is time-consuming to graph each system that you want to solve. Second (and a bigger problem), the graphical method is not precise. For instance, look at the graph of the system

$$x - 2y = 4$$
$$3x + 2y = 6$$

which is shown below.

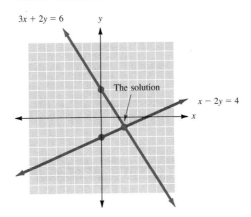

The exact solution for the system happens to be $\left(\dfrac{5}{2}, -\dfrac{3}{4}\right)$, but that would be difficult to read from the graph. Fortunately, there are algebraic methods that do not have this disadvantage and will allow you to find exact solutions for a system of equations.

Let's illustrate an algebraic method of finding a solution. It is called the *addition method.*

Example 1

This method uses the fact that if

$a = b$ and $c = d$

then

$a + c = b + d$

This is the *additive property* of equality. Note that by the additive property, if equals are added to equals, the resulting sums are equal.

Solve the system.

$$x + y = 8$$
$$x - y = 2$$

Note that the coefficients of the y terms are the opposites of each other (1 and -1) and that adding the two equations will "eliminate" the variable y. That addition step is shown on the next page.

This is also called *solution by elimination* for this reason.

$$\left.\begin{array}{r} x + y = 8 \\ x - y = 2 \end{array}\right\}$$
$$\begin{array}{r} 2x = 10 \\ x = 5 \end{array}$$

By adding, we eliminate the variable *y*. The resulting equation contains *only* the variable *x*.

We now know that 5 is the *x* coordinate of our solution. Substitute 5 for *x* into *either* of the original equations.

$$x + y = 8$$
$$5 + y = 8$$
$$y = 3$$

So $(5, 3)$ is the solution.

To check, replace *x* and *y* with these values in *both* of the original equations.

$$x + y = 8 \qquad\qquad x - y = 2$$
$$5 + 3 = 8 \qquad\qquad 5 - 3 = 2$$
$$8 = 8 \quad \text{(True)} \qquad\quad 2 = 2 \quad \text{(True)}$$

Since $(5, 3)$ satisfies both equations, it is the solution.

CHECK YOURSELF 1

Solve the system by adding.

$$x - y = -2$$
$$x + y = 6$$

Example 2

Solve the system.

$$-3x + 2y = 12$$
$$3x - y = -9$$

In this case, adding will eliminate the *x* terms.

Note that we don't care which variable is eliminated. Choose the one that requires the least work.

$$-3x + 2y = 12$$
$$3x - y = -9$$
$$y = 3$$

Now substitute 3 for y in either equation. From the first equation

$$-3x + 2 \cdot 3 = 12$$
$$-3x = 6$$
$$x = -2$$

and $(-2, 3)$ is the solution.

Show that you get the same x coordinate by substituting 3 for y in the second equation rather than in the first. Then check the solution.

CHECK YOURSELF 2

Solve the system by adding.

$$5x - 2y = 9$$
$$-5x + 3y = -11$$

Note that in both of the above examples we found an equation in a single variable by adding. We could do this because the coefficients of one of the variables were opposites. This gave 0 as a coefficient for one of the variables after we added the two equations. In some systems, you will not be able to directly eliminate either variable by adding. However, an equivalent system can always be written by multiplying one or both of the equations by a nonzero constant so that the coefficients of x (or of y) are opposites.

The following example illustrates.

Example 3

Solve the system.

$$2x + y = 13 \tag{1}$$
$$3x + y = 18 \tag{2}$$

Remember that multiplying both sides of an equation by some nonzero number does not change the solutions. So even though we have "altered" the equations, they are equivalent and will have the same solutions.

Note that adding the equations in this form will not eliminate either variable. You will still have terms in x and in y. However, look at what happens if we multiply both sides of equation (2) by -1 as the first step.

$$2x + y = 13 \xrightarrow{\hspace{3cm}} 2x + y = 13$$
$$3x + y = 18 \xrightarrow{\substack{\text{Multiply} \\ \text{by } -1}} -3x - y = -18$$

Now we can add.

$$
\begin{array}{r}
2x + y = 13 \\
-3x - y = -18 \\
\hline
-x \quad\;\; = -5 \\
x = \;\;\, 5
\end{array}
$$

Substitute 5 for x in equation (1).

$$
\begin{aligned}
2 \cdot 5 + y &= 13 \\
y &= \;\; 3
\end{aligned}
$$

$(5, 3)$ is the solution. We will leave it to the reader to check this solution.

CHECK YOURSELF 3

Solve the system by adding.

$$
\begin{aligned}
x - 2y &= \;\; 9 \\
x + 3y &= -1
\end{aligned}
$$

To summarize, multiplying both sides of one of the equations by a nonzero constant can yield an equivalent system in which the coefficients of the x terms or the y terms are opposites. This means that a variable can be eliminated by adding. Let's look at another example.

Example 4

Solve the system.

$$
\begin{aligned}
x + 4y &= \;\;\;\; 2 \qquad\qquad (1)\\
3x - 2y &= -22 \qquad\quad\; (2)
\end{aligned}
$$

One approach is to multiply both sides of equation (2) by 2. Do you see that the coefficients of the y terms will then be opposites?

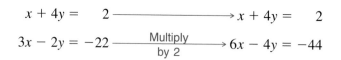

If we add the resulting equations, the variable y will be eliminated and we can solve for x.

Note that the coefficients of the y terms are opposites

$$
\begin{array}{r}
x + 4y = \;\;\;\; 2 \\
6x - 4y = -44 \\
\hline
7x \qquad\;\; = -42 \\
x = \;\; -6
\end{array}
$$

Now substitute -6 for x in equation (1) to find y.

Also -6 could be substituted for x in equation (2) to find y.

$$-6 + 4y = 2$$
$$4y = 8$$
$$y = 2$$

So $(-6, 2)$ is the solution.

Again you should check this result. As is often the case, there are several ways to solve the system. For example, what if we multiply both sides of equation (1) by -3? The coefficients of the x terms will then be opposites, and adding will eliminate the variable x so that we can solve for y. Try that for yourself in the following Check Yourself exercise.

CHECK YOURSELF 4

Solve the system by eliminating x.

$$x + 4y = 2$$
$$3x - 2y = -22$$

It may be necessary to multiply each equation separately so that one of the variables will be eliminated when the equations are added. The following example illustrates.

Example 5

Solve the system.

$$4x + 3y = 11 \tag{1}$$
$$3x - 2y = 4 \tag{2}$$

Do you see that multiplying in one equation will not help in this case? We will have to multiply in both equations.

The minus is used with the 4 so that the coefficients of the x term are opposites.

To eliminate x in this case, we can multiply both sides of equation (1) by 3 and both sides of equation (2) by -4. The coefficients of the x terms will then be opposites.

$$4x + 3y = 11 \xrightarrow[\text{by 3}]{\text{Multiply}} 12x + 9y = 33$$

$$3x - 2y = 4 \xrightarrow[\text{by } -4]{\text{Multiply}} -12x + 8y = -16$$

Adding the resulting equations gives

$$17y = 17$$
$$y = 1$$

Now substituting 1 for y in equation (1), we have

$$4x + 3 \cdot 1 = 11$$
$$4x = 8$$
$$x = 2$$

Check (2, 1) in both equations of the original system.

and (2, 1) is the solution.

Again there are different ways to approach the solution. You could choose to eliminate y. Try that in the exercise that follows.

CHECK YOURSELF 5

Solve the system by eliminating y.

$$4x + 3y = 11$$
$$3x - 2y = 4$$

Let's summarize the solution steps that we have illustrated.

TO SOLVE A SYSTEM OF LINEAR EQUATIONS BY ADDING

STEP 1   If necessary, multiply both sides of one or both equations by non-zero numbers to form an equivalent system in which the coefficients of one of the variables are opposites.

STEP 2   Add the equations of the new system.

STEP 3   Solve the resulting equation for the remaining variable.

STEP 4   Substitute the value found in step 3 into either of the original equations to find the value of the second variable.

STEP 5   Check your solution in both of the original equations.

In Section 8.1 we saw that certain systems had either no solution or infinitely many solutions. Let's see how this situation is indicated when we are using the addition method of solving equations.

Example 6

Solve

$$x + 3y = -2 \tag{1}$$
$$3x + 9y = -6 \tag{2}$$

We multiply both sides of equation (1) by -3.

$$x + 3y = -2 \xrightarrow[\text{by } -3]{\text{Multiply}} -3x - 9y = 6$$

$$3x + 9y = -6 \xrightarrow{} \underline{3x + 9y = -6}$$

$$0 = 0$$

Adding, we see that both variables have been eliminated, and we have the true statement $0 = 0$.

Look at the graph of the system.

The lines coincide. That will be the case whenever *adding eliminates both variables* and a true statement results.

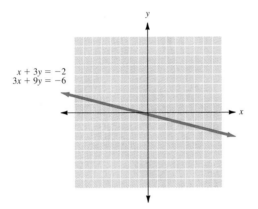

As we see, the two equations have the *same* graph. This means that the system is *dependent,* and there are *infinitely many solutions.* Any (x, y) that satisfies $x + 3y = -2$ will also satisfy $3x + 9y = -6$.

CHECK YOURSELF 6

Solve the system by adding.

$$x - 2y = 3$$
$$-2x + 4y = -6$$

In the previous section we also encountered systems that had *no* solutions. The final example illustrates what happens when we try to solve such a system with the algebraic methods of this section.

Example 7

Solve the system.

$$3x - y = 4 \qquad (1)$$
$$-6x + 2y = -5 \qquad (2)$$

We multiply both sides of equation (1) by 2.

Be sure to multiply the 4 by 2.

$3x - y = 4$ $\xrightarrow[\text{by 2}]{\text{Multiply}}$ $6x - 2y = 8$ We now add the two equations

$-6x + 2y = -5$ $\xrightarrow{\hspace{2cm}}$ $\underline{-6x + 2y = -5}$

$0 = 3$

Again both variables have been eliminated by addition. But this time we have the *false* statement $0 = 3$. This is because we tried to solve a system whose graph consists of two parallel lines, as we see in the graph below. Since the two lines do not intersect, there is *no* solution for the system. It is *inconsistent*.

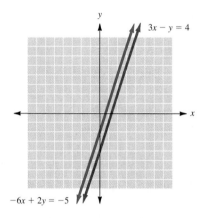

CHECK YOURSELF 7

Solve the system by adding.

$5x + 15y = 20$

$x + 3y = 3$

CHECK YOURSELF ANSWERS

1. $(2, 4)$.   **2.** $(1, -2)$.   **3.** $(5, -2)$.   **4.** $(-6, 2)$.   **5.** $(2, 1)$.
6. A dependent system.   **7.** There is no solution. The system is inconsistent.

Name _____

Date _____

Build Your Skills

Solve each of the following systems by adding.

1. $x + y = 6$
$x - y = 4$

2. $x - y = 8$
$x + y = 2$

3. $-x + y = 3$
$x + y = 5$

4. $x + y = 7$
$-x + y = 3$

5. $2x - y = 1$
$-2x + 3y = 5$

6. $x - 2y = 2$
$x + 2y = -14$

7. $x + 3y = 12$
$2x - 3y = 6$

8. $-3x + y = 8$
$3x - 2y = -10$

9. $x + 2y = -2$
$3x + 2y = -12$

10. $4x - 3y = 22$
$4x + 5y = 6$

11. $4x - 3y = 6$
$4x + 5y = 22$

12. $2x + 3y = 1$
$5x + 3y = 16$

13. $2x + y = 8$
$2x + y = 2$

14. $5x + 4y = 7$
$5x - 2y = 19$

15. $3x - 5y = 2$
$2x - 5y = -2$

16. $2x - y = 4$
$2x - y = 6$

17. $x + y = 3$
$3x - 2y = 4$

18. $x - y = -2$
$2x + 3y = 21$

19. $-5x + 2y = -3$
$x - 3y = -15$

20. $x + 5y = 10$
$-2x - 10y = -20$

ANSWERS

1. _____

2. _____

3. _____

4. _____

5. _____

6. _____

7. _____

8. _____

9. _____

10. _____

11. _____

12. _____

13. _____

14. _____

15. _____

16. _____

17. _____

18. _____

19. _____

20. _____

21. _____

22. _____

23. _____

24. _____

25. _____

26. _____

27. _____

28. _____

29. _____

30. _____

31. _____

32. _____

33. _____

34. _____

35. _____

36. _____

37. _____

38. _____

39. _____

40. _____

21. $7x + y = 10$
$2x + 3y = -8$

22. $3x - 4y = 2$
$4x - y = 20$

23. $5x + 2y = 28$
$x - 4y = -23$

24. $7x + 2y = 17$
$x - 5y = 13$

25. $3x - 4y = 2$
$-6x + 8y = -4$

26. $-x + 5y = 19$
$4x + 3y = -7$

27. $5x - 2y = 31$
$4x + 3y = 11$

28. $7x + 3y = -13$
$5x + 2y = -8$

29. $3x - 2y = 12$
$5x - 3y = 21$

30. $-4x + 5y = -6$
$5x - 2y = 16$

31. $-2x + 7y = 2$
$3x - 5y = -14$

32. $3x + 4y = 0$
$5x - 3y = -29$

33. $7x + 4y = 20$
$5x + 6y = 19$

34. $5x + 4y = 5$
$7x - 6y = 36$

35. $2x - 7y = 6$
$-4x + 3y = -12$

36. $3x + 2y = -18$
$7x - 6y = -42$

37. $5x - y = 20$
$4x + 3y = 16$

38. $3x + y = -5$
$5x - 4y = 20$

39. $3x + y = 1$
$5x + y = 2$

40. $2x - y = 2$
$2x + 5y = -1$

41. $3x + 4y = 3$
$6x - 2y = 1$

42. $3x + 3y = 1$
$2x + 4y = 2$

43. $5x - 2y = \dfrac{9}{5}$

$3x + 4y = -1$

44. $2x + 3y = -\dfrac{1}{12}$

$5x + 4y = \dfrac{2}{3}$

Think About These

Solve the following systems by adding.

45. $\dfrac{x}{3} - \dfrac{y}{4} = -\dfrac{1}{2}$

$\dfrac{x}{2} - \dfrac{y}{5} = \dfrac{3}{10}$

46. $\dfrac{1}{3}x - \dfrac{1}{2}y = \dfrac{5}{6}$

$\dfrac{1}{2}x - \dfrac{2}{5}y = \dfrac{9}{10}$

47. $0.4x - 0.2y = 0.6$
$0.5x - 0.6y = 9.5$

48. $0.2x + 0.37y = 0.8$
$-0.6x + 1.4y = 2.62$

49. In 1987, Japan released a total of 102 million tons of methane (CH_4) and CFCs. In the same year, the U.S.S.R. released 5 times as much methane as Japan and twice as many CFCs as Japan for a total of 240 million tons. The following equations can be derived from the given information.

$M + C = 102$ amount of gas released by Japan
$5M + 2C = 240$ amount of gas released by U.S.S.R.

Solve the system of equations to determine how much of each gas was released by each country. Your answer will be in millions of tons of each gas.

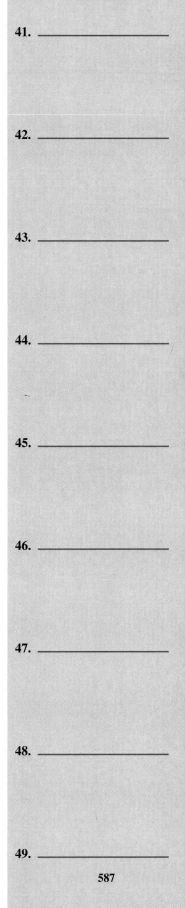

ANSWERS

41. _____

42. _____

43. _____

44. _____

45. _____

46. _____

47. _____

48. _____

49. _____

50. _____

a. _____

b. _____

c. _____

d. _____

e. _____

f. _____

50. In 1987, Japan released a total of 200 million tons of carbon dioxide (CO_2) and CFCs. In the same year, Poland released one-half as much CO_2 as Japan and one-sixth as many CFCs as Japan for a total of 70 million tons. The following equations can be derived from the given information.

$$D + C = 70 \qquad \text{amount of gas released by Poland}$$
$$2D + 6C = 200 \qquad \text{amount of gas released by Japan}$$

Solve the system of equations to determine how much of each gas was released by each country. Your answer will be in millions of tons of each gas.

Skillscan (Section 3.4)

Solve each of the following equations.

a. $2x + 3(x + 1) = 13$ **b.** $3(y - 1) + 4y = 18$ **c.** $x + 2(3x - 5) = 25$

d. $3x - 2(x - 7) = 12$ **e.** $2(3y + 1) - 4y = 7$ **f.** $3x - 2(3x - 2) = 5$

ANSWERS

1. $(5, 1)$ **3.** $(1, 4)$ **5.** $(2, 3)$ **7.** $(6, 2)$ **9.** $\left(-5, \dfrac{3}{2}\right)$ **11.** $(3, 2)$

13. Inconsistent   **15.** $(4, 2)$ **17.** $(2, 1)$ **19.** $(3, 6)$ **21.** $(2, -4)$ **23.** $\left(3, \dfrac{13}{2}\right)$

25. Dependent   **27.** $(5, -3)$ **29.** $(6, 3)$ **31.** $(-8, -2)$ **33.** $\left(2, \dfrac{3}{2}\right)$ **35.** $(3, 0)$

37. $(4, 0)$ **39.** $\left(\dfrac{1}{2}, -\dfrac{1}{2}\right)$ **41.** $\left(\dfrac{1}{3}, \dfrac{1}{2}\right)$ **43.** $\left(\dfrac{1}{5}, -\dfrac{2}{5}\right)$ **45.** $(3, 6)$

47. $(-11, -25)$ **49.** Japan: 12 million tons methane, 90 million tons CFCs; U.S.S.R.: 60 million tons methane, 180 million tons CFCs   **a.** 2   **b.** 3   **c.** 5   **d.** -2 **e.** $\dfrac{5}{2}$ **f.** $-\dfrac{1}{3}$

Systems of Linear Equations: Solving by Substitution

OBJECTIVE
To find the solution of a system of equations by substitution

In Sections 8.1 and 8.2, we looked at graphing and addition as methods of solving linear systems. A third method is called *solution by substitution.*

Example 1

Solve by substitution.

$$x + y = 12 \tag{1}$$
$$y = 3x \tag{2}$$

Notice that equation (2) says that y and $3x$ name the same quantity. So we may substitute $3x$ for y in equation (1). We then have

Replace y with $3x$ in equation (1).

$$\downarrow$$

The resulting equation contains only the variable x, so substitution is just another way of eliminating one of the variables from our system.

$$x + 3x = 12$$
$$4x = 12$$
$$x = 3$$

We can now substitute 3 for x in equation (1) to find the corresponding y coordinate of the solution.

$$3 + y = 12$$
$$y = 9$$

So $(3, 9)$ is the solution.

The solution for a system is written as an ordered pair.

This last step is identical to the one you saw in Section 8.2. As before, you can substitute the known coordinate value back into either of the original equations to find the value of the remaining variable. The check is also identical.

CHECK YOURSELF 1

Solve by substitution.

$$x - y = 9$$
$$y = 4x$$

The same technique can be readily used any time one of the equations is *already solved* for x or for y. The following example illustrates.

Example 2

Solve by substitution.

$$2x + 3y = 3 \tag{1}$$
$$y = 2x - 7 \tag{2}$$

Since Equation (2) tells us that y is $2x - 7$, we can replace y with $2x - 7$ in equation (1). This gives

Now y is eliminated from the equation, and we can proceed to solve for x.

$$2x + 3(2x - 7) = 3$$
$$2x + 6x - 21 = 3$$
$$8x = 24$$
$$x = 3$$

We now know that 3 is the x coordinate for the solution. So substituting 3 for x in equation (2), we have

$$y = 2 \cdot 3 - 7$$
$$= 6 - 7$$
$$= -1$$

And $(3, -1)$ is the solution. Once again you should verify this result by letting $x = 3$ and $y = -1$ in the original system.

CHECK YOURSELF 2

Solve by substitution.

$$2x - 3y = 6$$
$$x = 4y - 2$$

As we have seen, the substitution method works very well when one of the given equations is already solved for x or for y. It is also useful if you can readily solve for x or for y in one of the equations.

Example 3

Solve by substitution.

$$x - 2y = 5 \tag{1}$$
$$3x + y = 8 \tag{2}$$

Equation (2) could have been solved for *y* with the result substituted into equation (1).

Neither equation is solved for a variable. That is easily handled in this case. Solving for *x* in equation (1), we have

$x = 2y + 5$

Now substitute $2y + 5$ for *x* in equation (2).

$$3(2y + 5) + y = 8$$
$$6y + 15 + y = 8$$
$$7y = -7$$
$$y = -1$$

Substituting -1 for *y* in equation (2) yields

$$3x + (-1) = 8$$
$$3x = 9$$
$$x = 3$$

So $(3, -1)$ is the solution.

You should check this result by substituting 3 for *x* and -1 for *y* in the equations of the original system.

CHECK YOURSELF 3

Solve by substitution.

$$3x - y = 5$$
$$x + 4y = 6$$

In some systems, using substitution to solve for a specific variable results in fractional coefficients. The next example illustrates this possibility.

Example 4

Solve by substitution.

$$6x + 5y = 13 \tag{1}$$
$$3x + 2y = 4 \tag{2}$$

Neither equation is solved for a variable. Solving for *x* in equation (2), we have

$$3x = 4 - 2y$$
$$x = \frac{4}{3} - \frac{2}{3}y$$

Now substitute $\dfrac{4}{3} - \dfrac{2}{3}y$ for x in equation (1) to obtain

$$6\left(\dfrac{4}{3} - \dfrac{2}{3}y\right) + 5y = 13$$
$$8 - 4y + 5y = 13$$
$$y = 5$$

Now substitute 5 for y in equation (2).

$$3x + 2(5) = 4$$
$$3x + 10 = 4$$
$$3x = -6$$
$$x = -2$$

So $(-2, 5)$ is the solution.

This result can be checked by substituting -2 for x and 5 for y in both equations of the original system.

Note: Using the addition method in this case would have resulted in the same solution without the need to introduce fractional coefficients.

CHECK YOURSELF 4

Solve by substitution.

$$4x - 5y = 9$$
$$12x + 7y = 5$$

Inconsistent systems and dependent systems will show up in a fashion similar to that which we saw in Section 8.2. The following example illustrates.

Example 5

Solve the following systems by substitution.

(a) $4x - 2y = 6$ (1)

      $y = 2x - 3$ (2)

From equation (2) we can substitute $2x - 3$ for y in equation (1).

Don't forget to change both signs in the parentheses.

$$4x - 2(2x - 3) = 6$$
$$4x - 4x + 6 = 6$$
$$6 = 6$$

Both variables have been eliminated, and we have the true statement $6 = 6$.

Recall from the last section that a true statement tells us that the lines coincide. We call this system dependent. There are an infinite number of solutions.

(b) $3x - 6y = 9$ (3)

$x = 2y + 2$ (4)

Substitute $2y + 2$ for x in equation (3).

$3(2y + 2) - 6y = 9$
$6y + 6 - 6y = 9$ ⎤
$6 = 9$ ⎦ This time we have a false statement.

This means that the system is *inconsistent* and that the graphs of the two equations are parallel lines. There is no solution.

CHECK YOURSELF 5

Indicate whether the systems are inconsistent (no solution) or dependent (an infinite number of solutions).

1. $5x + 15y = 10$
$x = -3y + 1$

2. $12x - 4y = 8$
$y = 3x - 2$

The following summarizes our work in this section.

TO SOLVE A SYSTEM OF LINEAR EQUATIONS BY SUBSTITUTION

STEP 1   Solve one of the given equations for x or y. If this is already done, go on to step 2.

STEP 2   Substitute this expression for x or for y into the other equation.

STEP 3   Solve the resulting equation for the remaining variable.

STEP 4   Substitute the known value into either of the original equations to find the value of the second variable.

STEP 5   Check your solution in both of the original equations.

Strategies in Solving Systems of Equations

You have now seen three different ways to solve systems of linear equations: by graphing, adding, and substitution. The natural question is, Which method should I use in a given situation?

Graphing is the least exact of the methods, and solutions may have to be estimated.

The algebraic methods—addition and substitution—give exact solutions, and both will work for any system of linear equations. In fact, you may have noticed that several examples in this section could just as easily have been solved by adding (Example 3, for instance).

The choice of which of the algebraic methods (substitution or addition) to use is yours and depends largely on the given system. Here are some guidelines designed to help you choose an appropriate method for solving a linear system.

1. If one of the equations is already solved for x (or for y), then substitution is the preferred method.
2. If the coefficients of x (or of y) are the same, or opposites, in the two equations, then addition is the preferred method.
3. If solving for x (or for y) in either of the given equations will result in fractional coefficients, then addition is the preferred method.

Example 6

Select the most appropriate method for solving each of the following systems.

(a) $5x + 3y = 9$
   $2x - 7y = 8$

Addition is the most appropriate method since solving for a variable will result in fractional coefficients.

(b) $7x + 26 = 8$
   $x = 3y - 5$

Substitution is the most appropriate method since the second equation is already solved for x.

(c) $8x - 9y = 11$
   $4x + 9y = 15$

Addition is the most appropriate method since the coefficients of y are opposites.

CHECK YOURSELF 6

Select the most appropriate method for solving each of the following systems.

1. $2x + 5y = \quad 3$ 2. $4x - 3y = 2$
   $8x - 5y = -13$ $y = 3x - 4$
3. $3x - 5y = 2$ 4. $5x - 2y = 19$
   $x = 3y - 2$ $4x + 6y = 38$

CHECK YOURSELF ANSWERS

1. $(-3, -12)$. 2. $(6, 2)$. 3. $(2, 1)$. 4. $(1, -1)$. 5. (1) Inconsistent; (2) dependent.
6. (1) Addition; (2) substitution; (3) substitution; (4) addition.

Name _____

Date _____

Build Your Skills

Solve each of the following systems by substitution.

1. $x + y = 10$
  $y = 4x$

2. $x - y = 4$
  $x = 3y$

3. $2x - y = 10$
  $x = -2y$

4. $x + 3y = 10$
  $3x = y$

5. $3x + 2y = 12$
  $y = 3x$

6. $4x - 3y = 24$
  $y = -4x$

7. $x + y = 5$
  $y = x - 3$

8. $x + y = 9$
  $x = y + 3$

9. $x - y = 4$
  $x = 2y - 2$

10. $x - y = 7$
  $y = 2x - 12$

11. $2x + y = 7$
  $y - x = -8$

12. $3x - y = -15$
  $x = y - 7$

13. $2x - 5y = 10$
  $x - y = 8$

14. $4x - 3y = 0$
  $y = x + 1$

15. $3x + 4y = 9$
  $y - 3x = 1$

16. $5x - 2y = -5$
  $y - 5x = 3$

17. $3x - 18y = 4$
  $x = 6y + 2$

18. $4x + 5y = 6$
  $y = 2x - 10$

19. $5x - 3y = 6$
  $y = 3x - 6$

20. $8x - 4y = 16$
  $y = 2x - 4$

ANSWERS
1. _____
2. _____
3. _____
4. _____
5. _____
6. _____
7. _____
8. _____
9. _____
10. _____
11. _____
12. _____
13. _____
14. _____
15. _____
16. _____
17. _____
18. _____
19. _____
20. _____

21. _____

22. _____

23. _____

24. _____

25. _____

26. _____

27. _____

28. _____

29. _____

30. _____

31. _____

32. _____

33. _____

34. _____

35. _____

36. _____

37. _____

38. _____

21. $8x - 5y = 16$
$\qquad y = 4x - 5$

22. $6x - 5y = 27$
$\qquad x = 5y + 2$

23. $x + 3y = 7$
$\quad x - \ y = 3$

24. $2x - y = -4$
$\quad x + y = -5$

25. $\quad 6x - 3y = \ \ 9$
$\quad -2x + \ y = -3$

26. $5x - 6y = 21$
$\quad x - 2y = \ 5$

27. $\quad x - 7y = \ \ 3$
$\quad 2x - 5y = 15$

28. $\quad 4x - 12y = \ \ 5$
$\quad -x + \ 3y = -1$

29. $4x + 3y = -11$
$\quad 5x + \ y = -11$

30. $5x - 4y = \ \ 5$
$\quad 4x - \ y = -7$

Think About These

Solve each of the following systems by using either addition or substitution. If a unique solution does not exist, state whether the system is dependent or inconsistent.

31. $2x + 3y = -6$
$\qquad x = 3y + 6$

32. $7x + 3y = 31$
$\qquad y = -2x + 9$

33. $\quad 2x - \ y = 1$
$\quad -2x + 3y = 5$

34. $\quad x + 3y = 12$
$\quad 2x - 3y = \ 6$

35. $6x + 2y = 4$
$\qquad y = -3x + 2$

36. $\quad 3x - 2y = \ 15$
$\quad -x + 5y = -5$

37. $\quad x + 2y = \ \ -2$
$\quad 3x + 2y = -12$

38. $10x + 2y = 7$
$\qquad y = -5x + 3$

39. $2x - 3y = 14$
$4x + 5y = -5$

40. $2x + 3y = 1$
$5x + 3y = 16$

41. $4x - 2y = 0$
$x = \dfrac{3}{2}$

42. $4x - 3y = \dfrac{11}{2}$
$y = -\dfrac{3}{2}$

Solve each system.

43. $\dfrac{1}{3}x + \dfrac{1}{2}y = 5$
$\dfrac{x}{4} - \dfrac{y}{5} = -2$

44. $\dfrac{5x}{2} - y = \dfrac{9}{10}$
$\dfrac{3x}{4} + \dfrac{5y}{6} = \dfrac{2}{3}$

45. $0.4x - 0.2y = 0.6$
$2.5x - 0.3y = 4.7$

46. $0.4x - 0.1y = 5$
$6.4x + 0.4y = 60$

47. In 1987, human activities released a total of 8.550 billion metric tons of CO_2 and CH_4 into the atmosphere. The amount of CO_2 released was 29 times the amount of CH_4 released. Use the following system of equations to determine the amount of each gas that was released worldwide by human activities in 1987.

$C + M = 8.550$
$C = 29M$

48. Forty-five percent of the global warming effect comes from the CO_2 released by energy production activities and deforestation. Three and one-half times as much CO_2 is released by energy activities as by deforestation. Use the following system of equations to determine the percentage of global warming caused by each of the above activities.

$E + D = 45$
$E = 3.5D$

ANSWERS

39. _____

40. _____

41. _____

42. _____

43. _____

44. _____

45. _____

46. _____

47. _____

48. _____

49. _____

a. _____

b. _____

c. _____

d. _____

e. _____

f. _____

g. _____

h. _____

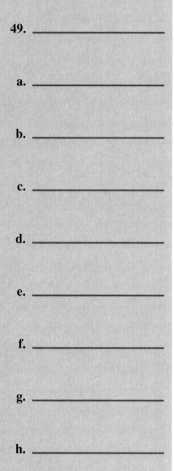

49. Methane and ozone contribute approximately 24 percent of the global warming effect. If methane contributes twice as much warming effect as ozone, use the following system of equations to determine the percentage of effect caused by each gas.

$$M + Z = 24$$
$$M = 2Z$$

Skillscan (Section 7.3)

Write an equation that describes each of the following relationships.

a. The sum of n and d is 55.

b. The value of L is 3 more than twice W.

c. Twice the sum of x and y is 36.

d. 3 times the sum of m and n is 72.

e. 4 times p plus 3 times q is 880.

f. The sum of 9 times x and 7 times y is 3100.

g. The larger number (y) is 8 less than twice the smaller number (x).

h. The larger number (m) is 2 more than 3 times the smaller number (n).

ANSWERS

1. (2, 8) **3.** (4, −2) **5.** $\left(\dfrac{4}{3}, 4\right)$ **7.** (4, 1) **9.** (10, 6) **11.** (5, −3) **13.** (10, 2)

15. $\left(\dfrac{1}{3}, 2\right)$ **17.** Inconsistent **19.** (3, 3) **21.** $\left(\dfrac{3}{4}, -2\right)$ **23.** (4, 1)

25. Dependent **27.** (10, 1) **29.** (−2, −1) **31.** (0, −2) **33.** (2, 3)

35. Dependent **37.** $\left(-5, \dfrac{3}{2}\right)$ **39.** $\left(\dfrac{5}{2}, -3\right)$ **41.** $\left(\dfrac{3}{2}, 3\right)$ **43.** (0, 10)

45. (2, 1) **47.** CH_4: 285 million tons, CO_2: 8.265 billion tons **49.** Ozone 8 percent, methane 16 percent **a.** $n + d = 55$ **b.** $L = 2W + 3$ **c.** $2(x + y) = 36$ **d.** $3(m + n) = 72$ **e.** $4p + 3q = 880$ **f.** $9x + 7y = 3100$ **g.** $y = 2x - 8$ **h.** $m = 3n + 2$

8.4

Systems of Linear Equations: Applications

OBJECTIVE

To solve applied or word problems by using systems of equations

In Chapter 3 we solved word problems by using equations in a single variable. Remember that all the unknowns in the problem had to be expressed in terms of that single variable.

Now that you have the background to use two equations in two variables to solve word problems, let's see how they can be applied. The five steps for solving word problems stay the same (in fact, we give them again for reference in our first example). Many students find that using two equations and two variables makes writing the necessary equations much easier.

Often problems can be solved by using either one or two variables. We'll start with an example done both ways so that you can compare.

Example 1

Here are the steps for using a single variable:

The sum of two numbers is 25. If the second number is 5 less than twice the first number, what are the two numbers?

1. Read the problem carefully. What do you want to find?

Step 1 You want to find the two unknown numbers.

2. Assign variables to the unknown quantities.

Step 2 Let $x =$ the first number.

Then $2x - 5$ is the second number.

Twice   5 less than

3. Write the equation for the solution.

Step 3

$$x + (2x - 5) = 25$$

The sum of the numbers is 25.

4. Solve the equation.

Step 4

$$x + (2x - 5) = 25$$
$$3x - 5 = 25$$
$$3x = 30$$
$$x = 10$$

x

10 is the first number.

$2x - 5$

15 is the second number.

5. Verify your result by returning to the original problem.

Step 5

$$10 + 15 = 25$$

The solution checks.

The sum of two numbers is 28. The second number is 4 more than twice the first number. Find the two numbers. Use an equation in a single variable for the solution.

Now, to compare the two approaches, let's look again at the problem of Example 1. This time we'll use two equations in two variables.

Example 2

The sum of two numbers is 25. If the second number is 5 less than twice the first number, what are the two numbers?

1. What do you want to find?

Step 1   You want to find the two unknown numbers.

2. Assign variables. This time we use two letters, x and y.

Step 2   Let $x =$ the first number and $y =$ the second number.

3. Write equations for the solution. Here two equations are needed because we have introduced two variables.

Step 3

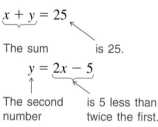

$$\underbrace{x + y}_{} = 25$$

The sum   is 25.

$$y = \underbrace{2x - 5}_{}$$

The second   is 5 less than
number   twice the first.

4. Solve the system of equations.

Step 4

$$x + y = 25 \tag{1}$$
$$y = 2x - 5 \tag{2}$$

We have used the substitution method because equation (2) is already solved for y.

Substitute $2x - 5$ for y in equation (1).

$$x + (2x - 5) = 25$$
$$3x - 5 = 25$$
$$x = 10$$

From equation (1),

$$10 + y = 25$$
$$y = 15$$

The two numbers are 10 and 15.

5. Check the result.

Step 5 The sum of the numbers is 25. The second number, 15, is 5 less than twice the first number, 10. The solution checks.

You should go back now and compare our two approaches to this problem. See which you prefer.

CHECK YOURSELF 2

Solve the problem of Check Yourself 1, this time using two variables for the solution.

There are some problems that cannot be readily solved by using a single variable, but can be solved rather easily in two variables. The following example illustrates.

Example 3

Ryan bought 8 pens and 7 pencils and paid a total of $7.40. Ashleigh purchased 2 pens and 10 pencils and paid $3.50. Find the cost for a single pen and a single pencil.

Step 1 You want to find the cost of a single pen and the cost of a single pencil.

Step 2 Let x be the cost of a pen and y be the cost of a pencil.

Step 3 Write the two necessary equations.

In the first equation, $8x$ is the total cost of the pens, $7y$ the total cost of the pencils. The second equation is formed in a similar fashion.

$$8x + 7y = 7.40 \qquad (1)$$

$$2x + 10y = 3.50 \qquad (2)$$

Step 4 Solve the system formed in step 3. We multiply equation (2) by -4. Adding will then eliminate the variable x.

$$8x + 7y = 7.40$$
$$-8x - 40y = -14.00$$

Now adding the equations, we have

$$-33y = -6.60$$
$$y = 0.20$$

Substituting 0.20 for y in equation (1), we have

$$8x + 7(0.20) = 7.40$$
$$8x + 1.40 = 7.40$$
$$8x = 6.00$$
$$x = 0.75$$

Step 5 From the results of step 4 we see that the pens are 75¢ each and that the pencils are 20¢ each.

To check these solutions, replace x with 0.75 and y with 0.20 in equation (1).

$$8(0.75) + 7(0.20) = 7.40$$
$$6.00 + 1.40 = 7.40$$
$$7.40 = 7.40 \qquad \text{(True)}$$

We will leave it to the reader to check these values in Equation (2).

CHECK YOURSELF 3

Annette bought three digital tapes and two compact disks on sale for $66. At the same sale, Terry bought three digital tapes and four compact disks for $96. Find the individual price for a tape and a disk.

The next example shows how sketches can be helpful in setting up a problem.

Example 4

An 18-ft board is cut into two pieces, one of which is 4 ft longer than the other. How long is each piece?

Step 1 You want to find the two lengths.

Step 2 Let x be the length of the longer piece and y the length of the shorter piece. You should always draw a sketch of the problem whenever it is appropriate.

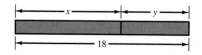

Step 3   Write the equations for the solution.

$$x + y = 18$$
$\uparrow$
The total length is 18.

Our second equation could also be written as

$$x = y + 4$$

in which case you would probably use substitution for the solution.

$$x - y = 4$$
$\uparrow$
The difference in lengths is 4.

Step 4   To solve the system, add:

$$
\begin{array}{ll}
x + y = 18 & \qquad (1) \\
\underline{x - y = 4} & \qquad (2) \\
2x = 22 & \\
x = 11 &
\end{array}
$$

Replace x with 11 in equation (1).

$$
\begin{array}{l}
11 + y = 18 \\
 y = 7
\end{array}
$$

The longer piece has length 11 ft, the shorter piece 7 ft.

Step 5   We'll leave it to you to check this result in the original problem.

CHECK YOURSELF 4

A 20-ft board is cut into two pieces, one of which is 6 ft longer than the other. How long is each piece?

Sketches will always be helpful in solving applications from geometry. Let's look at such an example.

Example 5

The length of a rectangle is 3 meters (m) more than twice its width. If the perimeter of the rectangle is 42 m, find the dimensions of the rectangle.

Step 1   You want to find the dimensions (length and width) of the rectangle.

We have used x and y as our two variables in the previous examples. Use whatever letters you want. The process is the same, and sometimes it helps you remember what letter stands for what. Here L = length and W = width.

Step 2   Let L be the length of the rectangle and W the width. Now draw a sketch of the problem.

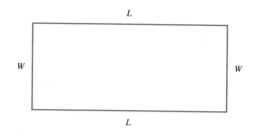

Step 3   Write the equations for the solution.

$$L = \underbrace{2W + 3}$$

3 more than twice
the width

$$\underbrace{2L + 2W} = 42$$

The perimeter

Step 4   Solve the system.

$$L = 2W + 3 \qquad\qquad (1)$$
$$2L + 2W = 42 \qquad\qquad (2)$$

Substitution is used since one equation is already solved for a variable.

From equation (1) we can substitute $2W + 3$ for L in equation (2).

$$2(2W + 3) + 2W = 42$$
$$4W + 6 + 2W = 42$$
$$6W = 36$$
$$W = 6$$

Replace W with 6 in equation (1) to find L.

$$L = 2 \cdot 6 + 3$$
$$= 12 + 3$$
$$= 15$$

The length is 15 m, the width is 6 m.

Step 5   Check these results. The perimeter is $2 \cdot 15 + 2 \cdot 6$ m, or 42 m. The length (15 m) is 3 m more than twice the width (6 m).

CHECK YOURSELF 5

The length of the two equal legs of an isosceles triangle is 5 in less than the length of the base. If the perimeter of the triangle is 50 in, find the lengths of the legs and the base.

Using two equations in two variables may also help you in writing equations to solve mixture problems.

Example 6

Winnifred has collected $4.50 in nickels and dimes. If she has 55 coins, how many of each kind of coin does she have?

Step 1   You want to find the number of nickels and the number of dimes.

Step 2   Let

Again we choose appropriate variables—*n* for nickels, *d* for dimes.

n = number of nickels

d = number of dimes

Step 3   Write the equations for the solution.

$$n + d = 55$$

There are 55 coins
in all.

Remember: The value of a number of coins is the value per coin times the number of coins: 5*n*, 10*d*, etc.

$$5n + 10d = 450$$

Value of   Value   Total value
nickels   of dimes   (in cents)

Step 4   We now have the system

$$n + d = 55 \qquad (1)$$
$$5n + 10d = 450 \qquad (2)$$

Addition is used to solve the system since the coefficients of the variables are not the same or opposites.

Let's solve this system by addition. Multiply equation (1) by -5. We then add the equation to eliminate the variable n.

$$
\begin{array}{rcr}
-5n - 5d & = & -275 \\
5n + 10d & = & 450 \\
\hline
5d & = & 175 \\
d & = & 35
\end{array}
$$

We now substitute d for 35 in equation (1).

$$n + 35 = 55$$
$$n = 20$$

There are 20 nickels and 35 dimes.

Step 5   We'll leave it to you to check this result. Just verify that the value of these coins is $4.50.

CHECK YOURSELF 6

Tickets for a play cost $8 or $6. If 350 tickets were sold in all and receipts were $2500, how many of each price ticket were sold?

We can also solve mixture problems which involve percentages by using two equations in two unknowns. Look at the following example.

Example 7

In a chemistry lab are a 20% acid solution and a 60% acid solution. How many milliliters of each should be mixed to produce 200 milliliters (mL) of a 44% acid solution?

Step 1 You need to know the amount of each solution to use.

Step 2 Let

x = amount of 20% acid solution
y = amount of 60% acid solution

Step 3 A drawing will help. Note that a 20% acid solution is 20% acid and 80% water.

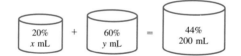

We can write equations from the total amount of the solution, here 200 mL, and from the amount of acid in that solution. Many students find a table helpful in organizing the information at this point. Here, for example, we might have

Note: The amount of acid is the amount of solution times the percentage of acid (as a decimal). That is the key to forming the third column of our table.

	AMOUNT OF SOLUTION	% ACID	AMOUNT OF ACID
	x	0.20	$0.20x$
	y	0.60	$0.60y$
TOTALS	200	0.44	(0.44)(200)

Now we are ready to form our system.

The total amount of the solution from the first column of our table.

$$x + y = 200 \tag{1}$$

The amount of acid from the third column of our table.

$$0.20x + 0.60y = 0.44(200) \tag{2}$$

Acid in 20% solution.　　Acid in 60% solution.　　Acid in mixture.

Step 4 If we multiply equation (2) by 100 to clear of decimals, we have

$$x + \quad y = \quad 200$$
$$20x + 60y = 8800$$

To complete the solution, multiply the first equation by -20 and add. We can then solve for y.

$$
\begin{array}{rcl}
-20x - 20y &=& -4000 \\
\underline{20x + 60y} &=& \underline{\quad 8800} \\
40y &=& \quad 4800 \\
y &=& \quad 120
\end{array}
$$

Substituting 120 for y in equation (1), we have

$$x + 120 = 200$$
$$x = \quad 80$$

The amounts to be mixed are 80 mL (20% acid solution) and 120 mL (60% acid solution).

Step 5 You can check this solution by verifying that the amount of acid from the 20% solution and the amount from the 60% solution are equal to the amount of acid in the mixture.

CHECK YOURSELF 7

You have a 30% alcohol solution and a 50% alcohol solution. How much of each solution should be combined to make 400 mL of a 45% alcohol solution?

A related kind of application involves interest. The key equation involves the *principal* (the amount invested), the annual *interest rate,* the *time* (in years) that the money is invested, and the amount of *interest* you receive.

$$I = P \cdot r \cdot t$$

Interest   Principal   Rate   Time

For 1 year we have

$$I = P \cdot r \qquad \text{since } t = 1$$

Example 8

Jeremy inherits $20,000 and invests part of the money in bonds with an interest rate of 11 percent. The remainder of the money is in savings at a 9 percent rate. What amount has he invested at each rate if he receives $2040 in interest for 1 year?

Step 1   You want to find the amounts invested at 11 percent and at 9 percent.

The amount invested at 11 percent could have been represented by y, and the amount invested at 9 percent could have been represented by x.

Step 2   Let $x =$ the amount invested at 11 percent and $y =$ the amount invested at 9 percent. Once again you may find a table helpful at this point. Here we might have

Note: $I = P \cdot r$—interest equals principal times rate—is the key to forming the third column of our table.

	PRINCIPAL	RATE	INTEREST
	x	11%	$0.11x$
	y	9%	$0.09y$
TOTALS	20,000		2040

Step 3   Form the equations for the solution, using the first and second columns of the above table.

$$x + y = 20,000$$

He has $20,000 invested in all.

Note the decimal form of 11 percent and 7 percent is used in the equation.

$$0.11x + 0.09y = 2040$$

The interest at 11% (rate · principal)   The interest at 9%   The total interest

Step 4   To solve the following system, use addition.

$$x + y = 20,000 \tag{1}$$
$$0.11x + 0.09y = 2,040 \tag{2}$$

To do this, multiply both sides of equation (1) by -9. Multiplying both sides of equation (2) by 100 will clear decimals. Adding the resulting equations will eliminate y.

$$
\begin{aligned}
-9x - 9y &= -180,000 \\
11x + 9y &= 204,000 \\
\hline
2x &= 24,000 \\
x &= 12,000
\end{aligned}
$$

Now substitute 12,000 for x in equation (1) and solve for y.

$$12,000 + y = 20,000$$
$$y = 8,000$$

Be sure to answer the question asked in the problem.

Jeremy has $12,000 invested at 11 percent and $8000 invested at 9 percent.

Step 5   To check, the interest at 11 percent is ($12,000)(0.11), or $1320. The interest at 9 percent is ($8000)(0.09), or $720. The total interest is $2040, and the solution is verified.

CHECK YOURSELF 8

Jan has $2000 more invested in a stock that pays 9 percent interest than in a savings account paying 8 percent. If her total interest for 1 year is $860, how much does she have invested at each rate?

Another group of applications is called *motion problems*. They involve a distance traveled, the rate, and the time of travel. Recall from Section 4.7 that these quantities are related by the equation

$$d = r \cdot t$$

Distance Rate Time

Our final example shows the use of this equation to form a system of equations to solve a motion problem.

Example 9

A boat can travel 36 mi downstream in 2 h. In coming back upstream, the trip takes 3 h. What is the rate of the boat in still water? What is the rate of the current?

Step 1 You want to find the two rates (of the boat and the current).

Step 2 Let

x = rate of boat in still water

y = rate of current

Step 3 To write the equations, think about the following: What is the effect of the current? Suppose the boat's rate in still water is 10 mi/h and the current is 2 mi/h.

The current *increases* the rate *downstream* to 12 mi/h ($10 + 2$). The current *decreases* the rate *upstream* to 8 mi/h ($10 - 2$). So here the rate downstream will be $x + y$, and the rate upstream will be $x - y$. At this point a table of information can again be helpful.

	DISTANCE	RATE	TIME
DOWNSTREAM	36	$x + y$	2
UPSTREAM	36	$x - y$	3

From the relationship $d = r \cdot t$ we can now use our table to write the system

From line 1 of our table $36 = 2(x + y)$

From line 2 $36 = 3(x - y)$

Step 4 Removing the parentheses in the equations of step 3, we have

$2x + 2y = 36$

$3x - 3y = 36$

By either of our earlier methods, this system gives values of 15 for x and 3 for y.

The rate in still water is 15 mi/h, and the rate of the current is 3 mi/h. We will leave the check to you.

CHECK YOURSELF 9

A plane flies 480 mi with the wind in 4 h. In returning against the wind, the trip takes 6 h. What is the rate of the plane in still air? What was the rate of the wind?

CHECK YOURSELF ANSWERS

1. The numbers are 8 and 20.   2. The numbers are 8 and 20.
3. Tape $12, disk $15.   4. 7 ft and 13 ft.
5. The legs have length 15 in; the base is 20 in.
6. 150 $6 tickets and 200 $8 tickets were sold.
7. 100 mL (30%), 300 mL (50%).
8. $4000 at 8 percent, $6000 at 9 percent.
9. Plane's rate in still air, 100 mi/h; wind's rate, 20 mi/h.

Name _____

Date _____

Build Your Skills

Solve each of the following problems. Be sure to show the equations used for the solution.

Find the two numbers in Exercises 1 to 8.

1. The sum of the numbers is 40. Their difference is 8.

2. The sum of the numbers is 100. The second is 3 times the first.

3. The sum of the numbers is 70. The second is 10 more than 3 times the first.

4. The sum of the numbers is 56. The second is 4 less than twice the first.

5. The difference of the numbers is 4. The larger is 8 less than twice the smaller.

6. The difference of the numbers is 22. The larger is 2 more than 3 times the smaller.

7. One number is 18 more than another, and the sum of the smaller number and twice the larger number is 45.

8. One number is 5 times another. The larger number is 9 more than twice the smaller.

9. Eight eagle stamps and two raccoon stamps cost $2.80. Three eagle stamps and four raccoon stamps cost $2.35. Find the cost of each kind of stamp.

10. Robin bought four chocolate bars and a pack of gum and paid $2.75. Meg bought two chocolate bars and three packs of gum and paid $2.25. Find the cost of each.

ANSWERS

1. _____

2. _____

3. _____

4. _____

5. _____

6. _____

7. _____

8. _____

9. _____

10. _____

11. _____

11. Xavier bought 5 Red Delicious apples and 4 Granny Smith apples at a cost of $4.81. Dean bought one of each of the two types at a cost of $1.08. Find the cost for each kind of apple.

12. _____

12. Four single-sided disks and two double-sided disks cost a total of $5.10. Two single-sided and four double-sided disks cost $5.40. Find the unit cost for each.

13. _____

13. A 30-m rope is cut into two pieces so that one piece is 6 m longer than the other. How long is each piece?

14. _____

14. An 18-ft board is cut into two pieces, one of which is twice as long as the other. How long is each piece?

15. _____

15. Two packages together weigh 32 kilograms (kg). The smaller package weighs 6 kg less than the larger. How much does each package weigh?

16. _____

16. A washer-dryer combination costs $400. If the washer costs $40 more than the dryer, what does each appliance cost separately?

17. In a town election, the winning candidate had 220 more votes than the loser. If 810 votes were cast in all, how many votes did each candidate receive?

17. _____

18. An office desk and chair together cost $250. If the desk cost $20 less than twice as much as the chair, what did each cost?

18. _____

19. The length of a rectangle is 2 in more than twice its width. If the perimeter of the rectangle is 34 in, find the dimensions of the rectangle.

19. _____

20. The perimeter of an isosceles triangle is 37 in. The lengths of the two equal legs are 6 in less than 3 times the length of the base. Find the lengths of the three sides.

20. _____

21. Jill has $3.50 in nickels and dimes. If she has 50 coins, how many of each type of coin does she have?

22. Richard has 22 coins with a total value of $4. If the coins are all quarters and dimes, how many of each type of coin does he have?

23. Theater tickets are $4 for general admission and $3 for students. During one evening 240 tickets were sold, and the receipts were $880. How many of each kind of ticket were sold?

24. 400 tickets were sold for a concert. The receipts from ticket sales were $3100, and the ticket prices were $7 and $9. How many of each price ticket were sold?

25. A coffee merchant has coffee beans which sell for $3 per pound and $5 per pound. The two types are to be mixed to create 100 lb of a mixture that will sell for $4.50 per pound. How much of each type of bean should be used in the mixture?

26. Peanuts are selling for $2 per pound, and cashews are selling for $5 per pound. How much of each type of nut would be needed to create 20 lb of a mixture that would sell for $2.75 per pound?

27. A chemist has a 25% and a 50% acid solution. How much of each solution should be used to form 200 milliliters (mL) of a 35% acid solution?

28. A pharmacist wishes to prepare 150 mL of a 20% alcohol solution. She has a 30% solution and a 15% solution in her stock. How much of each should be used in forming the desired mixture?

29. You have two alcohol solutions, one a 15% solution and one a 45% solution. How much of each solution should be used to obtain 300 mL of a 25% solution?

30. _____

31. _____

32. _____

33. _____

34. _____

35. _____

36. _____

37. _____

38. _____

30. If you combine a 10% acid solution and a 50% acid solution, how much of each should be used to make 40 centiliters (cL) of a 40% acid solution?

31. Otis has a total of $12,000 invested in two accounts. One account pays 8 percent and the other 9 percent. If his interest for 1 year is $1010, how much does he have invested at each rate?

32. Amy invests a part of $8000 in bonds paying 12 percent interest. The remainder is in a savings account at 8 percent. If she receives $840 in interest for 1 year, how much does she have invested at each rate?

33. Evander has $2000 more invested in an account paying 10 percent interest than in a second account that pays 8 percent. If his interest is $920 for 1 year, what does he have invested at each rate?

34. Raoul invested twice as much in an account paying 6 percent than he did in one paying 7 percent. He received $950 in interest for 1 year. What amount did he have invested at each rate?

35. David was able to row 16 mi downstream in 2 h. Returning upstream, he took 4 h to make the trip. How fast can he row in still water? What was the rate of the current?

36. A plane flies 450 mi with the wind in 3 h. Flying back against the wind, the plane takes 5 h to make the trip. What was the rate of the plane in still air? What was the rate of the wind?

37. An airliner made a trip of 1800 mi in 3 h, flying east across the country with the jetstream directly behind it. The return trip, against the jetstream, took 4 h. Find the speed of the plane in still air and the speed of the jetstream.

38. A boat traveled 60 mi upstream in 5 h. Returning downstream with the current, the boat took 3 h to make the trip. What was the boat's speed in still water, and what was the speed of the river's current?

Each of the following applications can be solved by the use of a system of linear equations. Match the application with the system on the right that could be used for its solution.

39. One number is 4 less than 3 times another. If the sum of the numbers is 36, what are the two numbers?

(a) $12x + 5y = 116$
$8x + 12y = 112$

40. Suppose that a movie theater sold 300 adult and student tickets for a showing with a revenue of $1440. If the adult tickets were $6 and the student tickets $4, how many of each type of ticket were sold?

(b) $x + y = 8000$
$0.06x + 0.09y = 600$

(c) $x + y = 200$
$0.20x + 0.60y = 90$

41. The length of a rectangle is 3 centimeters (cm) more than twice its width. If the perimeter of the rectangle is 36 cm, find the dimensions of the rectangle.

(d) $x + y = 36$
$y = 3x - 4$

42. An order of 12 dozen roller-ball pens and 5 dozen ballpoint pens cost $116. A later order for 8 dozen roller-ball pens and 12 dozen ballpoint pens cost $112. What was the cost of 1 dozen of each of the pens?

(e) $2(x + y) = 36$
$3(x - y) = 36$

(f) $x + y = 300$
$6x + 4y = 1440$

43. A candy merchant wishes to mix peanuts selling at $2 per pound with cashews selling for $5.50 per pound to form 140 lb of a mixed-nut blend which will sell for $3 per pound. What amount of each type of nut should be used?

(g) $L = 2W + 3$
$2L + 2W = 36$

(h) $x + y = 140$
$2x + 5.5y = 420$

44. Rolando has investments totaling $8000 in two accounts, one a savings account paying 6 percent interest and the other a bond paying 9 percent. If the annual interest from the two investments was $600, how much did he have invested at each rate?

45. A chemist wants to combine a 20% alcohol solution with a 60% solution to form 200 mL of a 45% solution. How much of each of the solutions should be used to form the mixture?

46. Marcus was able to make a downstream trip of 36 mi in 2 h. In returning upstream, it took 3 h to make the trip. How fast can his boat travel in still water? What was the rate of the river's current?

47. _____

47. In 1987 the United States released a total of 670 million tons of carbon dioxide heating equivalents in the form of CO_2 and CH_4. If there was 4 times as much CO_2 released as CH_4, how many million tons of each greenhouse gas were released by the United States that year?

48. _____

48. In 1987 the area then known as the U.S.S.R. released a total of 480 million tons of carbon dioxide heating equivalents in the form of CFCs and CH_4. If 3 times as many CFCs were released as CH_4, how many million tons of each greenhouse gas were released by the U.S.S.R. that year?

49. _____

49. A total of 2.7 million metric tons of methane was emitted in Canada from solid waste and livestock in 1987. If the United States emitted a total of 23 million metric tons of methane from the same sources and 8 times as much was caused by solid waste in the United States while 10 times as much was caused by livestock, find the amount of gas released by solid waste and the amount of gas released by livestock for each country.

50. _____

50. In 1987, the total CO_2 released by deforestation in Africa and South America was 630 million metric tons. If $1\frac{5}{8}$ times as much CO_2 was released in Africa as in South America, how much gas was released by each continent?

51. _____

51. The total 1987 increase in emissions of CH_4 and CFCs in South America was equivalent to 90 million metric tons of carbon. If North America released 600 million metric tons of carbon equivalent and there was 4 times as much CH_4 and 10 times as many CFCs released compared to South America, how many tons of carbon equivalent of each gas were released on each continent?

ANSWERS

1. 24, 16 **3.** 15, 55 **5.** 16, 12 **7.** 3, 21 **9.** 25¢, 40¢ **11.** 49¢, 59¢ **13.** 18 m, 12 m **15.** 13 kg, 19 kg **17.** 515, 295 **19.** 5 in by 12 in **21.** 30 nickels, 20 dimes **23.** 160 general, 80 student **25.** 25 lb at $3, 75 lb at $5 **27.** 120 mL of 25%, 80 mL of 50% **29.** 200 mL of 15%, 100 mL of 45% **31.** $7000 at 8%, $5000 at 9% **33.** $6000 at 10%, $4000 at 8% **35.** 6 mi/h, 2 mi/h **37.** 525 mi/h, 75 mi/h **39.** _d_ **41.** _g_ **43.** _h_ **45.** _c_ **47.** CO_2: 536 million tons, CH_4: 134 million tons **49.** Canada: solid waste 2 million tons, livestock 700,000 tons; United States: solid waste 16 million tons, livestock 7 million tons **51.** South America: 50 million tons CH_4, 40 million tons CFCs; North America: 200 million tons CH_4, 400 million tons CFCs

Summary

Systems of Linear Equations [8.1 to 8.3]

A System of Equations   Two or more equations considered together.

Solution   The solution of a system of two equations in two unknowns is an ordered pair that satisfies each equation of the system.

Solving by Graphing

1. Graph both equations on the same coordinate system.
2. The system may have
   (*a*) *One solution.* The lines intersect at one point (a consistent system). The solution is the ordered pair corresponding to that point.
   (*b*) *No solution.* The lines are parallel (an inconsistent system).
   (*c*) *Infinitely many solutions.* The two equations have the same graph (a dependent system). Any ordered pair corresponding to a point on the line is a solution.

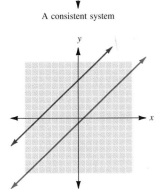

A consistent system

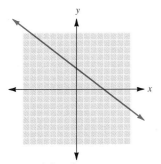

An inconsistent system

A dependent system

Solving by Adding

$2x - y = 4$ (1)
$3x + 2y = 13$ (2)
Multiply equation (1) by 2.

$4x - 2y = 8$
$3x + 2y = 13$

Add.

$7x = 21$
$x = 3$

1. If necessary, multiply both sides of one or both equations by nonzero numbers to form an equivalent system in which the coefficients of one of the variables are opposites.

2. Add the equations of the new system.

3. Solve the resulting equation for the remaining variable.

In equation (1),

$2 \cdot 3 - y = 4$

$y = 2$

(3, 2) is the solution.

$x - 2y = 3$ (1)

$2x + 3y = 13$ (2)

From equation (1),

$x = 2y + 3$

Substitute in equation (2):

$2(2y + 3) + 3y = 13$

$4y + 6 + 3y = 13$

$7y + 6 = 13$

$7y = 7$

$y = 1$

Continue as before.

4. Substitute the value found in step 3 into either of the original equations to find the value of the second variable.
5. Check your solution in both of the original equations.

Solving by Substitution

1. Solve one of the given equations for x or for y. If this is already done, go on to step 2.

2. Substitute this expression for x or for y into the other equation.

3. Solve the resulting equation for the remaining variable.
   Steps 4 and 5 are the same as above.

Solving Word Problems by Using Systems of Equations [8.4]

Applying Systems of Equations   Often word problems can be solved by using two variables and two equations to represent the unknowns and the given relationships in the problem.

The Solution Steps

1. Read the problem carefully. Then reread it to decide what you are asked to find.
2. Choose letters to represent the unknowns.
3. Translate the problem to the language of algebra to form a system of equations.
4. Solve the system.
5. Verify your solution in the original problem.

Summary Exercises Chapter 8

This summary exercise set is provided to give you practice with each of the objectives of the chapter. Each exercise is keyed to the appropriate chapter section. The answers are provided in the instructor's manual.

[8.1] Solve each of the following systems by graphing.

1. $x + y = 6$
$x - y = 2$

2. $x - y = 8$
$2x + y = 7$

3. $x + 2y = 4$
$x + 2y = 6$

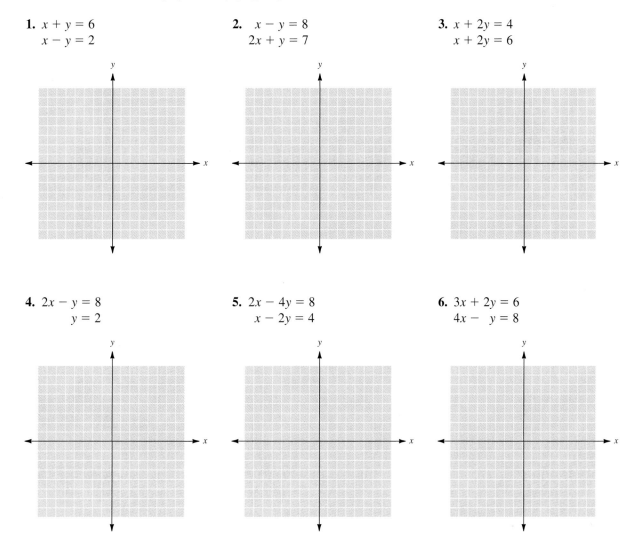

4. $2x - y = 8$
$y = 2$

5. $2x - 4y = 8$
$x - 2y = 4$

6. $3x + 2y = 6$
$4x - y = 8$

[8.2] Solve each of the following systems by adding.

7. $x + y = 8$
$x - y = 2$

8. $-x - y = 4$
$x - y = -8$

9. $2x - 3y = 16$
$5x + 3y = 19$

10. $2x + y = 7$
$3x - y = 3$

11. $3x - 5y = 14$
$3x + 2y = 7$

12. $2x - 4y = 8$
$x - 2y = 4$

13. $4x - 3y = -22$
$4x + 5y = -6$

14. $5x - 2y = 17$
$3x - 2y = 9$

15. $4x - 3y = 10$
$2x - 3y = 6$

16. $2x + 3y = -10$
$-2x + 5y = 10$

17. $3x + 2y = 3$
$6x + 4y = 5$

18. $3x - 2y = 23$
$x + 5y = -15$

19. $5x - 2y = -1$
$10x + 3y = 12$

20. $x - 3y = 9$
$5x - 15y = 45$

21. $2x - 3y = 18$
$5x - 6y = 42$

22. $3x + 7y = 1$
$4x - 5y = 30$

23. $5x - 4y = 12$
$3x + 5y = 22$

24. $6x + 5y = -6$
$9x - 2y = 10$

25. $4x - 3y = 7$
$-8x + 6y = -10$

26. $3x + 2y = 8$
$-x - 5y = -20$

27. $3x - 5y = -14$
$6x + 3y = -2$

[8.3] Solve each of the following systems by substitution.

28. $x + 2y = 10$
$y = 2x$

29. $x - y = 10$
$x = -4y$

30. $2x - y = 10$
$x = 3y$

31. $2x + 3y = 2$
$y = x - 6$

32. $4x + 2y = 4$
$y = 2 - 2x$

33. $x + 5y = 20$
$x = y + 2$

34. $6x + y = 2$
$y = 3x - 4$

35. $2x + 6y = 10$
$x = 6 - 3y$

36. $2x + y = 9$
$x - 3y = 22$

37. $x - 3y = 17$
$2x + y = 6$

38. $2x + 3y = 4$
$y = 2$

39. $4x - 5y = -2$
$x = -3$

40. $-6x + 3y = -4$
$y = -\dfrac{2}{3}$

41. $5x - 2y = -15$
$y = 2x + 6$

42. $3x + y = 15$
$x = 2y + 5$

[8.3] Solve each of the following systems by either addition or substitution.

43. $x - 4y = 0$
$4x + y = 34$

44. $2x + y = 2$
$y = -x$

45. $3x - 3y = 30$
$x = -2y - 8$

46. $5x + 4y = 40$
    $x + 2y = 11$

47. $x - 6y = -8$
    $2x + 3y = 4$

48. $4x - 3y = 9$
    $2x + y = 12$

49. $9x + y = 9$
    $x + 3y = 14$

50. $3x - 2y = 8$
    $-6x + 4y = -16$

51. $3x - 2y = 8$
    $2x - 3y = 7$

[8.4] Solve the following problems. Be sure to show the equations used.

52. The sum of two numbers is 40. If their difference is 10, find the two numbers.

53. The sum of two numbers is 17. If the second number is 1 more than 3 times the first, what are the two numbers?

54. The difference of two numbers is 8. The larger number is 2 less than twice the smaller. Find the numbers.

55. Five writing tablets and three pencils cost $8.25. Two tablets and two pencils cost $3.50. Find the cost for each item.

56. A cable which is 200 ft long is cut into two pieces so that one piece is 12 ft longer than the other. How long is each piece?

57. An amplifier and a pair of speakers cost $925. If the amplifier costs $75 more than the speakers, what does each cost?

58. A sofa and chair cost $850 as a set. If the sofa costs $100 more than twice as much as the chair, what is the cost of each?

59. The length of a rectangle is 4 cm more than its width. If the perimeter of the rectangle is 64 cm, find the dimensions of the rectangle.

60. The perimeter of an isosceles triangle is 29 in. The lengths of the two equal legs are 2 in more than twice the length of the base. Find the lengths of the three sides.

61. Darryl has 30 coins with a value of $5.50. If they are all nickels and quarters, how many of each kind of coin does he have?

62. Tickets for a concert sold for $11 and $8. If 600 tickets were sold for one evening and the receipts were $5550, how many of each kind of ticket were sold?

63. A laboratory has a 20% acid solution and a 50% acid solution. How much of each should be used to produce 600 mL of a 40% acid solution?

64. A service station wishes to mix 40 L of a 78% antifreeze solution. How many liters of a 75% solution and a 90% solution should be used in forming the mixture?

65. Martha has $18,000 invested. Part of the money is invested in a bond which yields 11 percent interest. The remainder is in her savings account, which pays 7 percent. If she earns $1660 in interest for 1 year, how much does she have invested at each rate?

66. A boat travels the 24 mi upstream in 3 h. It then takes the same 3 h to go 36 mi downstream. Find the speed of the boat in still water and the speed of the current.

67. A plane flying with the wind makes a trip of 2200 mi in 4 h. Returning against the wind, it can travel only 1800 mi in 4 h. What is the plane's rate in still air? What is the wind speed?

The purpose of this self-test is to help you check your progress and to review for a chapter test in class. Allow yourself about an hour to take the test. When you are done, check your answers in the back of the book. If you missed any problems, be sure to go back and review the appropriate sections in the chapter and the exercises that are provided.

ANSWERS

1. _____

2. _____

3. _____

4. _____

5. _____

6. _____

Solve each of the following systems by graphing.

1. $x + y = 5$
　$x - y = 3$

2. $x + 2y = 8$
　$x - y = 2$

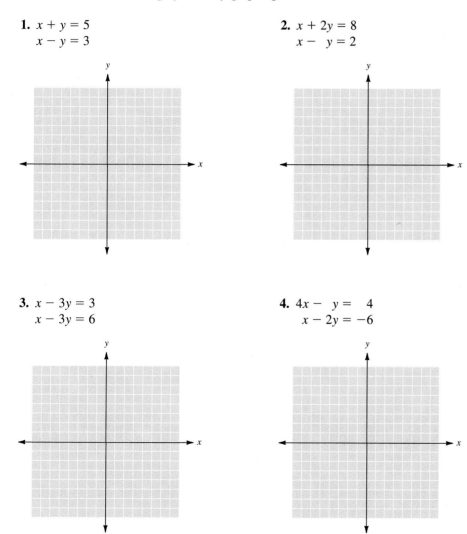

3. $x - 3y = 3$
　$x - 3y = 6$

4. $4x - y = 4$
　$x - 2y = -6$

Solve each of the following systems by adding.

5. $x + y = 5$
　$x - y = 3$

6. $x + 2y = 8$
　$x - y = 2$

7. _____

8. _____

9. _____

10. _____

11. _____

12. _____

13. _____

14. _____

15. _____

16. _____

17. _____

18. _____

19. _____

20. _____

21. _____

22. _____

23. _____

24. _____

25. _____

7. $3x + y = 6$
$-3x + 2y = 3$

8. $3x + 2y = 11$
$5x + 2y = 15$

9. $3x - 6y = 12$
$x - 2y = 4$

10. $4x + y = 2$
$8x - 3y = 9$

11. $2x - 5y = 2$
$3x + 4y = 26$

12. $x + 3y = 6$
$3x + 9y = 9$

Solve each of the following systems by substitution.

13. $x + y = 8$
$y = 3x$

14. $x - y = 9$
$x = -2y$

15. $2x - y = 10$
$x = y + 4$

16. $x - 3y = -7$
$y = x - 1$

17. $3x + y = -6$
$y = 2x + 9$

18. $4x + 2y = 8$
$y = 3 - 2x$

19. $5x + y = 10$
$x + 2y = -7$

20. $3x - 2y = 5$
$2x + y = 8$

Solve each of the following problems. Be sure to show the equations used.

21. The sum of two numbers is 30, and their difference is 6. Find the two numbers.

22. A rope 50 m long is cut into two pieces so that one piece is 8 m longer than the other. How long is each piece?

23. The length of a rectangle is 4 in less than twice its width. If the perimeter of the rectangle is 64 in, what are the dimensions of the rectangle?

24. Murray has 30 coins with a value of $5.70. If the coins are all dimes and quarters, how many of each coin does he have?

25. Jackson was able to travel 36 mi downstream in 2 h. In returning upstream, it took 3 h to make the trip. How fast can his boat travel in still water? What was the rate of the river current?

This test is provided to help you in the process of review of the previous chapters. Answers are provided in the back of the book. If you missed any problems, be sure to go back and review the appropriate chapter sections.

Determine which of the ordered pairs are solutions for the given equations.

1. $x + y = 8$ $(4, 4), (8, 0), (5, 2), (3, 5)$

2. $3x - y = 6$ $(0, -6), \left(\dfrac{1}{3}, 5\right), (2, 0), (1, -3)$

Complete the ordered pairs so that each pair is a solution for the given equation.

3. $2x - y = 8$ $(4, \ ), (\ , -8), \left(\dfrac{5}{2}, \ \right), (\ , -6)$

4. $3x + 4y = 12$ $(4, \ ), (8, \ ), (\ , 3), (\ , 6)$

Find four solutions for each of the given equations.

5. $3x - y = 6$ **6.** $x + 5y = 10$

Plot points with the given coordinates.

7. $A(0, -7)$

8. $B(4, -3)$

9. $C(-2, -5)$

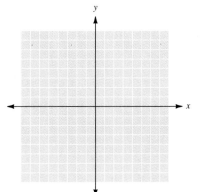

ANSWERS

10. _____

11. _____

12. _____

13. _____

14. _____

Graph each of the following equations.

10. $x - y = 5$

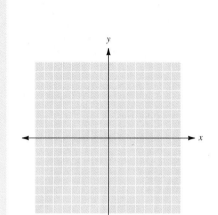

11. $y = \dfrac{2}{3}x + 3$

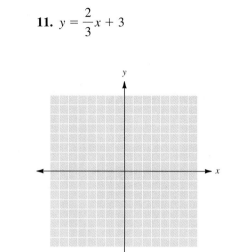

12. $x + 2y = 6$

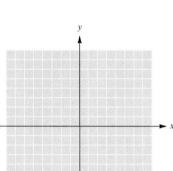

13. $2x - 5y = 10$

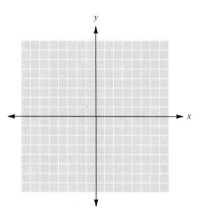

14. $y = -5$

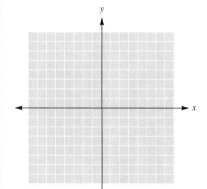

Find the slope of the line through the following pairs of points.

15. $(-2, -3)$ and $(5, 7)$

16. $(-3, 4)$ and $(2, -3)$

Find the slope and y intercept of the lines described by each of the following equations.

17. $y = -3x + 7$

18. $5x - 3y = 15$

Given the slope and y intercept for each of the following lines, write the equation of the line. Then graph the line.

19. $m = 2, b = -5$

20. $m = -\dfrac{3}{2}, b = 5$

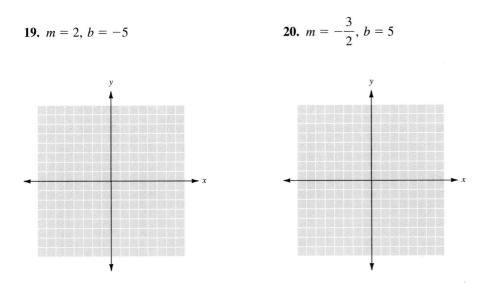

Graph each of the following inequalities.

21. $x + 2y < 6$

22. $3x - 4y \geq 12$

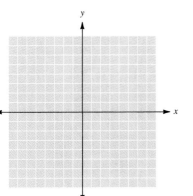

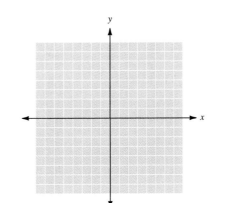

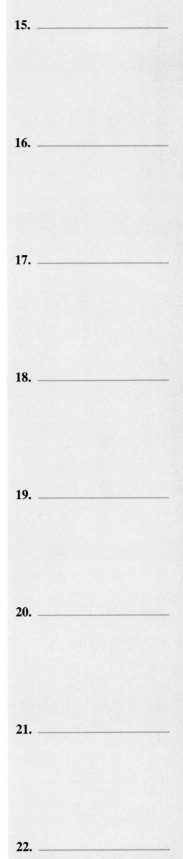

ANSWERS

15. _____

16. _____

17. _____

18. _____

19. _____

20. _____

21. _____

22. _____

Solve each of the following systems by graphing.

23. $x - y = 2$
$x + 3y = 6$

24. $3x + 2y = 6$
$x + 2y = -2$

Solve each of the following systems. If a unique solution does not exist, state whether the system is inconsistent or dependent.

25. $2x - 3y = 6$
$x - 3y = 2$

26. $2x + y = 4$
$y = 2x - 8$

27. $5x + 2y = 30$
$x - 4y = 17$

28. $2x - 6y = 8$
$x = 3y + 4$

29. $4x - 5y = 20$
$2x + 3y = 10$

30. $5x - 4y = -7$
$3x + 5y = -19$

31. $4x + 2y = 11$
$2x + y = 5$

32. $4x - 3y = 7$
$6x + 6y = 7$

33. $5x - 3y = 11$
$y = 10x - 7$

34. $2x - 5y = -17$
$x = \dfrac{3}{2}$

Solve each of the following applications. Be sure to show the system of equations used for your solution.

35. One number is 4 less than 5 times another. If the sum of the numbers is 26, what are the two numbers?

© 1993 McGraw-Hill, Inc.

36. Cynthia bought five blank VHS tapes and four cassette tapes for $28.50. Charlie bought four VHS tapes and two cassette tapes for $21.00. Find the cost of each type of tape.

37. Receipts for a concert, attended by 450 people, were $2775. If reserved-seat tickets were $7 and general-admission tickets were $4, how many of each type of ticket were sold?

38. Anthony invested part of his $12,000 inheritance in a bond paying 9 percent and the other part in a savings account paying 6 percent. If his interest from the two investments was $930 in the first year, how much did he have invested at each rate?

39. A chemist has a 30% acid solution and a 60% acid solution already prepared. How much of each of the two solutions should be mixed in order to form 300mL of a 50% acid solution?

40. Andrew was able to travel 75 mi downstream in 3 h. Returning upstream, he took 5 h to make the trip. How fast can his boat travel in still water, and what was the rate of the current?

CHAPTER 9

Exponents and Radicals

The ENVIRONMENT

World Population Growth

In 1990, the world's population was 5.3 billion people. The annual growth rate, or percentage increase, for the world's population was 1.8 percent. Although this rate of increase seems small, a 1.8 percent growth rate applied to 5.3 billion people means that 95 million more people were added to the earth's population in 1991 (0.018 × 5,300,000,000 = 95,000,000). Some rather startling figures result from this growth rate.

- It takes about 5 days to replace all the Americans killed in all U.S. wars.
- It takes about 9 months to replace the 75 million people killed in the world's greatest disaster ever, the bubonic plague epidemic of the 14th century.
- It takes less than 2 years (21 months) to add 165 million people. This is the number of people killed in all wars fought since the American Revolution.
- In less than 3 years (32 months) another population the size of the 1990 U.S. population (250 million) will be added to the world.

These figures illustrate what a small rate of growth and a very large population base can mean in actual numbers of new people.

It seems quite logical that this increase in human population should create a greater demand on the earth's resources. More people means more pressure for food, water, housing, clothes, energy, clean air, and other of life's necessities. While this relationship is accurate to some degree, it is not as simple as it might seem.

Each new person will certainly have some impact on the earth's resources, but not all people have the same amount of impact. Because the more developed countries (MDCs) use a greater share of the earth's resources, a child born in an MDC has a greater impact on the environment than a child born in a less developed country (LCD). The MDCs, which consist of the industrialized countries of Europe, North America, Japan, the Soviet Union, Australia, and New Zealand, contain only 23 percent of the world's population but consume about 80 percent of its resources. The LDCs, which are most of the countries of Africa, Asia, and Latin America, contain 77 percent of the world's population but use only 20 percent of its resources. These figures show that the culture into which a child is born greatly determines the impact of that child on the environment.

Much has been written about the large numbers of children being born in the LDCs. We certainly should be concerned about the large numbers of babies being produced by the poorer countries of the world, but we need to also recognize the greater impact of each child born in an MDC.

If current population growth rates continue, the world population is expected to be nearly 11 billion by 2050 and to almost triple, to 14 billion by 2100. Some scientists question whether the earth can support a population of this size. Many experts expect an increase in worldwide death rates to slow population growth as the environmental limits of clean air, water, or some other necessity are reached.

Not all researchers agree with these projections, however. Some social scientists contend that the population increase, especially in the MDCs, should not be restricted but should be encouraged. They contend that increased population will increase the most important resource, namely, human knowledge. According to this argument, the more people there are, the greater the chance of discovering a way to avoid, correct, or replace any loss of natural resources. These scientists believe that humans don't have the same environmental limits as the rest of life on earth because of the inventiveness of the human brain.

Whether the human population has natural limits or not, we know that there will continue to be resource allocation problems because not all people have equal access to the necessities of life. Currently, millions of people in the LDCs, and even large numbers of poor people in the MDCs, do not have clean water, sufficient food, or adequate housing. These numbers can only increase in the future because these are the areas of the world's population that are growing fastest.

Extending the Properties of Exponents

OBJECTIVE

To simplify expressions which require using the five properties of exponents

In Section 1.2, we introduced the first two properties of exponents. We stated there that the exponent notation indicates repeated multiplication and that the exponent tells us how many times the base is to be used as a factor.

Exponent
$$3^5 = \underbrace{3 \cdot 3 \cdot 3 \cdot 3 \cdot 3}_{5 \text{ factors}} = 243$$
Base

Our first property of exponents allowed us to multiply expressions that have the same base.

PROPERTY 1 OF EXPONENTS

For any real number a and positive integers m and n,

$$a^m \cdot a^n = a^{m+n}$$

In words, to multiply expressions with the same base, keep the base and add the exponents.

Example 1

Multiply.

(a) $x^6 \cdot x^7 = x^{6+7} = x^{13}$ Add the exponents.

(b) $2^3 \cdot 2^5 = 2^{3+5} = 2^8$ The base in the product is still 2.

CHECK YOURSELF 1

Multiply.

1. $y^5 \cdot y^6$ **2.** $3^2 \cdot 3^4$

Our second property allowed us to divide expressions with the same base.

PROPERTY 2 OF EXPONENTS

For any real number a, where a is not equal to 0, and positive integers m and n, where m is greater than n,

$$\frac{a^m}{a^n} = a^{m-n}$$

In words, to divide expressions with the same base, keep the base and subtract the exponents.

Example 2

Divide.

(a) $\dfrac{y^8}{y^4} = y^{8-4} = y^4$ Subtract the exponents.

(b) $\dfrac{3^6}{3^3} = 3^{6-3} = 3^3$ The base in the quotient is still 3.

CHECK YOURSELF 2

Divide.

1. $\dfrac{b^9}{b^5}$

2. $\dfrac{4^5}{4^2}$

In this section we want to expand our work with exponents to introduce three further properties. Consider the following expression:

Note that this means that the base, x^2, is used as a factor *4* times.

$(x^2)^4$

We can write this as

Using Property 1, we can *add* the exponents for the result x^8. However, note also that
$2 + 2 + 2 + 2 = 2 \times 4 = 8$.

$$(x^2)^4 = x^2 \cdot x^2 \cdot x^2 \cdot x^2 = x^8$$

and this leads us to our third property for exponents.

PROPERTY 3 OF EXPONENTS

For any real number a and positive integers m and n,

$$(a^m)^n = a^{m \cdot n}$$

In words, to raise a factor with an exponent to a power, keep the base and multiply the exponents.

The use of this new property is illustrated in our next example.

Example 3

Simplify each expression.

(a) $(x^4)^5 = x^{4\cdot5} = x^{20}$

(b) $(2^3)^4 = 2^{3\cdot4} = 2^{12}$ Multiply the exponents.

CAUTION

Be Careful! Be sure to distinguish between the correct use of Property 1 and Property 3.

$(x^4)^5 = x^{4\cdot5} = x^{20}$ This is a factor with an exponent raised to a *power,* so multiply the exponents.

but

$x^4 \cdot x^5 = x^{4+5} = x^9$ This is the product where each factor has the same base, so *add* the exponents.

CHECK YOURSELF 3

Simplify each expression.

1. $(m^5)^6$ **2.** $(m^5)(m^6)$ **3.** $(3^2)^4$ **4.** $(3^2)(3^4)$

Suppose we now have a product raised to a power. Consider an expression such as

Here the base is 3x.

$(3x)^4$

We know that

$(3x)^4 = (3x)(3x)(3x)(3x)$

Here we have applied the commutative and associative properties.

$= (3 \cdot 3 \cdot 3 \cdot 3)(x \cdot x \cdot x \cdot x)$

$= 3^4 \cdot x^4 = 81x^4$

Note that the power, here 4, has been applied to each factor, 3 and x. In general, we have

---

PROPERTY 4 OF EXPONENTS

For any real numbers a and b and positive integer m,

$(ab)^m = a^m b^m$

In words, to raise a product to a power, raise each factor to that same power.

The use of this property is shown in the following example.

Example 4

Simplify each expression.

Note that $(2x)^5$ and $2x^5$ are entirely different expressions. For $(2x)^5$, the base is $2x$, so we raise each factor to the fifth power. For $2x^5$, the base is x, and so the exponent applies only to x.

(a) $(2x)^5 = 2^5 \cdot x^5 = 32x^5$

(b) $(3ab)^4 = 3^4 \cdot a^4 \cdot b^4 = 81a^4b^4$

(c) $5(2r)^3 = 5 \cdot 2^3 \cdot r^3 = 40r^3$

CHECK YOURSELF 4

Simplify each expression.

1. $(3y)^4$ **2.** $(2mn)^6$ **3.** $3(4x)^2$ **4.** $5x^3$

We may have to use more than one of our properties in simplifying an expression involving exponents. Consider our next example.

Example 5

Simplify each expression.

To help you understand each step of the simplification we will provide reference to the property being applied. Make a list of the properties now to help you as you work through the remainder of this and the next section.

(a) $(r^4s^3)^3 = (r^4)^3 \cdot (s^3)^3$ Property 4

$= r^{12}s^9$ Property 3

(b) $(3x^2)^2 \cdot (2x^3)^3$

$= 3^2(x^2)^2 \cdot 2^3 \cdot (x^3)^3$ Property 4

$= 9x^4 \cdot 8x^9$ Property 3

$= 72x^{13}$ Multiply the coefficients and apply property 1.

(c) $\dfrac{(a^3)^5}{a^4} = \dfrac{a^{15}}{a^4}$ Property 3

$= a^{11}$ Property 2

CHECK YOURSELF 5

Simplify each expression.

1. $(m^5n^2)^3$ **2.** $(2p)^4(4p^2)^2$ **3.** $\dfrac{(s^4)^3}{s^5}$

We have one final exponent property to develop. Suppose we have a quotient raised to a power. Consider the following expression.

$$\left(\frac{x}{3}\right)^3$$

We know that

$$\left(\frac{x}{3}\right)^3 = \frac{x}{3} \cdot \frac{x}{3} \cdot \frac{x}{3} = \frac{x \cdot x \cdot x}{3 \cdot 3 \cdot 3} = \frac{x^3}{3^3}$$

Note that the power, here 3, has been applied to the numerator x and to the denominator 3. In general, we have

---

PROPERTY 5 OF EXPONENTS

For any real numbers a and b, where b is not equal to 0, and positive integer m,

$$\left(\frac{a}{b}\right)^m = \frac{a^m}{b^m}$$

In words, to raise a quotient to a power, raise the numerator and denominator to that same power.

Our final example illustrates the use of this property. Again note that the other properties may also have to be applied in simplifying an expression.

Example 6

Simplify each expression.

(a) $\left(\dfrac{3}{4}\right)^3 = \dfrac{3^3}{4^3} = \dfrac{27}{64}$ Property 5

(b) $\left(\dfrac{x^3}{y^2}\right)^4 = \dfrac{(x^3)^4}{(y^2)^4}$ Property 5

         $= \dfrac{x^{12}}{y^8}$ Property 3

(c) $\left(\dfrac{r^2 s^3}{t^4}\right)^2 = \dfrac{(r^2 s^3)^2}{(t^4)^2}$ Property 5

         $= \dfrac{(r^2)^2 (s^3)^2}{(t^4)^2}$ Property 4

         $= \dfrac{r^4 s^6}{t^8}$ Property 3

CHECK YOURSELF 6

Simplify each expression.

1. $\left(\dfrac{2}{3}\right)^4$ **2.** $\left(\dfrac{m^3}{n^4}\right)^5$ **3.** $\left(\dfrac{a^2 b^3}{c^5}\right)^2$

The following table summarizes the five properties of exponents that have been discussed in this section.

GENERAL FORM	EXAMPLE
1. $a^m a^n = a^{m+n}$	$x^2 \cdot x^3 = x^5$
2. $\dfrac{a^m}{a^n} = a^{m-n}$ $(m > n)$	$\dfrac{5^7}{5^3} = 5^4$
3. $(a^m)^n = a^{mn}$	$(z^5)^4 = z^{20}$
4. $(ab)^m = a^m b^m$	$(4x)^3 = 4^3 x^3 = 64x^3$
5. $\left(\dfrac{a}{b}\right)^m = \dfrac{a^m}{b^m}$	$\left(\dfrac{2}{3}\right)^6 = \dfrac{2^6}{3^6} = \dfrac{64}{729}$

CHECK YOURSELF ANSWERS

1. (1) y^{11}; (2) 3^6.

2. (1) b^4; (2) 4^3.

3. (1) m^{30}; (2) m^{11}; (3) 3^8; (4) 3^6.

4. (1) $81y^4$; (2) $64m^6 n^6$; (3) $48x^2$; (4) $5x^3$.

5. (1) $m^{15}n^6$; (2) $256p^8$; (3) s^7.

6. (1) $\dfrac{16}{81}$; (2) $\dfrac{m^{15}}{n^{20}}$; (3) $\dfrac{a^4 b^6}{c^{10}}$.

Name

Date

Build Your Skills

Use Properties 1 and 2 of exponents to simplify each of the following expressions.

1. $x^3 \cdot x^5$

2. $a^7 \cdot a^6$

3. $3^3 \cdot 3^4$

4. $5^2 \cdot 5^3$

5. $y^3 \cdot y^2 \cdot y^4$

6. $b^5 \cdot b^4 \cdot b$

7. $2^3 \cdot 2^2 \cdot 2$

8. $3^6 \cdot 3^4 \cdot 3^2$

9. $\dfrac{x^9}{x^4}$

10. $\dfrac{y^{10}}{y^3}$

11. $\dfrac{5^5}{5^2}$

12. $\dfrac{7^9}{7^6}$

Use Property 3 of exponents to simplify each of the following expressions.

13. $(x^2)^3$

14. $(a^5)^3$

15. $(m^4)^4$

16. $(p^7)^2$

17. $(2^4)^2$

18. $(3^3)^2$

19. $(5^3)^5$

20. $(7^2)^4$

1. _____

2. _____

3. _____

4. _____

5. _____

6. _____

7. _____

8. _____

9. _____

10. _____

11. _____

12. _____

13. _____

14. _____

15. _____

16. _____

17. _____

18. _____

19. _____

20. _____

Use Properties 4 and 5 of exponents to simplify each of the following expressions.

21. _____

22. _____

23. _____

24. _____

25. _____

26. _____

27. _____

28. _____

29. _____

30. _____

31. _____

32. _____

33. _____

34. _____

35. _____

36. _____

37. _____

38. _____

39. _____

40. _____

41. _____

42. _____

21. $(3x)^3$

22. $(4m)^2$

23. $(2xy)^4$

24. $(5pq)^3$

25. $5(3ab)^3$

26. $4(2rs)^4$

27. $\left(\dfrac{3}{4}\right)^2$

28. $\left(\dfrac{2}{3}\right)^3$

29. $\left(\dfrac{x}{5}\right)^3$

30. $\left(\dfrac{a}{2}\right)^5$

Use the properties of exponents to simplify each of the following expressions.

31. $(2x^3)^3$

32. $(3y^5)^2$

33. $(a^4b^3)^4$

34. $(p^2q^5)^3$

35. $(3x^3y)^2$

36. $(5m^2n^2)^3$

37. $(2m^3)^2(m^2)^3$

38. $(y^3)^5(3y^2)^2$

39. $\dfrac{(x^4)^3}{x^2}$

40. $\dfrac{(m^5)^3}{m^6}$

41. $\dfrac{(s^3)^2(s^2)^3}{(s^5)^2}$

42. $\dfrac{(y^5)^3(y^3)^2}{(y^4)^4}$

43. $\left(\dfrac{m^3}{n^2}\right)^3$

44. $\left(\dfrac{a^4}{b^3}\right)^4$

45. $\left(\dfrac{a^3b^2}{c^4}\right)^2$

46. $\left(\dfrac{x^5y^2}{z^4}\right)^3$

Think About These

47. Write x^{12} as a power of x^2.

48. Write y^{15} as a power of y^3.

49. Write a^{16} as a power of a^2.

50. Write m^{20} as a power of m^5.

51. Write each of the following as powers of 8. (Remember that $8 = 2^3$.)
2^{12}, 2^{18}, $(2^5)^3$, $(2^7)^6$

52. Write each of the following as powers of 9.
3^8, 3^{14}, $(3^5)^8$, $(3^4)^7$

The formula $(1 + R)^y = G$ gives us useful information about the growth of a population. Here R is the rate of growth expressed as a decimal, y is the time in years, and G is the growth factor. If a country has a 2 percent growth rate for 35 years, then it will double its population.

$(1.02)^{35} \approx 2$

53. (*a*) With this growth rate, how many doublings will occur in 105 years? How much larger will the country's population be?
   (*b*) The less developed countries of the world had an average growth rate of 2 percent in 1986. If their total population was 3.8 billion, what will their population be in 105 years if this rate remains unchanged?

54. The United States has a growth rate of 0.7 percent. What will be its growth factor after 35 years?

43. _____

44. _____

45. _____

46. _____

47. _____

48. _____

49. _____

50. _____

51. _____

52. _____

53. _____

_____

54. _____

Skillscan (Section 4.2)

Reduce each of the following fractions to simplest form.

a. $\dfrac{m^3}{m^5}$ b. $\dfrac{x^7}{x^{10}}$ c. $\dfrac{a^3}{a^9}$ d. $\dfrac{y^4}{y^8}$

e. $\dfrac{x^3}{x^3}$ f. $\dfrac{b^5}{b^5}$ g. $\dfrac{s^7}{s^7}$ h. $\dfrac{r^{10}}{r^{10}}$

ANSWERS

1. x^8 **3.** 3^7 **5.** y^9 **7.** 2^6 **9.** x^5 **11.** 5^3 **13.** x^6 **15.** m^{16} **17.** 2^8

19. 5^{15} **21.** $27x^3$ **23.** $16x^4y^4$ **25.** $135a^3b^3$ **27.** $\dfrac{9}{16}$ **29.** $\dfrac{x^3}{125}$ **31.** $8x^9$

33. $a^{16}b^{12}$ **35.** $9x^6y^2$ **37.** $4m^{12}$ **39.** x^{10} **41.** s^2 **43.** $\dfrac{m^9}{n^6}$ **45.** $\dfrac{a^6b^4}{c^8}$

47. $(x^2)^6$ **49.** $(a^2)^8$ **51.** $8^4, 8^6, 8^5, 8^{14}$ **53.** (a) Three doublings, 8 times as large;

(b) 30.4 billion   **a.** $\dfrac{1}{m^2}$ **b.** $\dfrac{1}{x^3}$ **c.** $\dfrac{1}{a^6}$ **d.** $\dfrac{1}{y^4}$ **e.** 1   **f.** 1   **g.** 1   **h.** 1

9.2 Zero and Negative Exponents

OBJECTIVES

1. To evaluate expressions involving zero or negative exponents
2. To simplify expressions involving zero or negative exponents

In the last section, we continued our discussion of the properties of exponents with the introduction of rules for raising expressions involving powers, products, and quotients to a power. We now want to extend our exponent notation to include 0 and negative integers as exponents.

First, what do we do with x^0? It will help to look at a problem that gives us x^0 as a result. What if the numerator and denominator of a fraction have the same base raised to the same power and we extend our division rule? For example,

By Property 2,

$$\frac{a^m}{a^n} = a^{m-n}$$

where $m > n$. Here m and n are *both* 5 so $m = n$.

$$\frac{a^5}{a^5} = a^{5-5} = a^0 \tag{1}$$

But from our experience with fractions we know that

$$\frac{a^5}{a^5} = 1 \tag{2}$$

By comparing equations (1) and (2), it seems reasonable to make the following definition:

As was the case with $\dfrac{0}{0}$, 0^0 will not be discussed until you study calculus.

> For any number a, $a \neq 0$,
>
> $a^0 = 1$
>
> In words, any expression, except 0, raised to the 0 power is 1.

The following example illustrates the use of this definition.

Example 1

Evaluate. Assume all variables are nonzero.

(a) $5^0 = 1$
(b) $27^0 = 1$
(c) $(x^2 y)^0 = 1$ if $x \neq 0$ and $y \neq 0$
(d) $6x^0 = 6 \cdot 1 = 6$ if $x \neq 0$

⟨CAUTION⟩

Be Careful! In part (d) the 0 exponent applies only to the x and *not* to the factor 6, since the base is x.

643

Evaluate. Assume all variables are nonzero.

1. 7^0 **2.** $(-8)^0$ **3.** $(xy^3)^0$ **4.** $3x^0$

The second property of exponents allows us to define a negative exponent. Suppose that the exponent in the denominator is *greater than* the exponent in the numerator. Consider the expression $\dfrac{x^2}{x^5}$.

Our previous work with fractions tells us that

Divide numerator and denominator by the two common factors of *x*.

$$\frac{x^2}{x^5} = \frac{x \cdot x}{x \cdot x \cdot x \cdot x \cdot x} = \frac{1}{x^3} \tag{1}$$

However, if we extend the second property to let n be greater than m, we have

Remember:

$$\frac{a^m}{a^n} = a^{m-n}$$

$$\frac{x^2}{x^5} = x^{2-5} = x^{-3} \tag{2}$$

Now, by comparing equations (1) and (2), it seems reasonable to define x^{-3} as $\dfrac{1}{x^3}$.

In general, we have this result:

John Wallis (1616–1703), an English mathematician, was the first to fully discuss the meaning of 0 and negative exponents.

> For any number a, $a \neq 0$, and any positive integer n,
>
> $$a^{-n} = \frac{1}{a^n}$$

Example 2

Rewrite each expression, using only positive exponents.

Negative exponent in numerator

(*a*) $x^{-4} = \dfrac{1}{x^4}$ Positive exponent in denominator

(*b*) $m^{-7} = \dfrac{1}{m^7}$

(*c*) $3^{-2} = \dfrac{1}{3^2}$ or $\dfrac{1}{9}$

(*d*) $10^{-3} = \dfrac{1}{10^3}$ or $\dfrac{1}{1000}$

(e) $2x^{-3} = 2 \cdot \dfrac{1}{x^3} = \dfrac{2}{x^3}$

The -3 exponent applies
only to x, since x is the base.

(f) $\dfrac{a^5}{a^9} = a^{5-9} = a^{-4} = \dfrac{1}{a^4}$

(g) $-4x^{-5} = -4 \cdot \dfrac{1}{x^5} = \dfrac{-4}{x^5}$

CHECK YOURSELF 2

Write, using only positive exponents.

1. a^{-10} **2.** 4^{-3} **3.** $3x^{-2}$ **4.** $\dfrac{x^5}{x^8}$

We will now allow negative integers as exponents in our first property for exponents. Consider our next example.

Example 3

Simplify each expression.

$a^m \cdot a^n = a^{m+n}$ for *any* integers m and n. So add the exponents.

(a) $x^5 x^{-2} = x^{5+(-2)} = x^3$

Note: An alternative approach would be

By definition

$x^{-2} = \dfrac{1}{x^2}$

$x^5 x^{-2} = x^5 \cdot \dfrac{1}{x^2} = \dfrac{x^5}{x^2} = x^3$

(b) $a^7 a^{-5} = a^{7+(-5)} = a^2$

(c) $y^5 y^{-9} = y^{5+(-9)} = y^{-4} = \dfrac{1}{y^4}$

CHECK YOURSELF 3

Simplify.

1. $x^7 x^{-2}$ **2.** $b^3 b^{-8}$

Our final example shows that all the properties of exponents introduced in the last section can be extended to expressions with negative exponents.

Example 4

Simplify each expression.

(a) $\dfrac{m^{-3}}{m^4} = m^{-3-4}$ Property 2

$= m^{-7} = \dfrac{1}{m^7}$

(b) $\dfrac{a^{-2}b^6}{a^5b^{-4}} = a^{-2-5}b^{6-(-4)}$ Apply Property 2 to each variable.

$= a^{-7}b^{10} = \dfrac{b^{10}}{a^7}$

This could also be done by using Property 4 first, so

$(2x^4)^{-3} = 2^{-3}x^{-12} = \dfrac{1}{2^3x^{12}} = \dfrac{1}{8x^{12}}$

(c) $(2x^4)^{-3} = \dfrac{1}{(2x^4)^3}$ Definition of the negative exponent

$= \dfrac{1}{2^3(x^4)^3}$ Property 4

$= \dfrac{1}{8x^{12}}$ Property 3

(d) $\dfrac{(y^{-2})^4}{(y^3)^{-2}} = \dfrac{y^{-8}}{y^{-6}}$ Property 3

$= y^{-8-(-6)}$ Property 2

$= y^{-2} = \dfrac{1}{y^2}$

CHECK YOURSELF 4

Simplify each expression.

1. $\dfrac{x^5}{x^{-3}}$ **2.** $\dfrac{m^3n^{-5}}{m^{-2}n^3}$ **3.** $(3a^3)^{-4}$ **4.** $\dfrac{(r^3)^{-2}}{(r^{-4})^2}$

CHECK YOURSELF ANSWERS

1. (1) 1; (2) 1; (3) 1; (4) 3.

2. (1) $\dfrac{1}{a^{10}}$; (2) $\dfrac{1}{4^3}$ or $\dfrac{1}{64}$; (3) $\dfrac{3}{x^2}$; (4) $\dfrac{1}{x^3}$.

3. (1) x^5; (2) $\dfrac{1}{b^5}$.

4. (1) x^8; (2) $\dfrac{m^5}{n^8}$; (3) $\dfrac{1}{81a^{12}}$; (4) r^2.

Name _____

Date _____

Build Your Skills

Evaluate (assume the variables are nonzero).

1. 4^0

2. $(-7)^0$

3. $(-29)^0$

4. 75^0

5. $(x^3y^2)^0$

6. $7m^0$

7. $9x^0$

8. $(3a^2b^6)^0$

9. $(-2p^5q^7)^0$

10. $-5x^0$

Write each of the following expressions, using positive exponents; simplify where possible.

11. b^{-8}

12. p^{-12}

13. 3^{-4}

14. 2^{-5}

15. 5^{-2}

16. 4^{-3}

17. 10^{-4}

18. 10^{-5}

19. $5x^{-1}$

20. $3a^{-2}$

21. $(5x)^{-1}$

22. $(3a)^{-2}$

© 1993 McGraw-Hill, Inc.

ANSWERS

1. _____
2. _____
3. _____
4. _____
5. _____
6. _____
7. _____
8. _____
9. _____
10. _____
11. _____
12. _____
13. _____
14. _____
15. _____
16. _____
17. _____
18. _____
19. _____
20. _____
21. _____
22. _____

23. $-2x^{-5}$

24. $3x^{-4}$

25. $(-2x)^{-5}$

26. $(3x)^{-4}$

Use Properties 1 and 2 to simplify each of the following expressions. Write your answers with positive exponents only.

27. a^5a^3

28. m^5m^7

29. x^9x^{-3}

30. $a^{10}a^{-6}$

31. b^5b^{-9}

32. y^3y^{-10}

33. x^0x^5

34. $r^{-3}r^0$

35. $\dfrac{a^8}{a^5}$

36. $\dfrac{m^9}{m^4}$

37. $\dfrac{x^7}{x^9}$

38. $\dfrac{a^3}{a^{10}}$

39. $\dfrac{r^{-3}}{r^5}$

40. $\dfrac{x^3}{x^{-5}}$

41. $\dfrac{x^{-4}}{x^{-5}}$

42. $\dfrac{p^{-6}}{p^{-3}}$

Simplify each of the following expressions. Write your answers with positive exponents only.

43. $\dfrac{m^5 n^{-3}}{m^{-4} n^5}$

44. $\dfrac{p^{-3} q^{-2}}{p^4 q^{-3}}$

45. $(2a^{-3})^4$

46. $(3x^2)^{-3}$

47. $(x^{-2} y^3)^{-2}$

48. $(a^5 b^{-3})^{-3}$

49. $\dfrac{(r^{-2})^3}{r^{-4}}$

50. $\dfrac{(y^3)^{-4}}{y^{-6}}$

51. $\dfrac{(x^{-3})^3}{(x^4)^{-2}}$

52. $\dfrac{(m^4)^{-3}}{(m^{-2})^4}$

53. $\dfrac{(a^{-3})^2 (a^4)}{(a^{-3})^{-3}}$

54. $\dfrac{(x^2)^{-3} (x^{-2})}{(x^2)^{-4}}$

Think About These

Decide which variables cannot be equal to zero.

55. $a^{-4} b^3$

56. $x^{-4} y^3$

57. $x^{-3} y^3 z^2$

58. $a^3 b^{-3} c^{-2}$

59. $\dfrac{a^4 b^{-3}}{c^3}$

60. $\dfrac{m^{-2} n^3}{r^2 s^3}$

ANSWERS
43. _____
44. _____
45. _____
46. _____
47. _____
48. _____
49. _____
50. _____
51. _____
52. _____
53. _____
54. _____
55. _____
56. _____
57. _____
58. _____
59. _____
60. _____

In 1975 the population of earth was approximately 4 billion and doubling every 35 years. The formula for the population P in year Y for this doubling rate is

$$P \text{ (in billions)} = 4 \times 2^{(Y-1975)/35}$$

61. What was the approximate population of earth in 1960?

62. What will the earth's population be in 2025?

The United States population in 1990 was approximately 250 million, and the average growth rate for the past 30 years gives a doubling time of 66 years. The above formula for the United States then becomes

$$P \text{ (in millions)} = 250 \times 2^{(Y-1990)/66}$$

63. What was the approximate population of the United States in 1960?

64. What will be the population of the United States in 2025 if this growth rate continues?

Skillscan (Section 2.4)

Evaluate each expression.

a. 3^2 **b.** $(-3)^2$ **c.** 5^3 **d.** $(-5)^3$

e. 2^4 **f.** $(-2)^4$ **g.** 3^5 **h.** $(-3)^5$

ANSWERS

1. 1   **3.** 1   **5.** 1   **7.** 9   **9.** 1   **11.** $\dfrac{1}{b^8}$ **13.** $\dfrac{1}{3^4}$ or $\dfrac{1}{81}$ **15.** $\dfrac{1}{5^2}$ or $\dfrac{1}{25}$

17. $\dfrac{1}{10^4}$ or $\dfrac{1}{10,000}$ **19.** $\dfrac{5}{x}$ **21.** $\dfrac{1}{5x}$ **23.** $-\dfrac{2}{x^5}$ **25.** $-\dfrac{1}{32x^5}$ **27.** a^8 **29.** x^6

31. $\dfrac{1}{b^4}$ **33.** x^5 **35.** a^3 **37.** $\dfrac{1}{x^2}$ **39.** $\dfrac{1}{r^8}$ **41.** x **43.** $\dfrac{m^9}{n^8}$ **45.** $\dfrac{16}{a^{12}}$

47. $\dfrac{x^4}{y^6}$ **49.** $\dfrac{1}{r^2}$ **51.** $\dfrac{1}{x}$ **53.** $\dfrac{1}{a^{11}}$ **55.** a **57.** x **59.** b, c

61. 2.97 billion   **63.** 182 million   **a.** 9   **b.** 9   **c.** 125   **d.** -125 **e.** 16
f. 16   **g.** 243   **h.** -243

Roots and Radicals

OBJECTIVES

1. To use the radical notation to represent roots
2. To distinguish between rational and irrational numbers

In the last section, we extended the properties of exponents to include the entire set of integers. Over the next four sections, we will be working with a new notation that "reverses" the process of raising to a power.

From our work in Chapter 1, we know that when we have a statement such as

$$x^2 = 9$$

it is read as "x squared equals 9."

Here we are concerned with the relationship between the variable x and the number 9. We call that relationship the *square root* and say, equivalently, that "x is the square root of 9."

We know from experience that x must be 3 (since $3^2 = 9$) or -3 [since $(-3)^2 = 9$]. We see that 9 has two square roots, 3 and -3. In fact, every positive number will have *two* square roots. In general, if $x^2 = a$, we call x the *square root* of a.

We are now ready for our new notation. The symbol $\sqrt{}$ is called a *radical sign*. We saw above that 3 was the positive square root of 9. We also call 3 the *principal square root* of 9 and can write

$$\sqrt{9} = 3$$

to indicate that 3 is the principal square root of 9.

To summarize:

$\sqrt{9}$ asks, "What positive number must we square (or multiply by itself) to get 9?"

$\sqrt{9}$ is read the "*square root*" of 9, and since $3^2 = 9$,

$$\sqrt{9} = 3$$

This leads us to the following definition.

The symbol $\sqrt{}$ first appeared in print in 1525. In Latin, "radix" means *root*, and this was contracted to a small *r*. The present symbol may have been used because it resembled the manuscript form of that small *r*.

DEFINITION

$\sqrt{a}$ is the *positive* (or *principal*) square root of *a*. It is the positive number whose square is *a*.

This definition is illustrated in our first example.

Example 1

Find the following square roots.

(a) $\sqrt{49} = 7$ Since 7 is the positive number, we must square to get 49.

(b) $\sqrt{81} = 9$

(c) $\sqrt{\dfrac{4}{9}} = \dfrac{2}{3}$ Since $\dfrac{2}{3}$ is the positive number,

we must square to get $\dfrac{4}{9}$.

CHECK YOURSELF 1

Find the following square roots.

1. $\sqrt{64}$ **2.** $\sqrt{144}$ **3.** $\sqrt{\dfrac{16}{25}}$

Note: When you use the radical sign, you will get only the *positive square root:*

$\sqrt{25} = 5$

Each positive number has two square roots. For instance, 25 has square roots of 5 and -5 because

$$5^2 = 25 \qquad \text{and} \qquad (-5)^2 = 25$$

If you want to indicate the negative square root, you must use a minus sign in front of the radical.

$$-\sqrt{25} = -5$$

Example 2

Find the following square roots.

(a) $\sqrt{100} = 10$ The principal root

(b) $-\sqrt{100} = -10$ The negative square root

(c) $-\sqrt{\dfrac{9}{16}} = -\dfrac{3}{4}$

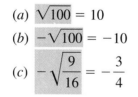

CHECK YOURSELF 2

Find the following square roots.

1. $\sqrt{16}$ **2.** $-\sqrt{16}$ **3.** $-\sqrt{\dfrac{16}{25}}$

The square roots of negative numbers are *not* real numbers. For instance, $\sqrt{-9}$ is *not* a real number because there is *no* real number x such that

$$x^2 = -9$$

Be Careful!   Do not confuse

$$-\sqrt{9} \quad \text{with} \quad \sqrt{-9}$$

The expression $-\sqrt{9}$ is -3, while $\sqrt{-9}$ is not a real number.
   The following example summarizes our discussion thus far.

Example 3

(a) $\sqrt{36} = 6$

(b) $\sqrt{121} = 11$

(c) $-\sqrt{64} = -8$

(d) $\sqrt{-64}$ is not a real number.

(e) $\sqrt{0} = 0$ (Because $0 \cdot 0 = 0$)

CHECK YOURSELF 3

Evaluate if possible.

1. $\sqrt{81}$ **2.** $\sqrt{49}$ **3.** $-\sqrt{49}$ **4.** $\sqrt{-49}$

   As we mentioned earlier, finding the square root of a number is the reverse of squaring a number. We can extend that idea to work with other roots of numbers. For instance, the *cube root* of a number is the number we must cube (or raise to the third power) to get that number.
   The cube root of 8 is 2 since $2^3 = 8$, and we write

$\sqrt[3]{8}$ is read "the cube root of 8."

$$\sqrt[3]{8} = 2$$

   We give special names to the parts of a radical expression. These are summarized as follows.

The index for $\sqrt[3]{a}$ is 3.

The index of 2 for square roots is generally not written.

We understand that $\sqrt{a}$ is the principal square root of a.

> Every radical expression contains three parts as shown below. The principal *n*th root of *a* is written as
>
> Index
> $$\sqrt[n]{a}$$
> Radical   Radicand
> sign

To illustrate, the *cube root* of 64 is written

Index $\longrightarrow$ $\sqrt[3]{64}$
of 3

and it represents the number we must cube (or raise to the third power) to get 64.

$$\sqrt[3]{64} = 4$$

because $4^3 = 64$. And

Index $\longrightarrow$ $\sqrt[4]{81}$
of 4

is the *fourth root* of 81. It represents the number we must raise to the fourth power to get 81.

$$\sqrt[4]{81} = 3$$

because $3^4 = 81$.

We can find roots of negative numbers as long as the index is *odd* (3, 5, etc.). For example,

$$\sqrt[3]{-64} = -4$$

because $(-4)^3 = -64$.

If the index is *even* (2, 4, etc.), roots of negative numbers are *not* real numbers. For example,

$$\sqrt[4]{-16}$$

The *even power* of a real number is always *positive* or *zero*.

is not a real number because there is no real number x such that $x^4 = -16$.

The following table shows the most common roots.

It would be helpful for your work here and in future mathematics classes to memorize these roots.

SQUARE ROOTS		CUBE ROOTS	FOURTH ROOTS
$\sqrt{1} = 1$	$\sqrt{49} = 7$	$\sqrt[3]{1} = 1$	$\sqrt[4]{1} = 1$
$\sqrt{4} = 2$	$\sqrt{64} = 8$	$\sqrt[3]{8} = 2$	$\sqrt[4]{16} = 2$
$\sqrt{9} = 3$	$\sqrt{81} = 9$	$\sqrt[3]{27} = 3$	$\sqrt[4]{81} = 3$
$\sqrt{16} = 4$	$\sqrt{100} = 10$	$\sqrt[3]{64} = 4$	$\sqrt[4]{256} = 4$
$\sqrt{25} = 5$	$\sqrt{121} = 11$	$\sqrt[3]{125} = 5$	$\sqrt[4]{625} = 5$
$\sqrt{36} = 6$	$\sqrt{144} = 12$		

You can use the table in the following example which summarizes the discussion above.

Example 4

Evaluate each of the following.

(*a*) $\sqrt[3]{125} = 5$ because $5^3 = 125$.

(*b*) $\sqrt[5]{32} = 2$ because $2^5 = 32$.

The cube root of a negative number will be negative.

(*c*) $\sqrt[3]{-125} = -5$ because $(-5)^3 = -125$.

The fourth root of a negative number is not a real number.

(*d*) $\sqrt[4]{-81}$ is not a real number.

CHECK YOURSELF 4

Evaluate if possible.

1. $\sqrt[3]{64}$ **2.** $\sqrt[4]{16}$ **3.** $\sqrt[4]{-256}$ **4.** $\sqrt[3]{-8}$

The radical notation allows us to distinguish between two important types of numbers.

First, a *rational number* can be represented by a fraction whose numerator and denominator are integers and whose denominator is nonzero. The form of a rational number is

$$\frac{a}{b} \qquad a \text{ and } b \text{ are integers, } b \neq 0$$

Some examples of rational numbers are $\dfrac{-2}{3}, \dfrac{4}{5}, 2,$ and $\dfrac{7}{3}$. Certain square roots are rational numbers also. For example,

Note that each radicand is a *perfect-square integer* (that is, an integer which is the square of another integer).

$$\sqrt{4} \qquad \sqrt{25} \qquad \text{and} \qquad \sqrt{64}$$

represent the rational numbers 2, 5, and 8, respectively.

There are numbers that represent a point on the number line but that *cannot* be written as the ratio of two integers. These are called *irrational numbers*. For example, the square root of any positive number which is not itself a perfect square is an irrational number. Because the radicands are *not* perfect squares, the expressions

$$\sqrt{2} \qquad \sqrt{3} \qquad \text{and} \qquad \sqrt{5}$$

Note: The fact that the square root of 2 is irrational will be proved in later mathematics courses and was known to Greek mathematicians over 2000 years ago.

represent irrational numbers.

Example 5

Which of the following numbers are rational and which are irrational?

$$\sqrt{\frac{2}{3}} \qquad \sqrt{\frac{4}{9}} \qquad \sqrt{7} \qquad \sqrt{16} \qquad \sqrt{25}$$

Here $\sqrt{7}$ and $\sqrt{\frac{2}{3}}$ are irrational numbers. And $\sqrt{16}$ and $\sqrt{25}$ are rational numbers because 16 and 25 are perfect squares. Also $\sqrt{\frac{4}{9}}$ is rational because $\sqrt{\frac{4}{9}} = \frac{2}{3}$, a ratio of two integers.

CHECK YOURSELF 5

Which of the following numbers are rational and which are irrational?

1. $\sqrt{26}$ 2. $\sqrt{49}$ 3. $\sqrt{\frac{6}{7}}$

4. $\sqrt{100}$ 5. $\sqrt{105}$ 6. $\sqrt{\frac{16}{9}}$

The decimal representation of a rational number always terminates or repeats. For instance,

$$\frac{3}{8} = 0.375$$

$$\frac{5}{11} = 0.454545\ldots$$

$$\approx 0.45$$

An important fact about the irrational numbers is that their decimal representations are always *nonterminating* and *nonrepeating*. We can therefore only approximate irrational numbers with a terminating decimal. A table of roots can be found at the end of this text, or a calculator can be used to find roots. However, note that the values found for the irrational roots are only approximations. For instance, $\sqrt{2}$ is approximately 1.414 (to three decimal places), and we can write

$$\sqrt{2} \approx 1.414$$

Read "approximately equal to"

With a calculator we find that

1.414 is an approximation to the number whose square is 2.

$$(1.414)^2 = 1.999396$$

The set of all rational numbers and the set of all irrational numbers together form the set of *real numbers*. The real numbers will represent every point that can be pictured on the number line. Some examples are shown below.

For this reason we will refer to the number line as the *real number line*.

The following diagram summarizes the relationships among the various numeric sets that have been introduced here and in Chapter 1.

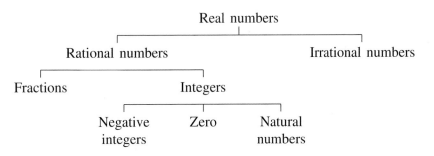

We conclude our work in this section by developing a general result that we will need later. Let's start by looking at two numerical examples.

$$\sqrt{2^2} = \sqrt{4} = 2 \tag{1}$$
$$\sqrt{(-2)^2} = \sqrt{4} = 2 \qquad \text{since } (-2)^2 = 4 \tag{2}$$

Consider the value of $\sqrt{x^2}$ where x is positive or negative.

This is because the principal square root of a number is always positive or zero.

In (1) where $x = 2$:   In (2) where $x = -2$:

$$\sqrt{2^2} = 2 \qquad\qquad \sqrt{(-2)^2} \neq -2$$
$$\text{Here } \sqrt{(-2)^2} = -(-2) = 2.$$

Comparing the results above, we see that $\sqrt{x^2}$ is x if x is positive (or 0) and $\sqrt{x^2}$ is $-x$ if x is negative. We can write

$$\sqrt{x^2} = \begin{cases} x & \text{where } x \geq 0 \\ -x & \text{where } x < 0 \end{cases}$$

From your earlier work with absolute values you will remember that

$$|x| = \begin{cases} x & \text{where } x \geq 0 \\ -x & \text{where } x < 0 \end{cases}$$

and we can summarize the discussion by writing

$$\sqrt{x^2} = |x| \qquad \text{for any real number } x$$

Example 6

Evaluate each of the following.

(a) $\sqrt{5^2} = 5$

(b) $\sqrt{(-4)^2} = |-4| = 4$

Note: Alternatively in (b), we could write

$$\sqrt{(-4)^2} = \sqrt{16} = 4$$

CHECK YOURSELF 6

Evaluate.

1. $\sqrt{6^2}$

2. $\sqrt{(-6)^2}$

Note: The case for roots with indices that are odd does *not* require the use of absolute value, as illustrated in the last example. For instance,

$$\sqrt[3]{3^3} = \sqrt[3]{27} = 3$$
$$\sqrt[3]{(-3)^3} = \sqrt[3]{-27} = -3$$

and we see that

$$\sqrt[n]{x^n} = x \qquad \text{where } n \text{ is odd}$$

To summarize, we can write

$$\sqrt[n]{x^n} = \begin{cases} |x| & \text{where } n \text{ is even} \\ x & \text{where } n \text{ is odd} \end{cases}$$

Example 7

Evaluate each of the following.

(a) $\sqrt{(-4)^2} = |-4| = 4 \qquad \sqrt{x} = |x|$

(b) $\sqrt[3]{5^3} = 5 \qquad \sqrt[3]{x} = x$ (the index is odd)

(c) $\sqrt[3]{(-5)^3} = -5$

(d) $\sqrt[4]{(-2)^4} = |-2| = 2 \qquad \sqrt[4]{x^4} = |x|$ (the index is even)

CHECK YOURSELF 7

Evaluate.

1. $\sqrt[3]{4^3}$

2. $\sqrt[4]{(-3)^4}$

3. $\sqrt[5]{(-2)^5}$

CHECK YOURSELF ANSWERS

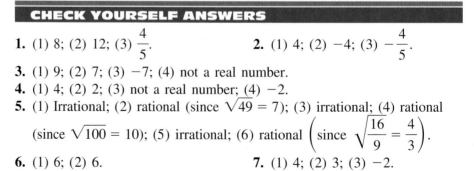

1. (1) 8; (2) 12; (3) $\dfrac{4}{5}$.   **2.** (1) 4; (2) -4; (3) $-\dfrac{4}{5}$.

3. (1) 9; (2) 7; (3) -7; (4) not a real number.

4. (1) 4; (2) 2; (3) not a real number; (4) -2.

5. (1) Irrational; (2) rational (since $\sqrt{49} = 7$); (3) irrational; (4) rational (since $\sqrt{100} = 10$); (5) irrational; (6) rational $\left(\text{since } \sqrt{\dfrac{16}{9}} = \dfrac{4}{3} \right)$.

6. (1) 6; (2) 6.   **7.** (1) 4; (2) 3; (3) -2.

Name _____

Date _____

Build Your Skills

Evaluate if possible.

1. $\sqrt{16}$

2. $\sqrt{121}$

3. $\sqrt{400}$

4. $\sqrt{64}$

5. $-\sqrt{100}$

6. $\sqrt{-100}$

7. $\sqrt{-81}$

8. $-\sqrt{81}$

9. $\sqrt{\dfrac{16}{9}}$

10. $-\sqrt{\dfrac{1}{25}}$

11. $\sqrt{-\dfrac{4}{5}}$

12. $\sqrt{\dfrac{4}{25}}$

13. $\sqrt[3]{27}$

14. $\sqrt[4]{81}$

15. $\sqrt[3]{-27}$

16. $\sqrt[4]{-16}$

17. $\sqrt[4]{-81}$

18. $-\sqrt[3]{64}$

19. $-\sqrt[3]{27}$

20. $-\sqrt[3]{-8}$

ANSWERS

1. _____

2. _____

3. _____

4. _____

5. _____

6. _____

7. _____

8. _____

9. _____

10. _____

11. _____

12. _____

13. _____

14. _____

15. _____

16. _____

17. _____

18. _____

19. _____

20. _____

ANSWERS

21. _____

22. _____

23. _____

24. _____

25. _____

26. _____

27. _____

28. _____

29. _____

30. _____

31. _____

32. _____

33. _____

34. _____

35. _____

36. _____

37. _____

38. _____

39. _____

40. _____

21. $\sqrt[4]{625}$

22. $\sqrt[3]{1000}$

23. $\sqrt[3]{\dfrac{1}{27}}$

24. $\sqrt[3]{-\dfrac{8}{27}}$

Which of the following roots are rational numbers and which are irrational numbers?

25. $\sqrt{19}$

26. $\sqrt{36}$

27. $\sqrt{100}$

28. $\sqrt{7}$

29. $\sqrt[3]{9}$

30. $\sqrt[3]{8}$

31. $\sqrt[4]{16}$

32. $\sqrt{\dfrac{4}{9}}$

33. $\sqrt{\dfrac{4}{7}}$

34. $\sqrt[3]{5}$

35. $\sqrt[3]{-27}$

36. $-\sqrt[4]{81}$

Evaluate each of the following expressions.

37. $\sqrt{5^2}$

38. $\sqrt{(-5)^2}$

39. $\sqrt[3]{4^3}$

40. $\sqrt[4]{(-3)^4}$

41. $\sqrt[4]{2^4}$

42. $\sqrt[3]{(-5)^3}$

43. $\sqrt[5]{3^5}$

44. $\sqrt[5]{(-2)^5}$

For Problems 45 to 50, find the two expressions that are equivalent.

45. $\sqrt{-16}, \ -\sqrt{16}, \ -4$

46. $-\sqrt{25}, \ -5, \ \sqrt{-25}$

47. $\sqrt[3]{-125}, \ -\sqrt[3]{125}, \ |-5|$

48. $\sqrt[5]{-32}, \ -\sqrt[5]{32}, \ |-2|$

49. $\sqrt[4]{10,000}, \ 100, \ \sqrt[3]{1000}$

50. $10^2, \ \sqrt{10,000}, \ \sqrt[3]{100,000}$

Suppose a population grew from x million in 1980 to y million in 1990. To estimate the population for 1985, we could assume that $x/z = z/y$, where z is the 1985 population. Then $z^2 = xy$, and $z = \sqrt{xy}$.

51. Use the equation $z = \sqrt{xy}$ to find the U.S. population in 1985 (z) if the 1980 census showed a population of 227.5 million (x) and the 1990 census showed a population of 249.2 million (y).

52. The world population was 3.02 billion in 1960. It had grown to 5.29 billion in 1990. Using the equation from the discussion above, determine the world population in 1975.

ANSWERS

41. _____

42. _____

43. _____

44. _____

45. _____

46. _____

47. _____

48. _____

49. _____

50. _____

51. _____

52. _____

Skillscan (Section 1.5)

Find each of the following products.

a. $(4x^2)(2x)$ **b.** $(9a^4)(5a)$ **c.** $(16m^2)(3m)$ **d.** $(8b^3)(2b)$

e. $(27p^6)(3p)$ **f.** $(81s^4)(s^3)$ **g.** $(100y^4)(2y)$ **h.** $(49m^6)(2m)$

ANSWERS

1. 4 **3.** 20 **5.** -10 **7.** Not a real number **9.** $\dfrac{4}{3}$ **11.** Not a real number

13. 3 **15.** -3 **17.** Not a real number **19.** -3 **21.** 5 **23.** $\dfrac{1}{3}$ **25.** Irrational

27. Rational **29.** Irrational **31.** Rational **33.** Irrational **35.** Rational **37.** 5
39. 4 **41.** 2 **43.** 3 **45.** $-\sqrt{16}, -4$ **47.** $\sqrt[3]{-125}, -\sqrt[3]{125}$ **49.** $\sqrt[4]{10,000},$
$\sqrt[3]{1000}$ **51.** 238.1 million **a.** $8x^3$ **b.** $45a^5$ **c.** $48m^3$ **d.** $16b^4$ **e.** $81p^7$
f. $81s^7$ **g.** $200y^5$ **h.** $98m^7$

Simplifying Radical Expressions

OBJECTIVE

To simplify expressions involving radicals

In Section 9.3, we introduced the radical notation. For most applications, we will want to make sure that all radical expressions are in *simplest form*. The following three conditions must be satisfied.

> An expression involving square roots is in *simplest form* if
>
> **1.** There are no perfect-square factors in a radical.
> **2.** No fraction appears inside a radical.
> **3.** No radical appears in the denominator.

For instance, considering condition 1,

$\sqrt{17}$ is in simplest form since 17 has *no* perfect-square factors

while

$\sqrt{12}$ is *not* in simplest form

because

$$\sqrt{12} = \sqrt{4 \cdot 3}$$

A perfect square

To simplify radical expressions, we'll need to develop two important properties. First, look at the following expressions:

$$\sqrt{4 \cdot 9} = \sqrt{36} = 6$$
$$\sqrt{4} \cdot \sqrt{9} = 2 \cdot 3 = 6$$

Since this tells us that $\sqrt{4 \cdot 9} = \sqrt{4} \cdot \sqrt{9}$, the following general rule for radicals is suggested.

> **PROPERTY 1 OF RADICALS**
>
> For any positive real numbers a and b,
>
> $$\sqrt{ab} = \sqrt{a} \cdot \sqrt{b}$$
>
> In words, the square root of a product is the product of the square roots.

Let's see how this property is applied in simplifying expressions when radicals are involved.

Example 1

Perfect-square factors are 1, 4, 9, 16, 25, 36, 49, 64, 81, 100, and so on.

Simplify each expression.

(a) $\sqrt{12} = \sqrt{4 \cdot 3}$

A perfect square

Apply Property 1.

$= \sqrt{4} \cdot \sqrt{3}$

Note that we have removed the perfect-square factor from inside the radical, so the expression is in simplest form.

$= 2\sqrt{3}$

It would not have helped to write

$\sqrt{45} = \sqrt{15 \cdot 3}$

since neither factor is a perfect square.

(b) $\sqrt{45} = \sqrt{9 \cdot 5}$

A perfect square

$= \sqrt{9} \cdot \sqrt{5}$

$= 3\sqrt{5}$

We look for the *largest* perfect-square factor, here 36.

Then apply Property 1.

(c) $\sqrt{72} = \sqrt{36 \cdot 2}$

A perfect square

$= \sqrt{36} \cdot \sqrt{2}$

$= 6\sqrt{2}$

(d) $5\sqrt{18} = 5\sqrt{9 \cdot 2}$

A perfect square

$= 5 \cdot \sqrt{9} \cdot \sqrt{2} = 5 \cdot 3\sqrt{2} = 15\sqrt{2}$

$\langle$CAUTION$\rangle$

Be Careful! Even though

$\sqrt{a \cdot b} = \sqrt{a} \cdot \sqrt{b}$

$\sqrt{a + b}$ is *not the same* as $\sqrt{a} + \sqrt{b}$

Let $a = 4$ and $b = 9$, and substitute.

$\sqrt{a + b} = \sqrt{4 + 9} = \sqrt{13}$
$\sqrt{a} + \sqrt{b} = \sqrt{4} + \sqrt{9} = 2 + 3 = 5$

Since $\sqrt{13} \neq 5$, we see that the expressions $\sqrt{a + b}$ and $\sqrt{a} + \sqrt{b}$ are not in general the same.

CHECK YOURSELF 1

Simplify.

1. $\sqrt{20}$ **2.** $\sqrt{75}$ **3.** $\sqrt{98}$ **4.** $\sqrt{48}$

In our remaining work with radicals we will assume that all variables represent positive real numbers.

The process is the same if variables are involved in a radical expression.

Example 2

Simplify each of the following radicals.

(a) $\sqrt{x^3} = \sqrt{x^2 \cdot x}$

                A perfect square

By our first rule for radicals.

Note: $\sqrt{x^2} = x$ (as long as x is positive).

$= \sqrt{x^2} \cdot \sqrt{x}$

$= x\sqrt{x}$

(b) $\sqrt{4b^3} = \sqrt{4 \cdot b^2 \cdot b}$

                Perfect squares

$= \sqrt{4b^2} \cdot \sqrt{b}$

$= 2b\sqrt{b}$

Note that we want the perfect-square factor to have the largest possible even exponent, here 4. Keep in mind that

$a^2 \cdot a^2 = a^4$

(c) $\sqrt{18a^5} = \sqrt{9 \cdot a^4 \cdot 2a}$

                Perfect squares

$= \sqrt{9a^4} \cdot \sqrt{2a}$

$= 3a^2\sqrt{2a}$

CHECK YOURSELF 2

Simplify.

1. $\sqrt{9x^3}$ **2.** $\sqrt{27m^3}$ **3.** $\sqrt{50b^5}$

To develop a second property for radicals, look at the following expressions:

$$\sqrt{\frac{16}{4}} = \sqrt{4} = 2$$

$$\frac{\sqrt{16}}{\sqrt{4}} = \frac{4}{2} = 2$$

Since $\sqrt{\dfrac{16}{4}} = \dfrac{\sqrt{16}}{\sqrt{4}}$, a second general rule for radicals is suggested.

PROPERTY 2 OF RADICALS

For any positive real numbers a and b,

$$\sqrt{\frac{a}{b}} = \frac{\sqrt{a}}{\sqrt{b}}$$

In words, the square root of a quotient is the quotient of the square roots.

This property is used in a fashion similar to Property 1 in simplifying radical expressions. Remember that our second condition for a radical expression to be in simplest form states that no fraction should appear inside a radical. The next example illustrates how expressions that violate that condition are simplified.

Example 3

Write each expression in simplest form.

Apply Property 2 to write the numerator and denominator as separate radicals.

(a) $\sqrt{\dfrac{9}{4}} = \dfrac{\sqrt{9}}{\sqrt{4}}$ $\begin{cases} \text{Remove any} \\ \text{perfect squares} \\ \text{from the radical.} \end{cases}$

$ = \dfrac{3}{2}$

Apply Property 2.

(b) $\sqrt{\dfrac{2}{25}} = \dfrac{\sqrt{2}}{\sqrt{25}}$

$ = \dfrac{\sqrt{2}}{5}$

Apply Property 2.

(c) $\sqrt{\dfrac{8x^2}{9}} = \dfrac{\sqrt{8x^2}}{\sqrt{9}}$

Factor $8x^2$ as $4x^2 \cdot 2$.

$ = \dfrac{\sqrt{4x^2 \cdot 2}}{3}$

Apply Property 1 in the numerator.

$ = \dfrac{\sqrt{4x^2} \cdot \sqrt{2}}{3}$

$ = \dfrac{2x\sqrt{2}}{3}$

CHECK YOURSELF 3

Simplify.

1. $\sqrt{\dfrac{25}{16}}$ **2.** $\sqrt{\dfrac{7}{9}}$ **3.** $\sqrt{\dfrac{12x^2}{49}}$

In our previous examples, the denominator of the fraction appearing in the radical was a perfect square, and we were able to write each expression in simplest radical form by removing that perfect square from the denominator.

If the denominator of the fraction in the radical is *not* a perfect square, we can still apply Property 2 of radicals. As we will see in our next example, the third condition for a radical to be in simplest form is then violated, and a new technique is necessary.

Example 4

Write each expression in simplest form.

We begin by applying Property 2.

(a) $\sqrt{\dfrac{1}{3}} = \dfrac{\sqrt{1}}{\sqrt{3}} = \dfrac{1}{\sqrt{3}}$

Do you see that $\dfrac{1}{\sqrt{3}}$ is still not in simplest form because of the radical in the denominator? To solve this problem, we multiply the numerator and denominator by $\sqrt{3}$. Note that the denominator will become

$$\sqrt{3} \cdot \sqrt{3} = \sqrt{9} = 3$$

In general, if $a \geq 0$, then $\sqrt{a} \cdot \sqrt{a} = a$. We then have

We can do this because we are multiplying the fraction by $\dfrac{\sqrt{3}}{\sqrt{3}}$ or 1, which does not change its value.

$$\dfrac{1}{\sqrt{3}} = \dfrac{1 \cdot \sqrt{3}}{\sqrt{3} \cdot \sqrt{3}} = \dfrac{\sqrt{3}}{3}$$

The expression $\dfrac{\sqrt{3}}{3}$ is now in simplest form since all three of our conditions are satisfied.

Note:

$\sqrt{2} \cdot \sqrt{5} = \sqrt{2 \cdot 5} = \sqrt{10}$
$\sqrt{5} \cdot \sqrt{5} = 5$

(b) $\sqrt{\dfrac{2}{5}} = \dfrac{\sqrt{2}}{\sqrt{5}}$

$= \dfrac{\sqrt{2} \cdot \sqrt{5}}{\sqrt{5} \cdot \sqrt{5}}$

$= \dfrac{\sqrt{10}}{5}$

and the expression is in simplest form since again our three conditions are satisfied.

We multiply numerator and denominator by $\sqrt{7}$ to "clear" the denominator of the radical.

(c) $\sqrt{\dfrac{3x}{7}} = \dfrac{\sqrt{3x}}{\sqrt{7}}$

$= \dfrac{\sqrt{3x} \cdot \sqrt{7}}{\sqrt{7} \cdot \sqrt{7}}$

$= \dfrac{\sqrt{21x}}{7}$

The expression is in simplest form.

Simplify.

1. $\sqrt{\dfrac{1}{2}}$ **2.** $\sqrt{\dfrac{2}{3}}$ **3.** $\sqrt{\dfrac{2y}{5}}$

Both of the properties of radicals given in this section are true for cube roots, fourth roots, and so on. Here we have limited ourselves to simplifying expressions involving square roots.

1. (1) $2\sqrt{5}$; (2) $5\sqrt{3}$; (3) $7\sqrt{2}$; (4) $4\sqrt{3}$.

2. (1) $3x\sqrt{x}$; (2) $3m\sqrt{3m}$; (3) $5b^2\sqrt{2b}$.

3. (1) $\dfrac{5}{4}$; (2) $\dfrac{\sqrt{7}}{3}$; (3) $\dfrac{2x\sqrt{3}}{7}$.

4. (1) $\dfrac{\sqrt{2}}{2}$; (2) $\dfrac{\sqrt{6}}{3}$; (3) $\dfrac{\sqrt{10y}}{5}$.

Build Your Skills

Use Property 1 to simplify each of the following radical expressions. Assume that all variables represent positive real numbers.

1. $\sqrt{18}$

2. $\sqrt{50}$

3. $\sqrt{28}$

4. $\sqrt{108}$

5. $\sqrt{45}$

6. $\sqrt{80}$

7. $\sqrt{48}$

8. $\sqrt{125}$

9. $\sqrt{200}$

10. $\sqrt{96}$

11. $\sqrt{147}$

12. $\sqrt{300}$

13. $3\sqrt{12}$

14. $5\sqrt{24}$

15. $\sqrt{5x^2}$

16. $\sqrt{7a^2}$

17. $\sqrt{3y^4}$

18. $\sqrt{10x^6}$

19. $\sqrt{2r^3}$

20. $\sqrt{5a^5}$

21. $\sqrt{27b^2}$

22. $\sqrt{98m^4}$

23. $\sqrt{24x^4}$

24. $\sqrt{72x^3}$

25. $\sqrt{54a^5}$

26. $\sqrt{200y^6}$

27. $\sqrt{x^3y^2}$

28. $\sqrt{a^2b^5}$

Use Property 2 to simplify each of the following radical expressions.

29. $\sqrt{\dfrac{4}{25}}$

30. $\sqrt{\dfrac{64}{9}}$

31. $\sqrt{\dfrac{9}{16}}$

32. $\sqrt{\dfrac{49}{25}}$

33. $\sqrt{\dfrac{3}{4}}$

34. $\sqrt{\dfrac{5}{9}}$

35. $\sqrt{\dfrac{5}{36}}$

36. $\sqrt{\dfrac{10}{49}}$

Use the properties for radicals to simplify each of the following expressions. Assume that all variables represent positive real numbers.

37. $\sqrt{\dfrac{8a^2}{25}}$

38. $\sqrt{\dfrac{12y^2}{49}}$

39. $\sqrt{\dfrac{1}{5}}$

40. $\sqrt{\dfrac{1}{7}}$

41. $\sqrt{\dfrac{3}{2}}$

42. $\sqrt{\dfrac{5}{3}}$

43. $\sqrt{\dfrac{3a}{5}}$

44. $\sqrt{\dfrac{2x}{7}}$

45. $\sqrt{\dfrac{2x^2}{3}}$

46. $\sqrt{\dfrac{5m^2}{2}}$

47. $\sqrt{\dfrac{8s^3}{7}}$

48. $\sqrt{\dfrac{12x^3}{5}}$

Think About These

Decide whether each of the following is already written in simplest form. If it is not, explain what needs to be done.

49. $\sqrt{10mn}$

50. $\sqrt{18ab}$

51. $\dfrac{\sqrt{98x^2y}}{7x}$

52. $\dfrac{\sqrt{6xy}}{3x}$

ANSWERS

37. _____

38. _____

39. _____

40. _____

41. _____

42. _____

43. _____

44. _____

45. _____

46. _____

47. _____

48. _____

49. _____

50. _____

51. _____

52. _____

Skillscan (Section 1.4)

Use the distributive property to combine the like terms in each of the following expressions.

a. $5x + 6x$ **b.** $8a - 3a$ **c.** $10y - 12y$ **d.** $7m + 10m$

e. $9a + 7a - 12a$ **f.** $5s - 8s + 4s$ **g.** $12m + 3n - 6m$ **h.** $8x + 5y - 4x$

ANSWERS

1. $3\sqrt{2}$ **3.** $2\sqrt{7}$ **5.** $3\sqrt{5}$ **7.** $4\sqrt{3}$ **9.** $10\sqrt{2}$ **11.** $7\sqrt{3}$ **13.** $6\sqrt{3}$
15. $x\sqrt{5}$ **17.** $y^2\sqrt{3}$ **19.** $r\sqrt{2r}$ **21.** $3b\sqrt{3}$ **23.** $2x^2\sqrt{6}$ **25.** $3a^2\sqrt{6a}$
27. $xy\sqrt{x}$ **29.** $\dfrac{2}{5}$ **31.** $\dfrac{3}{4}$ **33.** $\dfrac{\sqrt{3}}{2}$ **35.** $\dfrac{\sqrt{5}}{6}$ **37.** $\dfrac{2a\sqrt{2}}{5}$ **39.** $\dfrac{\sqrt{5}}{5}$
41. $\dfrac{\sqrt{6}}{2}$ **43.** $\dfrac{\sqrt{15a}}{5}$ **45.** $\dfrac{x\sqrt{6}}{3}$ **47.** $\dfrac{2s\sqrt{14s}}{7}$ **49.** Simplest form
51. Remove the perfect-square factors from the radical   **a.** $11x$ **b.** $5a$ **c.** $-2y$ **d.** $17m$
e. $4a$ **f.** s **g.** $6m + 3n$ **h.** $4x + 5y$

Adding and Subtracting Radicals

OBJECTIVE
To add or subtract expressions involving radicals

Like radicals have the same index and the same radicand (the expression inside the radical).

$2\sqrt{3}$ and $5\sqrt{3}$ are like radicals.

$\sqrt{2}$ and $\sqrt{5}$ are not like radicals—they have different radicands.

$\sqrt{2}$ and $\sqrt[3]{2}$ are not like radicals—they have different indices (2 and 3, representing a square root and a cube root).

"Indices" is the plural of "index."

Like radicals can be added (or subtracted) in the same way as like terms. We apply the distributive property and then combine the coefficients:

$$2x^2 + 3x^2 = (2 + 3)x^2 = 5x^2$$

and

$$2\sqrt{5} + 3\sqrt{5} = (2 + 3)\sqrt{5} = 5\sqrt{5}$$

Example 1

Simplify each expression.

Apply the distributive property, then combine the coefficients.

(a) $5\sqrt{2} + 3\sqrt{2} = (5 + 3)\sqrt{2} = 8\sqrt{2}$

(b) $7\sqrt{5} - 2\sqrt{5} = (7 - 2)\sqrt{5} = 5\sqrt{5}$

(c) $8\sqrt{7} - \sqrt{7} + 2\sqrt{7} = (8 - 1 + 2)\sqrt{7}$
$$= 9\sqrt{7}$$

CHECK YOURSELF 1

Simplify.

1. $2\sqrt{5} + 7\sqrt{5}$ **2.** $9\sqrt{7} - \sqrt{7}$ **3.** $5\sqrt{3} - 2\sqrt{3} + \sqrt{3}$

If a sum or difference involves terms that are *not* like radicals, we may be able to combine terms after simplifying the radicals according to our earlier methods.

Example 2

Simplify each expression.

(*a*) $3\sqrt{2} + \sqrt{8}$

We do not have like radicals, but we can simplify $\sqrt{8}$. Remember that

$$\sqrt{8} = \sqrt{4 \cdot 2} = 2\sqrt{2}$$

so

$$3\sqrt{2} + \overbrace{\sqrt{8}}^{\sqrt{8}} = 3\sqrt{2} + 2\sqrt{2}$$
$$= (3 + 2)\sqrt{2}$$
$$= 5\sqrt{2}$$

Simplify $\sqrt{12}$.

The radicals can now be combined. Do you see why?

(*b*) $5\sqrt{3} + \sqrt{12} = 5\sqrt{3} + \sqrt{4 \cdot 3}$
$$= 5\sqrt{3} + \sqrt{4}\,\sqrt{3}$$
$$= 5\sqrt{3} + 2\sqrt{3}$$
$$= (5 + 2)\sqrt{3}$$
$$= 7\sqrt{3}$$

Simplify both terms.

(*c*) $\sqrt{50} - 2\sqrt{18} = \sqrt{25 \cdot 2} - 2\sqrt{9 \cdot 2}$
$$= \sqrt{25}\,\sqrt{2} - 2\sqrt{9}\,\sqrt{2}$$
$$= 5\sqrt{2} - 2 \cdot 3\,\sqrt{2}$$
$$= 5\sqrt{2} - 6\sqrt{2}$$
$$= (5 - 6)\,\sqrt{2}$$
$$= -1\sqrt{2}$$
$$= -\sqrt{2}$$

(*d*) $3\sqrt{20} - 2\sqrt{45} + 4\sqrt{5} = 3\sqrt{4 \cdot 5} - 2\sqrt{9 \cdot 5} + 4\sqrt{5}$
$$= 3\sqrt{4}\,\sqrt{5} - 2\sqrt{9}\,\sqrt{5} + 4\sqrt{5}$$
$$= 3 \cdot 2\sqrt{5} - 2 \cdot 3\sqrt{5} + 4\sqrt{5}$$
$$= 6\sqrt{5} - 6\sqrt{5} + 4\sqrt{5}$$
$$= (6 - 6 + 4)\,\sqrt{5}$$
$$= 4\,\sqrt{5}$$

CHECK YOURSELF 2

Simplify.

1. $\sqrt{2} + \sqrt{18}$ **2.** $5\sqrt{3} - \sqrt{27}$ **3.** $\sqrt{80} - 2\sqrt{5} + 4\sqrt{20}$

There is an important point to keep in mind when radical expressions are combined. Combining terms requires the use of the distributive property, and that property can be applied *only* with like radicals. The following example illustrates.

Example 3

We begin by simplifying *each* term as before.

Simplify.

$$3\sqrt{50} - 2\sqrt{48} - \sqrt{72}$$
$$= 3\sqrt{25 \cdot 2} - 2\sqrt{16 \cdot 3} - \sqrt{36 \cdot 2}$$
$$= 3\sqrt{25}\sqrt{2} - 2\sqrt{16}\sqrt{3} - \sqrt{36}\sqrt{2}$$
$$= 3 \cdot 5\sqrt{2} - 2 \cdot 4\sqrt{3} - 6\sqrt{2}$$
$$= 15\sqrt{2} - 8\sqrt{3} - 6\sqrt{2}$$
$$= 9\sqrt{2} - 8\sqrt{3}$$

Now note that *only* the first and third terms are *like radicals* and can be combined.

CHECK YOURSELF 3

Simplify $2\sqrt{80} + \sqrt{54} - 3\sqrt{45}$.

If variables are involved in radical expressions, the process of combining terms proceeds in a fashion similar to that shown in previous examples. Consider our final example. We again assume that all variables represent positive real numbers.

Example 4

Simplify each expression.

Since like radicals are involved, we apply the distributive property and combine terms as before.

Simplify the first term.

The radicals can now be combined.

(a) $5\sqrt{3x} - 2\sqrt{3x} = (5 - 2)\sqrt{3x} = 3\sqrt{3x}$

(b) $2\sqrt{3a^3} + 5a\sqrt{3a}$
$$= 2\sqrt{a^2 \cdot 3a} + 5a\sqrt{3a}$$
$$= 2\sqrt{a^2} \cdot \sqrt{3a} + 5a\sqrt{3a}$$
$$= 2a\sqrt{3a} + 5a\sqrt{3a}$$
$$= (2a + 5a)\sqrt{3a} = 7a\sqrt{3a}$$

CHECK YOURSELF 4

Simplify each expression.

1. $2\sqrt{7y} + 3\sqrt{7y}$ **2.** $\sqrt{20a^2} - a\sqrt{45}$

CHECK YOURSELF ANSWERS

1. (1) $9\sqrt{5}$; (2) $8\sqrt{7}$; (3) $4\sqrt{3}$.
2. (1) $4\sqrt{2}$; (2) $2\sqrt{3}$; (3) $10\sqrt{5}$.
3. $3\sqrt{6} - \sqrt{5}$.
4. (1) $5\sqrt{7y}$; (2) $-a\sqrt{5}$.

Name

Date

Build Your Skills

Simplify by combining like terms.

1. $\sqrt{2} + 5\sqrt{2}$

2. $4\sqrt{3} + 2\sqrt{3}$

3. $9\sqrt{7} - 2\sqrt{7}$

4. $7\sqrt{2} - 3\sqrt{3}$

5. $5\sqrt{7} + 3\sqrt{6}$

6. $3\sqrt{5} - 5\sqrt{5}$

7. $2\sqrt{3} - 5\sqrt{3}$

8. $2\sqrt{11} + 5\sqrt{11}$

9. $2\sqrt{3x} + 5\sqrt{3x}$

10. $7\sqrt{2a} - 3\sqrt{2a}$

11. $2\sqrt{3} + \sqrt{3} + 3\sqrt{3}$

12. $3\sqrt{5} + 2\sqrt{5} + \sqrt{5}$

13. $5\sqrt{7} - 2\sqrt{7} + \sqrt{7}$

14. $3\sqrt{10} - 2\sqrt{10} + \sqrt{10}$

15. $2\sqrt{5x} + 5\sqrt{5x} - 2\sqrt{5x}$

16. $5\sqrt{3b} - 2\sqrt{3b} + 4\sqrt{3b}$

17. $2\sqrt{3} + \sqrt{12}$

18. $5\sqrt{2} + \sqrt{18}$

19. $\sqrt{20} - \sqrt{5}$

20. $\sqrt{98} - 3\sqrt{2}$

21. $2\sqrt{6} - \sqrt{54}$

22. $2\sqrt{3} - \sqrt{27}$

23. $\sqrt{72} + \sqrt{50}$

24. $\sqrt{27} - \sqrt{12}$

25. $3\sqrt{12} - \sqrt{48}$

26. $5\sqrt{8} + 2\sqrt{18}$

27. $4\sqrt{20} - 2\sqrt{45}$

28. $2\sqrt{72} - \sqrt{200}$

ANSWERS

1. _____
2. _____
3. _____
4. _____
5. _____
6. _____
7. _____
8. _____
9. _____
10. _____
11. _____
12. _____
13. _____
14. _____
15. _____
16. _____
17. _____
18. _____
19. _____
20. _____
21. _____
22. _____
23. _____
24. _____
25. _____
26. _____
27. _____
28. _____

29. _____

30. _____

31. _____

32. _____

33. _____

34. _____

35. _____

36. _____

37. _____

38. _____

39. _____

40. _____

a. _____

b. _____

c. _____

d. _____

e. _____

f. _____

g. _____

h. _____

29. $\sqrt{12} + \sqrt{27} - \sqrt{3}$

30. $\sqrt{50} + \sqrt{32} - \sqrt{8}$

31. $3\sqrt{24} - \sqrt{54} + \sqrt{6}$

32. $\sqrt{63} - 2\sqrt{28} + 5\sqrt{7}$

33. $2\sqrt{50} + 3\sqrt{18} - \sqrt{32}$

34. $3\sqrt{27} + 4\sqrt{12} - \sqrt{300}$

Simplify by combining like terms.

35. $a\sqrt{27} - 2\sqrt{3a^2}$

36. $5\sqrt{2y^2} - 3y\sqrt{8}$

37. $5\sqrt{3x^3} + 2\sqrt{27x}$

38. $7\sqrt{2a^3} - \sqrt{8a}$

39. $\sqrt{6} - \sqrt{\dfrac{2}{3}}$

40. $\sqrt{15} + \sqrt{\dfrac{3}{5}}$

Skillscan (Section 4.3)

Perform the indicated multiplication.

a. $2(x + 5)$ **b.** $3(a - 3)$ **c.** $m(m - 8)$ **d.** $y(y + 7)$

e. $(w + 2)(w - 2)$ **f.** $(x - 3)(x + 3)$ **g.** $(x + y)(x + y)$ **h.** $(b - 7)(b - 7)$

ANSWERS

1. $6\sqrt{2}$ 3. $7\sqrt{7}$ 5. Cannot be simplified   7. $-3\sqrt{3}$ 9. $7\sqrt{3x}$ 11. $6\sqrt{3}$
13. $4\sqrt{7}$ 15. $5\sqrt{5x}$ 17. $4\sqrt{3}$ 19. $\sqrt{5}$ 21. $-\sqrt{6}$ 23. $11\sqrt{2}$ 25. $2\sqrt{3}$
27. $2\sqrt{5}$ 29. $4\sqrt{3}$ 31. $4\sqrt{6}$ 33. $15\sqrt{2}$ 35. $a\sqrt{3}$ 37. $(5x + 6)\sqrt{3x}$
39. $\dfrac{2\sqrt{6}}{3}$ **a.** $2x + 10$ **b.** $3a - 9$ **c.** $m^2 - 8m$ **d.** $y^2 + 7y$ **e.** $w^2 - 4$
f. $x^2 - 9$ **g.** $x^2 + 2xy + y^2$ **h.** $b^2 - 14b + 49$

Multiplying and Dividing Radicals

OBJECTIVE
To multiply or divide expressions involving radicals

In Section 9.4 we stated the first property for radicals:

$$\sqrt{ab} = \sqrt{a} \cdot \sqrt{b} \qquad \text{where } a \text{ and } b \text{ are any positive real numbers}$$

That property has been used to simplify radical expressions up to this point. Suppose now that we want to find a product, such as $\sqrt{3} \cdot \sqrt{5}$.

We can use our first radical rule in the opposite manner.

The product of square roots is equal to the square root of the product of the radicands.

$$\sqrt{a} \cdot \sqrt{b} = \sqrt{ab}$$

so

$$\sqrt{3} \cdot \sqrt{5} = \sqrt{3 \cdot 5} = \sqrt{15}$$

Example 1

Multiply the radicals.

(a) $\sqrt{2}\,\sqrt{7} = \sqrt{2 \cdot 7} = \sqrt{14}$

(b) $\sqrt{5}\,\sqrt{3x} = \sqrt{5 \cdot 3x} = \sqrt{15x}$

(c) $\sqrt{5}\,\sqrt{3}\,\sqrt{2} = \sqrt{5 \cdot 3 \cdot 2} = \sqrt{30}$

CHECK YOURSELF 1

Multiply.

1. $\sqrt{3}\,\sqrt{10}$ **2.** $\sqrt{5}\,\sqrt{7a}$ **3.** $\sqrt{2}\,\sqrt{7}\,\sqrt{5}$

We may have to simplify after multiplying, as the following examples illustrate.

Example 2

Multiply then simplify each expression.

(a) $\sqrt{5}\sqrt{10} = \sqrt{5 \cdot 10} = \sqrt{50}$
$$= \sqrt{25 \cdot 2} = 5\sqrt{2}$$

(b) $\sqrt{12} \sqrt{6} = \sqrt{12 \cdot 6} = \sqrt{72}$
$$= \sqrt{36 \cdot 2} = \sqrt{36} \sqrt{2} = 6\sqrt{2}$$

Note: An alternative approach would be to simplify $\sqrt{12}$ first.

$\sqrt{12} \sqrt{6} = 2\sqrt{3} \sqrt{6} = 2\sqrt{18}$
$$= 2\sqrt{9 \cdot 2} = 2\sqrt{9} \sqrt{2}$$
$$= 2 \cdot 3\sqrt{2} = 6\sqrt{2}$$

(c) $\sqrt{10x} \sqrt{2x} = \sqrt{20x^2} = \sqrt{4x^2 \cdot 5}$
$$= \sqrt{4x^2} \sqrt{5} = 2x\sqrt{5}$$

CHECK YOURSELF 2

Simplify.

1. $\sqrt{3} \sqrt{6}$ 　　　　　　**2.** $\sqrt{3} \cdot \sqrt{18}$ 　　　　　　**3.** $\sqrt{8a} \sqrt{3a}$

If coefficients are involved in a product, we can use the commutative and associative properties to change the order and grouping of the factors. This is illustrated in our next example.

Example 3

Multiply.

Note: In practice, it is not necessary to show the intermediate steps.

$(2\sqrt{5})(3\sqrt{6}) = (2 \cdot 3)(\sqrt{5} \sqrt{6})$
$$= 6\sqrt{5 \cdot 6}$$
$$= 6\sqrt{30}$$

CHECK YOURSELF 3

Multiply $(3\sqrt{7})(5\sqrt{3})$.

The distributive property can also be applied in multiplying radical expressions. Consider the following.

Example 4

Multiply.

(a) $\sqrt{3} (\sqrt{2} + \sqrt{3})$
$$= \sqrt{3} \cdot \sqrt{2} + \sqrt{3} \cdot \sqrt{3} \qquad \text{The distributive property}$$
$$= \sqrt{6} + 3 \qquad\qquad\qquad \text{Multiply the radicals.}$$

(b) $\sqrt{5}\,(2\sqrt{6} + 3\sqrt{3})$

$\quad = \sqrt{5} \cdot 2\sqrt{6} + \sqrt{5} \cdot 3\sqrt{3}$ The distributive property

$\quad = 2 \cdot \sqrt{5} \cdot \sqrt{6} + 3 \cdot \sqrt{5} \cdot \sqrt{3}$ The commutative property

$\quad = 2\sqrt{30} + 3\sqrt{15}$

CHECK YOURSELF 4

Multiply.

1. $\sqrt{5}\,(\sqrt{6} + \sqrt{5})$ **2.** $\sqrt{3}\,(2\sqrt{5} + 3\sqrt{2})$

Our earlier FOIL pattern for multiplying binomials can also be applied in multiplying radical expressions. This is shown in our next example.

Example 5

Multiply.

(a) $(\sqrt{3} + 2)\,(\sqrt{3} + 5)$

$\quad = \sqrt{3} \cdot \sqrt{3} + 5\sqrt{3} + 2\sqrt{3} + 2 \cdot 5$

$\quad = 3 + 5\sqrt{3} + 2\sqrt{3} + 10$ Combine like terms.

$\quad = 13 + 7\sqrt{3}$

Be Careful!   This result *cannot* be further simplified: 13 and $7\sqrt{3}$ are *not* like terms.

(b) $(\sqrt{7} + 2)(\sqrt{7} - 2) = \sqrt{7} \cdot \sqrt{7} - 2\sqrt{7} + 2\sqrt{7} - 4$

$\quad\quad\quad\quad\quad\quad\quad = 7 - 4 = 3$

(c) $(\sqrt{3} + 5)^2 = (\sqrt{3} + 5)(\sqrt{3} + 5)$

$\quad\quad\quad\quad\quad = \sqrt{3}\,\sqrt{3} + 5\sqrt{3} + 5\sqrt{3} + 5 \cdot 5$

$\quad\quad\quad\quad\quad = 3 + 5\sqrt{3} + 5\sqrt{3} + 25$

$\quad\quad\quad\quad\quad = 28 + 10\sqrt{3}$

$\langle$ CAUTION $\rangle$

Note: You can use the pattern $(a + b)(a - b) = a^2 - b^2$, where $a = \sqrt{7}$ and $b = 2$, for the same result. $\sqrt{7} + 2$ and $\sqrt{7} - 2$ are called *conjugates* of each other. Note that their product is the rational number 3. The product of conjugates will *always be rational*.

CHECK YOURSELF 5

Multiply.

1. $(\sqrt{5} + 3)(\sqrt{5} - 2)$ **2.** $(\sqrt{3} + 4)(\sqrt{3} - 4)$ **3.** $(\sqrt{2} - 3)^2$

We can also use our second property for radicals in the opposite manner.

The quotient of square roots is equal to the square root of the quotient of the radicands.

$$\frac{\sqrt{a}}{\sqrt{b}} = \sqrt{\frac{a}{b}}$$

One use of this property to divide radical expressions is illustrated in our next example.

Example 6

Simplify.

The clue to recognizing when to use this approach is in noting that 48 is divisible by 3.

(a) $\dfrac{\sqrt{48}}{\sqrt{3}} = \sqrt{\dfrac{48}{3}} = \sqrt{16} = 4$

(b) $\dfrac{\sqrt{200}}{\sqrt{2}} = \sqrt{\dfrac{200}{2}} = \sqrt{100} = 10$

(c) $\dfrac{\sqrt{125x^2}}{\sqrt{5}} = \sqrt{\dfrac{125x^2}{5}} = \sqrt{25x^2} = 5x$

CHECK YOURSELF 6

Simplify.

1. $\dfrac{\sqrt{75}}{\sqrt{3}}$

2. $\dfrac{\sqrt{81s^2}}{\sqrt{9}}$

Another approach is called for if the quotient cannot be simplified as in the previous examples. Consider the following.

Example 7

Simplify each of the following.

(a) $\dfrac{\sqrt{3}}{\sqrt{5}}$

Note that this approach to simplifying a quotient is identical to our earlier work in Section 9.4.

Multiply the numerator and denominator by $\sqrt{5}$. The value of the fraction is unchanged, and we will have a rational number in the denominator.

$$\dfrac{\sqrt{3}}{\sqrt{5}} = \dfrac{\sqrt{3} \cdot \sqrt{5}}{\sqrt{5} \cdot \sqrt{5}} = \dfrac{\sqrt{15}}{\sqrt{25}} = \dfrac{\sqrt{15}}{5}$$

The expression is now in simplest radical form (there is no radical in the denominator). This process is called *rationalizing the denominator*.

Multiply numerator and denominator by $\sqrt{7}$.

(b) $\dfrac{2}{\sqrt{7}} = \dfrac{2\sqrt{7}}{\sqrt{7}\sqrt{7}} = \dfrac{2\sqrt{7}}{\sqrt{49}} = \dfrac{2\sqrt{7}}{7}$

(c) $\dfrac{\sqrt{3x^3y^2}}{\sqrt{5}} = \dfrac{\sqrt{3x^3y^2} \cdot \sqrt{5}}{\sqrt{5} \cdot \sqrt{5}}$

$= \dfrac{\sqrt{15x^3y^2}}{5}$ We must now simplify the
                  radical in the numerator.

$= \dfrac{\sqrt{x^2y^2 \cdot 15x}}{5}$ Apply Property 1.

$= \dfrac{\sqrt{x^2y^2} \cdot \sqrt{15x}}{5}$

$= \dfrac{xy\sqrt{15x}}{5}$

CHECK YOURSELF 7

Simplify by rationalizing the denominators.

1. $\dfrac{3}{\sqrt{6}}$ **2.** $\dfrac{\sqrt{10}}{\sqrt{3}}$ **3.** $\dfrac{\sqrt{2a^2b}}{\sqrt{5}}$

We will now look at another type of division problem in which the denominator will be a binomial. In simplifying such an expression we will use the idea of conjugates introduced in Example 5.

Example 8

Rationalize the denominator in the expression

$$\frac{5}{\sqrt{7} + 2}$$

Recall that $\sqrt{7} + 2$ and $\sqrt{7} - 2$ are conjugates of each other and that the product of conjugates will *always be a rational number*. Therefore, to rationalize the denominator in this expression, we can multiply numerator and denominator by $\sqrt{7} - 2$.

If an expression involves a binomial with a radical in the denominator, multiply the numerator and denominator by the *conjugate* of the denominator to rationalize.

See Example 5(*b*) for details of the multiplication in the denominator.

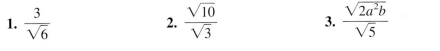

$\dfrac{5}{\sqrt{7} + 2} = \dfrac{5(\sqrt{7} - 2)}{(\sqrt{7} + 2)(\sqrt{7} - 2)}$

$= \dfrac{5\sqrt{7} - 5 \cdot 2}{3}$

$= \dfrac{5\sqrt{7} - 10}{3}$ The quotient is now in simplest form.

CHECK YOURSELF 8

Rationalize the denominator of the expression $\dfrac{7}{\sqrt{10} - 2}$.

There is one final quotient form which you may encounter in simplifying expressions, and it will be extremely important in our work with quadratic equations in the next chapter. This form is shown in our final example.

Example 9

Simplify the expression

$$\frac{3 + \sqrt{72}}{3}$$

First, we must simplify the radical in the numerator.

<CAUTION>

Be Careful! Students are sometimes tempted to write

$$\frac{\cancel{3} + 6\sqrt{2}}{\cancel{3}} = 1 + 6\sqrt{2}$$

This is *not* correct. We must divide *both terms* of the numerator by the common factor.

$$\frac{3 + \sqrt{72}}{3} = \frac{3 + \sqrt{36 \cdot 2}}{3}$$

Use Property 1 to simplify $\sqrt{72}$.

$$= \frac{3 + \sqrt{36} \cdot \sqrt{2}}{3} = \frac{3 + 6\sqrt{2}}{3}$$

$$= \frac{3(1 + 2\sqrt{2})}{3} = 1 + 2\sqrt{2}$$

Factor the numerator—then divide by the *common* factor of 3.

CHECK YOURSELF 9

Simplify $\dfrac{15 + \sqrt{75}}{5}$.

CHECK YOURSELF ANSWERS

1. (1) $\sqrt{30}$; (2) $\sqrt{35a}$; (3) $\sqrt{70}$.
2. (1) $3\sqrt{2}$; (2) $3\sqrt{6}$; (3) $2a\sqrt{6}$.
3. $15\sqrt{21}$.
4. (1) $\sqrt{30} + 5$; (2) $2\sqrt{15} + 3\sqrt{6}$.
5. (1) $-1 + \sqrt{5}$; (2) -13; (3) $11 - 6\sqrt{2}$.
6. (1) 5; (2) $3s$.
7. (1) $\dfrac{\sqrt{6}}{2}$; (2) $\dfrac{\sqrt{30}}{3}$; (3) $\dfrac{a\sqrt{10b}}{5}$.
8. $\dfrac{7\sqrt{10} + 14}{6}$.   **9.** $3 + \sqrt{3}$.

Name _____

Date _____

Build Your Skills

Perform the indicated multiplication. Then simplify each radical expression.

1. $\sqrt{7}\,\sqrt{5}$

2. $\sqrt{3}\,\sqrt{7}$

3. $\sqrt{5}\,\sqrt{11}$

4. $\sqrt{13}\,\sqrt{5}$

5. $\sqrt{3}\,\sqrt{10m}$

6. $\sqrt{7a}\,\sqrt{13}$

7. $\sqrt{2x}\,\sqrt{15}$

8. $\sqrt{17}\,\sqrt{2b}$

9. $\sqrt{3}\,\sqrt{7}\,\sqrt{2}$

10. $\sqrt{5}\,\sqrt{7}\,\sqrt{3}$

11. $\sqrt{3}\sqrt{12}$

12. $\sqrt{7}\,\sqrt{7}$

13. $\sqrt{10}\,\sqrt{10}$

14. $\sqrt{5}\,\sqrt{15}$

15. $\sqrt{18}\,\sqrt{6}$

16. $\sqrt{8}\,\sqrt{10}$

17. $\sqrt{2x}\,\sqrt{6x}$

18. $\sqrt{3a}\,\sqrt{15a}$

19. $2\sqrt{3}\,\sqrt{7}$

20. $3\sqrt{2}\,\sqrt{5}$

21. $(3\sqrt{3})(5\sqrt{7})$

22. $(2\sqrt{5})(3\sqrt{11})$

ANSWERS

1. _____

2. _____

3. _____

4. _____

5. _____

6. _____

7. _____

8. _____

9. _____

10. _____

11. _____

12. _____

13. _____

14. _____

15. _____

16. _____

17. _____

18. _____

19. _____

20. _____

21. _____

22. _____

23. _____

24. _____

25. _____

26. _____

27. _____

28. _____

29. _____

30. _____

31. _____

32. _____

33. _____

34. _____

35. _____

36. _____

37. _____

38. _____

39. _____

40. _____

41. _____

42. _____

43. _____

44. _____

45. _____

46. _____

23. $(3\sqrt{5})(2\sqrt{10})$ **24.** $(4\sqrt{3})(3\sqrt{6})$

25. $\sqrt{5}\,(\sqrt{2} + \sqrt{5})$ **26.** $\sqrt{3}(\sqrt{5} - \sqrt{3})$

27. $\sqrt{3}\,(2\sqrt{5} - 3\sqrt{3})$ **28.** $\sqrt{7}\,(2\sqrt{3} + 3\sqrt{7})$

29. $(\sqrt{3} + 5)(\sqrt{3} + 3)$ **30.** $(\sqrt{5} - 2)(\sqrt{5} - 1)$

31. $(\sqrt{5} - 1)(\sqrt{5} + 3)$ **32.** $(\sqrt{2} + 3)(\sqrt{2} - 7)$

33. $(\sqrt{5} - 2)(\sqrt{5} + 2)$ **34.** $(\sqrt{7} + 5)(\sqrt{7} - 5)$

35. $(\sqrt{10} + 5)(\sqrt{10} - 5)$ **36.** $(\sqrt{11} - 3)(\sqrt{11} + 3)$

37. $(\sqrt{x} + 3)(\sqrt{x} - 3)$ **38.** $(\sqrt{a} - 4)(\sqrt{a} + 4)$

39. $(\sqrt{3} + 2)^2$ **40.** $(\sqrt{5} - 3)^2$

41. $(\sqrt{y} - 5)^2$ **42.** $(\sqrt{x} + 4)^2$

Perform the indicated division. Rationalize the denominator if necessary. Then simplify each radical expression.

43. $\dfrac{\sqrt{98}}{\sqrt{2}}$ **44.** $\dfrac{\sqrt{108}}{\sqrt{3}}$

45. $\dfrac{\sqrt{72a^2}}{\sqrt{2}}$ **46.** $\dfrac{\sqrt{48m^2}}{\sqrt{3}}$

47. $\dfrac{\sqrt{5}}{\sqrt{2}}$

48. $\dfrac{\sqrt{3}}{\sqrt{5}}$

49. $\dfrac{3}{\sqrt{7}}$

50. $\dfrac{2}{\sqrt{5}}$

51. $\dfrac{4}{\sqrt{6}}$

52. $\dfrac{2}{\sqrt{10}}$

53. $\dfrac{2}{\sqrt{8}}$

54. $\dfrac{3}{\sqrt{27}}$

55. $\dfrac{\sqrt{2a^2b}}{\sqrt{3}}$

56. $\dfrac{\sqrt{5x^3y^2}}{\sqrt{2}}$

57. $\dfrac{\sqrt{5x^4y}}{\sqrt{7}}$

58. $\dfrac{\sqrt{7m^2n^3}}{\sqrt{3}}$

59. $\dfrac{3}{\sqrt{6}+2}$

60. $\dfrac{4}{\sqrt{7}-2}$

61. $\dfrac{12}{\sqrt{10}-2}$

62. $\dfrac{6}{\sqrt{11}-3}$

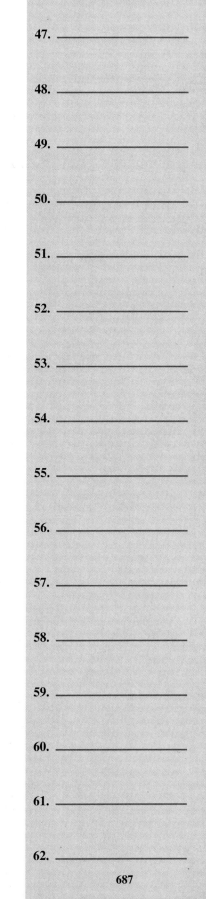

ANSWERS

47. _____

48. _____

49. _____

50. _____

51. _____

52. _____

53. _____

54. _____

55. _____

56. _____

57. _____

58. _____

59. _____

60. _____

61. _____

62. _____

Think About These

63. $\dfrac{4}{\sqrt{5} + \sqrt{3}}$

64. $\dfrac{10}{\sqrt{7} - \sqrt{2}}$

65. $\dfrac{6 + \sqrt{18}}{3}$

66. $\dfrac{6 - \sqrt{20}}{2}$

67. $\dfrac{15 - \sqrt{75}}{5}$

68. $\dfrac{8 + \sqrt{48}}{4}$

69. $\dfrac{\sqrt{2} + \sqrt{3}}{\sqrt{2} - \sqrt{3}}$

70. $\dfrac{\sqrt{3} - \sqrt{5}}{\sqrt{3} + \sqrt{5}}$

ANSWERS

63. _____

64. _____

65. _____

66. _____

67. _____

68. _____

69. _____

70. _____

ANSWERS

1. $\sqrt{35}$ **3.** $\sqrt{55}$ **5.** $\sqrt{30m}$ **7.** $\sqrt{30x}$ **9.** $\sqrt{42}$ **11.** 6   **13.** 10   **15.** $6\sqrt{3}$

17. $2x\sqrt{3}$ **19.** $2\sqrt{21}$ **21.** $15\sqrt{21}$ **23.** $30\sqrt{2}$ **25.** $\sqrt{10} + 5$ **27.** $2\sqrt{15} - 9$

29. $18 + 8\sqrt{3}$ **31.** $2 + 2\sqrt{5}$ **33.** 1   **35.** -15 **37.** $x - 9$ **39.** $7 + 4\sqrt{3}$

41. $y - 10\sqrt{y} + 25$ **43.** 7   **45.** $6a$ **47.** $\dfrac{\sqrt{10}}{2}$ **49.** $\dfrac{3\sqrt{7}}{7}$ **51.** $\dfrac{2\sqrt{6}}{3}$

53. $\dfrac{\sqrt{2}}{2}$ **55.** $\dfrac{a\sqrt{6b}}{3}$ **57.** $\dfrac{x^2\sqrt{35y}}{7}$ **59.** $\dfrac{3\sqrt{6} - 6}{2}$ **61.** $2\sqrt{10} + 4$

63. $2\sqrt{5} - 2\sqrt{3}$ **65.** $2 + \sqrt{2}$ **67.** $3 - \sqrt{3}$ **69.** $-5 - 2\sqrt{6}$

The Distance Between Two Points

OBJECTIVE
To find the distance between two points

The concept of irrational numbers was formulated by the Greeks. It was an extension of the search to find the distance between two points. If two points lie on an axis, it is a simple matter of subtraction.

Example 1

Find the distance between $(-3, 0)$ and $(5, 0)$.

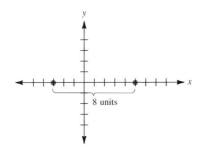

There are 8 units between the points, so the distance is 8.

CHECK YOURSELF 1

Find the distance between $(0, 7)$ and $(0, -6)$.

If two points lie on the same vertical or horizontal line, it is equally straightforward to find the distance between them.

Example 2

Find the distance between $(-2, -3)$ and $(-2, 2)$.

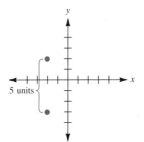

Again, by inspection we can see that there are 5 units between the two points. The distance is 5.

CHECK YOURSELF 2

Find the distance between $(2, -5)$ and $(-8, -5)$.

In order to find the distance between any two points on a plane, we use a formula derived from the Pythagorean theorem.

PYTHAGOREAN THEOREM

Given a right triangle, the square of the length of the hypotenuse is equal to the sum of the squares of the two sides.

We write $c^2 = a^2 + b^2$

We can rewrite the formula as

$c = \sqrt{a^2 + b^2}$

A distance is always positive, so we use only the principal square root.

We'll use the Pythagorean theorem in the next example.

Example 3

Find the distance from $(2, 3)$ to $(5, 7)$.

The distance can be seen as the hypotenuse of a right triangle.

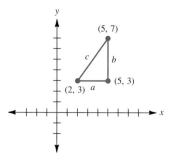

The lengths of the two legs can be found by finding the difference of the two x coordinates and the difference of the two y coordinates. So

$a = 5 - 2 = 3$ and $b = 7 - 3 = 4$

The distance, c, can then be found using the formula

$$c = \sqrt{a^2 + b^2}$$

or, in this case

$$c = \sqrt{3^2 + 4^2}$$
$$c = \sqrt{9 + 16}$$
$$= \sqrt{25}$$
$$= 5$$

The distance is 5 units.

CHECK YOURSELF 3

Find the distance between $(0, 2)$ and $(5, 14)$

If we call our points (x_1, y_1) and (x_2, y_2), we can state the *distance formula.*

DISTANCE FORMULA

The distance between points (x_1, y_1) and (x_2, y_2) can be found using the formula

$$d = \sqrt{(x_2 - x_1)^2 + (y_2 - y_1)^2}$$

Example 4

Find the distance between $(-2, 5)$ and $(2, -3)$.

Using the formula,

$$d = \sqrt{[2 - (-2)]^2 + [(-3) - 5]^2}$$
$$= \sqrt{(4)^2 + (-8)^2}$$
$$= \sqrt{16 + 64}$$
$$= \sqrt{80}$$
$$= 4\sqrt{5}$$

$\sqrt{80} = \sqrt{16 \cdot 5} = 4\sqrt{5}$

CHECK YOURSELF 4

Find the distance between $(2, 5)$ and $(-5, 2)$

CHECK YOURSELF ANSWERS

1. 13.   **2.** 10.

3. 13.   **4.** $\sqrt{58}$.

Name _____

Date _____

Build Your Skills

Find the distance between each two points.

1. $(2, 0)$ and $(-4, 0)$

2. $(-3, 0)$ and $(4, 0)$

3. $(0, -2)$ and $(0, -9)$

4. $(0, 8)$ and $(0, -4)$

5. $(2, 5)$ and $(5, 2)$

6. $(3, 3)$ and $(5, 7)$

7. $(5, 1)$ and $(3, 8)$

8. $(2, 9)$ and $(7, 4)$

9. $(-2, 8)$ and $(1, 5)$

10. $(2, 6)$ and $(-3, 4)$

11. $(6, -1)$ and $(2, 2)$

12. $(2, -8)$ and $(1, 0)$

13. $(-1, -1)$ and $(2, 5)$

14. $(-2, -2)$ and $(3, 3)$

15. $(-2, 9)$ and $(-3, 3)$

16. $(4, -1)$ and $(0, -5)$

17. $(-1, -4)$ and $(-3, 5)$

18. $(-2, 3)$ and $(-7, -1)$

19. $(-2, -4)$ and $(-4, 1)$

20. $(-1, -1)$ and $(4, -2)$

ANSWERS

1. _____

2. _____

3. _____

4. _____

5. _____

6. _____

7. _____

8. _____

9. _____

10. _____

11. _____

12. _____

13. _____

14. _____

15. _____

16. _____

17. _____

18. _____

19. _____

20. _____

21. _____

22. _____

23. _____

24. _____

25. _____

26. _____

21. $(-4, -2)$ and $(-1, -5)$

22. $(-2, -2)$ and $(-4, -4)$

23. $(-2, 0)$ and $(-4, -1)$

24. $(-5, -2)$ and $(-7, -1)$

Think About These

Use the distance formula to show that each set of points describes an isosceles triangle (a triangle with two sides of equal length).

25. $(-3, 0)$, $(2, 3)$ and $(7, 6)$

26. $(-2, 4)$, $(2, 7)$, and $(5, 3)$

ANSWERS

1. 6   **3.** 7   **5.** $3\sqrt{2}$ **7.** $\sqrt{53}$ **9.** $3\sqrt{2}$ **11.** 5   **13.** $3\sqrt{5}$ **15.** $\sqrt{37}$
17. $\sqrt{85}$ **19.** $\sqrt{29}$ **21.** $3\sqrt{2}$ **23.** $\sqrt{5}$ **25.** Sides have length $\sqrt{34}$, $\sqrt{34}$, and $\sqrt{136}$

Summary

Properties of Exponents [9.1]

For any nonzero real numbers a and b and integers m and n:

$x^5 \cdot x^7 = x^{5+7} = x^{12}$

Property 1 $a^m \cdot a^n = a^{m+n}$

$\dfrac{x^7}{x^5} = x^{7-5} = x^2$

Property 2 $\dfrac{a^m}{a^n} = a^{m-n}$

$(x^5)^3 = x^{5\cdot3} = x^{15}$

Property 3 $(a^m)^n = a^{m \cdot n}$

$(2xy)^3 = 2^3 x^3 y^3$
$\qquad = 8x^3 y^3$

Property 4 $(ab)^m = a^m b^m$

$\left(\dfrac{x^2}{3}\right)^2 = \dfrac{(x^2)^2}{3^2}$
$\qquad = \dfrac{x^4}{9}$

Property 5 $\left(\dfrac{a}{b}\right)^m = \dfrac{a^m}{b^m}$

Zero and Negative Exponents [9.2]

$5^0 = 1$

$(-3)^0 = 1$

$2x^0 = 2 \cdot 1 = 2$

Zero as an Exponent $a^0 = 1$ $(a \neq 0)$

$x^{-3} = \dfrac{1}{x^3}$

$3x^{-5} = \dfrac{3}{x^5}$

Negative Integers as Exponents $a^{-n} = \dfrac{1}{a^n}$ $(a \neq 0)$

Radicals [9.3]

$\sqrt{49} = 7$

Square Roots $\sqrt{x}$ is the principal (or positive) square root of x. It is the positive number we must square to get x.

$-\sqrt{49} = -7$
$\sqrt{-49}$ is not a real number.

     $-\sqrt{x}$ is the negative square root of x.
     The square root of a negative number is not a real number.

$\sqrt[3]{64} = 4$ because $4^3 = 64$.
$\sqrt[4]{81} = 3$ because $3^4 = 81$.
$\dfrac{2}{3}, \dfrac{-7}{12}, 5, \sqrt{36},$ and $\sqrt[3]{64}$ are rational numbers.
$\sqrt{5}, \sqrt{37},$ and $\sqrt[3]{65}$ are irrational numbers.

Other Roots $\sqrt[3]{x}$ is the cube root of x.
     $\sqrt[4]{x}$ is the fourth root of x.

Rational and Irrational Numbers   Rational numbers can be expressed as the quotient of two integers with a nonzero denominator.
     Irrational numbers cannot be expressed as the quotient of two integers.

Real Numbers   The real numbers are the set of rational numbers and the set of irrational numbers together.

$\sqrt{5^2} = 5 \qquad \sqrt{(-3)^2} = 3$
$\sqrt[3]{2^3} = 2 \qquad \sqrt[3]{(-3)^3} = -3$

Definitions $\sqrt{x^2} = |x|$ for any real number x
              $\sqrt[3]{x^3} = x$ for any real number x

Simplifying Radical Expressions [9.4]

> An expression involving square roots is in *simplest form* if
>
> **1.** There are no perfect-square factors in a radical.
> **2.** No fraction appears inside a radical.
> **3.** No radical appears in the denominator.

$$\sqrt{40} = \sqrt{4 \cdot 10}$$
$$= \sqrt{4} \cdot \sqrt{10}$$
$$= 2\sqrt{10}$$
$$\sqrt{12x^3} = \sqrt{4x^2 \cdot 3x}$$
$$= \sqrt{4x^2}\,\sqrt{3x}$$
$$= 2x\sqrt{3x}$$

To simplify a radical expression, use one of the following properties. The square root of a product is the product of the square roots.

$$\sqrt{ab} = \sqrt{a} \cdot \sqrt{b}$$

$$\sqrt{\frac{5}{16}} = \frac{\sqrt{5}}{\sqrt{16}} = \frac{\sqrt{5}}{4}$$

The square root of a quotient is the quotient of the square roots.

$$\sqrt{\frac{2y}{3}} = \frac{\sqrt{2y}}{\sqrt{3}} = \frac{\sqrt{2y} \cdot \sqrt{3}}{\sqrt{3} \cdot \sqrt{3}}$$
$$= \frac{\sqrt{6y}}{\sqrt{9}} = \frac{\sqrt{6y}}{3}$$

$$\sqrt{\frac{a}{b}} = \frac{\sqrt{a}}{\sqrt{b}}$$

Adding and Subtracting Radicals [9.5]

$3\sqrt{5}$ and $2\sqrt{5}$ are like radicals.

$$2\sqrt{3} + 3\sqrt{3} = (2 + 3)\,\sqrt{3}$$
$$= 5\sqrt{3}$$
$$5\sqrt{7} - 2\sqrt{7} = (5 - 2)\,\sqrt{7}$$
$$= 3\sqrt{7}$$
$$\sqrt{12} + \sqrt{3} = 2\sqrt{3} + \sqrt{3}$$
$$= (2 + 1)\,\sqrt{3}$$
$$= 3\sqrt{3}$$

Like radicals have the same index and the same radicand (the expression inside the radical).

Like radicals can be added (or subtracted) in the same way as like terms. Apply the distributive law and combine the coefficients.

Certain expressions can be combined after one or more of the terms involving radicals are simplified.

Multiplying and Dividing Radicals [9.6]

$$\sqrt{7}\,\sqrt{5} = \sqrt{7 \cdot 5}$$
$$= \sqrt{35}$$
$$\sqrt{6}\,\sqrt{15} = \sqrt{6 \cdot 15} = \sqrt{90}$$
$$= \sqrt{9 \cdot 10}$$
$$= 3\sqrt{10}$$
$$\sqrt{5}(\sqrt{3} + 2\sqrt{5})$$
$$= \sqrt{5} \cdot \sqrt{3} + \sqrt{5} \cdot 2\sqrt{5}$$
$$= \sqrt{15} + 10$$

Multiplying   To multiply radical expressions, use the first property of radicals in the following way:

$$\sqrt{a}\,\sqrt{b} = \sqrt{ab}$$

The distributive property can also be applied in multiplying radical expressions.

$(\sqrt{5} + 2)(\sqrt{5} - 1)$
$= \sqrt{5} \cdot \sqrt{5} - \sqrt{5} + 2\sqrt{5} - 2$
$= 3 + \sqrt{5}$
$(\sqrt{10} + 3)(\sqrt{10} - 3)$
$= 10 - 9 = 1$

The FOIL pattern allows us to find the product of binomial radical expressions.

Dividing   To divide radical expressions, use the second property of radicals in the following way:

$\dfrac{\sqrt{50}}{\sqrt{2}} = \sqrt{\dfrac{50}{2}}$
$= \sqrt{25}$
$= 5$

$$\dfrac{\sqrt{a}}{\sqrt{b}} = \sqrt{\dfrac{a}{b}}$$

$\dfrac{\sqrt{3}}{\sqrt{5}} = \dfrac{\sqrt{3} \cdot \sqrt{5}}{\sqrt{5} \cdot \sqrt{5}}$
$= \dfrac{\sqrt{15}}{\sqrt{25}} = \dfrac{\sqrt{15}}{5}$

When necessary, multiply the numerator and denominator of the expression by the same root to rationalize the denominator.

$\sqrt{5} - 2$ is the conjugate of $\sqrt{5} + 2$. To simplify,

If the divisor (the denominator) is a binomial, multiply the numerator and denominator by the *conjugate* of the denominator.

$\dfrac{3}{\sqrt{5} - 2} =$
$\dfrac{3(\sqrt{5} + 2)}{(\sqrt{5} - 2)(\sqrt{5} + 2)}$
$= \dfrac{3(\sqrt{5} + 2)}{5 - 4}$
$= 3\sqrt{5} + 6$

The Distance Between Two Points [9.7]

The distance between $(2, -3)$ and $(5, -4)$

$d = \sqrt{(5 - 2)^2 + [(-4) - (-3)]^2}$
$= \sqrt{3^2 + (-1)^2}$
$= \sqrt{9 + 1}$
$= \sqrt{10}$

Given two points (x_1, y_1) and (x_2, y_2), the distance between them can be found using the distance formula

$$d = \sqrt{(x_2 - x_1)^2 + (y_2 - y_1)^2}$$

Summary Exercises Chapter 9

This summary exercise set is provided to give you practice with each of the objectives of the chapter. Each exercise is keyed to the appropriate chapter section. The answers are provided in the instructor's manual. Your instructor will give you guidelines on how to best use these exercises in your instructional setting.

[9.1] Use the properties of exponents to simplify each of the following expressions. Assume that all variables represent nonzero real numbers.

1. $2^3 \cdot 2^4$ **2.** $\dfrac{x^{12}}{x^8}$ **3.** $(a^4)^3$ **4.** $(3xy)^4$

5. $(r^4 s^5)^3$ **6.** $(3x^2)^2(x^3)^3$ **7.** $\dfrac{(b^4)^4}{b^{10}}$ **8.** $\left(\dfrac{x^5}{y^3}\right)^2$

[9.2] Evaluate each of the following expressions.

9. 5^0 **10.** $(5m)^0$ **11.** $5m^0$ **12.** $(4x^3y)^0$

[9.2] Write, using positive exponents.

13. a^{-7} **14.** 2^{-4} **15.** 10^{-3} **16.** $4x^{-3}$

17. $\dfrac{a^5}{a^8}$ **18.** $p^6 p^{-8}$ **19.** $\dfrac{m^{-3}}{m^{-8}}$ **20.** $\dfrac{x^3 y^{-4}}{x^{-4} y^3}$

21. $(3s^{-2})^3$ **22.** $\dfrac{(y^3)^{-4}}{(y^{-2})^{-3}}$

[9.3] Evaluate if possible.

23. $\sqrt{81}$ **24.** $-\sqrt{49}$ **25.** $\sqrt{-49}$ **26.** $\sqrt[3]{64}$

27. $\sqrt[3]{-64}$ **28.** $\sqrt[4]{81}$ **29.** $\sqrt[4]{-81}$

[9.4] Simplify each of the following radical expressions. Assume that all variables represent positive real numbers.

30. $\sqrt{50}$ **31.** $\sqrt{45}$ **32.** $\sqrt{7a^3}$ **33.** $\sqrt{20x^4}$

34. $\sqrt{49m^5}$ **35.** $\sqrt{200b^3}$ **36.** $\sqrt{147r^3s^2}$ **37.** $\sqrt{108a^2b^5}$

38. $\sqrt{\dfrac{10}{81}}$ **39.** $\sqrt{\dfrac{18x^2}{25}}$ **40.** $\sqrt{\dfrac{12m^5}{49}}$

41. $\sqrt{\dfrac{3}{7}}$ **42.** $\sqrt{\dfrac{3a}{2}}$ **43.** $\sqrt{\dfrac{8x^2}{7}}$

[9.5] Simplify by combining like terms.

44. $\sqrt{3} + 4\sqrt{3}$ **45.** $9\sqrt{5} - 3\sqrt{5}$ **46.** $3\sqrt{2} + 2\sqrt{3}$

47. $3\sqrt{3a} - \sqrt{3a}$ **48.** $7\sqrt{6} - 2\sqrt{6} + \sqrt{6}$ **49.** $5\sqrt{3} + \sqrt{12}$

50. $3\sqrt{18} - 5\sqrt{2}$ **51.** $\sqrt{32} - \sqrt{18}$ **52.** $\sqrt{27} - \sqrt{3} + 2\sqrt{12}$

53. $\sqrt{8} + 2\sqrt{27} - \sqrt{75}$ **54.** $x\sqrt{18} - 3\sqrt{8x^2}$

[9.6] Simplify each radical expression.

55. $\sqrt{6}\sqrt{5}$ **56.** $\sqrt{3}\sqrt{6}$ **57.** $\sqrt{3x}\sqrt{2}$

58. $\sqrt{2}\sqrt{8}\sqrt{3}$ **59.** $\sqrt{5a} \cdot \sqrt{10a}$ **60.** $\sqrt{2}(\sqrt{3} + \sqrt{5})$

61. $\sqrt{7}(2\sqrt{3} - 3\sqrt{7})$ **62.** $(\sqrt{3} + 5)(\sqrt{3} - 3)$ **63.** $(\sqrt{15} - 3)(\sqrt{15} + 3)$

64. $(\sqrt{2} + 3)^2$ **65.** $\dfrac{\sqrt{7x^3}}{\sqrt{3}}$ **66.** $\dfrac{5}{\sqrt{11} - 2}$

[9.7] Find the distance between each pair of points.

67. $(-3, 2)$ and $(-7, 2)$ **68.** $(2, 0)$ and $(5, 9)$

69. $(-2, 7)$ and $(-5, -1)$ **70.** $(5, -1)$ and $(-2, 3)$

The purpose of this self-test is to help you check your progress and to review for a chapter test in class. Allow yourself about an hour to take the test. When you are done, check your answers in the back of the book. If you missed any problems, be sure to go back and review the appropriate sections in the chapter and the exercises that are provided.

Use the properties of exponents to simplify each of the following expressions. Assume that all variables represent nonzero real numbers.

1. $(x^5)^3$

2. $(4a^2b)^2$

3. $\dfrac{(y^5)^2}{y^7}$

4. $\left(\dfrac{m^5}{n^2}\right)^3$

Evaluate (assume the variables are nonzero).

5. 9^0

6. $7x^0$

Write, using positive exponents.

7. x^{-7}

8. $5a^{-4}$

9. x^5x^{-7}

10. $\dfrac{m^{-4}}{m^4}$

Evaluate if possible.

11. $\sqrt{121}$

12. $\sqrt[3]{27}$

13. $\sqrt{-144}$

ANSWERS

1. _____

2. _____

3. _____

4. _____

5. _____

6. _____

7. _____

8. _____

9. _____

10. _____

11. _____

12. _____

13. _____

14. _____

15. _____

16. _____

17. _____

18. _____

19. _____

20. _____

21. _____

22. _____

23. _____

24. _____

25. _____

Simplify each of the following radical expressions.

14. $\sqrt{75}$

15. $\sqrt{24a^3}$

16. $\sqrt{\dfrac{16}{25}}$

17. $\sqrt{\dfrac{5}{9}}$

Simplify by combining like terms.

18. $2\sqrt{10} - 3\sqrt{10} + 5\sqrt{10}$

19. $3\sqrt{8} - \sqrt{18}$

Simplify each of the following radical expressions.

20. $\sqrt{3x} \cdot \sqrt{6x}$

21. $(\sqrt{5} + 3)(\sqrt{5} + 2)$

22. $\dfrac{\sqrt{7}}{\sqrt{2}}$

23. $\dfrac{4}{\sqrt{11} + 3}$

Find the distance between the two points.

24. $(-3, 7)$ and $(-12, 7)$

25. $(-2, 5)$ and $(-9, -1)$

Retain Your Skills

This review covers selected topics from the first eight chapters.

Perform the indicated operation.

1. $3x^2y^2 - 5xy - 2x^2y^2 + 2xy$

2. $\dfrac{36m^5n^2}{27m^2n}$

3. $(x^2 - 3x + 5) - (x^2 - 2x - 4)$

4. $(5z^2 - 3z) - (2z^2 - 5)$

Multiply.

5. $(2x - 3)(x + 7)$

6. $(2a - 2b)(a + 4b)$

Divide.

7. $(x^2 + 3x + 2) \div (x - 3)$

8. $(x^4 - 2x) \div (x + 2)$

Solve each equation and check your results.

9. $5x - 2 = 2x - 6$

10. $3(x - 2) = 2(3x + 1) - 2$

Factor each polynomial completely.

11. $x^2 - x - 56$

12. $4x^3y - 2x^2y^2 + 8x^4y$

13. $8a^3 - 18ab^2$

14. $15x^2 - 21xy + 6y^2$

Find the slope of the line through the following pairs of points.

15. $(2, -4)$ and $(-3, -9)$

16. $(-1, 7)$ and $(3, -2)$

Graph each of the following equations.

17. $x + y = 4$

18. $y = 2x - 5$

Solve each of the following systems.

19. $2x - y = 9$
    $3x - 6y = 9$

20. $2x - 5y = 1$
        $y = 3x + 5$

CHAPTER 10

Quadratic Equations

The ENVIRONMENT

Energy Conservation

War in the Middle East, oil slicks on the beaches of Alaska and the Persian Gulf, acid rain in the northeast, global warming, and the greenhouse effect—what do these recent hot news topics have in common? None of these topics ever seems to be discussed without energy use in the United States also being mentioned. U.S. energy consumption is directly related to these problems and so must be considered when seemingly unrelated world events are investigated.

The United States has used increasing amounts of oil and gas for the past 100 years. Combined with coal, these three fossil fuels have supported most of the economic growth of the United States, and the world, since 1950. There has been a steady parallel increase between energy consumption and GNP in the United States since the end of World War II.

Production of the energy necessary to fuel the modern U.S. lifestyle is expensive, in terms of both the monetary cost and the environmental impact. The United States spends hundreds of billions of dollars per year on its energy needs. In the production of this energy, the United States emits thousands of tons of air and water pollutants, is a principal cause of much of the ocean pollution caused by oil spills around the world, and must maintain a military presence around the globe to protect its energy supply lines. Many people are beginning to question the need for these levels of energy costs.

What can be done to reduce the costs associated with energy production from fossil fuels? Nuclear energy is available but is expensive and entails its own environmental costs. Solar power and other alternative energy sources are being developed, but only hydroelectric power is currently a major contributor to U.S. power supplies. Hydropower is restricted by the availability of usable sites, and most good hydroelectric sites are already being exploited. While research is continuing on many different aspects of alternative energy, the funding cutbacks of the 1980s have delayed any substantial developments in most of these fields.

Conservation, however, is showing a very high potential for reducing our energy costs. Many experts believe conservation is the area with the highest promise for energy savings in the foreseeable future. Generally included with conservation is the increased efficiency in using the amount of energy we currently have. This means getting more done with the energy we already produce.

Conservation is already making a large contribution to the U.S. energy supply. From 1973 to 1986 the U.S. economy grew by 35 percent while there was no increase in total energy use. Conservation measures such as increased fuel economy in automobiles, plugged steam leaks, increased home insulation, more efficient appliances and industrial motors, and new lighting technologies have allowed the economy to grow without a corresponding growth in energy demand.

Most conservative experts feel much more can be done, such as producing cars that travel over 100 miles on 1 gallon of gas, making fluorescent light bulbs that last for years instead of months, and further improving home insulation and appliances. Over three times as much energy is available from conservation and improved efficiency as from developing remaining fossil-fuel and nuclear resources.

Another benefit of conservation is the reduction of pollutants emitted into the environment. Since 1979, emissions of air pollutants alone are 40 percent below what they would have been without the conservation and efficiency improvements. It is abundantly clear that energy conservation is an energy source that should be exploited to the fullest extent possible.

More on Quadratic Equations

OBJECTIVES

1. To solve quadratic equations of the form

$$ax^2 = k$$

2. To solve quadratic equations of the form

$$(x - h)^2 = k \quad \text{and} \quad a(x - h)^2 = k$$

We now have more tools for solving quadratic equations. In this section and the next we will be using the ideas of Sections 9.3 and 9.4 to extend our solution techniques.

In Section 5.5 we identified all equations of the form

$$ax^2 + bx + c = 0$$

as quadratic equations in standard form. In that section, we discussed solving these equations whenever the quadratic expression was factorable. In this chapter, we want to extend our equation-solving techniques so that we can find solutions for all such quadratic equations.

Let's first review the factoring method of solution that we introduced in Chapter 5.

Example 1

Solve each quadratic equation by factoring.

(*a*) $x^2 = -7x - 12$

First, we write the equation in standard form.

Add 7x and 12 to both sides of the equation. The equation must be *set equal to 0*.

$$x^2 + 7x + 12 = 0$$

Once the equation is in standard form, we can factor the quadratic member.

$$(x + 3)(x + 4) = 0$$

Finally, using the zero product rule, we solve the equations $x + 3 = 0$ and $x + 4 = 0$ as follows:

These solutions can be checked as before by substitution into the original equation.

$$x = -3 \quad \text{or} \quad x = -4$$

(*b*) $x^2 = 16$

Again, we write the equation in standard form.

Here we factor the quadratic member of the equation as a difference of squares.

$$x^2 - 16 = 0$$

Factoring, we have

$$(x + 4)(x - 4) = 0$$

and finally the solutions are

$$x = -4 \quad \text{or} \quad x = 4$$

CHECK YOURSELF 1

Solve each of the following quadratic equations.

1. $x^2 - 4x = 45$ **2.** $w^2 = 25$

    Certain quadratic equations can be solved by other methods. Let's return to the equation of Example 1(*b*).
    Beginning with

$$x^2 = 16$$

we can take the square root of each side, to write

$$\sqrt{x^2} = \sqrt{16}$$

From Section 9.3, we know that this is equivalent to

$$\sqrt{x^2} = 4 \tag{1}$$

or

Recall that by definition $\sqrt{x^2} = |x|$

$$|x| = 4 \tag{2}$$

Values for x of 4 or -4 will both satisfy equation (2), and so we have the two solutions

$$x = 4 \quad \text{or} \quad x = -4$$

We usually write the solutions as

$x = \pm 4$ is simply a convenient "shorthand" for indicating the two solutions, and we generally will go directly to this form.

$$x = \pm 4$$

    Let's look at two more equations solved by this method in the next example.

Example 2

Solve each of the following equations by the square root method shown above.

(*a*) $x^2 = 9$

By taking the square root of each side, we have

$$x = \pm\sqrt{9}$$

or

$$x = \pm 3$$

(*b*) $x^2 = 5$

Again, we take the square root of each side to write our two solutions as

$$x = \pm\sqrt{5}$$

CHECK YOURSELF 2

Solve.

1. $x^2 = 100$ **2.** $t^2 = 15$

You may have to add or subtract on both sides of the equation to write an equation in the form of those in the previous example. Our next example illustrates.

Example 3

Solve $x^2 - 8 = 0$.

First, add 8 to both sides of the equation. We have

$$x^2 = 8$$

Now take the square root of both sides.

$$x = \pm\sqrt{8}$$

Recall that
$$\sqrt{8} = \sqrt{4 \cdot 2}$$
$$= \sqrt{4} \cdot \sqrt{2}$$
$$= 2\sqrt{2}$$

Normally the solution should be written in the simplest form. In this case we have

$$x = \pm 2\sqrt{2}$$

CHECK YOURSELF 3

Solve.

1. $x^2 - 18 = 0$ **2.** $x^2 + 1 = 7$

In the form

$ax^2 = k$

a is the coefficient of x^2 and k is some number.

To solve a quadratic equation of the form $ax^2 = k$, divide both sides of the equation by a as the first step. This is shown in our next example.

Example 4

Solve $4x^2 = 3$.

Divide both sides of the equation by 4.

$$x^2 = \frac{3}{4}$$

Now take the square root of both sides.

$$x = \pm\sqrt{\frac{3}{4}}$$

Recall that
$$\sqrt{\frac{3}{4}} = \frac{\sqrt{3}}{\sqrt{4}}$$
$$= \frac{\sqrt{3}}{2}$$

Again write your result in the simplest form, so

$$x = \pm\frac{\sqrt{3}}{2}$$

CHECK YOURSELF 4

Solve $9x^2 = 5$.

Equations of the form $(x - h)^2 = k$ can also be solved by taking the square root of both sides. Consider the next example.

Example 5

Solve $(x - 1)^2 = 6$.

Again, take the square root of both sides of the equation.

$$x - 1 = \pm\sqrt{6}$$

Now add 1 to both sides of the equation to isolate x.

$$x = 1 \pm \sqrt{6}$$

CHECK YOURSELF 5

Solve $(x + 2)^2 = 12$.

Equations of the form $a(x - h)^2 = k$ can also be solved if each side of the equation is divided by a first. This is shown in our next example.

Example 6

Solve $3(x - 2)^2 = 5$.

$$\sqrt{\frac{5}{3}} = \frac{\sqrt{5}}{\sqrt{3}} \cdot \frac{\sqrt{3}}{\sqrt{3}} = \frac{\sqrt{15}}{3}$$

$$(x - 2)^2 = \frac{5}{3}$$

$$x - 2 = \pm\sqrt{\frac{5}{3}} = \frac{\pm\sqrt{15}}{3}$$

$$x = 2 \pm \frac{\sqrt{15}}{3}$$

$$x = \frac{6}{3} \pm \frac{\sqrt{15}}{3}$$

$$x = \frac{6 \pm \sqrt{15}}{3}$$

CHECK YOURSELF 6

Solve $5(x + 3)^2 = 2$.

What about an equation such as the following?

$$x^2 + 5 = 0$$

If we apply the above methods, we first subtract 5 from both sides, to write

$$x^2 = -5$$

Taking the square root of both sides gives

$$x = \pm\sqrt{-5}$$

But we know there are no square roots of -5 in the real numbers, so this equation has *no real number solutions.* You'll work with this type of equation in your next algebra course.

CHECK YOURSELF ANSWERS

1. (1) -5, 9; (2) -5, 5.

2. (1) ± 10; (2) $\pm \sqrt{15}$.

3. (1) $\pm 3\sqrt{2}$; (2) $\pm \sqrt{6}$.

4. $\pm \dfrac{\sqrt{5}}{3}$.

5. $-2 \pm 2\sqrt{3}$.

6. $\dfrac{-15 \pm \sqrt{10}}{5}$.

Name

Date

Build Your Skills

Solve each of the equations for x.

1. $x^2 = 3$ **2.** $x^2 = 11$

3. $x^2 = 29$ **4.** $x^2 = 37$

5. $x^2 - 7 = 0$ **6.** $x^2 - 13 = 0$

7. $x^2 - 20 = 0$ **8.** $x^2 = 28$

9. $x^2 = 40$ **10.** $x^2 - 54 = 0$

11. $x^2 + 3 = 12$ **12.** $x^2 - 7 = 18$

13. $x^2 + 5 = 8$ **14.** $x^2 - 4 = 17$

15. $x^2 - 2 = 16$ **16.** $x^2 + 6 = 30$

17. $9x^2 = 25$ **18.** $16x^2 = 9$

ANSWERS

1. _____

2. _____

3. _____

4. _____

5. _____

6. _____

7. _____

8. _____

9. _____

10. _____

11. _____

12. _____

13. _____

14. _____

15. _____

16. _____

17. _____

18. _____

19. _____

20. _____

21. _____

22. _____

23. _____

24. _____

25. _____

26. _____

27. _____

28. _____

29. _____

30. _____

31. _____

32. _____

33. _____

34. _____

35. _____

36. _____

19. $36x^2 = 5$

20. $25x^2 = 7$

21. $5x^2 = 3$

22. $2x^2 = 7$

23. $(x - 1)^2 = 5$

24. $(x - 3)^2 = 10$

25. $(x + 1)^2 = 12$

26. $(x + 2)^2 = 32$

27. $(x - 3)^2 = 24$

28. $(x - 5)^2 = 27$

29. $(x + 5)^2 = 25$

30. $(x + 2)^2 = 16$

31. $3(x - 5)^2 = 7$

32. $2(x - 5)^2 = 3$

33. $4(x + 5)^2 = 9$

34. $16(x + 2)^2 = 25$

35. $-2(x + 2)^2 = -6$

36. $-5(x + 4)^2 = -10$

37. $-4(x - 1)^2 = -7$

38. $-2(x - 1)^2 = -9$

39. $(3x - 1)^2 = 8$

40. $(2x - 1)^2 = 12$

Think About These

Solve each equation for x.

41. $x^2 - 2x + 1 = 7$
(*Hint:* Factor the left-hand side.)

42. $x^2 + 4x + 4 = 7$
(*Hint:* Factor the left-hand side.)

43. $(2x + 11)^2 - 9 = 0$

44. $(3x + 14)^2 - 25 = 0$

Windmills are often considered as an alternative form of power generation because wind is considered a renewable resource. For a constant windspeed, the relationship between power output and windmill diameter is given by

$P = kD^2$

where P is power in watts (W), D is the diameter of the windmill in meters (m), and k is a constant of proportionality.

45. For $k = 10$, find the diameter necessary to produce 120 watts (W) of power.

46. For $k = 8$, find the diameter necessary to produce 160 W of power.

47. For $k = 10$, by how much does a 4-m windmill need to be increased in order to generate 250 W of power?

48. For $k = 15$, by how much does a 6-m windmill need to be increased in order to generate 600 W of power?

ANSWERS

37. _____

38. _____

39. _____

40. _____

41. _____

42. _____

43. _____

44. _____

45. _____

46. _____

47. _____

48. _____

ANSWERS

a. _____

b. _____

c. _____

d. _____

e. _____

f. _____

g. _____

h. _____

Skillscan (Section 4.4)

Multiply each of the following expressions.

a. $(x + 1)^2$ **b.** $(x + 5)^2$

c. $(x - 2)^2$ **d.** $(x - 7)^2$

e. $(x + 4)^2$ **f.** $(x - 3)^2$

g. $(2x + 5)^2$ **h.** $(2x - 1)^2$

ANSWERS

1. $\pm\sqrt{3}$ **3.** $\pm\sqrt{29}$ **5.** $\pm\sqrt{7}$ **7.** $\pm 2\sqrt{5}$ **9.** $\pm 2\sqrt{10}$ **11.** ± 3

13. $\pm\sqrt{3}$ **15.** $\pm 3\sqrt{2}$ **17.** $\pm\dfrac{5}{3}$ **19.** $\dfrac{\pm\sqrt{5}}{6}$ **21.** $\dfrac{\pm\sqrt{15}}{5}$ **23.** $1 \pm \sqrt{5}$

25. $-1 \pm 2\sqrt{3}$ **27.** $3 \pm 2\sqrt{6}$ **29.** $-10, 0$ **31.** $\dfrac{15 \pm \sqrt{21}}{3}$ **33.** $\dfrac{-13}{2}, \dfrac{-7}{2}$

35. $-2 \pm \sqrt{3}$ **37.** $\dfrac{2 \pm \sqrt{7}}{2}$ **39.** $\dfrac{1 \pm 2\sqrt{2}}{3}$ **41.** $1 \pm \sqrt{7}$ **43.** $-4, -7$

45. $\approx 3.46\,\text{m}$ **47.** $1\,\text{m}$ **a.** $x^2 + 2x + 1$ **b.** $x^2 + 10x + 25$ **c.** $x^2 - 4x + 4$
d. $x^2 - 14x + 49$ **e.** $x^2 + 8x + 16$ **f.** $x^2 - 6x + 9$ **g.** $4x^2 + 20x + 25$
h. $4x^2 - 4x + 1$

© 1993 McGraw-Hill, Inc.

10.2 Completing the Square

OBJECTIVE
To solve a quadratic equation by completing the square

We can solve a quadratic equation such as

$$x^2 - 2x + 1 = 5$$

very easily if we notice that the expression on the left is a perfect-square trinomial. Factoring, we have

$$(x - 1)^2 = 5$$

so

$$x - 1 = \pm\sqrt{5} \qquad \text{or} \qquad x = 1 \pm \sqrt{5}$$

The solutions for the original equation are then $1 + \sqrt{5}$ and $1 - \sqrt{5}$.

It is true that every quadratic equation can be written in the form above (with a perfect-square trinomial on the left). That is the basis for the *completing-the-square* method for solving quadratic equations.

First, let's look at two perfect-square trinomials.

$$x^2 + 6x + 9 = (x + 3)^2 \tag{1}$$
$$x^2 - 8x + 16 = (x - 4)^2 \tag{2}$$

There is an important relationship between the coefficient of the middle term (the x term) and the constant.

In equation (1),

$$\left(\frac{1}{2} \cdot 6\right)^2 = 3^2 = 9$$

The x coefficient The constant

In equation (2),

$$\left[\frac{1}{2}(-8)\right]^2 = (-4)^2 = 16$$

The x coefficient The constant

It is always true that, in a perfect-square trinomial with a coefficient of 1 for x^2, the square of one-half of the x coefficient is equal to the constant term.

Example 1

(*a*) Find the term that should be added to $x^2 + 4x$ so that the expression is a perfect-square trinomial.

The coefficient of x^2 must be 1 before the added term is found.

To complete the square of $x^2 + 4x$, add the square of one-half of 4 (the x coefficient).

$$x^2 + 4x + \left(\frac{1}{2} \cdot 4\right)^2$$

or $x^2 + 4x + 2^2$ or $x^2 + 4x + 4$

The trinomial $x^2 + 4x + 4$ is a perfect square because

$$x^2 + 4x + 4 = (x + 2)^2$$

(b) Find the term that should be added to $x^2 - 10x$ so that the expression is a perfect-square trinomial.

To complete the square of $x^2 - 10x$, add the square of one-half of -10 (the x coefficient).

$$x^2 - 10x + \left[\frac{1}{2}(-10)\right]^2$$

or $x^2 - 10x + (-5)^2$ or $x^2 - 10x + 25$

Check for yourself, by factoring, that this is a perfect-square trinomial.

CHECK YOURSELF 1

Complete the square and factor.

1. $x^2 + 2x$ **2.** $x^2 - 12x$

We can now use the above process along with the solution methods of Section 10.1 to solve a quadratic equation.

Example 2

Solve $x^2 + 4x - 2 = 0$ by completing the square.

Add 2 to both sides to remove -2 from the left side.

$$x^2 + 4x = 2$$

We find the term needed to complete the square by squaring one-half of the x coefficient.

$$\left(\frac{1}{2} \cdot 4\right)^2 = 2^2 = 4$$

We now add 4 to both sides of the equation.

This *completes the square* on the left.

$$x^2 + 4x + 4 = 2 + 4$$

Now factor on the left and simplify on the right.

$$(x + 2)^2 = 6$$

Now solving as before, we have

$$x + 2 = \pm\sqrt{6}$$
$$x = -2 \pm \sqrt{6}$$

CHECK YOURSELF 2

Solve by completing the square.

$$x^2 + 6x - 4 = 0$$

For the completing-the-square method to work, the coefficient of x^2 must be 1. The following example illustrates the solution process when that coefficient is not equal to 1.

Example 3

Solve $2x^2 - 4x - 5 = 0$ by completing the square.

$$2x^2 - 4x - 5 = 0 \qquad \text{Add 5 to both sides.}$$
$$2x^2 - 4x = 5 \qquad \text{Since the coefficient of } x^2 \text{ is not 1 (here it is 2), divide every term by 2. This will make the new leading coefficient equal to 1.}$$
$$x^2 - 2x = \frac{5}{2}$$

We now complete the square and solve as before.

$$x^2 - 2x + 1 = \frac{5}{2} + 1$$

$$(x - 1)^2 = \frac{7}{2}$$

$$x - 1 = \pm\sqrt{\frac{7}{2}}$$

Simplify the radical on the right.

$$x - 1 = \pm\frac{\sqrt{14}}{2}$$

$$x = 1 \pm \frac{\sqrt{14}}{2}$$

or

We have combined the terms on the right with the common denominator of 2.

$$x = \frac{2 \pm \sqrt{14}}{2}$$

CHECK YOURSELF 3

Solve by completing the square.

$3x^2 - 6x + 2 = 0$

Let's summarize by listing the steps to solve a quadratic equation by completing the square.

SOLVING A QUADRATIC EQUATION BY COMPLETING THE SQUARE

STEP 1 Write the equation in the form

$ax^2 + bx = k$

so that the variable terms are on the left side and the constant is on the right side.

STEP 2 If the coefficient of x^2 is not 1, divide both sides of the equation by that coefficient.

STEP 3 Add the square of one-half the coefficient of x to both sides of the equation.

STEP 4 The left side of the equation is now a perfect-square trinomial. Factor and solve as before.

CHECK YOURSELF ANSWERS

1. (1) $x^2 + 2x + 1 = (x + 1)^2$; (2) $x^2 - 12x + 36 = (x - 6)^2$.

2. $-3 \pm \sqrt{13}$.   **3.** $\dfrac{3 \pm \sqrt{3}}{3}$.

Name _____

Date _____

Build Your Skills

Determine whether each of the following trinomials are perfect squares.

1. $x^2 + 4x + 4$

2. $x^2 + 6x - 9$

3. $x^2 - 10x - 25$

4. $x^2 - 12x + 36$

5. $x^2 - 18x + 81$

6. $x^2 - 24x + 48$

Find the constant term that should be added to make each of the following expressions perfect-square trinomials.

7. $x^2 + 6x$

8. $x^2 - 8x$

9. $x^2 - 10x$

10. $x^2 + 5x$

11. $x^2 + 9x$

12. $x^2 - 20x$

Solve each of the following quadratic equations by completing the square.

13. $x^2 + 4x - 12 = 0$

14. $x^2 - 6x + 8 = 0$

15. $x^2 - 2x - 5 = 0$

16. $x^2 + 4x - 7 = 0$

17. $x^2 + 3x - 27 = 0$

18. $x^2 + 5x - 3 = 0$

19. $x^2 + 6x - 1 = 0$

20. $x^2 + 4x - 4 = 0$

21. $x^2 - 5x + 6 = 0$

22. $x^2 - 6x - 3 = 0$

23. $x^2 + 6x - 5 = 0$

24. $x^2 - 2x = 1$

ANSWERS

1. _____
2. _____
3. _____
4. _____
5. _____
6. _____
7. _____
8. _____
9. _____
10. _____
11. _____
12. _____
13. _____
14. _____
15. _____
16. _____
17. _____
18. _____
19. _____
20. _____
21. _____
22. _____
23. _____
24. _____

25. _____

26. _____

27. _____

28. _____

29. _____

30. _____

31. _____

32. _____

33. _____

34. _____

35. _____

36. _____

a. _____

b. _____

c. _____

d. _____

e. _____

f. _____

25. $x^2 = 8x + 3$

26. $x^2 = 2 - 5x$

27. $x^2 = 3 - x$

28. $x^2 = x + 1$

29. $2x^2 - 6x + 1 = 0$

30. $2x^2 + 10x + 11 = 0$

31. $2x^2 - 4x + 1 = 0$

32. $2x^2 - 8x + 5 = 0$

33. $4x^2 - 2x - 1 = 0$

34. $3x^2 - x - 2 = 0$

Think About These

Solve each of the quadratic equations by completing the square.

35. $3x^2 - 4x + 7x - 9 = 2x^2 + 5x - 4$

36. $-4x^2 - 8x + 4x + 5 = -5x^2 + 2x + 16$

Skillscan (Section 2.6)

Evaluate the expression $b^2 - 4ac$ for each set of values.

a. $a = 1, b = 1, c = -3$

b. $a = 1, b = -1, c = -1$

c. $a = 1, b = -8, c = -3$

d. $a = 1, b = -2, c = -1$

e. $a = -2, b = 4, c = -2$

f. $a = 2, b = -3, c = 4$

ANSWERS

1. Yes   **3.** No   **5.** Yes   **7.** 9   **9.** 25   **11.** $\dfrac{81}{4}$ **13.** $-6, 2$ **15.** $1 \pm \sqrt{6}$

17. $\dfrac{-3 \pm 3\sqrt{13}}{2}$ **19.** $-3 \pm \sqrt{10}$ **21.** 2, 3   **23.** $-3 \pm \sqrt{14}$ **25.** $4 \pm \sqrt{19}$

27. $\dfrac{-1 \pm \sqrt{13}}{2}$ **29.** $\dfrac{3 \pm \sqrt{7}}{2}$ **31.** $\dfrac{2 \pm \sqrt{2}}{2}$ **33.** $\dfrac{1 \pm \sqrt{5}}{4}$ **35.** $1 \pm \sqrt{6}$ **a.** 13

b. 5   **c.** 76   **d.** 8   **e.** 0   **f.** -23

The Quadratic Formula

OBJECTIVE

To solve a quadratic equation by using the quadratic formula

We are now ready to derive and use the *quadratic formula,* which will allow us to solve all quadratic equations. We derive the formula by using the method of completing the square.

The derivation of the quadratic formula is shown below.

To use the quadratic formula, the quadratic equation you want to solve must be in *standard form.* That form is

$$ax^2 + bx + c = 0 \qquad \text{where } a \neq 0$$

Example 1

Write each equation in standard form.

(*a*) $2x^2 - 5x + 3 = 0$ is in standard form.

$$a = 2 \qquad b = -5 \qquad \text{and} \qquad c = 3$$

(*b*) $5x^2 + 3x = 5$ is *not* in standard form. Rewrite the equation by subtracting 5 from both sides.

$5x^2 + 3x - 5 = 0$ is in standard form.

$$a = 5 \qquad b = 3 \qquad \text{and} \qquad c = -5$$

CHECK YOURSELF 1

Rewrite each quadratic in standard form.

1. $x^2 - 3x = 5$ **2.** $3x^2 = 7 - 2x$

Once a quadratic equation is written in standard form, we will be able to find both solutions to the equation. Remember that a solution is a value for x that will make the equation true.

What follows is the derivation of the formula that can be used to solve quadratic equations.

We call it the quadratic formula.

TO DERIVE THE QUADRATIC FORMULA

Let $ax^2 + bx + c = 0$, where $a \neq 0$.

$ax^2 + bx = -c$ Subtract c from both sides.

$x^2 + \dfrac{b}{a}x = -\dfrac{c}{a}$ Divide both sides by a.

$x^2 + \dfrac{b}{a}x + \dfrac{b^2}{4a^2} = \dfrac{b^2}{4a^2} - \dfrac{c}{a}$ Add $\dfrac{b^2}{4a^2}$ to both sides.

$\left(x + \dfrac{b}{2a}\right)^2 = \dfrac{b^2 - 4ac}{4a^2}$ Factor on the left, and add the fractions on the right.

$x + \dfrac{b}{2a} = \pm\sqrt{\dfrac{b^2 - 4ac}{4a^2}}$ Take the square root of both sides.

$x + \dfrac{b}{2a} = \pm\dfrac{\sqrt{b^2 - 4ac}}{2a}$ Simplify the radical on the right.

$x = -\dfrac{b}{2a} \pm \dfrac{\sqrt{b^2 - 4ac}}{2a}$ Subtract $\dfrac{b}{2a}$ from both sides.

This is the completing-the-square step that makes the left-hand side a perfect square.

$$x = \frac{-b \pm \sqrt{b^2 - 4ac}}{2a}$$

THE QUADRATIC FORMULA

Example 2

Solve $x^2 - 5x + 4 = 0$ by formula.

The equation is in standard form, so first identify a, b, and c.

The leading coefficient is 1, so $a = 1$.

$x^2 - 5x + 4 = 0$

$a = 1 \quad b = -5 \quad c = 4$

We now substitute the values for a, b, and c into the formula.

$$x = \frac{-b \pm \sqrt{b^2 - 4ac}}{2a}$$

$$= \frac{-(-5) \pm \sqrt{(-5)^2 - 4(1)(4)}}{2(1)}$$

Simplify the expression.

$$= \frac{5 \pm \sqrt{25 - 16}}{2}$$

$$= \frac{5 \pm \sqrt{9}}{2}$$

$$= \frac{5 \pm 3}{2}$$

Now,

$$x = \frac{5 + 3}{2} \quad \text{or} \quad x = \frac{5 - 3}{2}$$

$$= 4 \qquad\qquad\qquad = 1$$

The solutions are 4 and 1.

Note: These results could also have been found by factoring the original equation. You should check that for yourself.

CHECK YOURSELF 2

Solve $x^2 - 2x - 8 = 0$ by formula. Check your result by factoring.

Of course, the main use of the formula is to solve equations that cannot be factored.

Example 3

Solve $2x^2 = x + 4$ by formula.

First, the equation *must be written* in standard form to find a, b, and c.

$$2x^2 - x - 4 = 0$$
$$a = 2 \qquad b = -1 \qquad c = -4$$

Substitute the values for a, b, and c into the formula.

$$x = \frac{-b \pm \sqrt{b^2 - 4ac}}{2a}$$

$$= \frac{-(-1) \pm \sqrt{(-1)^2 - 4(2)(-4)}}{2(2)}$$

$$= \frac{1 \pm \sqrt{1 + 32}}{4}$$

$$= \frac{1 \pm \sqrt{33}}{4}$$

CHECK YOURSELF 3

Solve $3x^2 = 3x + 4$ by formula.

Example 4

Solve $x^2 - 2x = 4$ by formula.

In standard form, the equation is

$$x^2 - 2x - 4 = 0$$

$a = 1 \qquad b = -2 \qquad c = -4$

Again substitute the values into the quadratic formula.

$$x = \frac{-(-2) \pm \sqrt{(-2)^2 - 4(1)(-4)}}{2(1)}$$

$$= \frac{2 \pm \sqrt{20}}{2}$$

You should always write your solution in simplest form.

Since 20 has a perfect-square factor,

$\sqrt{20} = \sqrt{4 \cdot 5}$

$\quad = 2\sqrt{5}$

Now factor the numerator and divide by the common factor of 2.

$$x = \frac{2 \pm 2\sqrt{5}}{2}$$

$$= \frac{2(1 \pm \sqrt{5})}{2}$$

$$= 1 \pm \sqrt{5}$$

CHECK YOURSELF 4

Solve $3x^2 = 2x + 4$ by formula.

There is another point that should be made about the use of the quadratic formula. This is illustrated in our next example.

Example 5

Solve $3x^2 - 6x - 3 = 0$ by formula. Since the equation is in standard form, we could use

$a = 3 \qquad b = -6 \qquad$ and $\qquad c = -3$

in the quadratic formula. There is, however, a better approach.

Note the common factor of 3 in the quadratic member of the original equation. Factoring, we have

$$3(x^2 - 2x - 1) = 0$$

and dividing both sides of the equation by 3 gives

$$x^2 - 2x - 1 = 0$$

The advantage to this approach is that these values will require much less simplification after we substitute into the quadratic formula.

Now let $a = 1$, $b = -2$, and $c = -1$. Then

$$x = \frac{-(-2) \pm \sqrt{(-2)^2 - 4(1)(-1)}}{2 \cdot 1}$$

$$= \frac{2 \pm \sqrt{8}}{2}$$

$$= \frac{2 \pm 2\sqrt{2}}{2}$$

$$= \frac{2(1 \pm \sqrt{2})}{2}$$

$$= 1 \pm \sqrt{2}$$

CHECK YOURSELF 5

Solve $4x^2 - 20x = 12$ by formula.

In applications that lead to quadratic equations, you may want to find approximate values for the solutions.

Example 6

Solve $x^2 - 5x + 5 = 0$ by formula, and write your solutions in approximate decimal form.

Substituting $a = 1$, $b = -5$, and $c = 5$ gives

$$x = \frac{-(-5) \pm \sqrt{(-5)^2 - 4(1)(5)}}{2(1)}$$

$$= \frac{5 \pm \sqrt{5}}{2}$$

From the table of square roots in Appendix 2 or a calculator we have, $\sqrt{5} \approx 2.236$, so

$$x \approx \frac{5 + 2.236}{2} \qquad \text{or} \qquad x \approx \frac{5 - 2.236}{2}$$

$$= \frac{7.236}{2} \qquad\qquad\qquad = \frac{2.764}{2}$$

$$= 3.618 \qquad\qquad\qquad\quad = 1.382$$

CHECK YOURSELF 6

Solve $x^2 - 3x - 5 = 0$ by formula, and approximate the solutions in decimal form.

You may be wondering whether the quadratic formula can be used to solve all quadratic equations. It can, but not all quadratic equations will have real solutions, as the next example shows.

Example 7

Make sure the quadratic equation is in standard form.

Solve $x^2 - 3x + 5 = 0$ by formula.

Substituting $a = 1$, $b = -3$, and $c = 5$, we have

$$x = \frac{-(-3) \pm \sqrt{(-3)^2 - 4(1)(5)}}{2(1)}$$

$$= \frac{3 \pm \sqrt{-11}}{2}$$

In this case there are no real number solutions because of the negative number in the radical.

CHECK YOURSELF 7

Solve $x^2 - 3x = -3$ by formula.

Let's review the steps used for solving equations by the use of the quadratic formula.

SOLVING EQUATIONS WITH THE QUADRATIC FORMULA

STEP 1 Rewrite the equation in standard form.

$ax^2 + bx + c = 0$

STEP 2 If a common factor exists, divide both sides of the equation by that common factor.

STEP 3 Identify the coefficients a, b, and c.

STEP 4 Substitute values for a, b, and c into the formula

$$x = \frac{-b \pm \sqrt{b^2 - 4ac}}{2a}$$

STEP 5 Simplify the right side of the expression formed in step 4 to write the solutions for the original equation.

Often, applied problems will lead to quadratic equations that must be solved by the methods of this or the previous section. Our final example illustrates such an application.

Example 8

The word "numerically" is used because we cannot compare units of area to units of length.

The perimeter of a square is numerically 6 less than its area. Find the length of one side of the square.

Step 1   You want to find the length of one side of the square.

Step 2   Let x represent the length of one side. A sketch of the problem will help.

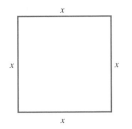

Step 3   The perimeter is $4x$. The area is x^2. So

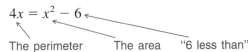

$$4x = x^2 - 6$$

The perimeter   The area   "6 less than"

Step 4   Writing the equation in standard form, we solve as before.

$$x^2 - 4x - 6 = 0$$

has solutions

$$x = 2 + \sqrt{10} \approx 5.162$$

We reject the solution of -1.162; a length cannot be negative.

$$x = 2 - \sqrt{10} \approx -1.162$$

The length of a side is approximately 5.162.

Step 5   If you have a calculator handy, you can easily check this result. Letting x be 5.162, find the perimeter and area to verify (approximately) the result in the original problem.

CHECK YOURSELF 8

The perimeter of a square is numerically 2 more than its area. Find the length of one side.

You have now studied four methods for solving quadratic equations:

1. Factoring
2. Extracting roots
3. Completing the square
4. The quadratic formula

The choice of which method to use depends largely on the equation you want to solve. Factoring is usually easiest and should be tried first.

Extracting roots is used only for equations in the particular form $(x - h)^2 = k$.

For this reason, it is important that you commit the quadratic formula to memory.

Both the completing-the-square method and the quadratic formula are applicable to all quadratic equations and can always be used. Many students seem to find the quadratic formula quicker and easier to apply.

CHECK YOURSELF ANSWERS

1. (1) $x^2 - 3x - 5 = 0$; (2) $3x^2 + 2x - 7 = 0$.

2. $x = 4, -2$.   **3.** $x = \dfrac{3 \pm \sqrt{57}}{6}$.   **4.** $x = \dfrac{1 \pm \sqrt{13}}{3}$.

5. $x = \dfrac{5 \pm \sqrt{37}}{2}$.   **6.** $x \approx 4.193$ or -1.193.

7. $\dfrac{3 \pm \sqrt{-3}}{2}$, no real solutions.

8. Approximately 3.414 or 0.586.

Name

Date

Build Your Skills

Solve each of the following quadratic equations by formula.

1. $x^2 + 3x + 2 = 0$

2. $x^2 - 4x + 3 = 0$

3. $x^2 - 6x + 5 = 0$

4. $x^2 - 7x + 6 = 0$

5. $3x^2 + 2x - 1 = 0$

6. $x^2 - 8x + 16 = 0$

7. $x^2 + 5x = -4$

8. $4x^2 + 5x = 6$

9. $x^2 = 6x - 9$

10. $2x^2 - 5x = 3$

11. $2x^2 - 3x - 7 = 0$

12. $x^2 - 5x + 2 = 0$

13. $x^2 + 2x - 4 = 0$

14. $x^2 - 4x + 2 = 0$

15. $2x^2 - 3x = 3$

16. $3x^2 - 2x + 1 = 0$

17. $3x^2 - 2x = 6$

18. $4x^2 = 4x + 5$

19. $2x^2 + 2x + 1 = 0$

20. $3x^2 - x = 1$

ANSWERS

1. _____
2. _____
3. _____
4. _____
5. _____
6. _____
7. _____
8. _____
9. _____
10. _____
11. _____
12. _____
13. _____
14. _____
15. _____
16. _____
17. _____
18. _____
19. _____
20. _____

21. _____

22. _____

23. _____

24. _____

25. _____

26. _____

27. _____

28. _____

29. _____

30. _____

31. _____

32. _____

33. _____

34. _____

35. _____

36. _____

37. _____

38. _____

39. _____

40. _____

21. $5x^2 = 8x - 2$

22. $5x^2 - 2 = 2x$

23. $2x^2 - 9 = 4x$

24. $3x^2 - 6x = 2$

25. $3x - 5 = \dfrac{1}{x}$

26. $x + 3 = \dfrac{1}{x}$

27. $(x - 2)(x + 1) = 3$

28. $(x - 3)(x + 2) = 5$

Solve the following quadratic equations by factoring or by any of the techniques of this chapter.

29. $(x - 1)^2 = 7$

30. $(2x + 3)^2 = 5$

31. $x^2 - 5x - 14 = 0$

32. $3x^2 + 2x - 1 = 0$

33. $6x^2 - 23x + 10 = 0$

34. $x^2 + 7x - 18 = 0$

35. $2x^2 - 8x + 3 = 0$

36. $x^2 + 2x - 1 = 0$

37. $x^2 - 8x + 16 = 9$

38. $3x^2 + 6x + 1 = 1$

39. $2x^2 - 16x + 32 = 3$

40. $x^2 - 3x = 10$

Use your calculator or a table of square roots for the following problems.

41. The perimeter of a square is numerically 3 less than its area. Find the length of one side.

42. The perimeter of a square is numerically 1 more than its area. Find the length of one side.

43. A picture frame is 15 in by 12 in. The area of the picture that shows is 140 in². What is the width of the frame?

44. A garden area is 30 ft long by 20 ft wide. A path of uniform width is set around the edge. If the remaining garden area is 400 ft², what is the width of the path?

45. $\dfrac{3}{x} + \dfrac{5}{x^2} = 9$

46. $\dfrac{8}{x} - \dfrac{3}{x^2} = -6$

Think About These

Solve the following equations.

47. $\dfrac{x}{x+1} + \dfrac{10x}{x^2+4x+3} = \dfrac{15}{x+3}$

48. $x - \dfrac{9x}{x-2} = \dfrac{-10}{x-2}$

49. A solar collector is 2.0 m long by 1.5 m wide. The exposed collector area is held in place by a frame of uniform width around its outside edge. If this exposed area is 2.5 m², what is the width of the frame?

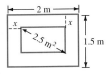

50. A solar collector is 2.5 m long by 2.0 m wide. The exposed collector area is held in place by a frame of uniform width around its outside edge. If this exposed area is 4.0 m², what is the width of the frame?

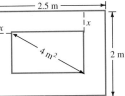

41. _____

42. _____

43. _____

44. _____

45. _____

46. _____

47. _____

48. _____

49. _____

50. _____

A solar cooker is a three-dimensional, parabolic surface which reflects sunlight onto a focal point where a grill is located to cook food. If a cooker has a grill located 0.25 m high, the equation describing the size of the cooker itself is

$$y = x^2$$

where y is the depth of the cooker and $2x$ is the width.

 51. Find the width of a cooker which is 0.35 m deep.

52. Find the width of a cooker which is 0.49 m deep.

Skillscan (Section 9.3)

Find the value of the expression $\sqrt{a^2 + b^2}$ for the following pairs of values. Where necessary, approximate the value by using a calculator or our table of square roots.

a. $a = 3$, $b = 4$ **b.** $a = 5$, $b = 12$ **c.** $a = 5$, $b = 8$ **d.** $a = 6$, $b = 9$

e. $a = 7$, $b = 8$ **f.** $a = 1$, $b = 2$ **g.** $a = 7$, $b = 24$ **h.** $a = 10$, $b = 15$

ANSWERS

1. $-2, -1$ **3.** $1, 5$ **5.** $-1, \dfrac{1}{3}$ **7.** $-4, -1$ **9.** 3 **11.** $\dfrac{3 \pm \sqrt{65}}{4}$ **13.** $-1 \pm \sqrt{5}$

15. $\dfrac{3 \pm \sqrt{33}}{4}$ **17.** $\dfrac{1 \pm \sqrt{19}}{3}$ **19.** No real solutions   **21.** $\dfrac{4 \pm \sqrt{6}}{5}$ **23.** $\dfrac{2 \pm \sqrt{22}}{2}$

25. $\dfrac{5 \pm \sqrt{37}}{6}$ **27.** $\dfrac{1 \pm \sqrt{21}}{2}$ **29.** $1 \pm \sqrt{7}$ **31.** $-2, 7$ **33.** $\dfrac{1}{2}, \dfrac{10}{3}$ **35.** $\dfrac{4 \pm \sqrt{10}}{2}$

37. $1, 7$ **39.** $\dfrac{8 \pm \sqrt{6}}{2}$ **41.** Approximately 4.646   **43.** Approximately 0.787 in

45. $\dfrac{1 \pm \sqrt{21}}{6}$ **47.** 5   **49.** ≈ 7.5 cm   **51.** 1.2 m   **a.** 5   **b.** 13   **c.** 9.434

d. 10.817   **e.** 10.630   **f.** 2.236   **g.** 25   **h.** 18.028

Graphing Quadratic Equations

OBJECTIVE

To graph a quadratic equation with the form

$$y = ax^2 + bx + c \qquad a \neq 0$$

In Section 7.3 you learned to graph first-degree equations. Similar methods will allow you to graph quadratic equations of the form

$$y = ax^2 + bx + c \qquad a \neq 0$$

The first thing you will notice is that the graph of an equation in this form is not a straight line. The graph is always the curve called a *parabola*.

Here are some examples:

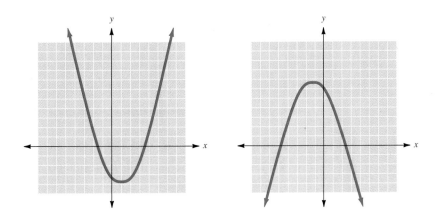

We can start as before by finding solutions for the equation. Choose any convenient values for x. Then use the given equation to compute the corresponding values for y.

Example 1

If $y = x^2$, complete the ordered pairs to form solutions. Then show these results in a table of values.

$$(-2, \ \ ), (-1, \ \ ), (0, \ \ ), (1, \ \ ), (2, \ \ )$$

For example, to complete the pair $(-2, \ \ )$, substitute -2 for x in the given equation.

$$y = (-2)^2 = 4$$

Remember that a solution is a pair of values that make the equation a true statement.

So $(-2, 4)$ is a solution.

Substituting the other values for x in the same manner, we have the following table of values for $y = x^2$:

x	y
-2	4
-1	1
0	0
1	1
2	4

CHECK YOURSELF 1

If $y = x^2 + 2$, complete the ordered pairs to form solutions and form a table of values.

$(-2, \ ), (-1, \ ), (0, \ ), (1, \ ), (2, \ )$

We can now plot points in the cartesian coordinate system that correspond to solutions to the equation.

Example 2

Plot the points from the table of values corresponding to $y = x^2$ from Example 1.

x	y
-2	4
-1	1
0	0
1	1
2	4

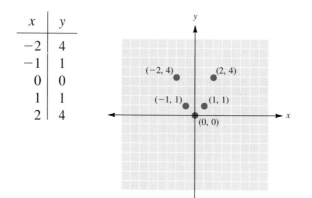

CHECK YOURSELF 2

Plot the points from the table of values formed in Check Yourself 1.

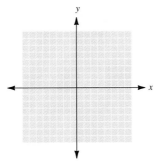

The graph of the equation can be drawn by joining the points with a smooth curve.

Example 3

Draw the graph of $y = x^2$.

We can now draw a smooth curve between the points found in Example 2 to form the graph of $y = x^2$.

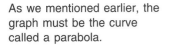

As we mentioned earlier, the graph must be the curve called a parabola.

Notice that a parabola *does not* come to a point.

CHECK YOURSELF 3

Draw a smooth curve between the points plotted in the Check Yourself 2 exercise.

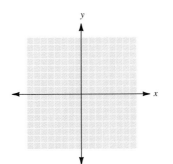

You can use any convenient values for x in forming your table of values. You should use as many pairs as are necessary to get the correct shape of the graph (a parabola).

Example 4

Graph $y = x^2 - 2x$. Use values of x between -1 and 3.

First, determine solutions for the equation. For instance, if $x = -1$,

$$y = (-1)^2 - 2(-1)$$
$$= 1 + 2$$
$$= 3$$

and $(-1, 3)$ is a solution for the given equation.

Substituting the other values for x, we can form the table of values shown below. We then plot the corresponding points and draw a smooth curve to form our graph.

Any values can be substituted for x in the original equation.

x	y
-1	3
0	0
1	-1
2	0
3	3

The graph of $y = x^2 - 2x$.

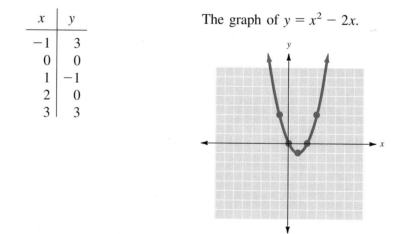

CHECK YOURSELF 4

Graph $y = x^2 + 4x$. Use values of x between -4 and 0.

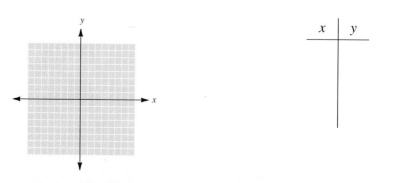

Choosing values for x is also a valid method of graphing a quadratic equation that contains a constant term.

Example 5

Graph $y = x^2 - x - 2$. Use values of x between -2 and 3. We'll show the computation for two of the solutions.

If $x = -2$,

$$y = (-2)^2 - (-2) - 2$$
$$= 4 + 2 - 2$$
$$= 4$$

If $x = 3$,

$$y = 3^2 - 3 - 2$$
$$= 9 - 3 - 2$$
$$= 4$$

You should substitute the remaining values for x into the given equation to verify the other solutions shown in the table of values below.

x	y
-2	4
-1	0
0	-2
1	-2
2	0
3	4

The graph of $y = x^2 - x - 2$.

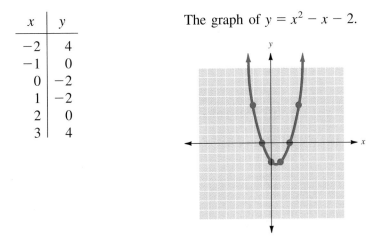

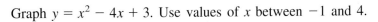

CHECK YOURSELF 5

Graph $y = x^2 - 4x + 3$. Use values of x between -1 and 4.

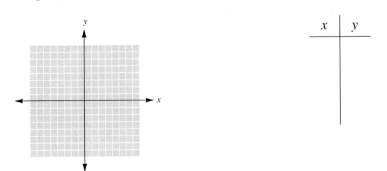

x	y

In the next example the graph looks significantly different from previous graphs.

Example 6

Graph $y = -x^2 + 3$. Use x values between -2 and 2. Again we'll show two computations.

If $x = -2$,

Note: $-(-2)^2 = -4$

$$y = -(-2)^2 + 3$$
$$= -4 + 3$$
$$= -1$$

If $x = 1$,

$$y = -(1)^2 + 3$$
$$= -1 + 3$$
$$= 2$$

Verify the remainder of the solutions shown in the table of values below for yourself.

x	y
-2	-1
-1	2
0	3
1	2
2	-1

The graph of $y = -x^2 + 3$.

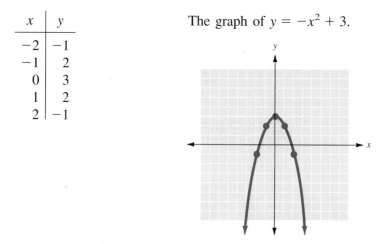

There is an important difference between this graph and the others we have seen. This time the parabola opens downward! Can you guess why? The answer is in the coefficient of the x^2 term.

If the coefficient of x^2 is *positive,* the parabola opens *upward.*

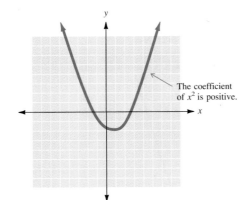

If the coefficient of x^2 is *negative,* the parabola opens *downward.*

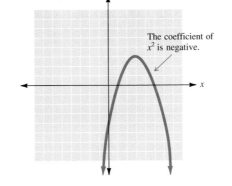

Graph $y = -x^2 - 2x$. Use x values between -3 and 1.

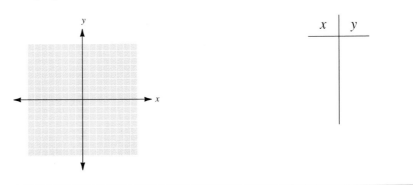

There are two other terms we would like to introduce before closing this section on graphing quadratic equations. As you may have noticed, all the parabolas that we graphed are symmetric about a vertical line. This is called the *axis of symmetry* for the parabola.

The point at which the parabola intersects that vertical line (this will be the lowest—or the highest—point on the parabola) is called the *vertex*. You'll learn more about finding the axis of symmetry and the vertex of a parabola in your next course in algebra.

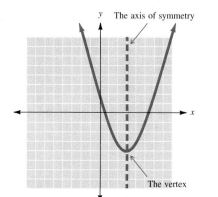

CHECK YOURSELF ANSWERS

1.

x	y
-2	6
-1	3
0	2
1	3
2	6

2.

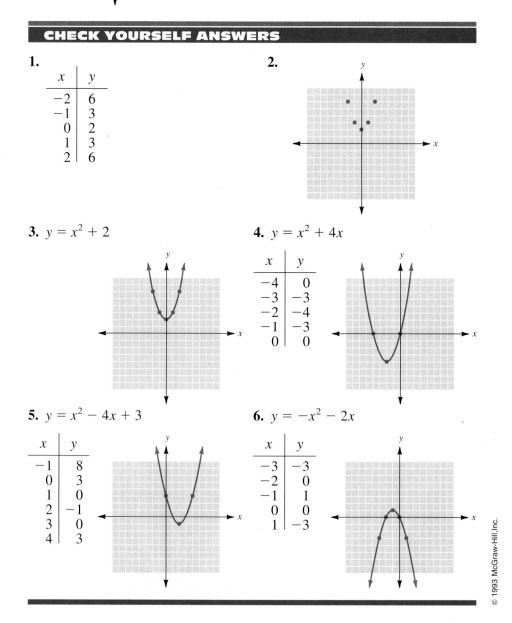

3. $y = x^2 + 2$

4. $y = x^2 + 4x$

x	y
-4	0
-3	-3
-2	-4
-1	-3
0	0

5. $y = x^2 - 4x + 3$

x	y
-1	8
0	3
1	0
2	-1
3	0
4	3

6. $y = -x^2 - 2x$

x	y
-3	-3
-2	0
-1	1
0	0
1	-3

Name

Date

Build Your Skills

Graph each of the following quadratic equations after completing the given table of values.

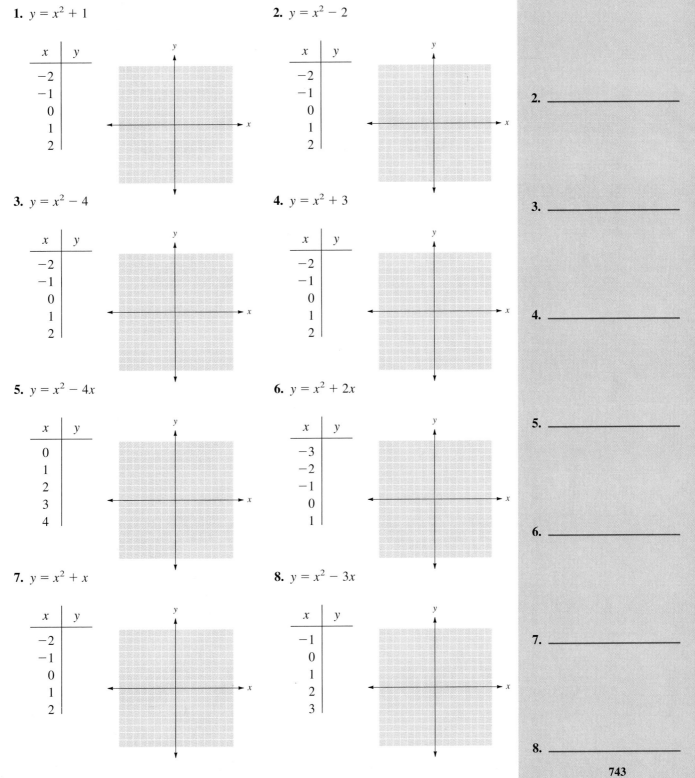

1. $y = x^2 + 1$

x	y
−2	
−1	
0	
1	
2	

2. $y = x^2 − 2$

x	y
−2	
−1	
0	
1	
2	

3. $y = x^2 − 4$

x	y
−2	
−1	
0	
1	
2	

4. $y = x^2 + 3$

x	y
−2	
−1	
0	
1	
2	

5. $y = x^2 − 4x$

x	y
0	
1	
2	
3	
4	

6. $y = x^2 + 2x$

x	y
−3	
−2	
−1	
0	
1	

7. $y = x^2 + x$

x	y
−2	
−1	
0	
1	
2	

8. $y = x^2 − 3x$

x	y
−1	
0	
1	
2	
3	

ANSWERS

1. _____

2. _____

3. _____

4. _____

5. _____

6. _____

7. _____

8. _____

9. _____

10. _____

11. _____

12. _____

13. _____

14. _____

15. _____

16. _____

9. $y = x^2 - 2x - 3$

x	y
-1	
0	
1	
2	
3	

10. $y = x^2 - 5x + 6$

x	y
0	
1	
2	
3	
4	

11. $y = x^2 - x - 6$

x	y
-1	
0	
1	
2	
3	

12. $y = x^2 + 3x - 4$

x	y
-4	
-3	
-2	
-1	
0	

13. $y = -x^2 + 2$

x	y
-2	
-1	
0	
1	
2	

14. $y = -x^2 - 2$

x	y
-2	
-1	
0	
1	
2	

15. $y = -x^2 - 4x$

x	y
-4	
-3	
-2	
-1	
0	

16. $y = -x^2 + 2x$

x	y
-1	
0	
1	
2	
3	

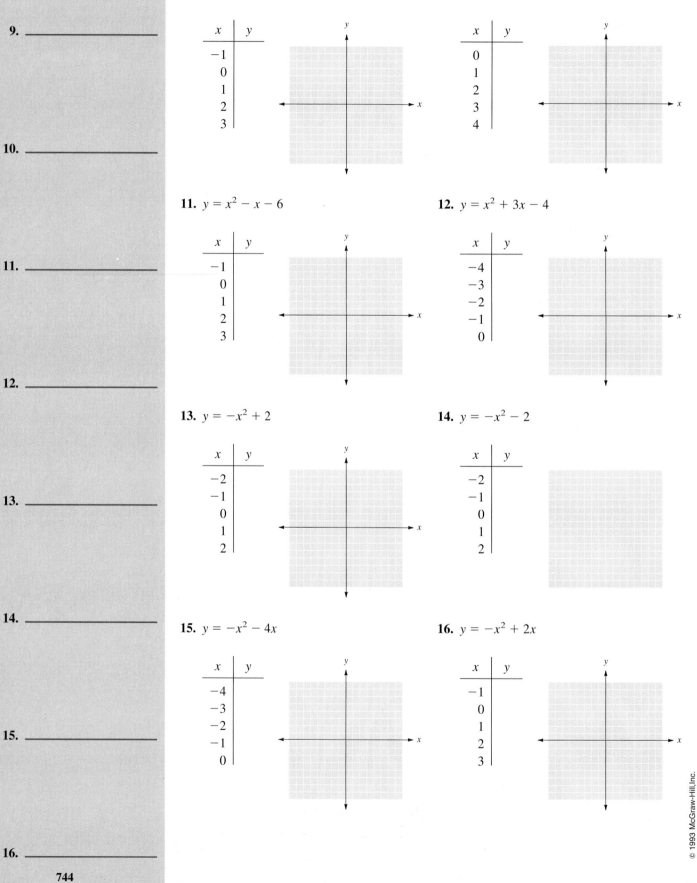

Think About These

Match each graph with the correct equation on the right.

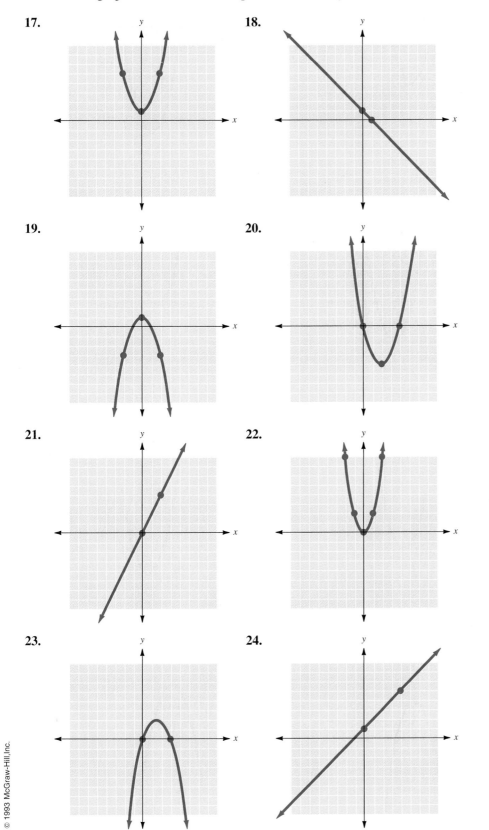

17.

18.

19.

20.

21.

22.

23.

24.

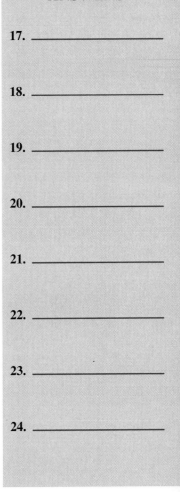

ANSWERS

17. _____

18. _____

19. _____

20. _____

21. _____

22. _____

23. _____

24. _____

(a) $y = -x^2 + 1$

(b) $y = 2x$

(c) $y = x^2 - 4x$

(d) $y = -x + 1$

(e) $y = -x^2 + 3x$

(f) $y = x^2 + 1$

(g) $y = x + 1$

(h) $y = 2x^2$

745

ANSWERS

1. $y = x^2 + 1$

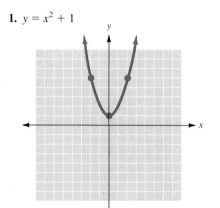

3. $y = x^2 - 4$

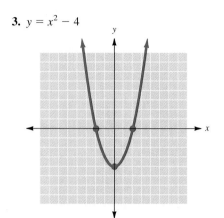

5. $y = x^2 - 4x$

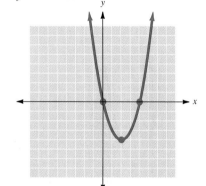

7. $y = x^2 + x$

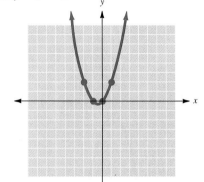

9. $y = x^2 - 2x - 3$

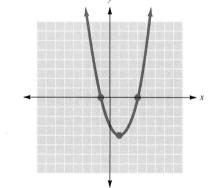

11. $y = x^2 - x - 6$

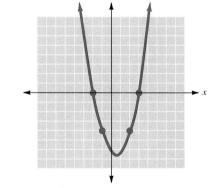

13. $y = -x^2 + 2$

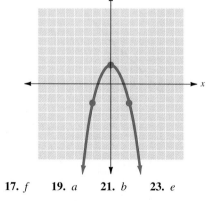

15. $y = -x^2 - 4x$

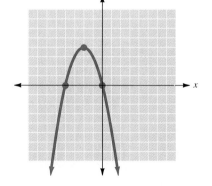

17. f **19.** a **21.** b **23.** e

746

10.5 Applications of The Pythagorean Theorem

OBJECTIVE
To apply the Pythagorean theorem in solving problems

One very important application of our work with quadratic equations is the *Pythagorean theorem*. The theorem was named for the Greek mathematician Pythagoras, born in 572 B.C. Pythagoras was the founder of the Greek society the Pythagoreans. Although the theorem bears Pythagoras' name, his own work on this theorem is uncertain as the Pythagoreans credited new discoveries to their founder.

You may recall that we used this theorem to find the distance between two points in Chapter 9.

The theorem states that for every right triangle, the following is true.

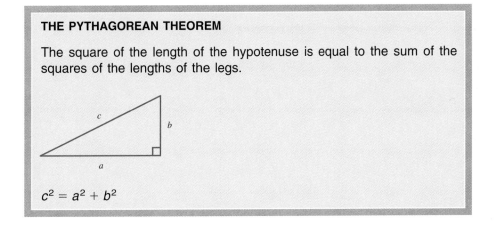

THE PYTHAGOREAN THEOREM

The square of the length of the hypotenuse is equal to the sum of the squares of the lengths of the legs.

$$c^2 = a^2 + b^2$$

Example 1

Verify the Pythagorean theorem for the given triangles.

(*a*) For the right triangle below,

$$5^2 \stackrel{?}{=} 3^2 + 4^2$$
$$25 \stackrel{?}{=} 9 + 16$$
$$25 = 25$$

(*b*) For the right triangle below.

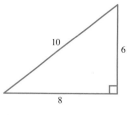

$$13^2 \stackrel{?}{=} 12^2 + 5^2$$
$$169 \stackrel{?}{=} 144 + 25$$
$$169 = 169$$

CHECK YOURSELF 1

Verify the Pythagorean theorem for the right triangle shown.

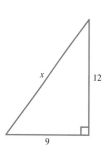

The Pythagorean theorem can be used to find the length of one side of a right triangle when the lengths of the two other sides are known.

Example 2

Find length x.

Note x will be longer than the given sides since it is the hypotenuse.

$$x^2 = 9^2 + 12^2$$
$$= 81 + 144$$
$$= 225$$

so

$$x = 15 \qquad \text{or} \qquad \underbrace{x = -15}$$

We reject this solution because a length must be positive.

CHECK YOURSELF 2

Find length x.

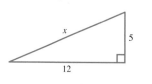

One or more of the lengths of the sides may be represented by an irrational number.

Example 3

Find length x.

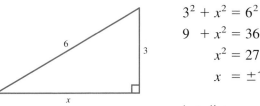

$$3^2 + x^2 = 6^2$$
$$9 + x^2 = 36$$
$$x^2 = 27$$
$$x = \pm\sqrt{27}$$

but distance cannot be negative, so

$$x = \sqrt{27} \quad \text{or} \quad x = 3\sqrt{3}$$

So x is approximately 5.2.

Note: You can approximate $3\sqrt{3}$ (or $\sqrt{27}$) with the use of a calculator or by the table in Appendix 2.

CHECK YOURSELF 3

Find length x.

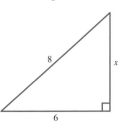

The Pythagorean theorem can be applied to solve a variety of geometric problems.

Example 4

Find the length of the diagonal of a rectangle which is 8 cm long and 5 cm wide. Let x be the unknown length of the diagonal:

Always draw and label a sketch showing the information from a problem when geometric figures are involved.

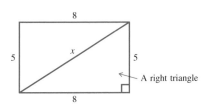

A right triangle

So

Again, distance cannot be negative, so we eliminate $x = -\sqrt{89}$.

$$x^2 = 5^2 + 8^2$$
$$= 25 + 64$$
$$= 89$$
$$x = \sqrt{89} \text{ cm} \qquad \text{so } x \approx 9.4 \text{ cm}$$

CHECK YOURSELF 4

The diagonal of a rectangle is 12 in, and its width is 6 in. Find its length.

The next application makes use of the Pythagorean theorem in obtaining the final solution.

Example 5

How long must a guy wire be to reach from the top of a 30-ft pole to a point on the ground 20 ft from the base of the pole?

Again be sure to draw a sketch of the problem.

Always check to see if your final answer is reasonable.

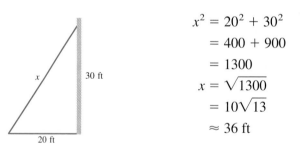

$$x^2 = 20^2 + 30^2$$
$$= 400 + 900$$
$$= 1300$$
$$x = \sqrt{1300}$$
$$= 10\sqrt{13}$$
$$\approx 36 \text{ ft}$$

CHECK YOURSELF 5

A 16-ft ladder leans against a wall with its base 4 ft from the wall. How far off the floor is the top of the ladder?

In the next example one of the legs is the missing part of the triangle.

Example 6

The length of one leg of a right triangle is 2 cm more than the other. If the length of the hypotenuse is 6 cm, what are the lengths of the two legs?

Draw a sketch of the problem, labeling the known and unknown lengths. Here, if one leg is represented by x, the other must be represented by $x + 2$.

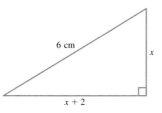

Use the Pythagorean theorem to form an equation.

The sum of the squares of the lengths of the unknown legs is equal to the square of the length of the hypotenuse.

$$x^2 + (x + 2)^2 = 6^2$$
$$x^2 + x^2 + 4x + 4 = 36$$
$$2x^2 + 4x - 32 = 0$$
$$x^2 + 2x - 16 = 0 \quad \left\{ \begin{array}{l} \text{Divide both} \\ \text{sides by 2.} \end{array} \right.$$

We apply the quadratic formula as before:

$$x = \frac{-2 \pm \sqrt{2^2 - 4(1)(-16)}}{2(1)}$$
$$= -1 \pm \sqrt{17}$$

Now,

$$x = -1 + \sqrt{17} \qquad \text{or} \qquad x = -1 - \sqrt{17}$$
$$\approx 3.123 \qquad\qquad\qquad \approx -5.123$$

Reject the negative solution in a geometric problem.

Make sure to include the units with the final answer.

If $x \approx 3.123$, then $x + 2 \approx 5.123$. The lengths of the legs are approximately 3.123 and 5.123 cm.

CHECK YOURSELF 6

The length of one leg of a right triangle is 1 in more than the other. If the length of the hypotenuse is 3 in, what are the lengths of the legs?

CHECK YOURSELF ANSWERS

1. $10^2 \overset{?}{=} 8^2 + 6^2$; $100 \overset{?}{=} 64 + 36$; $100 = 100$.
2. 13.
3. $2\sqrt{7}$; or approximately 5.3.
4. Length: $6\sqrt{3}$ in. or approximately 10.4 in.
5. The height is approximately 15.5 ft.
6. Approximately 1.561 and 2.561 in.

Find the length *x* in each triangle.

1.
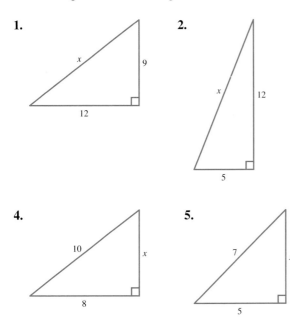

2.
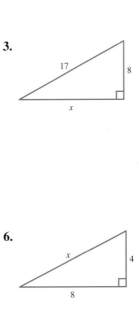

3.

4.

5.

6.

7. Find the length of the diagonal of a rectangle with 10 cm length and 7 cm width.

8. Find the length of the diagonal of a rectangle with 5 in width and 7 in length.

9. Find the width of a rectangle whose diagonal is 12 ft and whose length is 10 ft.

10. Find the length of a rectangle whose diagonal is 9 in and whose width is 6 in.

11. How long must a guy wire be to run from the top of a 20-ft pole to a point on the ground 8 ft from the base of the pole?

12. The base of a 15-ft ladder is 5 ft away from a wall. How high from the floor is the top of the ladder?

13. The length of one leg of a right triangle is 3 in more than the other. If the length of the hypotenuse is 8 in, what are the lengths of the two legs?

14. The length of a rectangle is 1 cm longer than its width. If the diagonal of the rectangle is 4 cm, what are the dimensions (the length and width) of the rectangle?

ANSWERS

1. _____

2. _____

3. _____

4. _____

5. _____

6. _____

7. _____

8. _____

9. _____

10. _____

11. _____

12. _____

13. _____

14. _____

Think About These

Find the altitude of each triangle.

15.

16.

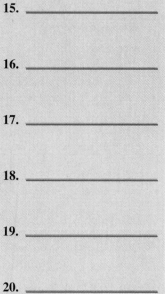

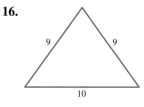

A homeowner wishes to insulate her attic with fiberglass insulation in order to conserve energy. The insulation comes in 40-cm-wide rolls which are cut to fit between the rafters in the attic.

17. If the roof is 6 m from peak to eave and the attic space is 2 m high at the peak, how long does each of the pieces of insulation need to be?

18. If the roof is 7 m from peak to eave and the attic space is 3 m high at the peak, how long does each of the pieces of insulation need to be?

19. A solar collector and its stand are in the shape of a right triangle. The collector is 5.00 m long, the upright leg is 3.00 m long, and the base leg is 4.00 m long. Due to inefficiencies in the collector's position, it needs to be raised by 0.50 m on the upright leg. How long will the new base leg be?

20. A solar collector and its stand are in the shape of a right triangle. The collector is 5.00 m long, the upright leg is 3.00 m long, and the base leg is 4.00 m long. Due to inefficiencies in the collector's position, it needs to be lowered by 0.50 m on the upright leg. How long will the new base leg be?

ANSWERS

1. 15　　**3.** 15　　**5.** $2\sqrt{6}$　　**7.** ≈ 12.207 cm　　**9.** ≈ 6.633 ft　　**11.** ≈ 21.541 ft
13. ≈ 3.954 in, 6.954 in　　**15.** 4　　**17.** $4\sqrt{2} \approx 5.7$ m　　**19.** ≈ 3.6 m
17. $4\sqrt{2} \approx 5.7$ m　　**19.** ≈ 3.6 m

Summary

Solving Quadratic Equations [10.1 to 10.3]

Solve $4x^2 = 13$.

$$x^2 = \frac{13}{4}$$

$$x = \sqrt{\frac{13}{4}} \text{ or } -\sqrt{\frac{13}{4}}$$

Simplifying gives

$$x = \frac{\sqrt{13}}{2} \text{ or } -\frac{\sqrt{13}}{2}$$

Solving Equations of the Form $ax^2 = k$

Divide both sides of the equation by a. The equation can then be solved by taking the square root of both sides.

$$x^2 = \frac{k}{a}$$

$$x = \sqrt{\frac{k}{a}} \qquad \text{or} \qquad x = -\sqrt{\frac{k}{a}}$$

To solve:

$$2x^2 + 2x - 1 = 0$$
$$2x^2 + 2x = 1$$
$$x^2 + x = \frac{1}{2}$$
$$x^2 + x + \left(\frac{1}{2}\right)^2 = \frac{1}{2} + \left(\frac{1}{2}\right)^2$$
$$\left(x + \frac{1}{2}\right)^2 = \frac{3}{4}$$
$$x + \frac{1}{2} = \pm\sqrt{\frac{3}{4}} = \pm\frac{\sqrt{3}}{2}$$
$$x = \frac{-1 \pm \sqrt{3}}{2}$$

Completing the Square   To solve a quadratic equation by completing the square:

1. Write the equation in the form

   $$ax^2 + bx = k$$

   so that the variable terms are on the left side and the constant is on the right side.
2. If the leading coefficient (of x^2) is not 1, divide both sides by that coefficient.
3. Add the square of one-half the middle (x) coefficient to both sides of the equation.
4. The left side of the equation is now a perfect-square trinomial. Factor and solve as before.

To solve:

$$x^2 - 2x = 4$$

Write the equation as

$$x^2 - 2x - 4 = 0$$

$a = 1 \qquad b = -2 \qquad c = -4$

$$x =$$
$$\frac{-(-2) \pm \sqrt{(-2)^2 - 4(1)(-4)}}{2(1)}$$
$$= \frac{2 \pm \sqrt{20}}{2}$$
$$= \frac{2 \pm 2\sqrt{5}}{2} = \frac{2(1 \pm \sqrt{5})}{2}$$
$$= 1 \pm \sqrt{5}$$

The Quadratic Formula   To solve an equation by formula:

1. Rewrite the equation in standard form.

   $$ax^2 + bx + c = 0$$

2. If a common factor exists, divide both sides of the equation by that common factor.
3. Identify the coefficients a, b, and c.
4. Substitute the values for a, b, and c into the quadratic formula.

   $$x = \frac{-b \pm \sqrt{b^2 - 4ac}}{2a}$$

5. Simplify the right side of the expression formed in step 4 to write the solutions for the original equation.

Graphing Quadratic Equations [10.4]

To graph equations of the form

$$y = ax^2 + bx + c$$

1. Form a table of values by choosing convenient values for x and finding the corresponding values for y.
2. Plot the points from the table of values.
3. Draw a smooth curve between the points.

The graph of a quadratic equation will always be a parabola. The parabola opens upward if a, the coefficient of the x^2 term, is positive.

$$y = x^2 - 4x$$

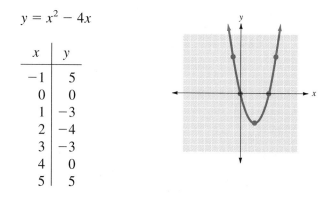

x	y
-1	5
0	0
1	-3
2	-4
3	-3
4	0
5	5

The parabola opens downward if a, the coefficient of the x^2 term, is negative.

$$y = -x^2 + 2x$$

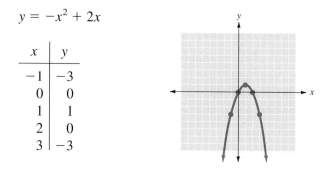

x	y
-1	-3
0	0
1	1
2	0
3	-3

The Pythagorean Theorem [10.5]

Find length x:

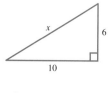

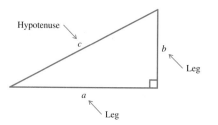

$$x^2 = 10^2 + 6^2$$
$$= 100 + 36$$
$$= 136$$
$$x = \sqrt{136} \text{ or } 2\sqrt{34}$$

In words, given a right triangle, the square of the length of the hypotenuse is equal to the sum of the squares of the lengths of the legs.

Summary Exercises Chapter 10

This summary exercise set is provided to give you practice with each of the objectives of the chapter. Each exercise is keyed to the appropriate chapter section. The answers are provided in the instructor's manual. Your instructor will give you guidelines on how to best use these exercises in your instructional setting.

[10.1] Solve each of the following equations for x by the square root method.

1. $x^2 = 10$

2. $x^2 = 48$

3. $x^2 - 20 = 0$

4. $x^2 + 2 = 8$

5. $(x - 1)^2 = 5$

6. $(x + 2)^2 = 8$

7. $(x + 3)^2 = 5$

8. $64x^2 = 25$

9. $4x^2 = 27$

10. $9x^2 = 20$

11. $25x^2 = 7$

12. $7x^2 = 3$

[10.2] Solve each of the following equations by completing the square.

13. $x^2 - 3x - 10 = 0$

14. $x^2 - 8x + 15 = 0$

15. $x^2 - 5x + 2 = 0$

16. $x^2 - 2x - 2 = 0$

17. $x^2 - 4x - 4 = 0$

18. $x^2 + 3x = 7$

19. $x^2 - 4x = -2$

20. $x^2 + 3x = 5$

21. $x^2 - x = 7$

22. $2x^2 + 6x = 12$

23. $2x^2 - 4x - 7 = 0$

24. $3x^2 + 5x + 1 = 0$

[10.3] Solve each of the following equations by using the quadratic formula.

25. $x^2 - 5x - 14 = 0$

26. $x^2 - 8x + 16 = 0$

27. $x^2 + 5x - 3 = 0$

28. $x^2 - 7x - 1 = 0$

29. $x^2 - 6x + 1 = 0$

30. $x^2 - 3x + 5 = 0$

31. $3x^2 - 4x = 2$

32. $2x - 3 = \dfrac{3}{x}$

33. $(x - 1)(x + 4) = 3$

34. $x^2 - 5x + 7 = 5$

35. $2x^2 - 8x = 12$

36. $5x^2 = 15 - 15x$

Solve by factoring or by any of the methods of this chapter.

37. $5x^2 = 3x$

38. $(2x - 3)(x + 5) = -11$

39. $(x - 1)^2 = 10$

40. $2x^2 = 7$

41. $2x^2 = 5x + 4$

42. $2x^2 - 4x = 30$

43. $2x^2 = 5x + 7$

44. $3x^2 - 4x = 2$

45. $3x^2 + 6x - 15 = 0$

46. $x^2 - 3x = 2(x + 5)$

47. $x - 2 = \dfrac{2}{x}$

48. The perimeter of a square is numerically 2 less than its area. Find the length of one side. (Approximate your answer to three decimal places, using a calculator or a table of square roots.)

[10.4] Graph each quadratic equation after completing the table of values.

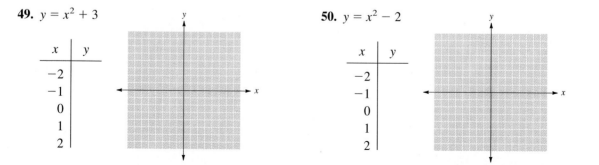

49. $y = x^2 + 3$

x	y
-2	
-1	
0	
1	
2	

50. $y = x^2 - 2$

x	y
-2	
-1	
0	
1	
2	

51. $y = x^2 - 3x$

x	y
−1	
0	
1	
2	
3	

52. $y = x^2 + 4x$

x	y
−4	
−3	
−2	
−1	
0	

53. $y = x^2 - x - 2$

x	y
−1	
0	
1	
2	
3	

54. $y = x^2 - 4x + 3$

x	y
0	
1	
2	
3	
4	

55. $y = x^2 + 2x - 3$

x	y
−3	
−2	
−1	
0	
1	

56. $y = 2x^2$

x	y
−2	
−1	
0	
1	
2	

57. $y = 2x^2 - 3$

x	y
−2	
−1	
0	
1	
2	

58. $y = -x^2 + 3$

x	y
−2	
−1	
0	
1	
2	

59. $y = -x^2 - 2$

x	y
−2	
−1	
0	
1	
2	

60. $y = -x^2 + 4x$

x	y
0	
1	
2	
3	
4	

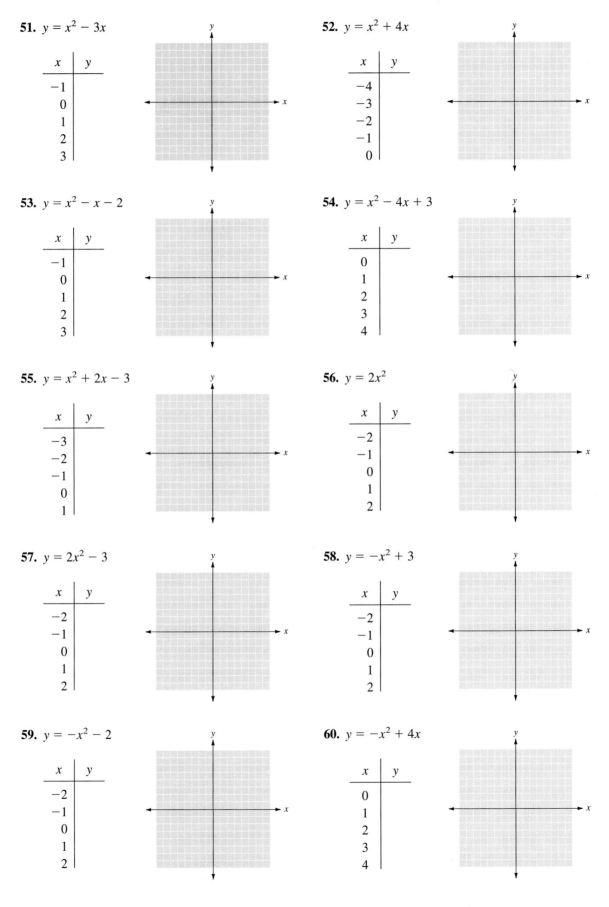

[10.5] Find length x in each triangle.

61.   **62.**   **63.**

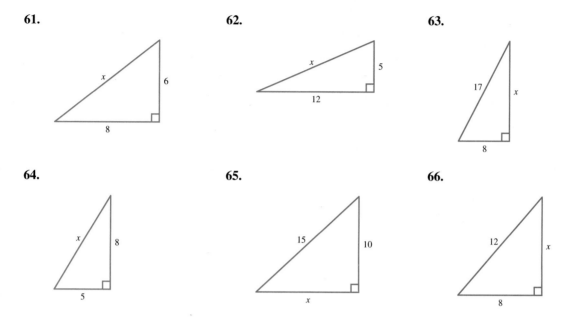

[10.5] Solve each of the following word problems. Approximate your answer to one decimal place where necessary.

67. Find the length of the diagonal of a rectangle whose length is 12 in and whose width is 9 in.

68. Find the length of a rectangle whose diagonal has a length of 10 cm and whose width is 5 cm.

69. How long must a guy wire be to run from the top of an 18-ft pole to a point on level ground 16 ft away from the base of the pole?

70. The length of one leg of a right triangle is 1 in more than the length of the other. If the length of the hypotenuse of the triangle is 6 in, what are the lengths of the two legs?

ANSWERS

The purpose of this self-test is to help you check your progress and to review for a chapter test in class. Allow yourself about an hour to take the test. When you are done, check your answers in the back of the book. If you missed any problems, be sure to go back and review the appropriate sections in the chapter and the exercises that are provided.

Solve each of the following equations for x.

1. $x^2 = 15$

2. $x^2 - 8 = 0$

3. $(x - 1)^2 = 7$

4. $9x^2 = 10$

Solve each of the following equations by completing the square.

5. $x^2 - 2x - 8 = 0$

6. $x^2 + 3x - 1 = 0$

7. $x^2 + 2x - 5 = 0$

8. $2x^2 - 5x + 1 = 0$

Solve each of the following equations by using the quadratic formula.

9. $x^2 - 2x - 3 = 0$

10. $x^2 - 6x + 9 = 0$

11. $x^2 - 5x = 2$

12. $2x^2 = 2x + 5$

13. $2x - 1 = \dfrac{4}{x}$

14. $(x - 1)(x + 3) = 2$

Graph each quadratic equation after completing the given table of values.

15. $y = x^2 + 4$

x	y
-2	
-1	
0	
1	
2	

16. $y = x^2 - 2x$

x	y
-1	
0	
1	
2	
3	

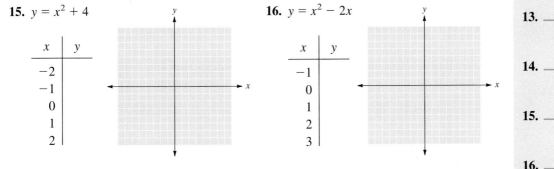

1. _____

2. _____

3. _____

4. _____

5. _____

6. _____

7. _____

8. _____

9. _____

10. _____

11. _____

12. _____

13. _____

14. _____

15. _____

16. _____

17. $y = x^2 - 3$

x	y
-2	
-1	
0	
1	
2	

18. $y = x^2 + x - 2$

x	y
-2	
-1	
0	
1	
2	

19. $y = -x^2 + 4$

x	y
-2	
-1	
0	
1	
2	

20. $y = -x^2 + 2x$

x	y
-1	
0	
1	
2	
3	

Find length x in each triangle.

21.

22.

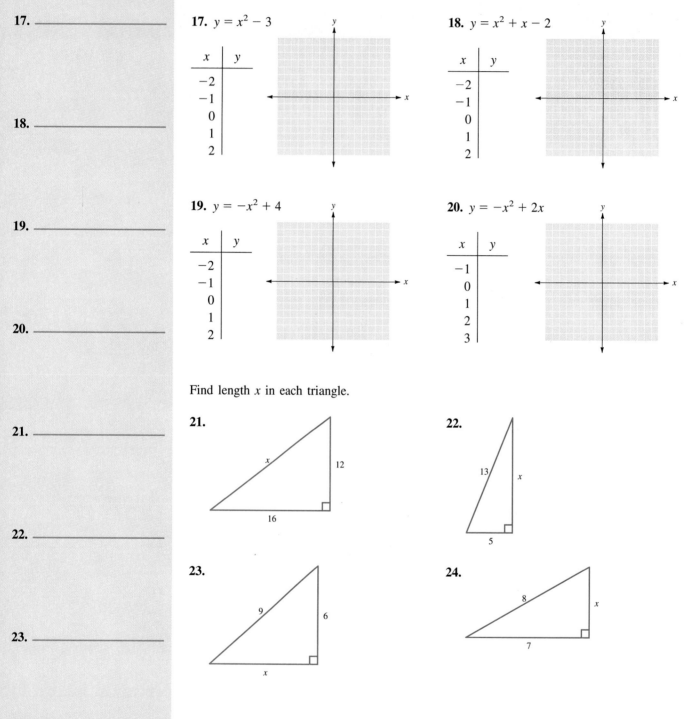

23.

24.

Solve the following word problem.

25. If the length of the diagonal of a rectangle is 12 cm and the width of the rectangle is 7 cm, what is the length of the rectangle?

ANSWERS

This test is provided to help you in the process of review of the previous chapters. Answers are provided in the back of the book. If you missed any problems, be sure to go back and review the appropriate chapter sections.

1. _____

Use the properties of exponents to simplify each of the following expressions. Assume that all variables represent nonzero real numbers.

2. _____

1. $(m^4)^5$ **2.** $(3xy^4)^2$

3. _____

3. $(2a^3)^2(a^2)^4$ **4.** $\left(\dfrac{r^4}{s^3}\right)^2$

4. _____

Evaluate each of the following expressions.

5. _____

5. $5x^0$ **6.** $(5m^3n)^0$

6. _____

Simplify each expression, using positive exponents only.

7. x^{-8} **8.** $4a^{-5}$

7. _____

8. _____

9. $r^4 r^{-7}$ **10.** $\dfrac{y^{-5}}{y^5}$

9. _____

Evaluate each root if possible.

11. $\sqrt{144}$ **12.** $-\sqrt{144}$

10. _____

13. $\sqrt{-144}$ **14.** $\sqrt[3]{-27}$

11. _____

12. _____

13. _____

14. _____

Simplify each of the following radical expressions. Assume that all variables represent positive real numbers.

15. $\sqrt{98x^4}$

16. $\sqrt{150m^3n^2}$

17. $\sqrt{\dfrac{12a^2}{25}}$

18. $\sqrt{\dfrac{5y^3}{3}}$

Simplify each of the following radical expressions by combining like terms.

19. $\sqrt{12} + 3\sqrt{27} - \sqrt{75}$

20. $a\sqrt{20} - 2\sqrt{45a^2}$

Simplify each of the following radical expressions.

21. $3\sqrt{2a} \cdot 5\sqrt{6a}$

22. $\sqrt{5}\,(2\sqrt{3} - 3\sqrt{5})$

23. $(\sqrt{2} - 5)(\sqrt{2} + 3)$

24. $\dfrac{\sqrt{8x^3}}{\sqrt{3}}$

25. $\dfrac{12}{\sqrt{10} - 2}$

Solve each of the following equations by factoring or by any of the methods of Chapter 10.

26. $x^2 - 72 = 0$

27. $x^2 + 6x - 3 = 0$

28. $(x - 2)^2 = 7$

29. $3x^2 = 6x + 45$

30. $2x^2 - 3x = 2(x + 1)$

31. $x^2 - 2x = 24$

32. $4x^2 - 10x = 8$

33. $2x - 2 = \dfrac{3}{x}$

Graph each of the following quadratic equations after completing the table of values.

34. $y = x^2 - 2$

x	y
-2	
-1	
0	
1	
2	

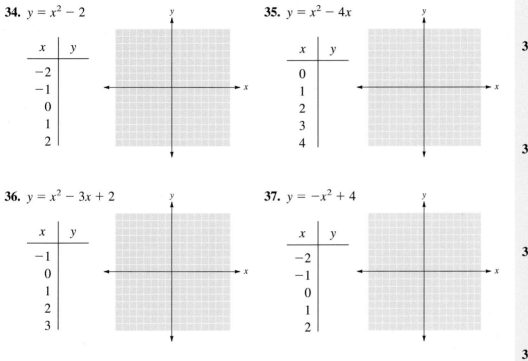

35. $y = x^2 - 4x$

x	y
0	
1	
2	
3	
4	

36. $y = x^2 - 3x + 2$

x	y
-1	
0	
1	
2	
3	

37. $y = -x^2 + 4$

x	y
-2	
-1	
0	
1	
2	

Solve the following application. Approximate your answer to the nearest tenth of an inch.

38. The length of a rectangle is 2 in less than 3 times its width. If the area of the rectangle is 10 in^2, find the dimensions of the rectangle.

ANSWERS

30. _____

31. _____

32. _____

33. _____

34. _____

35. _____

36. _____

37. _____

38. _____

39. ⎯⎯⎯⎯⎯⎯⎯⎯⎯⎯

40. ⎯⎯⎯⎯⎯⎯⎯⎯⎯⎯

Find length x in the triangle below, using the Pythagorean theorem.

39.

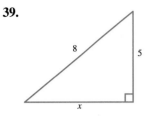

Solve the following application by using the Pythagorean theorem. Approximate your answer to the nearest tenth of a foot.

40. The length of one leg of a right triangle is 2 ft more than the length of the other. If the length of the hypotenuse of the triangle is 6 ft, find the lengths of the two legs. Approximate your answer to the nearest hundredth of a foot.

Appendix 1   The Arithmetic of Rational Numbers

This review is provided as an aid for students who wish to refresh their background in the arithmetic of fractions. You will find that much of what you will be asked to do with fractions or rational expressions in algebra has its basis in the methods you used in arithmetic.

Let's start the review with some terminology.

The Language of Fractions

Fractions name a number of equal parts of a unit or whole. A fraction is written in the form $\frac{a}{b}$, where a *and* b *are whole numbers and* b *cannot be zero.*

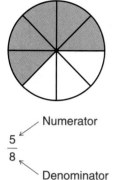

Numerator

$\frac{5}{8}$

Denominator

Denominator   The number of equal parts the whole is divided into.

Numerator   The number of equal parts of the whole that are used.

Proper fraction   A fraction whose numerator is less than its denominator. It names a number less than 1.

$\frac{2}{3}$ and $\frac{11}{15}$ are proper fractions.

$\frac{7}{5}$, $\frac{21}{20}$, and $\frac{8}{8}$ are improper fractions.

$2\frac{1}{3}$ and $5\frac{7}{8}$ are mixed numbers. Note that $2\frac{1}{3}$ means $2 + \frac{1}{3}$.

Improper fraction   A fraction whose numerator is greater than or equal to its denominator. It names a number greater than or equal to 1.

Mixed number   The sum of a whole number and a proper fraction.

Converting Mixed Numbers and Improper Fractions

To Change an Improper Fraction to a Mixed Number

1. Divide the numerator by the denominator. The quotient is the whole-number portion of the mixed number.
2. If there is a remainder, write the remainder over the original denominator. This gives the fractional portion of the mixed number.

Example 1

Change $\frac{22}{5}$ to a mixed number.

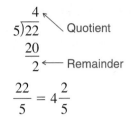

$$\frac{22}{5} = 4\frac{2}{5}$$

To Change a Mixed Number to an Improper Fraction

1. Multiply the denominator of the fraction by the whole-number portion of the mixed number.
2. Add the numerator of the fraction to that product.
3. Write that sum over the original denominator to form the improper fraction.

Example 2

Change $5\dfrac{3}{4}$ to an improper fraction.

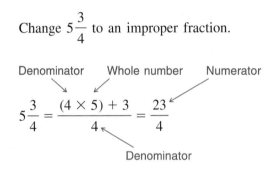

Equivalent Fractions

Equivalent fractions   Two fractions are equivalent (have equal value) if they are different names for the same number.

Cross products

$$\frac{a}{b} = \frac{c}{d} \qquad a \times d \text{ and } b \times c \text{ are called the cross products.}$$

$\dfrac{2}{3} = \dfrac{4}{6}$ because $2 \times 6 =$ 3×4.

If their cross products are equal, two fractions are equivalent.

$\dfrac{1}{2} = \dfrac{1 \times 5}{2 \times 5} = \dfrac{5}{10}$

The fundamental principle   For the fraction $\dfrac{a}{b}$, and any nonzero number c,

$\dfrac{1}{2}$ and $\dfrac{5}{10}$ are equivalent fractions.

$$\frac{a}{b} = \frac{a \times c}{b \times c}$$

In words: We can multiply the numerator and denominator of a fraction by the same nonzero number. The result will be an equivalent fraction.

$\dfrac{2}{3}$ is in simplest form.

$\dfrac{12}{18}$ is *not* in simplest form.

The numerator and denominator have the common factor 6.

$\dfrac{10}{15} = \dfrac{10 \div 5}{15 \div 5} = \dfrac{2}{3}$

$\dfrac{3}{4} = \dfrac{3 \times 2}{4 \times 2} = \dfrac{6}{8}$

Simplest form   A fraction is in simplest form, or in lowest terms, if the numerator and denominator have no common factors other than 1. This means that the fraction has the smallest possible numerator and denominator.

To write a fraction in simplest form   Divide the numerator and denominator by any common factors greater than 1 to write a fraction as an equivalent fraction in simplest form.

To build a fraction   Multiply the numerator and denominator by any whole number greater than 1 to build a fraction to an equivalent fraction with a specified denominator.

Multiplying Fractions

To Multiply Fractions

1. Multiply numerator by numerator. This gives the numerator of the product.
2. Multiply denominator by denominator. This gives the denominator of the product.
3. Simplify the resulting fraction if possible.

Example 3

Multiply $\dfrac{5}{8} \times \dfrac{3}{7}$.

$$\dfrac{5}{8} \times \dfrac{3}{7} = \dfrac{5 \times 3}{8 \times 7} = \dfrac{15}{56}$$

In multiplying fractions, it is usually easiest to divide by any common factors in the numerator and denominator *before* multiplying.

Example 4

Multiply $\dfrac{7}{15} \times \dfrac{10}{21}$.

We divide numerator and denominator by the common factors of 7 and 5.

$$\dfrac{7}{15} \times \dfrac{10}{21} = \dfrac{7 \times 10}{15 \times 21}$$

$$= \dfrac{\overset{1}{7} \times \overset{2}{10}}{\underset{3}{15} \times \underset{3}{21}}$$

$$= \dfrac{2}{9}$$

Dividing Fractions

© 1993 McGraw-Hill, Inc.

Remember: The divisor *follows* the division sign. That is the fraction that is inverted or "turned."

To divide fractions   Invert the divisor and multiply.

Example 5

Divide $\dfrac{3}{7} \div \dfrac{4}{5}$.

Invert

$$\frac{3}{7} \div \frac{4}{5} = \frac{3}{7} \times \frac{5}{4} = \frac{15}{28}$$

You can divide by common factors *only* after the divisor has been inverted.

Example 6

Note: Since a fraction bar also indicates division,

$$\frac{\dfrac{3}{5}}{\dfrac{9}{10}} = \frac{3}{5} \div \frac{9}{10}$$

and the quotient is simplified as in our example.

Divide $\dfrac{3}{5} \div \dfrac{9}{10}$.

$$\frac{3}{5} \div \frac{9}{10} = \frac{3}{5} \times \frac{10}{9} = \frac{\overset{1}{\cancel{3}} \times \overset{2}{\cancel{10}}}{\underset{1}{\cancel{5}} \times \underset{3}{\cancel{9}}} = \frac{2}{3}$$

Multiplying or Dividing Mixed Numbers

To multiply or divide mixed numbers   Convert any mixed or whole numbers to improper fractions. Then multiply or divide the fractions as before.

Example 7

Multiply $6\dfrac{2}{3} \times 3\dfrac{1}{5}$.

$$6\frac{2}{3} \times 3\frac{1}{5} = \frac{20}{3} \times \frac{16}{5} = \frac{20 \times 16}{3 \times \underset{1}{\cancel{5}}}^{4} = \frac{64}{3} = 21\frac{1}{3}$$

Example 8

Divide $\dfrac{7}{8} \div 5\dfrac{1}{4}$.

$$\frac{7}{8} \div 5\frac{1}{4} = \frac{7}{8} \div \frac{21}{4} = \frac{\overset{1}{\cancel{7}}}{\underset{2}{\cancel{8}}} \times \frac{\overset{1}{\cancel{4}}}{\underset{3}{\cancel{21}}} = \frac{1}{6}$$

Finding the Least Common Denominator

To Find the LCD

1. Write the prime factorization for each of the denominators.
2. Find all the prime factors that appear in any one of the prime factorizations.
3. Form the product of those prime factors, using each factor the greatest number of times it occurs in any one factorization.

Example 9

Find the LCD of fractions with denominators 4, 6, and 15.

$$
\begin{array}{r}
4 = 2 \times 2 \\
6 = 2 \quad\ \times 3 \\
\underline{15 = \qquad\quad\ 3 \times 5} \\
2 \times 2 \times 3 \times 5
\end{array}
$$

To form the LCD, use two factors of 2, one of 3, and one of 5.

The LCD $= 2 \times 2 \times 3 \times 5$, or 60.

Adding Fractions

To Add Like Fractions

1. Add the numerators.
2. Place the sum over the common denominator.
3. Simplify the resulting fraction if necessary.

Example 10

Add $\dfrac{5}{18} + \dfrac{7}{18}$.

$$\frac{5}{18} + \frac{7}{18} = \frac{12}{18} = \frac{\overset{2}{\cancel{12}}}{\underset{3}{\cancel{18}}} = \frac{2}{3}$$

To Add Unlike Fractions

1. Find the LCD of the fractions.
2. Change each fraction to equivalent fractions that have the LCD as a common denominator.
3. Add the resulting like fractions.

Example 11

Add $\dfrac{3}{4} + \dfrac{7}{10}$.

To convert to equivalent fractions with denominator 20, multiply numerator and denominator of the first fraction by 5, and of the second fraction by 2.

1. The LCD for 4 and 10 is 20.

2. $\dfrac{3}{4} = \dfrac{15}{20} \qquad \dfrac{7}{10} = \dfrac{14}{20}$

3. $\dfrac{3}{4} + \dfrac{7}{10} = \dfrac{15}{20} + \dfrac{14}{20} = \dfrac{29}{20} = 1\dfrac{9}{20}$

Subtracting Fractions

To Subtract Like Fractions

1. Subtract the numerators.
2. Place the difference over the common denominator.
3. Simplify the resulting fraction if necessary.

Example 12

$\dfrac{17}{20} - \dfrac{7}{20} = \dfrac{10}{20} = \dfrac{\overset{1}{\cancel{10}}}{\underset{2}{\cancel{20}}} = \dfrac{1}{2}$

To Subtract Unlike Fractions

1. Find the LCD of the fractions.
2. Change each fraction to equivalent fractions that have the LCD as a common denominator.
3. Subtract the resulting like fractions.

Example 13

Subtract $\dfrac{8}{9} - \dfrac{5}{6}$.

1. The LCD for 9 and 6 is 18.

2. $\overset{\times 2}{\overbrace{\dfrac{8}{9} = \dfrac{16}{18}}}$ $\overset{\times 3}{\overbrace{\dfrac{5}{6} = \dfrac{15}{18}}}$
$\underset{\times 2}{}$ $\underset{\times 3}{}$

3. $\dfrac{8}{9} - \dfrac{5}{6} = \dfrac{16}{18} - \dfrac{15}{18} = \dfrac{1}{18}$

Adding or Subtracting Mixed Numbers

To Add or Subtract Mixed Numbers

1. Add or subtract the whole-number parts.
2. Add or subtract the fractional parts.
   Note: Subtracting may require renaming the first mixed number.
3. Combine the results as a mixed number.

Example 14

Add $1\dfrac{2}{3} + 2\dfrac{3}{4}$.

First note that the LCD for the fractional portions of the mixed numbers is 12. Then convert the fractional portions of the mixed numbers to fractions with that LCD.

$$1\dfrac{2}{3} + 2\dfrac{3}{4} = 1\dfrac{8}{12} + 2\dfrac{9}{12} = 3\dfrac{17}{12} = 4\dfrac{5}{12}$$

with annotations $(1 + 2)$ and $\left(\dfrac{8}{12} + \dfrac{9}{12}\right)$

Example 15

Subtract $5\dfrac{1}{2} - 3\dfrac{3}{4}$.

To rename the first fraction, borrow 1 from 5 and think of that 1 as $\dfrac{4}{4}$.

$$5\dfrac{1}{2} - 3\dfrac{3}{4} = 5\dfrac{2}{4} - 3\dfrac{3}{4}$$

Rename

$$= 4\dfrac{6}{4} - 3\dfrac{3}{4} = 1\dfrac{3}{4}$$

$$(4 - 3) \qquad \left(\dfrac{6}{4} - \dfrac{3}{4}\right)$$

Name _____

Date _____

1. Give the fractions that name the shaded portions of the following diagrams. Indicate the numerator and the denominator.

a.

Fraction _____

Numerator _____

Denominator _____

b.

Fraction _____

Numerator _____

Denominator _____

1. _____

2. _____

3. _____

4. _____

2. If your English class has 23 students and 13 are women:

a. What fraction names the portion that are women? _____

b. What fraction names the portion that are not women? _____

3. You are given the following group of numbers:

$$\frac{2}{3}, \frac{5}{4}, 2\frac{3}{7}, \frac{45}{8}, \frac{7}{7}, 3\frac{4}{5}, \frac{9}{1}, \frac{7}{10}, \frac{12}{5}, 5\frac{2}{9}$$

a. List the proper fractions.

b. List the improper fractions.

c. List the mixed numbers.

4. Convert to mixed or whole numbers.

a. $\dfrac{41}{6}$

b. $\dfrac{32}{8}$

5. _____

5. Convert to improper fractions.

    a. $3\frac{1}{6}$

                                        b. $4\frac{3}{8}$

6. _____

6. Find out whether each pair of fractions is equivalent.

    a. $\frac{5}{8}, \frac{7}{12}$

                                          b. $\frac{8}{15}, \frac{32}{60}$

7. _____

7. Reduce each fraction to lowest terms.

    a. $\frac{15}{18}$ **b.** $\frac{24}{36}$ **c.** $\frac{140}{180}$ **d.** $\frac{210}{294}$

8. Find the missing numerators.

    a. $\frac{5}{8} = \frac{?}{24}$

                                          b. $\frac{4}{5} = \frac{?}{40}$

8. _____

9. Arrange the fractions in order from smallest to largest.

    a. $\frac{5}{9}, \frac{4}{7}$

                                          b. $\frac{5}{6}, \frac{4}{5}, \frac{7}{10}$

9. _____

10. Write as equivalent fractions with the LCD as a common denominator.

    a. $\frac{1}{6}, \frac{7}{8}$

                                          b. $\frac{3}{10}, \frac{5}{8}, \frac{7}{12}$

10. _____

11. Multiply.

    a. $\frac{5}{8} \times \frac{3}{4}$ **b.** $\frac{3}{5} \times \frac{4}{9}$ **c.** $\frac{10}{27} \times \frac{9}{20}$ **d.** $4 \times \frac{3}{8}$

    e. $\frac{4}{7} \times 2\frac{3}{8}$ **f.** $5\frac{1}{3} \times 1\frac{4}{5}$ **g.** $2\frac{2}{5} \times 1\frac{7}{8}$ **h.** $1\frac{5}{12} \times 8$

11. _____

12. Divide.

a. $\dfrac{3}{5} \div \dfrac{1}{4}$ b. $\dfrac{5}{12} \div \dfrac{5}{8}$ c. $1\dfrac{7}{9} \div \dfrac{4}{9}$

d. $3\dfrac{3}{8} \div 2\dfrac{1}{4}$ e. $\dfrac{9}{10} \div 3$ f. $4 \div 2\dfrac{2}{3}$

13. A kitchen measures $5\dfrac{1}{3}$ yards (yd) by $4\dfrac{1}{4}$ yd. If you purchase linoleum which costs $12 per square yard, what will it cost to cover the floor?

14. If you drive 117 mi in $2\dfrac{1}{4}$ hours, what is your average speed?

15. Add.

a. $\dfrac{1}{8} + \dfrac{3}{8}$ b. $\dfrac{6}{7} + \dfrac{5}{7}$

16. Add.

a. $\dfrac{1}{5} + \dfrac{3}{4}$ b. $\dfrac{5}{8} + \dfrac{5}{6}$ c. $\dfrac{5}{18} + \dfrac{7}{12}$ d. $\dfrac{3}{5} + \dfrac{1}{4} + \dfrac{5}{6}$

17. Subtract.

a. $\dfrac{5}{8} - \dfrac{3}{8}$ b. $\dfrac{3}{5} - \dfrac{1}{6}$ c. $\dfrac{7}{10} - \dfrac{7}{12}$ d. $\dfrac{11}{20} - \dfrac{7}{25}$

18. Add or subtract as indicated.

a. $5\dfrac{2}{5} + 4\dfrac{1}{5}$ **b.** $6\dfrac{5}{7} + 3\dfrac{4}{7}$ **c.** $4\dfrac{1}{8} + 3\dfrac{5}{12}$

d. $5\dfrac{7}{10} + 3\dfrac{11}{12}$ **e.** $2\dfrac{1}{2} + 3\dfrac{5}{6} + 4\dfrac{3}{8}$ **f.** $7\dfrac{7}{9} - 3\dfrac{4}{9}$

g. $5\dfrac{1}{7} - 3\dfrac{3}{7}$ **h.** $7\dfrac{1}{6} - 3\dfrac{1}{8}$ **i.** $6\dfrac{5}{12} - 3\dfrac{5}{8}$

j. $4 - 2\dfrac{2}{3}$

19. Jan ran $3\dfrac{2}{3}$ mi on Monday, $1\dfrac{3}{4}$ mi on Wednesday, and $4\dfrac{1}{2}$ mi on Friday. How far did she run during the week?

20. At the beginning of a year Miguel was $51\dfrac{3}{4}$ in tall. In June he measured $53\dfrac{1}{8}$ in. How much did he grow during that period?

21. Amelia buys an 8-yd roll of wallpaper on sale. After measuring, she finds that she needs the following amounts of the paper: $2\dfrac{1}{3}$, $1\dfrac{1}{2}$, and $3\dfrac{3}{4}$ yd. Does she have enough for the job? If so, how much will be left over?

ANSWERS

1. (*a*) Fraction, $\dfrac{3}{8}$; numerator, 3; denominator, 8; (*b*) fraction, $\dfrac{5}{6}$; numerator, 5; denominator, 6

3. (*a*) $\dfrac{2}{3}$, $\dfrac{7}{10}$; (*b*) $\dfrac{5}{4}$, $\dfrac{45}{8}$, $\dfrac{7}{7}$, $\dfrac{9}{1}$, $\dfrac{12}{5}$; (*c*) $2\dfrac{3}{7}$, $3\dfrac{4}{5}$, $5\dfrac{2}{9}$ **5.** (*a*) $\dfrac{19}{6}$; (*b*) $\dfrac{35}{8}$ **7.** (*a*) $\dfrac{5}{6}$; (*b*) $\dfrac{2}{3}$; (*c*) $\dfrac{7}{9}$; (*d*) $\dfrac{5}{7}$ **9.** (*a*) $\dfrac{5}{9}$, $\dfrac{4}{7}$; (*b*) $\dfrac{7}{10}$, $\dfrac{4}{5}$, $\dfrac{5}{6}$ **11.** (*a*) $\dfrac{15}{32}$; (*b*) $\dfrac{4}{15}$; (*c*) $\dfrac{1}{6}$; (*d*) $1\dfrac{1}{2}$; (*e*) $1\dfrac{5}{14}$; (*f*) $9\dfrac{3}{5}$; (*g*) $4\dfrac{1}{2}$; (*h*) $11\dfrac{1}{3}$ **13.** \$272   **15.** (*a*) $\dfrac{1}{2}$; (*b*) $1\dfrac{4}{7}$ **17.** (*a*) $\dfrac{1}{4}$; (*b*) $\dfrac{13}{30}$; (*c*) $\dfrac{7}{60}$; (*d*) $\dfrac{27}{100}$ **19.** $9\dfrac{11}{12}$ mi   **21.** Yes, $\dfrac{5}{12}$ yd

Appendix 2 A Table of Squares and Square Roots

N	N^2	$\sqrt{N}$	N	N^2	$\sqrt{N}$
1	1	1.000	36	1,296	6.000
2	4	1.414	37	1,369	6.083
3	9	1.732	38	1,444	6.164
4	16	2.000	39	1,521	6.245
5	25	2.236	40	1,600	6.325
6	36	2.449	41	1,681	6.403
7	49	2.646	42	1,764	6.481
8	64	2.828	43	1,849	6.557
9	81	3.000	44	1,936	6.633
10	100	3.162	45	2,025	6.708
11	121	3.317	46	2,116	6.782
12	144	3.464	47	2,209	6.856
13	169	3.606	48	2,304	6.928
14	196	3.742	49	2,401	7.000
15	225	3.873	50	2,500	7.071
16	256	4.000	51	2,601	7.141
17	289	4.123	52	2,704	7.211
18	324	4.243	53	2,809	7.280
19	361	4.359	54	2,916	7.348
20	400	4.472	55	3,025	7.416
21	441	4.583	56	3,136	7.483
22	484	4.690	57	3,249	7.550
23	529	4.796	58	3,364	7.616
24	576	4.899	59	3,481	7.681
25	625	5.000	60	3,600	7.746
26	676	5.099	61	3,721	7.810
27	729	5.196	62	3,844	7.874
28	784	5.292	63	3,969	7.937
29	841	5.385	64	4,096	8.000
30	900	5.477	65	4,225	8.062
31	961	5.568	66	4,356	8.124
32	1,024	5.657	67	4,489	8.185
33	1,089	5.745	68	4,624	8.246
34	1,156	5.831	69	4,761	8.307
35	1,225	5.916	70	4,900	8.367

N	N²	√N	N	N²	√N
71	5,041	8.426	116	13,456	10.770
72	5,184	8.485	117	13,689	10.817
73	5,329	8.544	118	13,924	10.863
74	5,476	8.602	119	14,161	10.909
75	5,625	8.660	120	14,400	10.954
76	5,776	8.718	121	14,641	11.000
77	5,929	8.775	122	14,884	11.045
78	6,084	8.832	123	15,129	11.091
79	6,241	8.888	124	15,376	11.136
80	6,400	8.944	125	15,625	11.180
81	6,561	9.000	126	15,876	11.225
82	6,724	9.055	127	16,129	11.269
83	6,889	9.110	128	16,384	11.314
84	7,056	9.165	129	16,641	11.358
85	7,225	9.220	130	16,900	11.402
86	7,396	9.274	131	17,161	11.446
87	7,569	9.327	132	17,424	11.489
88	7,744	9.381	133	17,689	11.533
89	7,921	9.434	134	17,956	11.576
90	8,100	9.487	135	18,225	11.619
91	8,281	9.539	136	18,496	11.662
92	8,464	9.592	137	18,769	11.705
93	8,649	9.644	138	19,044	11.747
94	8,836	9.695	139	19,321	11.790
95	9,025	9.747	140	19,600	11.832
96	9,216	9.798	141	19,881	11.874
97	9,409	9.849	142	20,164	11.916
98	9,604	9.899	143	20,449	11.958
99	9,801	9.950	144	20,736	12.000
100	10,000	10.000	145	21,025	12.042
101	10,201	10.050	146	21,316	12.083
102	10,404	10.100	147	21,609	12.124
103	10,609	10.149	148	21,904	12.166
104	10,816	10.198	149	22,201	12.207
105	11,025	10.247	150	22,500	12.247
106	11,236	10.296	151	22,801	12.288
107	11,449	10.344	152	23,104	12.329
108	11,664	10.392	153	23,409	12.369
109	11,881	10.440	154	23,716	12.410
110	12,100	10.488	155	24,025	12.450
111	12,321	10.536	156	24,336	12.490
112	12,544	10.583	157	24,649	12.530
113	12,769	10.630	158	24,964	12.570
114	12,996	10.677	159	25,281	12.610
115	13,225	10.724	160	25,600	12.649

N	N²	√N	N	N²	√N
161	25,921	12.689	181	32,761	13.454
162	26,244	12.728	182	33,124	13.491
163	26,569	12.767	183	33,489	13.528
164	26,896	12.806	184	33,856	13.565
165	27,225	12.845	185	34,225	13.601
166	27,556	12.884	186	34,596	13.638
167	27,889	12.923	187	34,969	13.675
168	28,224	12.961	188	35,344	13.711
169	28,561	13.000	189	35,721	13.748
170	28,900	13.038	190	36,100	13.784
171	29,241	13.077	191	36,481	13.820
172	29,584	13.115	192	36,864	13.856
173	29,929	13.153	193	37,249	13.892
174	30,276	13.191	194	37,636	13.928
175	30,625	13.229	195	38,025	13.964
176	30,976	13.266	196	38,416	14.000
177	31,329	13.304	197	38,809	14.036
178	31,684	13.342	198	39,204	14.071
179	32,041	13.379	199	39,601	14.107
180	32,400	13.416	200	40,000	14.142

Environmental Bibliography

Berreby, David. "The Numbers Game," *Discover,* April 1990, pp. 42–43.

Brown, Lester R., and John E. Young. "Feeding the World in the Nineties," in Linda Starke (ed.), *State of the World 1990, A Worldwatch Institute Report on Progress Toward a Sustainable Society.* Norton, New York, 1990.

Caufield, Catherine. "The Ancient Forest," *The New Yorker,* May 14, 1990, pp. 46–84.

Chiles, James R. "Tomorrow's Energy Today (Possibilities for a Nonfossil Fuel Economy)," *Audubon,* January 1990, pp. 58–73.

The Council on Economic Priorities. *The Price of Power—Electric Utilities and the Environment.* MIT Press, Cambridge, Mass. 1972.

Environmental Education Association of Oregon. *THREE R's as a Guide to the Best Use of Our Finite Resources,* Eugene, Oreg., 1977.

Fickett, Arnold P., Clark W. Gellings, and Amory B. Lovins. "Efficient Use of Electricity," *Scientific American,* September 1990, pp. 64–74.

Gagliardi, R. *The Mathematics of the Energy Crisis.* Intergalactic, Westmont, N.J., 1978.

Miller, G. Tyler. *Environmental Science,* 3d ed. Wadsworth, Belmont, Calif., 1991.

Metro Recycling Information Center. *Monthly Report,* March 1990 and April 1990, Portland, Oreg.

Myers, Norman. "Tropical-forest Species: Going, Going, Going. . . ," *Scientific American,* December 1988, p. 132.

National Council of Teachers of Mathematics. *A Sourcebook of Applications of School Mathematics,* Reston, Va., 1980.

"New Thinking on Old Growth," *Science,* Apr. 14, 1989, pp. 141–143.

Pimental, D., et al. "World Agriculture and Soil Erosion," *Bioscience,* April 1987, pp. 277–283.

Ryan, John C. "Timber's Last Stand," *World · watch,* July/August 1990, pp. 27–34.

Satchell, Michael. "The Endangered Logger: Big Business and a Little Bird Threaten a Northwest Way of Life," *U.S. News and World Report,* June 25, 1990, pp. 27–29.

Wilson, Edward O. "Threats to Biodiversity," *Scientific American,* September 1989, pp. 108–116.

World Resources Institute. *World Resources 1990–91.* Oxford University Press, New York, 1990.

Answers to Self-Tests and Cumulative Review Tests

Chapter 1 Self-Test

1. $a - 5$ **2.** $6m$ **3.** $4(m + n)$ **4.** $\dfrac{a + b}{3}$ **5.** 4^4 **6.** $7b^3$ **7.** 3   **8.** 65   **9.** 144

10. Commutative property of multiplication   **11.** Distributive property   **12.** Associative property of addition

13. 21   **14.** $20x + 12$ **15.** $15a$ **16.** $3x^2y$ **17.** $19x + 5y$ **18.** $8a^2$ **19.** a^{14} **20.** $15x^3y^7$ **21.** $2x^3$

22. $4ab^3$ **23.** 6   **24.** 10   **25.** 3

Chapter 2 Self-Test

1.
2. 7   **3.** 7   **4.** 11   **5.** 11   **6.** -13 **7.** -3 **8.** -21 **9.** 1

10. -6 **11.** -24 **12.** 9   **13.** 0   **14.** -40 **15.** 63   **16.** -27 **17.** -24 **18.** -25 **19.** 3

20. -5 **21.** Undefined   **22.** -4 **23.** 80   **24.** 144   **25.** 5

Chapter 3 Self-Test

1. No   **2.** Yes   **3.** 11   **4.** 12   **5.** 7   **6.** 7   **7.** -12 **8.** 25   **9.** 3   **10.** 4   **11.** $-\dfrac{2}{3}$ **12.** -5

13. $\dfrac{C}{2\pi}$ **14.** $\dfrac{3V}{B}$ **15.** $\dfrac{6 - 3x}{2}$ **16.** $x \geq 9$ **17.** $x < -3$

18. $x \leq 14$ **19.** $x < -4$ **20.** $x \geq \dfrac{4}{3}$

21. $x > -1$ **22.** 7   **23.** 21, 22, 23   **24.** Steve, 6; Jan, 12; Rick, 17   **25.** 10 in, 21 in

Chapters 1 to 3 Cumulative Test

1. $3(r + s)$ **2.** $\dfrac{x - 5}{3}$ **3.** 5^5 **4.** $8x^3y^2$ **5.** Associative property of addition   **6.** Distributive property

7. $5a^2b$ **8.** $12a^2 + 3a$ **9.** $15m^5n^3$ **10.** $5xy$ **11.** 2   **12.** 80   **13.** 7   **14.** 7   **15.** -16 **16.** 4

17. 63   **18.** -16 **19.** 0   **20.** -9 **21.** 13   **22.** 3   **23.** 5   **24.** -24 **25.** $\dfrac{5}{4}$ **26.** $-\dfrac{2}{5}$ **27.** 5

28. $\dfrac{I}{Pt}$ **29.** $\dfrac{2A}{b}$ **30.** $\dfrac{c - ax}{b}$ **31.** $x < 3$ **32.** $x \leq -\dfrac{3}{2}$

33. $x > 4$ **34.** $x \geq \dfrac{4}{3}$ **35.** 13   **36.** 42, 43   **37.** 7   **38.** \$420

39. 5 cm, 17 cm   **40.** 8 in, 13 in, 16 in

Chapter 4 Self-Test

1. Binomial   **2.** Trinomial   **3.** $8x^4 - 3x^2 - 7$; 8, -3, -7; 4   **4.** $10x^2 - 12x - 7$ **5.** $7a^3 + 11a^2 - 3a$

6. $3x^2 + 11x - 12$ **7.** $b^2 - 7b - 5$ **8.** $7a^2 - 10a$ **9.** $4x^2 + 5x - 6$ **10.** $2x^2 - 7x + 5$

11. $15a^3b^2 - 10a^2b^2 + 20a^2b^3$ **12.** $3x^2 + x - 14$ **13.** $a^2 - 49b^2$ **14.** $9m^2 + 12mn + 4n^2$

15. $2x^3 + 7x^2y - xy^2 - 2y^3$ **16.** $2x^2 - 3y$ **17.** $4c^2 - 6 + 9cd$ **18.** $x - 6$ **19.** $x + 2 + \dfrac{10}{2x - 3}$

20. $2x^2 - 3x + 2 + \dfrac{7}{3x + 1}$ **21.** 4   **22.** -2 **23.** 6, 11   **24.** 20 dimes, 25 quarters   **25.** 175 mi/h, 225 mi/h

Chapter 5 Self-Test

1. $6(2b + 3)$ **2.** $3p^2(3p - 4)$ **3.** $5(x^2 - 2x + 4)$ **4.** $6ab(a - 3 + 2b)$ **5.** $(a + 5)(a - 5)$
6. $(8m + n)(8m - n)$ **7.** $(7x + 4y)(7x - 4y)$ **8.** $2b(4a + 5b)(4a - 5b)$ **9.** $(a - 7)(a + 2)$
10. $(b + 3)(b + 5)$ **11.** $(x - 4)(x - 7)$ **12.** $(y + 10z)(y + 2z)$ **13.** $(2x - 1)(x + 8)$ **14.** $(3w + 7)(w + 1)$

15. $(4x - 3y)(2x + y)$ **16.** $3x(2x + 5)(x - 2)$ **17.** $3, 5$ **18.** $-1, 4$ **19.** $-1, \dfrac{2}{3}$ **20.** $0, 3$ **21.** $5, 7$

22. 3 cm by 11 cm   **23.** $\dfrac{P - 2L}{2}$ **24.** $\dfrac{b}{a - 1}$ **25.** 18

Chapter Six Self-Test

1. 4   **2.** $-3, 3$ **3.** $\dfrac{-3x^4}{4y^2}$ **4.** $\dfrac{4}{a}$ **5.** $\dfrac{x + 1}{x - 2}$ **6.** $\dfrac{4p^2}{7q}$ **7.** $\dfrac{2}{x - 1}$ **8.** $\dfrac{3}{4y}$ **9.** $\dfrac{3}{m}$ **10.** a **11.** 2

12. 5   **13.** $\dfrac{17x}{15}$ **14.** $\dfrac{3s - 2}{s^2}$ **15.** $\dfrac{4x + 17}{(x - 2)(x + 3)}$ **16.** $\dfrac{15}{w - 5}$ **17.** $\dfrac{2}{3x}$ **18.** $\dfrac{n}{2n + m}$ **19.** 36

20. 9   **21.** 2, 6   **22.** 4, 12   **23.** 50 mi/h, 45 mi/h   **24.** 6   **25.** 20 ft, 35 ft

Chapters 4 to 6 Cumulative Test

1. $8x^2 - 7x - 4$ **2.** $-a^2 - 3a - 7$ **3.** $w^2 - 8w - 4$ **4.** $28x^3y^2 - 14x^2y^2 + 21x^2y^3$ **5.** $15s^2 - 23s - 28$

6. $6a^3 + a^2b - 4ab^2 + b^3$ **7.** $-x^2 + 2xy - 3y$ **8.** $2x + 4$ **9.** $x^2 - 2x + 2 + \dfrac{5}{3x + 6}$ **10.** 9

11. 27 $5 bills, 56 $10 bills   **12.** 120 mi/h, 140 mi/h   **13.** $8a^2(3a - 2)$ **14.** $7mn(m - 3 - 7n)$
15. $(a + 8b)(a - 8b)$ **16.** $5p(p + 4q)(p - 4q)$ **17.** $(a - 6)(a - 8)$ **18.** $2w(w - 7)(w + 3)$

19. $(3r - 7s)(r + 4s)$ **20.** 4, 5   **21.** $-4, 4$ **22.** $-\dfrac{2}{3}, 2$ **23.** 7   **24.** 5 in by 17 in   **25.** $\dfrac{2S_n - na_n}{n}$

26. $-1, \dfrac{3}{2}$ **27.** $\dfrac{m}{3}$ **28.** $\dfrac{a - 7}{3a + 1}$ **29.** $\dfrac{3}{x}$ **30.** $\dfrac{1}{3w}$ **31.** $\dfrac{8r + 3}{6r^2}$ **32.** $\dfrac{x + 33}{3(x - 3)(x + 3)}$ **33.** $\dfrac{12}{y - 4}$

34. $\dfrac{x - 1}{2x + 1}$ **35.** $\dfrac{n}{3n + m}$ **36.** $\dfrac{6}{5}$ **37.** $-\dfrac{9}{2}, 7$ **38.** 2   **39.** 52 mi/h, 48 mi/h   **40.** 125 min

Chapter Seven Self-Test

1. $(3, 6), (9, 0)$ **2.** $(4, 0), (5, 4)$ **3.** $(3, 3), (6, 2), (9, 1)$ **4.** $(3, 0), (0, 4), \left(\dfrac{3}{4}, 3\right), \left(\dfrac{3}{4}, 3\right)$ **5.** Different answers

are possible   **6.** Different answers are possible   **7.** $(4, 2)$ **8.** $(-4, 6)$ **9.** $(0, -7)$ **10–12.**

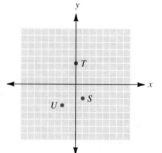

13. $x + y = 4$

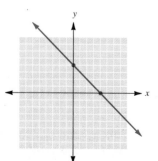

14. $y = 3x$

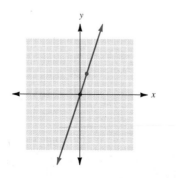

15. $y = \dfrac{3}{4}x - 4$

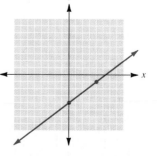

16. $x + 3y = 6$

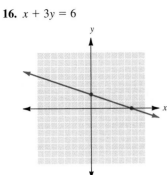

17. $2x + 5y = 10$

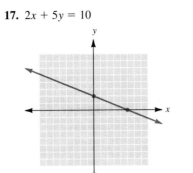

18. $y = -4$

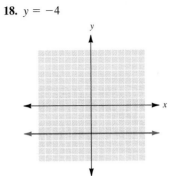

19. 1

20. $\dfrac{3}{4}$

21. $y = -3x + 6$

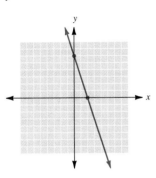

22. $y = \dfrac{2}{5}x - 3$

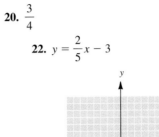

23. $x + y < 3$

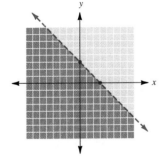

24. $3x + y \geq 9$

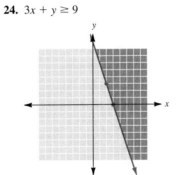

25. $x \leq 7$

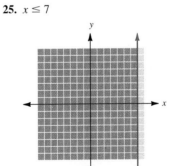

Chapter Eight Self-Test

1. $x + y = 5$
　　$x - y = 3$

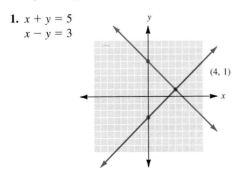

(4, 1)

2. $x + 2y = 8$
　　$x - \ y = 2$

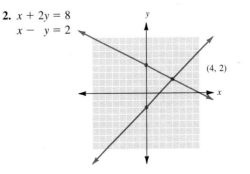

(4, 2)

3. $x - 3y = 3$
$x - 3y = 6$

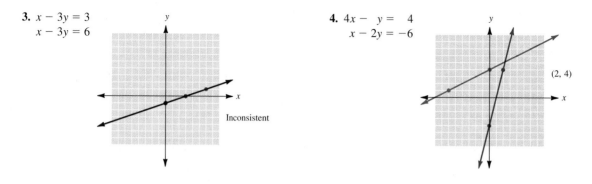

Inconsistent

4. $4x - y = 4$
$x - 2y = -6$

$(2, 4)$

5. $(4, 1)$ **6.** $(4, 2)$ **7.** $(1, 3)$ **8.** $\left(2, \dfrac{5}{2}\right)$ **9.** Dependent   **10.** $\left(\dfrac{3}{4}, -1\right)$ **11.** $(6, 2)$ **12.** Inconsistent

13. $(2, 6)$ **14.** $(6, -3)$ **15.** $(6, 2)$ **16.** $(5, 4)$ **17.** $(-3, 3)$ **18.** Inconsistent   **19.** $(3, -5)$ **20.** $(3, 2)$

21. 12, 18   **22.** 21 m, 29 m   **23.** 12 in by 20 in   **24.** 12 dimes, 18 quarters   **25.** Boat 15 mi/h, current 3 mi/h

Chapters 7 and 8 Cumulative Test

1. $(4, 4), (8, 0), (3, 5)$ **2.** $(0, -6), (2, 0), (1, -3)$ **3.** $0, 0, -3, 1$ **4.** $0, -3, 0, -4$

5. $(2, 0), (1, -3), (0, -6), (-1, -9)$ **6.** $(0, 2), (5, 1), (10, 0), (15, -1)$ **7–9.**

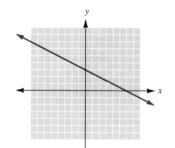

10. $x - y = 5$

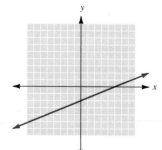

11. $y = \dfrac{2}{3}x + 3$

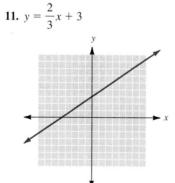

12. $x + 2y = 6$

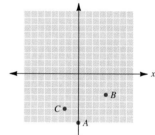

13. $2x - 5y = 10$

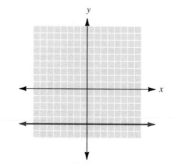

14. $y = -5$

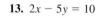

15. $\dfrac{10}{7}$ **16.** $-\dfrac{7}{5}$

17. $m = -3, b = 7$

18. $m = \dfrac{5}{3}, b = -5$

19. $y = 2x - 5$ **20.** $y = -\dfrac{3}{2}x + 5$ **21.** $x + 2y < 6$

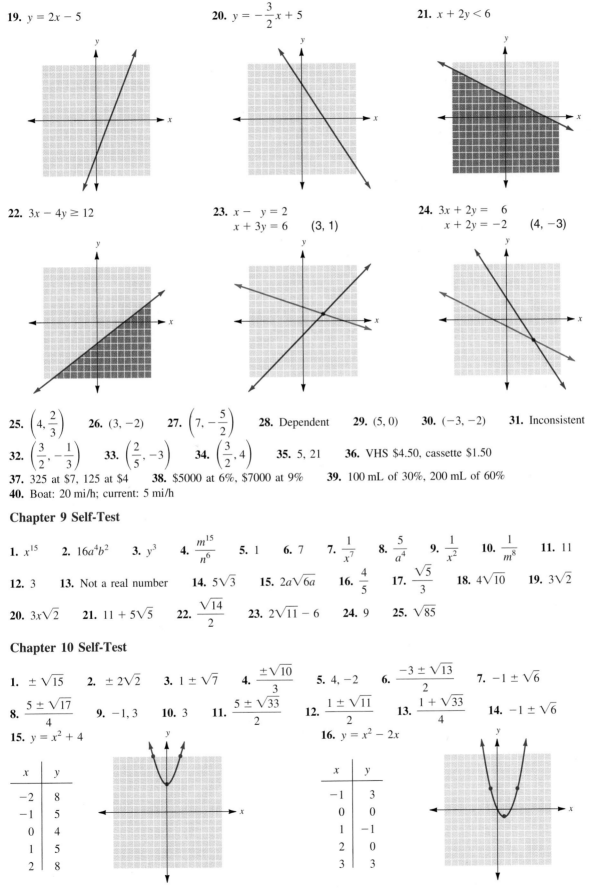

22. $3x - 4y \geq 12$ **23.** $x - y = 2$
$x + 3y = 6$ **(3, 1)**   **24.** $3x + 2y = 6$
$x + 2y = -2$ **(4, −3)**

25. $\left(4, \dfrac{2}{3}\right)$ **26.** $(3, -2)$ **27.** $\left(7, -\dfrac{5}{2}\right)$ **28.** Dependent   **29.** $(5, 0)$ **30.** $(-3, -2)$ **31.** Inconsistent

32. $\left(\dfrac{3}{2}, -\dfrac{1}{3}\right)$ **33.** $\left(\dfrac{2}{5}, -3\right)$ **34.** $\left(\dfrac{3}{2}, 4\right)$ **35.** 5, 21   **36.** VHS $4.50, cassette $1.50

37. 325 at $7, 125 at $4   **38.** $5000 at 6%, $7000 at 9%   **39.** 100 mL of 30%, 200 mL of 60%

40. Boat: 20 mi/h; current: 5 mi/h

Chapter 9 Self-Test

1. x^{15} **2.** $16a^4b^2$ **3.** y^3 **4.** $\dfrac{m^{15}}{n^6}$ **5.** 1   **6.** 7   **7.** $\dfrac{1}{x^7}$ **8.** $\dfrac{5}{a^4}$ **9.** $\dfrac{1}{x^2}$ **10.** $\dfrac{1}{m^8}$ **11.** 11

12. 3   **13.** Not a real number   **14.** $5\sqrt{3}$ **15.** $2a\sqrt{6a}$ **16.** $\dfrac{4}{5}$ **17.** $\dfrac{\sqrt{5}}{3}$ **18.** $4\sqrt{10}$ **19.** $3\sqrt{2}$

20. $3x\sqrt{2}$ **21.** $11 + 5\sqrt{5}$ **22.** $\dfrac{\sqrt{14}}{2}$ **23.** $2\sqrt{11} - 6$ **24.** 9   **25.** $\sqrt{85}$

Chapter 10 Self-Test

1. $\pm\sqrt{15}$ **2.** $\pm 2\sqrt{2}$ **3.** $1 \pm \sqrt{7}$ **4.** $\dfrac{\pm\sqrt{10}}{3}$ **5.** $4, -2$ **6.** $\dfrac{-3 \pm \sqrt{13}}{2}$ **7.** $-1 \pm \sqrt{6}$

8. $\dfrac{5 \pm \sqrt{17}}{4}$ **9.** $-1, 3$ **10.** 3   **11.** $\dfrac{5 \pm \sqrt{33}}{2}$ **12.** $\dfrac{1 \pm \sqrt{11}}{2}$ **13.** $\dfrac{1 + \sqrt{33}}{4}$ **14.** $-1 \pm \sqrt{6}$

15. $y = x^2 + 4$

x	y
-2	8
-1	5
0	4
1	5
2	8

16. $y = x^2 - 2x$

x	y
-1	3
0	0
1	-1
2	0
3	3

17. $y = x^2 - 3$

x	y
-2	1
-1	-2
0	-3
1	-2
2	1

18. $y = x^2 + x - 2$

x	y
-2	0
-1	-2
0	-2
1	0
2	4

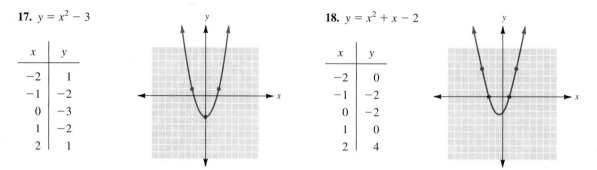

19. $y = -x^2 + 4$

x	y
-2	0
-1	3
0	4
1	3
2	0

20. $y = -x^2 + 2x$

x	y
-1	-3
0	0
1	1
2	0
3	-3

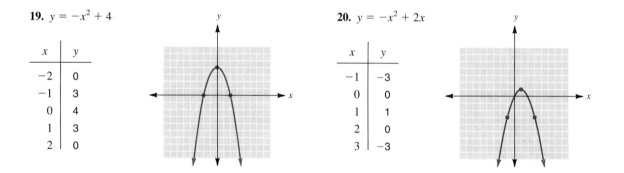

21. 20   **22.** 12   **23.** $3\sqrt{5}$ **24.** $\sqrt{15}$ **25.** Approximately 9.747 cm

Chapters 9 and 10 Cumulative Test

1. m^{20} **2.** $9x^2y^8$ **3.** $4a^{14}$ **4.** $\dfrac{r^8}{s^6}$ **5.** 5   **6.** 1   **7.** $\dfrac{1}{x^8}$ **8.** $\dfrac{4}{a^5}$ **9.** $\dfrac{1}{r^3}$ **10.** $\dfrac{1}{y^{10}}$ **11.** 12

12. -12 **13.** Not a real number   **14.** -3 **15.** $7x^2\sqrt{2}$ **16.** $5mn\sqrt{6m}$ **17.** $\dfrac{2a\sqrt{3}}{5}$ **18.** $\dfrac{y\sqrt{15y}}{3}$

19. $6\sqrt{3}$ **20.** $-4a\sqrt{5}$ **21.** $30a\sqrt{3}$ **22.** $2\sqrt{15} - 15$ **23.** $-13 - 2\sqrt{2}$ **24.** $\dfrac{2x\sqrt{6x}}{3}$ **25.** $2\sqrt{10} + 4$

26. $\pm 6\sqrt{2}$ **27.** $-3 \pm 2\sqrt{3}$ **28.** $2 \pm \sqrt{7}$ **29.** 5, -3 **30.** $\dfrac{5 \pm \sqrt{41}}{4}$ **31.** -4, 6   **32.** $\dfrac{5 \pm \sqrt{57}}{4}$

33. $\dfrac{1 \pm \sqrt{7}}{2}$

34. $y = x^2 - 2$

x	y
-2	2
-1	-1
0	-2
1	-1
2	2

35. $y = x^2 - 4x$

x	y
0	0
1	-3
2	-4
3	-3
4	0

36. $y = x^2 - 3x + 2$

x	y
-1	6
0	2
1	0
2	0
3	2

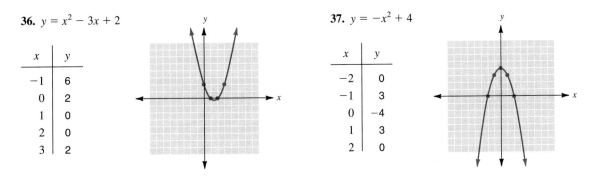

37. $y = -x^2 + 4$

x	y
-2	0
-1	3
0	-4
1	3
2	0

38. Width 2.2 in, length 4.6 in   **39.** $\sqrt{39}$ **40.** 3.12 ft, 5.12 ft